Main Version

D1298377

Laboratory Manual for

Human Anatomy & Physiology

Second Edition

TERRY R. MARTIN

Kishwaukee College

Mc Graw Hill

Connect
Learn
Succeed™

LABORATORY MANUAL FOR HUMAN ANATOMY & PHYSIOLOGY: MAIN VERSION, SECOND EDITION

Published by McGraw-Hill, a business unit of The McGraw-Hill Companies, Inc., 1221 Avenue of the Americas, New York, NY 10020. Copyright © 2013 by The McGraw-Hill Companies, Inc. All rights reserved. Printed in the United States of America. Previous edition © 2010. No part of this publication may be reproduced or distributed in any form or by any means, or stored in a database or retrieval system, without the prior written consent of The McGraw-Hill Companies, Inc., including, but not limited to, in any network or other electronic storage or transmission, or broadcast for distance learning.

Some ancillaries, including electronic and print components, may not be available to customers outside the United States.

This book is printed on acid-free paper.

1 2 3 4 5 6 7 8 9 0 QDB/QDB 1 0 9 8 7 6 5 4 3 2

ISBN 978–0–07–735306–3
MHID 0–07–735306–4

Vice President, Editor-in-Chief: *Marty Lange*
Vice President, EDP: *Kimberly Meriwether David*
Senior Director of Development: *Kristine Tibbetts*
Publisher: *Michael S. Hackett*
Executive Editor: *James F. Connely*
Senior Developmental Editor: *Fran Simon*
Marketing Manager: *Chris Loewenberg*
Lead Project Manager: *Peggy J. Selle*
Senior Buyer: *Sandy Ludovissy*
Senior Media Project Manager: *Tammy Juran*
Senior Designer: *Laurie B. Janssen*
Cover Designer: *Ron Bissell*
Cover Illustration: *Ross Martin*
Senior Photo Research Coordinator: *John C. Leland*
Photo Research: *Danny Meldung/Photo Affairs, Inc*
Compositor: *Precision Graphics*
Typeface: *10/12 Times LT Std*
Printer: *Quad/Graphics*

All credits appearing on page or at the end of the book are considered to be an extension of the copyright page.

Some of the laboratory experiments included in this text may be hazardous if materials are handled improperly or if procedures are conducted incorrectly. Safety precautions are necessary when you are working with chemicals, glass test tubes, hot water baths, sharp instruments, and the like, or for any procedures that generally require caution. Your school may have set regulations regarding safety procedures that your instructor will explain to you. Should you have any problems with materials or procedures, please ask your instructor for help.

www.mhhe.com

Contents

Supplemental Laboratory Exercises*

*These laboratory exercises are available online only at **www.mhhe.com/martinseries2** (click on this laboratory manual cover).

Preface

In Touch
with Anatomy & Physiology Lab Courses

Author Terry Martin's forty years of teaching anatomy and physiology courses, authorship of three laboratory manuals, and active involvement in the Human Anatomy and Physiology Society (HAPS) drove his determination to create a laboratory manual with an innovative approach that would benefit students. The *Laboratory Manual for Human Anatomy & Physiology* includes a cat version, a fetal pig version, and for the first time, a rat version. Each of these versions includes sixty-one laboratory exercises, three supplemental labs found online, and six cat, fetal pig, or rat dissection labs. A main version with no dissection exercises is also available. All four versions are written to work well with any anatomy and physiology text.

Martin Lab Manual Series . . .
IN TOUCH with Anatomy & Physiology Lab Courses

▶ NEW! Available in **4 Versions:** main (no dissection), cat dissection, fetal pig dissection, and rat dissection.

▶ Incorporates **learning outcomes and assessments** to help students master important material!

▶ NEW! **Pre-Lab** assignments are printed in the lab manual. They will help students be more prepared for lab and save instructors time during lab.

▶ **Clear, concise** writing style facilitates more thorough understanding of lab exercises.

▶ **BIOPAC**© exercises use hardware and software for data acquisition, analysis, and recording.

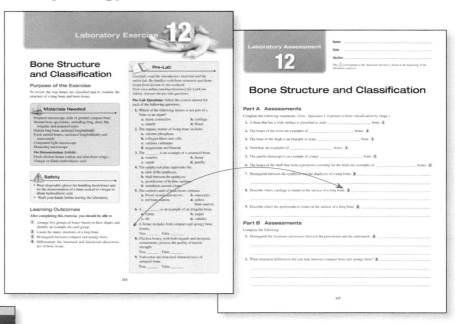

▶ NEW! **Ph.I.L.S. 4.0** included and physiology lab simulations interspersed throughout make otherwise difficult and expensive experiments a breeze through digital simulations.

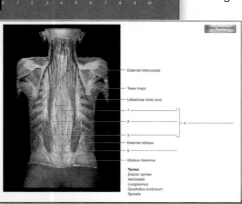

▶ Cadaver images from **Anatomy & Physiology Revealed**® **(APR)** are incorporated throughout the lab. Cadaver images help students make the connection from specimen to cadaver.

▶ **Micrographs** incorporated throughout the lab aid students' visual understanding of difficult topics.

▶ **Instructor's Guide** is annotated for quick and easy use by adjuncts and is available online at www.mhhe.com/martinseries2.

In Touch
with Student Needs

▶ The procedures are clear, concise, and easy to follow. Relevant lists and summary tables present the contents efficiently. Histology micrographs and cadaver photos are incorporated in the appropriate locations within the associated labs.

▶ NEW! The pre-lab section now includes quiz questions. It also directs the student to carefully read the introductory material and the entire lab to become familiar with its contents. If necessary, a textbook or lecture notes might be needed to supplement the concepts. A visit to **www.mhhe.com/martinseries2** will provide a list of animations from **Anatomy & Physiology Revealed® (APR)** and LabCam videos to review before answering five or more fundamental laboratory questions for that particular lab.

▶ **Terminologia Anatomica** is used as the source for universal terminology in this laboratory manual. Alternative names are included when a term is introduced for the first time.

▶ Laboratory assessments immediately follow each laboratory exercise.

▶ Histology photos are placed within the appropriate laboratory exercise.

▶ A section called "Study Skills for Anatomy and Physiology" is located in the front of this laboratory manual. This section was written by students enrolled in a Human Anatomy and Physiology course.

▶ Critical Thinking Activities are incorporated within most of the laboratory exercises to enhance valuable critical thinking skills that students need throughout their lives.

▶ Cadaver images are incorporated with dissection labs.

In Touch
with Instructor Needs

▶ The instructor will find digital assets for use in creating customized lectures, visually enhanced tests and quizzes, and other printed support material.

▶ A correlation guide for **Anatomy & Physiology Revealed® (APR)** and the entire lab manual is located on the lab manual's website at **www.mhhe.com/martinseries2**. Cadaver images from APR are included within many of the laboratory exercises.

▶ Some unique labs included are "Scientific Method and Measurements," "Chemistry of Life," "Fetal Skeleton," "Surface Anatomy," "Diabetic Physiology," and "Genetics."

▶ The annotated instructor's guide for *Laboratory Manual for Human Anatomy and Physiology* describes the purpose of the laboratory manual and its special features, provides suggestions for presenting the laboratory exercises to students, instructional approaches, a suggested time schedule, and annotated figures and assessments. It contains a "Student Safety Contract" and a "Student Informed Consent Form."

▶ Each laboratory exercise can be completed during a single laboratory session.

In Touch
with Educational Needs

▶ Learning outcomes with icons ⃝ have matching assessments with icons 𝖠 so students can be sure they have accomplished the laboratory exercise content. Outcomes and assessments include all levels of learning skills: remember, understand, apply, analyze, evaluate, and create.

▶ Assessment rubrics for entire laboratory assessments are included in Appendix 2.

In Touch
with Technology

▶ Physiology Interactive Lab Simulations (Ph.I.L.S. 4.0) is included with the lab manual. Eleven lab simulations are interspersed throughout the lab manual. The correlation guide for all of the simulations is included in Appendix 3.

▶ **BIOPAC** Systems, Inc. BIOPAC© exercises are included on four different body systems. BIOPAC© systems use hardware and software for data acquisition, analysis, and recording of information for an individual.

Engaging Presentation Materials for Lecture and Lab

New! All content in Connect is correlated to HAPS Learning Outcomes.

McGraw-Hill *ConnectPlus Anatomy & Physiology* is a web-based assignment and assessment platform that gives students the means to better connect with their coursework, with their instructors, and with the important concepts that they will need to know for success now and in the future. With Connect Anatomy & Physiology, instructors can deliver assignments, quizzes and tests easily online. Stu-

dents can practice important skills at their own pace and on their own schedule. With Connect Anatomy & Physiology Plus, students also get 24/7 online access to an eBook—an online edition of the text—to aid them in successfully completing their work, wherever and whenever they choose at www.mhhe.com/martinseries2

McGraw-Hill Higher Education and Blackboard® have teamed up. What does this mean for you?

1. **Your life, simplified.** Now you and your students can access McGraw-Hill's Connect™ and Create™ right from within your Blackboard course – all with one single sign-on. Say goodbye to the days of logging in to multiple applications.

2. **Deep integration of content and tools.** Not only do you get single sign-on with Connect™ and Create™, you also get deep integration of McGraw-Hill content and content engines right in Blackboard. Whether you're choosing a book for your course or building Connect™ assignments, all the tools you need are right where you want them – inside of Blackboard.

3. **Seamless Gradebooks.** Are you tired of keeping multiple gradebooks and manually synchronizing grades into Blackboard? We thought so. When a student completes an integrated Connect™ assignment, the grade for that assignment automatically (and instantly) feeds your Blackboard grade center.

4. **A solution for everyone.** Whether your institution is already using Blackboard or you just want to try Blackboard on your own, we have a solution for you. McGraw-Hill and Blackboard can now offer you easy access to industry leading technology and content, whether your campus hosts it, or we do. Be sure to ask your local McGraw-Hill representative for details.

Guided Tour Through A Lab Exercise

The laboratory exercises include a variety of special features that are designed to stimulate interest in the subject matter, to involve students in the learning process, and to guide them through the planned activities. These features include the following:

Purpose of the Exercise
The purpose provides a statement about the intent of the exercise—that is, what will be accomplished.

Materials Needed This section lists the laboratory materials that are required to complete the exercise and to perform the demonstrations and learning extensions.

Safety A list of safety guidelines is included inside the front cover. Each lab session that requires special safety guidelines has a safety section. Your instructor might require some modifications of these guidelines.

Learning Outcomes The learning outcomes list what a student should be able to do after completing the exercise. Each learning outcome will have matching assessments indicated by the corresponding icon A in the laboratory exercise or the laboratory assessment.

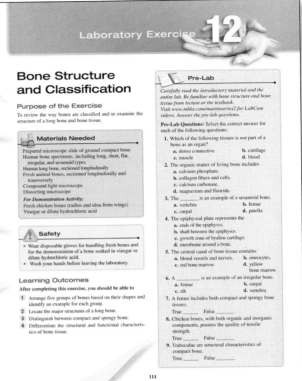

Pre-Lab The pre-lab includes quiz questions and directs the student to carefully read introductory material and examine the entire laboratory contents after becoming familiar with the topics from a textbook or lecture. Students will also be directed to visit **www.mhhe.com/martinseries2** to obtain a list of correlated **Anatomy and Physiology Revealed®** animations and LabCam videos. After successfully answering the pre-lab questions, the student is prepared to become involved in the laboratory exercise.

Introduction The introduction describes the subject of the exercise or the ideas that will be investigated. It includes all of the information needed to perform the laboratory exercise.

Procedure The procedure provides a set of detailed instructions for accomplishing the planned laboratory activities. Usually these instructions are presented in outline form so that a student can proceed efficiently through the exercise in stepwise fashion.

The procedures include a wide variety of laboratory activities and, from time to time, direct the student to complete various tasks in the laboratory assessments.

There are also separate procedures in 11 labs that utilize Ph.I.L.S. 4.0.

Demonstration Activities Demonstration activities appear in separate boxes. They describe specimens, specialized laboratory equipment, or other materials of interest that an instructor may want to display to enrich the student's laboratory experience.

Demonstration Activity

Examine a fresh chicken bone and a chicken bone that has been soaked for several days in vinegar or overnight in dilute hydrochloric acid. Wear disposable gloves for handling these bones. This acid treatment removes the inorganic salts from the bone extracellular matrix. Rinse the bones in water and note the texture and flexibility of each (fig. 12.7a). The bone becomes soft and flexible without the support of the inorganic salts with calcium.

Examine the specimen of chicken bone that has been exposed to high temperature (baked at 121°C/250°F for 2 hours). This treatment removes the ... from the bone ... bone comes ... the collagen ... of the quali-... rovides ten-... the chicken

Learning Extension Activities Learning extension activities also appear in separate boxes. They encourage students to extend their laboratory experiences. Some of these activities are open-ended in that they suggest the student plan an investigation or experiment and carry it out after receiving approval from the laboratory instructor. Some of the figures are illustrated as line art or in grayscale. This will allow colored pencils to be used as a visual learning activity to distinguish various structures.

Learning Extension Activity

Repeat the demonstration of diffusion using a petri dish filled with ice-cold water and a second dish filled with very hot water. At the same moment, add a crystal of potassium permanganate to each dish and observe the circle as before. What difference do you note in the rate of diffusion in the two dishes? How do you explain this difference? _____

_____

_____

Illustrations Diagrams similar to those in a textbook often are used as aids for reviewing subject matter. Other illustrations provide visual instructions for performing steps in procedures or are used to identify parts of instruments or specimens. Micrographs are included to help students identify microscopic structures or to evaluate student understanding of tissues.

In some exercises, the figures include line drawings suitable for students to color with colored pencils. This activity may motivate students to observe the illustrations more carefully and help them to locate the special features represented in the figures.

Laboratory Assessments A laboratory assessment form to be completed by the student immediately follows each exercise. These assessments include various types of review activities, spaces for sketches of microscopic objects, tables for recording observations and experimental results, and questions dealing with the analysis of such data.

As a result of these activities, students will develop a better understanding of the structural and functional characteristics of their bodies and will increase their skills in gathering information by observation and experimentation. By completing all of the assessments, students will be able to determine if they were able to accomplish all of the learning outcomes.

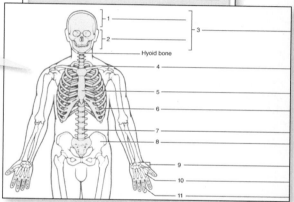

Laboratory Assessment

12

Name _____
Date _____
Section _____

The △ corresponds to the indicated outcome(s) found at the beginning of the laboratory exercise.

Bone Structure and Classification

Part A Assessments

Complete the following statements: (*Note:* Questions 1–6 pertain to bone classification by shape.)

1. A bone that has a wide surface is classified as a(an) _____ bone. △

2. The bones of the wrist are examples of _____ bones. △

3. The bone of the thigh is an example of a(an) _____ bone. △

4. Vertebrae are examples of _____ bones. △

5. The patella (kneecap) is an example of a large _____ bone. △

6. The bones of the skull that form a protective covering for the brain are examples of _____ bones. △

7. Distinguish between the epiphysis and the diaphysis of a long bone. △ _____

8. Describe where cartilage is found on the surface of a long bone. △ _____

9. Describe where the periosteum is found on the surface of a long bone. △ _____

... tween the periosteum and the endosteum. △ _____

... between compact bone and spongy bone? △ _____

Osteon

Lamella

Central canal

Lacuna (occupied by osteocyte in living bone)

Bone extracellular matrix

Canaliculus

Histology Histology photos placed within the appropriate exercise.

Changes to This Edition

Global Changes

- Introductory materials expanded; introductory material precedes most procedures.
- Pre-Lab questions expanded and placed in the laboratory manual rather than online.
- BIOPAC exercises rewritten.
- Ph.I.L.S. laboratory simulations updated.
- Ph.I.L.S. 4.0 online included with lab manual.
- Structural lists have functions and descriptions added.
- Muscle tables added with origins, insertions, and actions.

- New design and sequence of items placed on the introductory page of the laboratory exercise.
- Laboratory Reports changed to Laboratory Assessments.
- Matching assessments for the learning outcomes are all in the Laboratory Assessments.
- Laboratory exercises contain fully labeled figures.
- Laboratory Assessments expanded and contain figures to label.
- All micrographs contain magnifications.

Laboratory Exercise	Topic	Change
1	Laboratory Assessment	Improved directions
2	Procedures A, B, and C Structural lists Laboratory Assessment	Added introductory material Functions and descriptions added Added content
3	Fig. 3.1 (pH values) Procedure A	Improved depth Added introductory material
4	Fig. 4.3 (microscope) Laboratory Assessment	Added figure Added content; improved accuracy in Part B
5	Fig. 5.1 (composite cell) Introductory material Ph.I.L.S. Lesson 2	Improved depth Updated and expanded content Clarity added
6	Procedures B, C, and D Ph.I.L.S. Lesson 1	Added introductory material Clarity added
7	Fig. 7.2 (interphase) Fig. 7.5 (mitosis) Fig. 7.6a (human chromosomes) Introductory material	Added micrograph Improved depth New micrograph Improved depth
8	Fig. 8.1a, b, d, g, and h (epithelial tissues) Fig. 8.2 (sectional cuts) Table 8.1 (epithelial tissues)	New micrographs Added comparisons to body tube Added table with descriptions, functions, and locations
9	Fig. 9.1b and h (connective tissues) Table 9.1 (connective tissues) Table 9.2 (connective tissues) Introductory material	New micrographs Added table with descriptions and functions Improved design Improved depth
10	Fig. 10.1a and c (muscle tissues) Table 10.1 (muscle and nervous tissues)	New micrographs Added table with descriptions, functions, and locations
11	Fig. 11.1 (skin layers) Fig. 11.4b (skin structures) Table 11.1 (epidermal layers) Procedure	Added figure New micrograph Added table with locations and descriptions Reworked
12	Fig. 12.1 (bone classification) Fig. 12.4 (compact and spongy bone) Procedure Demonstration activity	Added figure Added figure Added introductory material Rewritten information
13	Fig. 13.1a–b (skeleton) Fig. 13.2a–h (bone features) Introductory material	Redrawn Added figure Improved depth
14	Figs. 14.1 and 14.2 (skulls) Fig. 14.7 (paranasal sinuses) Table 14.1 (skull passageways) Procedure	Redrawn Added figure Added table with locations and contents Expanded depth
15	Fig. 15.5 (rib) Procedures A and B Structural lists	Added figure Added introductory material Added functions and descriptions

Laboratory Exercise	Topic	Change
16	Fig. 16.2*b* (scapula) Fig. 16.5 (hand bones) Procedures A and B Structural lists	Added figure New figure Added introductory material Added functions and descriptions
17	Fig. 17.2*a* (hip bone) Fig. 17.2*b* (hip bone) Fig. 17.5 (foot bones) Procedures A and B Table 17.1 (male and female pelves) Critical Thinking Activities	Revised figure Added figure New figure Added introductory material Added comparison table Two added
18	Introductory material	Improved depth
19	Fig. 19.2 (synovial joint) Fig. 19.3*b* (cadaver knee) Procedures A and B Critical Thinking Activity	New figure Added figure Added introductory material One added
20	Fig. 20.1 (neuromuscular junctions) Fig. 20.3 (fascicle) Fig. 20.5 (sarcomere) Table 20.1 (muscle descriptions) Structural list Ph.I.L.S. Lesson 5	Added micrograph Added micrograph Added micrograph Added table Functions and descriptions added Clarity added
21	BIOPAC Exercise (Electromyography)	Rewritten
22	Tables 22.1, 22.2, 22.3, and 22.4 (head and neck muscles)	Added tables with origins, insertions, and actions
23	Fig. 23.5*a–b* (forearm muscles) Tables 23.1, 23.2, 23.3, and 23.4 (chest, shoulder, and upper limb muscles) Procedure	New figure Added tables with origins, insertions, and actions Reworked
24	Title and two procedures Fig. 24.4*a–c* (pelvic floor muscles) Tables 24.1, 24.2, and 24.3 (vertebral column, abdominal wall, and pelvic floor muscles)	Improved topics, clarity, and depth New figure Added tables with origins, insertions, and actions
25	Fig. 25.5*b* (leg muscles) Fig. 25.7 (leg muscles) Tables 25.1, 25.2, and 25.3 (hip and lower limb muscles)	Added figure Added figure Added tables with origins, insertions, and actions
26	Procedure Laboratory Assessment Part C	Reworked sequence and clarity Improved design and directions
27	Procedures A and B Fig. 27.1 (structural neurons) Fig. 27.6 (neuroglia) Fig. 27.8 (Purkinje cell) Fig. 27.9 Tables 27.1 and 27.2 (neurons and neuroglia)	New organization Added figure Added figure Added micrograph Added figure Added tables with characteristics, locations, and functions
28	Title and three procedures Fig. 28.3 (spinal cord) Fig. 28.4 (spinal nerves) Fig. 28.7 (meninges) Fig. 28.8 (spinal cord)	Expanded topics, clarity, and depth Expanded content Added figure Added figure New micrograph
29	Fig. 29.1 (withdrawal reflex arc) Fig. 29.2 (stretch reflex arc) Laboratory Assessment Part A table	Expanded content Added figure Expanded components
30	Figs. 30.1 and 30.2 (ventricles of brain) Fig. 30.7 (cerebellum and brainstem) Tables 30.1, 30.2, and 30.3 (brain and cranial nerves) Procedure A	Added figures Added figure Added tables with descriptions and functions Added introductory material

Changes to This Edition

Laboratory Exercise	Topic	Change
31	BIOPAC (Electroencephalography)	Rewritten
32	Fig. 32.3 (sheep brain) Fig. 32.6 (sheep brain)	Redrawn figure Added figure
33	Fig. 33.3 (two-point test) Table 33.1 (skin receptors)	Added figure Added table
34	Fig. 34.1 (smell receptors) Fig. 34.4 (taste bud)	Revised figure Revised orientation
35	Fig. 35.1 (lacrimal apparatus) Fig. 35.6 (eye exam) Fig. 35.12 (sectioned eye) Table 35.1 (eye muscles) Structural list	New figure New figure New micrograph Added table with actions and nerves Descriptions and functions added
36	Fig. 36.1 (refractive defects) Procedure A	Added figure Clarified directions
37	Structural list	Descriptions and functions added
38	Laboratory exercise title	Better reflects content of lab
39	Fig. 39.1 (major endocrine glands) Fig. 39.6 (thyroid gland) Fig. 39.12 (pancreas) Ph.I.L.S. Lesson 19	New figure New micrograph New micrograph Clarity added
40	Procedure A	Added introductory material
41	Figs. 41.2 and 41.5 (blood cells) Table 41.1 (blood components) Introductory material	Added micrographs Updated and expanded content Improved depth
42	Procedure D (cholesterol test) Ph.I.L.S. Lesson 34 Introductory material	Added procedure Clarity added Improved depth
43	Fig. 43.4 (blood test results) Table 43.2 (blood typing reactions)	Added figure Updated
44	Fig. 44.3 (sectioned heart) Fig. 44.6 (blood circuits) Structural list Terminology	Added figure Added figure Added descriptions and functions Updated
45	Fig. 45.1 (heart sound locations) Fig. 45.4 (cardiac cycle) Procedure B	Updated Added figure Added introductory material
46	BIOPAC (Electrocardiography)	Rewritten
47	Fig. 47.1 (blood vessel wall structure) Fig. 47.2 (artery and vein) Fig. 47.6 (cerebral arterial circle) Introductory material Procedures A, C, and D	Added figure New micrograph Added figure Improved depth Added introductory material
48	Fig. 48.2 (taking pulse rate) Fig. 48.5 (taking blood pressure) Introductory material Procedure B Ph.I.L.S. Lesson 40	Added figure Added figure Improved depth Added introductory material Clarity added
49	Fig. 49.1 (fluid movements) Fig. 49.2 (lymph drainage areas) Fig. 49.6a (cadaver lymph node) Fig. 49.6c (lymph node) Fig. 49.7 (thymus) Fig. 49.8 (spleen) Introductory material	Added figure Added figure Added figure New micrograph New micrograph New micrograph Improved depth

Laboratory Exercise	Topic	Change
50	Fig. 50.1 (respiratory organs) Fig. 50.5 (respiratory organs) Fig. 50.6 (trachea wall) Structural list	New figure Added figure New micrograph Added descriptions and functions
51	Fig. 51.1 (respiratory muscles) Fig. 51.2 (model for air movements) Procedure A Ph.I.L.S. Lesson 38	Added figure Added figure Added introductory material Clarity added
52	BIOPAC (Spirometry)	Rewritten
53	Figs. 53.1 and 53.2 (respiratory organs) Fig. 53.3 (peripheral chemoreceptors) Introductory material	New figures Added figure Updated and improved depth
54	Fig. 54.7 (stomach wall) Structural lists	New micrograph Added descriptions and functions
55	Fig. 55.1 (lock-and-key model) Introductory material	Added figure Improved depth
56	Laboratory exercise title Fig. 56.2 (kidney section) Fig. 56.6 (urethra of female and male) Fig. 56.9 (urethra) Procedure C Structural lists	Reflects expanded content Revised labels Added figure Added micrograph Added introductory material and urethra Added descriptions and functions
57	Procedures A and B	Reflects new organization of contents
58	Fig. 58.1 (male reproductive system) Fig. 58.2 (testis of cadaver) Fig. 58.3 (testis) Fig. 58.4 (seminiferous tubule) Fig. 58.5 (epididymis) Fig. 58.6 (ductus deferens) Structural lists Laboratory Assessments Part B	New figure Added figure Expanded labels New figure New figure Added figure Added descriptions and functions New content arrangement
59	Fig. 59.3 (female cadaver organs) Fig. 59.8 (uterine tube) Fig. 59.9 (uterine wall) Structural lists Laboratory Assessments Part B	Added figure New micrograph New micrograph Added descriptions and functions New content arrangement
60	Structural list	Added descriptions and functions
61	Introductory material	Improved depth

Acknowledgments

I value all the support and encouragement by the staff at McGraw-Hill, including Michelle Watnick, Jim Connely, Fran Simon, and Peggy Selle. Special recognition is granted to Colin Wheatley for his insight, confidence, wisdom, warmth, and friendship

I am grateful for the professional talent of Lurana Bain, Phillip Snider, and Janet Brodsky for their significant involvement in the development of some of the laboratory exercises. The main contents of the surface anatomy lab were provided by Lurana Bain, Ph.I.L.S. lab simulations components by Phillip Snider, and BIOPAC® labs by Janet Brodsky. I also want to thank Phillip Snider, Sue Caley-Opsal, and Kash Dutta for their contributions to the digital content included with this lab manual.

I am particularly thankful to Dr. Norman Jenkins and Dr. David Louis, retired presidents of Kishwaukee College, and Dr. Thomas Choice, president of Kishwaukee College, for their support, suggestions, and confidence in my endeavors. I am appreciative for the expertise of Womack Photography for numerous contributions. The professional reviews of the nursing procedures were provided by Kathy Schnier. I am also grateful to Laura Anderson, Rebecca Doty, Michele Dukes, Troy Hanke, Jenifer Holtzclaw, Stephen House, Shannon Johnson, Brian Jones, Marissa Kannheiser, Morgan Keen, Marcie Martin, Angele Myska, Sparkle Neal, Bonnie Overton, Susan Rieger, Eric Serna, Robert Stockley, Shatina Thompson, Nancy Valdivia, Marla Van Vickle, Jana Voorhis, Joyce Woo, and DeKalb Clinic for their contributions. There have been valuable contributions from my students, who have supplied thoughtful suggestions and assisted in clarification of details.

To my son Ross, an art instructor, I owe gratitude for his keen eye, creative suggestions, and cover illustration. Foremost, I am appreciative to Sherrie Martin, my spouse and best friend, for advice, understanding, and devotion throughout the writing and revising.

Terry R. Martin
Kishwaukee College
21193 Malta Road
Malta, IL 60150

Reviewers

I would like to express my sincere gratitude to all reviewers of the laboratory manual who provided suggestions for its improvement. Their thoughtful comments and valuable suggestions are greatly appreciated. They include the following:

M. Abdel-Sayed
  MEC/CUNY
Sharon I. Allen
  Reading Area Community College
Frank Ambriz
  University of Texas–Pan American
Pamela Anderson Cole
  Shelton State Community College
Bert Atsma
  Union County College
Jerry D. Barton II
  Tarrant County College–South Campus
Rachel Beecham
  Mississippi Valley State University
Moges Bizuneh
  Ivy Tech Community College
Danita Bradshaw-Ward
  Eastfield College
Gary Brady
  Spokane Falls Community College
Janet Brodsky
  Ivy Tech Community College
Beth Campbell
  Itawamba Community College
Ronald Canterbury
  University of Cincinnati

Claire Carpenter
  Yakima Valley Community College
Roger Choate
  Oklahoma City Community College
Ana Christensen
  Lamar University
Larissa Clark
  Arkansas State University-Newport
Harry Davis
  Gloucester County College
Mary Beth Davison
  Chatham University
Frank J. DeMaria
  Suffolk County Community College
William E. Dunscombe
  Union County College
Scott Elliott
  Whatcom Community College
Marirose T. Ethington
  Genesee Community College
David L. Evans
  Pennsylvania College of Technology
Amy Fenech Sandy
  Columbus Technical College
Tammy Filliater
  MSU–Great Falls College of Technology

Vanessa A. Fitsanakis
  King College
Andrew Flick
  Blue Ridge Community College
Katelijne Flies
  Central New Mexico Community College
Maria Florez
  Lone Star College–Cy Fair
Mary C. Fox
  University of Cincinnati
Sharon Fugate
  Madisonville Community College
Tammy R. Gamza
  Arkansas State University Beebe
Joseph D. Gar
  West Kentucky Community and Technical College
Kristine Garner
  University of Arkansas–Fort Smith
Anthony J. Gaudin
  Ivy Tech Community College
Samita Ghoshal
  Lone Star College, Texas
Louis A. Giacinti
  Milwaukee Area Technical College

Tejendra Gill
University of Houston
Gary Glaser
Erie Community College
Brent M. Graves
Northern Michigan University
Dale Harrington
Caldwell Community College
Gillian Hart
Trinidad State Junior College
Clare Hays
Metro State College of Denver
Gerald A. Heins
Northeast Wisconsin Technical College
Todd Heldreth
Charleston Southern University
D.J. Hennager
Kirkwood Community College
Kerrie Hoar
University of Wisconsin–La Crosse
Dale R. Horeth
Tidewater Community College
Michele Iannuzzi Sucich
SUNY Orange
Mark Jaffe
Nova Southeastern University
Dena Johnson
Tarrant County College NW
Edward Johnson
Central Oregon Community College
Jody Johnson
Arapahoe Comm. College
Geeta Joshi
Southeastern Community College
Susanne Kalup
Westmoreland County Community College
Lisa A. E. Kaplan
Quinnipiac University
Robert Keeton
University of Arkansas Community College
Suzanne Kempke
St. John's River Community College
L. Henry Kermott
St. Olaf College
Steven Kish
Zane State College
Will Kleinelp
Middlesex County College
Leigh Kleinert
Grand Rapids Community College
Michael S. Kopenitis
Amarillo College
Gina Langley
ENMU-Ruidoso

J. Ellen Lathrop-Davis
CCBC-Catonsville
Steven Leadon
Durham Technical Community College
Leigh Levitt
Union County College
Jerri K. Lindsey
Tarrant County College-Northeast Campus
Lynn B. Littman
Gwynedd-Mercy College
Community College of Philadelphia
Kenneth Long
California Lutheran University
Sarah Lovern
Concordia University
Jaime Malcore Tjossem
Rochester Community and Technical College
Peter Malo
Richard J Daley College
Anita Mandal
Edward Waters College
Barry Markillie
Cape Fear Community College
Deborah McCool
Mount Aloysius College
Lori McGrew
Belmont
Cherie McKeever
MSU College of Technology
Jonathan McMemamin-Balano
Norwalk Community College
Mark E. Meade
Jacksonville State University
Judy Megaw
Indian River State College
Claire Miller
Community College of Denver
Howard K. Motoike
LaGuardia Community College
Necia Nicholas
Calhoun Community College
Nancy O'Keefe
Purdue University Calumet
Sidney Palmer
Brigham Young University–Idaho
Karen Payne
Chattanooga State Technical Community College
Roger D. Peffer
Montana State University–Great Falls College of Technology
Andrew J. Petto
University of Wisconsin–Milwaukee

Danny M. Pincivero
The University of Toledo
Sara Reed Houser
Jefferson College of Health Sciences
Jackie Reynolds
Richland College
Rebecca Roush
Sandhills Community College
Tim Roye
San Jacinto College
Charlotte Russell
Simmons College
Radmila Sarac
Purdue University Calumet
Kathleen Sellers
Columbus State University
Kelly J. Sexton
Dallas County Community College District
Donald Shaw
University of Tennessee at Martin
Mark A. Shoop
Tennessee Wesleyan College
Crystal Sims
Cossatot Community College UA
Michelle Slover
Clarke College
Phillip Snider
Gadsden State Community College
Olufemi Sodeinde
Medgar Evers College, CUNY
Asha Stephens
College of the Mainland
Lydia Thebeau
Missouri Baptist University
Janis Thompson
Lorain County Community College
Diane G. Tice
Morrisville State College
Terri L. Tillen
Madisonville Community College
Pete Van Dyke
Walla Walla Community College
Delon Washo-Krupps
Arizona State University
Clement Yedjou
Jackson State University
Ruth A. Young
Northern Essex Community College
Paul Zillgitt
University of Wisconsin-Waukesha
David J. Zimmer
Erie Community College
Jay Zimmer
South Florida Community College
Jeffrey Zuiderveen
Columbus State University

About the Author

This laboratory manual series is by Terry R. Martin of Kishwaukee College. Terry's teaching experience of over forty years, his interest in students and love for college instruction, and his innovative attitude and use of technology-based learning enhance the solid tradition of his other well-established laboratory manuals. Among Terry's awards are the Kishwaukee College Outstanding Educator, Phi Theta Kappa Outstanding Instructor Award, Kishwaukee College ICCTA Outstanding Educator Award, Who's Who Among America's Teachers, Kishwaukee College Faculty Board of Trustees Award of Excellence, and Continued Excellence Award for Phi Theta Kappa Advisors. Terry's professional memberships include the National Association of Biology Teachers, Illinois Association of Community College Biologists, Human Anatomy and Physiology Society, former Chicago Area Anatomy and Physiology Society (founding member), Phi Theta Kappa (honorary member), and Nature Conservancy. In addition to writing many publications, he co-produced with Hassan Rastegar a videotape entitled *Introduction to the Human Cadaver and Prosection,* published by Wm. C. Brown Publishers. Terry revised the *Laboratory Manual to Accompany Hole's Human Anatomy and Physiology,* Thirteenth Edition, revised the *Laboratory Manual to Accompany Hole's Essentials of Human Anatomy and Physiology,* Eleventh Edition, and authored *Human Anatomy and Physiology Laboratory Manual, Fetal Pig Dissection*, Third Edition. A series of seven LabCam videos of anatomy and physiology laboratory processes were produced at Kishwaukee College and published by McGraw-Hill Higher Education. Cadaver dissection experiences have been provided for

his students for over twenty-five years. Terry teaches portions of EMT and paramedic classes and serves as a Faculty Consultant for Advanced Placement Biology examination readings. Terry has also been a faculty exchange member in Ireland. The author locally supports historical preservation, natural areas, scouting, and scholarship. We are pleased to have Terry continue the tradition of authoring laboratory manuals for McGraw-Hill Higher Education.

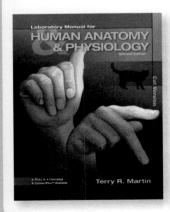
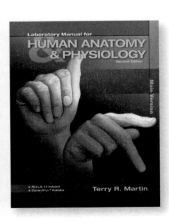
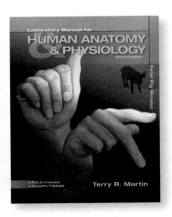
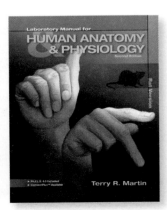

The exercises in this laboratory manual will provide you with opportunities to observe various anatomical structures and to investigate certain physiological phenomena. Such experiences should help you relate specimens, models, microscope slides, and your body to what you have learned in the lecture and read about in the textbook.

Frequent variations exist in anatomical structures among humans. The illustrations in the laboratory manual represent normal (normal means the most common variation) anatomy. Variations from normal anatomy do not represent abnormal anatomy unless some function is impaired.

The following list of suggestions and study skills may make your laboratory activities more effective and profitable.

1. Prepare yourself before attending the laboratory session by reading the assigned exercise and reviewing the related sections of the textbook and lecture notes as indicated in the pre-lab section of the laboratory exercise. Answer the pre-lab questions. It is important to have some understanding of what will be done in the lab before you come to class.

2. Visit the website **www.mhhe.com/martinseries2** and view suggested *Anatomy & Physiology Revealed* animations and view LabCam videos.

3. Be on time. During the first few minutes of the laboratory meeting, the instructor often will provide verbal instructions. Make special note of any changes in materials to be used or procedures to be followed. Also listen carefully for information about special techniques to be used and precautions to be taken.

4. Keep your work area clean and your materials neatly arranged so that you can locate needed items. This will enable you to efficiently proceed and will reduce the chances of making mistakes.

5. Pay particular attention to the purpose of the exercise, which states what you are to accomplish in general terms, and to the learning outcomes, which list what you should be able to do as a result of the laboratory experience. Then, before you leave the class, review the outcomes and make sure that you can perform all of the assessments.

6. Precisely follow the directions in the procedure and proceed only when you understand them clearly. Do not improvise procedures unless you have the approval of the laboratory instructor. Ask questions if you do not understand exactly what you are supposed to do and why you are doing it.

7. Handle all laboratory materials with care. These materials often are fragile and expensive to replace. When-

ever you have questions about the proper treatment of equipment, ask the instructor.

8. Treat all living specimens humanely and try to minimize any discomfort they might experience.

9. Although at times you might work with a laboratory partner or a small group, try to remain independent when you are making observations, drawing conclusions, and completing the activities in the laboratory reports.

10. Record your observations immediately after making them. In most cases, such data can be entered in spaces provided in the laboratory assessments.

11. Read the instructions for each section of the laboratory assessment before you begin to complete it. Think about the questions before you answer them. Your responses should be based on logical reasoning and phrased in clear and concise language.

12. At the end of each laboratory period, clean your work area and the instruments you have used. Return all materials to their proper places and dispose of wastes, including glassware or microscope slides that have become contaminated with human blood or body fluids, as directed by the laboratory instructor. Wash your hands thoroughly before leaving the laboratory.

Study Skills for Anatomy and Physiology

Students have found that certain study skills worked well for them while enrolled in Human Anatomy and Physiology. Although everyone has his or her learning style, there are techniques that work well for most students. Using some of the skills listed here could make your course more enjoyable and rewarding.

1. **Time management:** Prepare monthly, weekly, and daily schedules. Include dates of quizzes, exams, and projects on the calendar. On your daily schedule, budget several short study periods. Daily repetition alleviates cramming for exams. Prioritize your time so that you still have time for work and leisure activities. Find an appropriate study atmosphere with minimum distractions.

2. **Note taking:** Look for the main ideas and briefly express them in your own words. Organize, edit, and review your notes soon after the lecture. Add textbook information to your notes as you reorganize them. Underline or highlight with different colors the important points,

To the Student

major headings, and key terms. Study your notes daily, as they provide sequential building blocks of the course content.

3. **Chunking:** Organize information into logical groups or categories. Study and master one chunk of information at a time. For example, study the bones of the upper limb, lower limb, trunk, and head as separate study tasks.

4. **Mnemonic devices:** An *acrostic* is a combination of association and imagery to aid your memory. It is often in the form of a poem, rhyme, or jingle in which the first letter of each word corresponds to the first letters of the words you need to remember. **S**o **L**ong **T**op **P**art, **H**ere **C**omes **T**he **T**humb is an example of such a mnemonic device for remembering the eight carpals in a correct sequence. *Acronyms* are words formed by the first letters of the items to remember. *IPMAT* is an example of this type of mnemonic device to help you remember the phases of the cell cycle in the correct sequence. Try to create some of your own.

5. **Note cards/flash cards:** Make your own. Add labels and colors to enhance the material. Keep them with you in your pocket or purse. Study them often and for short periods. Concentrate on a small number of cards at one time. Shuffle your cards and have someone quiz you on their content. As you become familiar with the material, you can set aside cards that don't require additional mastery.

6. **Recording and recitation:** An auditory learner can benefit by recording lectures and review sessions with a cassette recorder. Many students listen to the taped sessions as they drive or just before going to bed. Reading your notes aloud can help also. Explain the material to anyone (even if there are no listeners). Talk about anatomy and physiology in everyday conversations.

7. **Study groups:** Small study groups that meet periodically to review course material and compare notes have helped and encouraged many students. However, keep the group on the task at hand. Work as a team and alternate leaders. This group often becomes a support group.

Practice sound study skills during your anatomy and physiology endeavor.

The Use of Animals in Biology Education*

The National Association of Biology Teachers (NABT) believes that the study of organisms, including nonhuman animals, is essential to the understanding of life on Earth. NABT recommends the prudent and responsible use of animals in the life science classroom. NABT believes that biology teachers should foster a respect for life. Biology teachers also should teach about the interrelationship and interdependency of all things.

Classroom experiences that involve nonhuman animals range from observation to dissection. NABT supports these experiences so long as they are conducted within the long-established guidelines of proper care and use of animals, as developed by the scientific and educational community.

As with any instructional activity, the use of nonhuman animals in the biology classroom must have sound educational objectives. Any use of animals, whether for observation or dissection, must convey substantive knowledge of biology. NABT believes that biology teachers are in the best position to make this determination for their students.

NABT acknowledges that no alternative can substitute for the actual experience of dissection or other use of animals and urges teachers to be aware of the limitations of alternatives. When the teacher determines that the most effective means to meet the objectives of the class do not require dissection, NABT accepts the use of alternatives to dissection, including models and the various forms of multimedia. The Association encourages teachers to be sensitive to substantive student objections to dissection and to consider providing appropriate lessons for those students where necessary.

To implement this policy, NABT endorses and adopts the "Principles and Guidelines for the Use of Animals in Precollege Education" of the Institute of Laboratory Animals Resources (National Research Council). Copies of the "Principles and Guidelines" may be obtained from the ILAR (2101 Constitution Avenue, NW, Washington, DC 20418; 202-334-2590).

*Adopted by the Board of Directors in October 1995. This policy supersedes and replaces all previous NABT statements regarding animals in biology education.

Scientific Method and Measurements

Purpose of the Exercise

To become familiar with the scientific method of investigation, to learn how to formulate sound conclusions, and to provide opportunities to use the metric system of measurements.

Materials Needed

Meterstick
Calculator
Human skeleton

Learning Outcomes

After completing this exercise, you should be able to

1. Convert English measurements to the metric system, and vice versa.

2. Measure and record upper limb lengths and heights of ten subjects.

3. Apply the scientific method to test the validity of a hypothesis concerning the direct, linear relationship between human upper limb length and height.

4. Design an experiment, formulate a hypothesis, and test it using the scientific method.

Pre-Lab

Carefully read the introductory material and examine the entire lab. Be familiar with the scientific method from lecture or the textbook. Answer the pre-lab questions.

Pre-Lab Questions: Select the correct answer for each of the following questions:

1. To explain scientific biological phenomena, scientists use a technique called
   a. the scientific method.   **b.** the scientific law.
   c. conclusions.   **d.** measurements.

2. Which of the following represents the correct sequence of the scientific method?
   a. analysis of data, conclusions, observations, experiment, hypothesis
   b. conclusions, experiment, hypothesis, analysis of data, observations
   c. observations, hypothesis, experiment, analysis of data, conclusions
   d. hypothesis, observations, experiment, analysis of data, conclusions

3. A theory, verified continuously from experiments, might become known as the
   a. conclusions.   **b.** hypothesis.
   c. valid results.   **d.** scientific law.

4. The most likely scientific unit for measuring the height of a person would be
   a. feet.   **b.** centimeters.
   c. inches.   **d.** kilometers.

5. Which of the following is *not* a unit of the metric system of measurements?
   a. centimeters   **b.** liters
   c. inches   **d.** millimeters

6. The hypothesis is formulated from the results of the experiment.
   True _____   False _____

7. A centimeter represents an example of a metric unit of length.
   True _____   False _____

Scientific investigation involves a series of logical steps to arrive at explanations for various biological phenomena. This technique, called the *scientific method,* is used in all disciplines of science. It allows scientists to draw logical and reliable conclusions about phenomena.

The scientific method begins with *observations* related to the topic under investigation. This step commonly involves the accumulation of previously acquired information and/or your observations of the phenomenon. These observations are used to formulate a tentative explanation known as the *hypothesis.* An important attribute of a hypothesis is that it must be testable. The testing of the hypothesis involves performing a carefully controlled *experiment* to obtain data that can be used to support, reject, or modify the hypothesis. An *analysis of data* is conducted using sufficient information collected during the experiment. Data analysis may include organization and presentation of data as tables, graphs, and drawings. From the interpretation of the data analysis, *conclusions* are drawn. (If the data do not support the hypothesis, you must reexamine the experimental design and the data, and if needed develop a new hypothesis.) The final presentation of the information is made from the conclusions. Results and conclusions are presented to the scientific community for evaluation through peer reviews, presentations at professional meetings, and published articles. If many investigators working independently can validate the hypothesis by arriving at the same conclusions, the explanation becomes a **theory.** A theory verified continuously over time and accepted by the scientific community becomes known as a **scientific law** or **principle.** A scientific law serves as the standard explanation for an observation unless it is disproved by new information. The five components of the scientific method are summarized as

Observations

↓

Hypothesis

↓

Experiment

↓

Analysis of data

↓

Conclusions

Metric measurements are characteristic tools of scientific investigations. The English system of measurements is often used in the United States, so the investigator must make conversions from the English system to the metric system. Use table 1.1 for the conversion of English units of measure to metric units for length, mass, volume, time, and temperature.

Procedure A—Using the Steps of the Scientific Method

To familiarize you with the components of the scientific method, this procedure represents a specific example of the order of the steps utilized. Each of the steps for this procedure will guide you through the proper sequence in an efficient pathway.

1. Many people have observed a correlation between the length of the upper and lower limbs and the height (stature) of an individual. For example, a person who has long upper limbs (the arm, forearm, and hand combined) tends to be tall. Make some visual observations of other people in your class to observe a possible correlation.

2. From such observations, the following hypothesis is formulated: The length of a person's upper limb is equal to 0.4 (40%) of the height of the person. Test this hypothesis by performing the following experiment.

3. In this experiment, use a meterstick (fig. 1.1) to measure an upper limb length of ten subjects. For each measurement, place the meterstick in the axilla (armpit) and record the length in centimeters to the end of the longest finger (fig. 1.2). Obtain the height of each person in centimeters by measuring them without shoes against a wall (fig. 1.3). The height of each person can be calculated by multiplying each individual's height in inches by 2.54 to obtain his/her height in centimeters. Record all your measurements in Part A of Laboratory Assessment 1.

4. The data collected from all of the measurements can now be analyzed. The expected (predicted) correlation between upper limb length and height is determined using the following equation:

Height × 0.4 = expected upper limb length

FIGURE 1.1 Metric ruler with metric lengths indicated. A meterstick length would be 100 centimeters. (The image size is approximately to scale.)

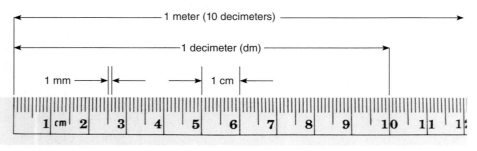

Metric ruler

TABLE 1.1 Metric Measurement System and Conversions

Measurement	Unit & Abbreviation	Metric Equivalent	Conversion Factor Metric to English (approximate)	Conversion Factor English to Metric (approximate)
Length	1 kilometer (km)	1,000 (10^3) m	1 km = 0.62 mile	1 mile = 1.61 km
	1 meter (m)	100 (10^2) cm 1,000 mm	1 m = 1.1 yards = 3.3 feet = 39.4 inches	1 yard = 0.9 m 1 foot = 0.3 m
	1 decimeter (dm)	0.1 (10^{-1}) m	1 dm = 3.94 inches	1 inch = 0.25 dm
	1 centimeter (cm)	0.01 (10^{-2}) m	1 cm = 0.4 inches	1 foot = 30.5 cm 1 inch = 2.54 cm
	1 millimeter (mm)	0.001 (10^{-3}) m 0.1 cm	1 mm = 0.04 inches	
	1 micrometer (μm)	0.000001 (10^{-6}) m 0.001 mm		
Mass	1 metric ton (t)	1,000 kg	1 t = 1.1 ton	1 ton = 0.91 t
	1 kilogram (kg)	1,000 g	1 kg = 2.2 pounds	1 pound = 0.45 kg
	1 gram (g)	1,000 mg	1 g = 0.04 ounce	1 pound = 454 g 1 ounce = 28.35 g
	1 milligram (mg)	0.001 g		
Volume (liquids and gases)	1 liter (L)	1,000 mL	1 L = 1.06 quarts	1 gallon = 3.78 L 1 quart = 0.95 L
	1 milliliter (mL)	0.001 L 1 cubic centimeter (cc or cm^3)	1 mL = 0.03 fluid ounce 1 mL = 1/4 teaspoon 1 mL = 15–16 drops	1 quart = 946 mL 1 fluid ounce = 29.6 mL 1 teaspoon = 5 mL
Time	1 second (s)	1/60 minute	same	same
	1 millisecond (ms)	0.001 s	same	same
Temperature	Degrees Celsius (°C)		°F = 9/5 °C + 32	°C = 5/9 (°F − 32)

FIGURE 1.2 Measurement of upper limb length.

The observed (actual) correlation to be used to test the hypothesis is determined by

Length of upper limb/height = actual % of height

5. A graph is an excellent way to display a visual representation of the data. Plot the subjects' data in Part A of the laboratory assessment. Plot the upper limb length of each subject on the x-axis and the height of each person on the y-axis. A line is already located on the graph that represents a hypothetical relationship of 0.4 (40%) upper limb length compared to height. This is a graphic representation of the original hypothesis.

6. Compare the distribution of all of the points (actual height and upper limb length) that you placed on the graph with the distribution of the expected correlation represented by the hypothesis.

7. Complete Part A of the laboratory assessment.

Procedure B—Design an Experiment

You have completed the steps of the scientific method with guidance directions in Procedure A. This procedure will allow for less guidance and more flexibility using the scientific method.

FIGURE 1.3 Measurement of height.

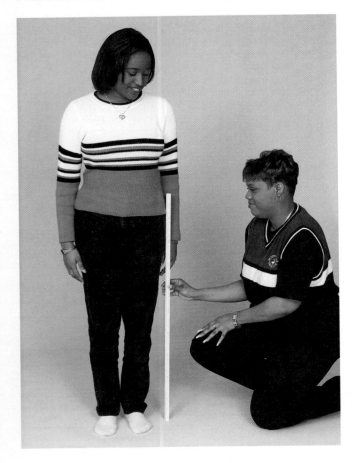

 Critical Thinking Activity

You have probably concluded that there is some correlation of the length of body parts to height. Often when a skeleton is found, it is not complete, especially when paleontologists discover a skeleton. It is occasionally feasible to use the length of a single bone to estimate the height of an individual. Observe human skeletons and locate the radius bone in the forearm. Use your observations to identify a mathematical relationship between the length of the radius and height. Formulate a hypothesis that can be tested. Make measurements, analyze data, and develop a conclusion from your experiment. Complete Part B of the laboratory assessment.

Name _____

Date _____

Section _____

The ⚠ corresponds to the indicated outcome(s) found at the beginning of the laboratory exercise.

Scientific Method and Measurements

Part A Assessments

1. Record measurements for the upper limb length and height of ten subjects. Use a calculator to determine the expected upper limb length and the actual percentage (as a decimal or a percentage) of the height for the ten subjects. Record your results in the following table. ⚠2

Subject	Measured Upper Limb Length (cm)	Height* (cm)	Height × 0.4 = Expected Upper Limb Length (cm)	Actual % of Height = Measured Upper Limb Length (cm)/Height (cm)
1.				
2.				
3.				
4.				
5.				
6.				
7.				
8.				
9.				
10.				

*The height of each person can be calculated by multiplying each individual's height in inches by 2.54 to obtain his/her height in centimeters. ⚠1

2. Plot the distribution of data (upper limb length and height) collected for the ten subjects on the following graph. The line located on the graph represents the *expected* 0.4 (40%) ratio of upper limb length to measured height (the original hypothesis). (The x-axis represents upper limb length, and the y-axis represents height.) Draw a line of *best fit* through the distribution of points of the plotted data of the ten subjects. Compare the two distributions (expected line and the distribution line drawn for the ten subjects). **3**

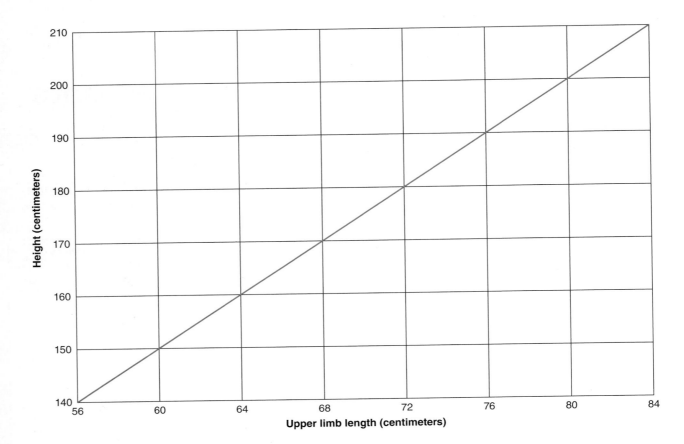

3. Does the distribution of the ten subjects' measured upper limb lengths support or reject the original hypothesis? _____ Explain your answer. **3**

Part B Assessments

1. Describe your observations of a possible correlation between the radius length and height. ◢4◣

2. Write a hypothesis based on your observations. ◢4◣

3. Describe the design of the experiment that you devised to test your hypothesis. ◢4◣

4. Place your analysis of the data in this space in the form of a table and a graph. ◢4◣

    a. Table:

    b. Graph:

5. Based on an analysis of your data, what conclusions can you make? Did these conclusions confirm or refute your original hypothesis? **4**

6. Discuss your results and conclusions with classmates. What common conclusion can the class formulate about the correlation between radius length and height? **4**

Body Organization, Membranes, and Terminology

Purpose of the Exercise

To review the organizational pattern of the human body, to review its organ systems and the organs included in each system, and to become acquainted with the terms used to describe the relative position of body parts, body sections, and body regions.

Materials Needed

Dissectible human torso model (manikin)
Variety of specimens or models sectioned along various planes

Learning Outcomes

After completing this exercise, you should be able to

1. Locate and name the major body cavities and identify the membranes associated with each cavity.

2. Associate the organs and functions included within each organ system and locate the organs in a dissectible human torso model.

3. Select the terms used to describe the relative positions of body parts.

4. Differentiate the terms used to identify body sections and identify the plane along which a particular specimen is cut.

5. Label body regions and associate the terms used to identify body regions.

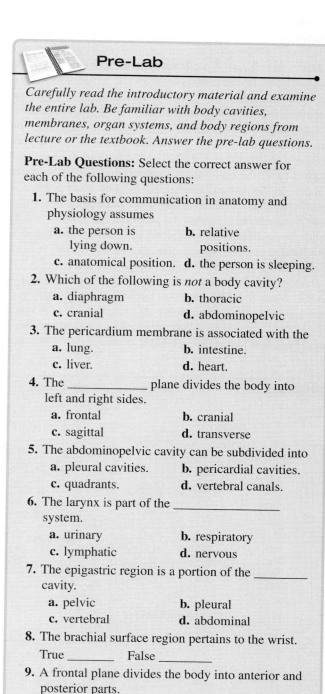
Pre-Lab

Carefully read the introductory material and examine the entire lab. Be familiar with body cavities, membranes, organ systems, and body regions from lecture or the textbook. Answer the pre-lab questions.

Pre-Lab Questions: Select the correct answer for each of the following questions:

1. The basis for communication in anatomy and physiology assumes
   a. the person is lying down.
   b. relative positions.
   c. anatomical position.
   d. the person is sleeping.

2. Which of the following is *not* a body cavity?
   a. diaphragm
   b. thoracic
   c. cranial
   d. abdominopelvic

3. The pericardium membrane is associated with the
   a. lung.
   b. intestine.
   c. liver.
   d. heart.

4. The _____ plane divides the body into left and right sides.
   a. frontal
   b. cranial
   c. sagittal
   d. transverse

5. The abdominopelvic cavity can be subdivided into
   a. pleural cavities.
   b. pericardial cavities.
   c. quadrants.
   d. vertebral canals.

6. The larynx is part of the _____ system.
   a. urinary
   b. respiratory
   c. lymphatic
   d. nervous

7. The epigastric region is a portion of the _____ cavity.
   a. pelvic
   b. pleural
   c. vertebral
   d. abdominal

8. The brachial surface region pertains to the wrist.
   True _____ False _____

9. A frontal plane divides the body into anterior and posterior parts.
   True _____ False _____

The major features of the organization of the human body include certain body cavities. The dorsal body cavity includes a cranial cavity containing the brain and a vertebral canal (spinal cavity) containing the spinal cord. The ventral body cavity includes the thoracic cavity, which is subdivided into a mediastinum containing primarily the heart, esophagus, and trachea. The thoracic cavity also includes two pleural cavities, each surrounding a lung. Also included in the ventral body cavity is the abdominopelvic cavity, composed of an abdominal cavity and pelvic cavity. The entire abdominopelvic cavity is further subdivided into either nine regions or four quadrants. The large size of the abdominopelvic cavity, with its many visceral organs, warrants these further subdivisions into regions or quadrants for convenience and for accuracy in describing organ locations, injury sites, and pain locations. Because early anatomical references do not address dorsal and ventral body cavities, some anatomists prefer not to use those broader terms.

Located within the ventral body cavities are thin serous membranes containing small amounts of a lubricating serous fluid. The double-layered membranes that secrete this fluid include the pericardium, pleura, and peritoneum. The pericardium is associated with the heart; the pleura is associated with a lung; the peritoneum is associated with the viscera located in the abdominopelvic cavity. The inner portion of each membrane attached to the organ is the visceral component, whereas the outer parietal portion forms an outer cavity wall. Several abdominopelvic organs, such as the kidneys, are located just behind (are retroperitoneal to) the parietal peritoneum, thus lacking a mesentery.

Although the human body functions as one entire unit, it is customary to divide the body into eleven body organ systems. In order to communicate effectively with each other about the body, scientists have devised anatomical terminology. Foremost in this task we use *anatomical position* as our basis for communication, including directional terms, body regions, and planes of the body. A person in anatomical position is standing erect, facing forward, with upper limbs at the sides and palms forward. This standard position allows us to describe relative positions of various body parts using such directional terms as left-right, anterior (ventral)-posterior (dorsal), proximal-distal, and superior-inferior. Body regions include certain surface areas, portions of limbs, and portions of body cavities. In order to study internal structures, often the body is depicted as sectioned into a sagittal plane, frontal plane, or transverse plane.

Procedure A—Body Cavities and Membranes

This procedure outlines the body cavities by location. Certain cavities contain thin layers of cells, called membranes, that line the cavity and cover the organs within the cavity. The figures included in Procedure A will allow you to locate and identify the associated membrane location and name appropriate for the cavity and the organs involved.

1. Study figures 2.1, 2.2, and 2.3 to become familiar with body cavities and associated membranes.
2. Locate the following features on the dissectible human torso model (fig. 2.4):

   body cavities
   - cranial cavity—houses brain
   - vertebral canal (spinal cavity)—houses spinal cord
   - thoracic cavity
     - mediastinum—region between the lungs; includes pericardial cavity
     - pleural cavities (2)

FIGURE 2.1 Major body cavities, left lateral view.

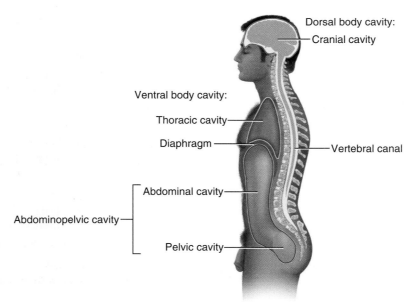

FIGURE 2.2 Thoracic membranes and cavities associated with (a) the lungs and (b) the heart.

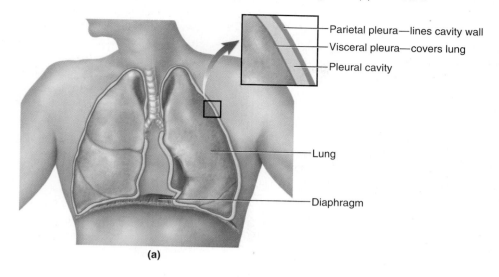

Parietal pleura—lines cavity wall
Visceral pleura—covers lung
Pleural cavity
Lung
Diaphragm

(a)

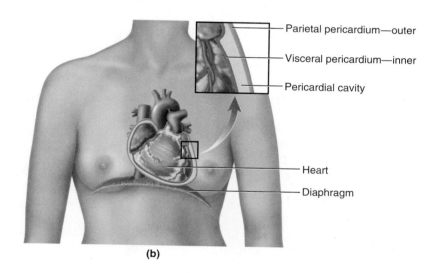

Parietal pericardium—outer
Visceral pericardium—inner
Pericardial cavity
Heart
Diaphragm

(b)

- abdominopelvic cavity
  - abdominal cavity
  - pelvic cavity

diaphragm—separates thoracic and abdominopelvic cavities

smaller cavities within the head

- oral cavity
- nasal cavity with connected sinuses
- orbital cavity
- middle ear cavity

membranes and cavities

- pleural cavity—associated with lungs
  - parietal pleura—lines cavity wall
  - visceral pleura—covers lungs
- pericardial cavity—associated with heart
  - parietal pericardium—covered by fibrous pericardium
  - visceral pericardium—epicardium
- peritoneal cavity—associated with abdominal organs
  - parietal peritoneum—lines cavity wall
  - visceral peritoneum—covers organs

3. Complete Part A of Laboratory Assessment 2.

FIGURE 2.3 Serous membranes of the abdominal cavity are shown in this left lateral view.

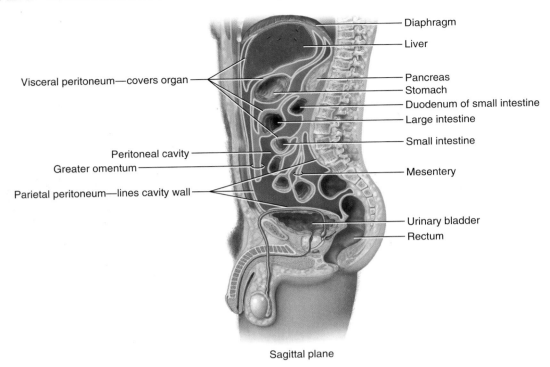

Diaphragm

Liver

Visceral peritoneum—covers organ

Pancreas
Stomach
Duodenum of small intestine
Large intestine
Small intestine

Peritoneal cavity
Greater omentum
Parietal peritoneum—lines cavity wall

Mesentery

Urinary bladder
Rectum

Sagittal plane

FIGURE 2.4 Dissectible human torso model with body cavities, abdominopelvic quadrants, body planes, and major organs indicated.

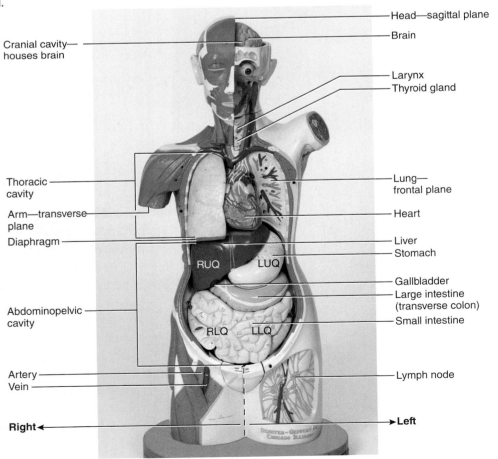

Head—sagittal plane

Brain

Cranial cavity—houses brain

Larynx
Thyroid gland

Thoracic cavity

Arm—transverse plane

Diaphragm

Lung—frontal plane

Heart

Liver
Stomach

RUQ   LUQ

Gallbladder
Large intestine (transverse colon)
Small intestine

Abdominopelvic cavity

RLQ   LLQ

Artery
Vein

Lymph node

Right ◄   ► **Left**

12

Procedure B—Organ Systems

The eleven body systems are listed in this procedure, and a major function of each system is included. Representative major organs are listed within each of the systems. By using models and charts to locate each major organ included within the system, you will have an early overview of each system of the human body. More detailed organs and functions are covered within each system during later laboratory exercises.

1. Use the dissectible human torso model (fig. 2.4) to locate the following systems and their major organs:

    integumentary system—protection
    - skin
    - accessory organs such as hair and nails

    skeletal system—support and protection
    - bones
    - ligaments

    muscular system—movement
    - skeletal muscles
    - tendons

    nervous system—detects changes; interprets sensory information; stimulates muscles and glands
    - brain
    - spinal cord
    - nerves

    endocrine system—secretes hormones
    - pituitary gland
    - thyroid gland
    - parathyroid glands
    - adrenal glands
    - pancreas
    - ovaries—in females
    - testes—in males
    - pineal gland
    - thymus

    cardiovascular system—transports gases, nutrients, and wastes
    - heart
    - arteries
    - veins

    lymphatic system—immune cell production and return tissue fluid to blood
    - lymphatic vessels
    - lymph nodes
    - thymus
    - spleen

    respiratory system—gas exchange between air and blood
    - nasal cavity
    - pharynx
    - larynx
    - trachea
    - bronchi
    - lungs

    digestive system—food breakdown and absorption
    - mouth
    - tongue
    - teeth
    - salivary glands
    - pharynx
    - esophagus
    - stomach
    - liver
    - gallbladder
    - pancreas
    - small intestine
    - large intestine

    urinary system—removes liquids and wastes from blood
    - kidneys
    - ureters
    - urinary bladder
    - urethra

    male reproductive system—sperm production
    - scrotum
    - testes
    - penis
    - urethra

    female reproductive system—egg production and fetal development
    - ovaries
    - uterine tubes (oviducts; fallopian tubes)
    - uterus
    - vagina

2. Complete Part B of the laboratory assessment.

Procedure C—Relative Positions, Planes, and Regions

This procedure illustrates and incorporates planes (sections), abdominopelvic subdivisions, and surface regions using the anatomical position of the human body. Communications and directions for anatomical study assume the person is in an anatomical position.

Anatomical planes (sections; cuts) of the entire body and individual organs allow a better understanding of the internal structure and function. Sectional anatomy is accomplished by cutting the body in standard ways to maximize the view of the particular organ. The anatomical planes include the sagittal, frontal, and transverse sections or cuts. Radiologic images by such techniques as a CT (computed tomography) scan or an MRI (magnetic resonance imaging) are examples of the advances and the importance of sectional anatomy for the diagnosis and treatment of medical situations.

The abdominopelvic area of the body is rather large and contains numerous viscera representing structural and functional parts of several body systems. People with abdominopelvic discomfort will often complain about it using terms like a "stomach ache." This is too general a term to be used for medical tests and procedures. Therefore, this abdominopelvic portion of the body is further subdivided into four quadrants and/or nine regions. Some medical disciplines prefer the use of quadrants, while others prefer the more specific use of the regions.

The general surface of the entire body is subdivided into regions using anatomical terms appropriate for that particular portion of the body. This becomes helpful for descriptions of injuries or pain locations that are more specific than using broader general terms such as head, neck, and upper limb, for example.

1. Observe the person standing in anatomical position (fig. 2.5). Anatomical terminology assumes the body is in anatomical position even though a person is often observed differently.
2. Study figures 2.6, 2.7, and 2.8 to become familiar with anatomical planes, abdominopelvic quadrants and regions, and body surface regions.
3. Examine the sectioned specimens on the demonstration table and identify the plane along which each is cut.
4. Complete Parts C, D, and E of the laboratory assessment.

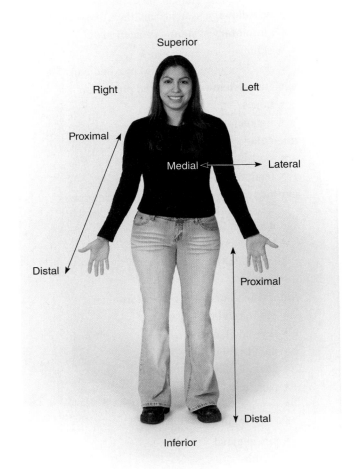

FIGURE 2.5 Anatomical position with directional terms indicated. The body is standing erect, face forward, with upper limbs at the sides and palms forward. When the palms are forward (supinated) the radius and ulna in the forearm are nearly parallel. This results in an anterior view of the body. The relative positional terms are used to describe a body part's location in relation to other body parts.

FIGURE 2.6   Anatomical planes (sections) of the body.

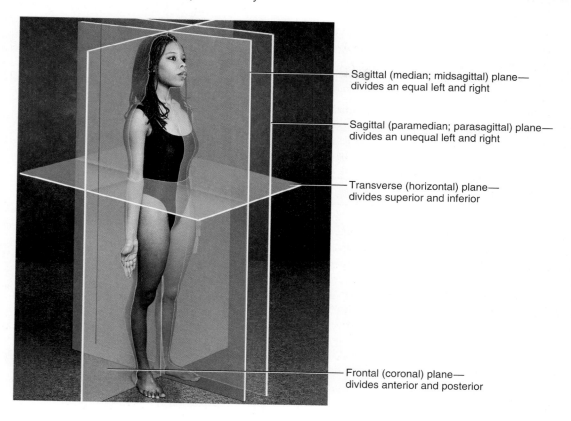

Sagittal (median; midsagittal) plane—
divides an equal left and right

Sagittal (paramedian; parasagittal) plane—
divides an unequal left and right

Transverse (horizontal) plane—
divides superior and inferior

Frontal (coronal) plane—
divides anterior and posterior

FIGURE 2.7   Abdominopelvic (a) quadrants and (b) regions. An area as large as the abdominopelvic cavity is subdivided into either four quadrants or nine regions for purposes of locating organs, injuries, or pain, or for performing medical procedures.

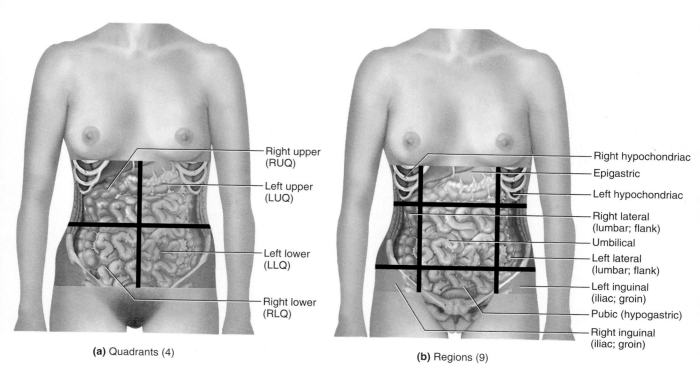

Right upper
(RUQ)

Left upper
(LUQ)

Left lower
(LLQ)

Right lower
(RLQ)

Right hypochondriac

Epigastric

Left hypochondriac

Right lateral
(lumbar; flank)

Umbilical

Left lateral
(lumbar; flank)

Left inguinal
(iliac; groin)

Pubic (hypogastric)

Right inguinal
(iliac; groin)

(a) Quadrants (4)

(b) Regions (9)

FIGURE 2.8 Diagrams of the body surface regions (a) anterior regions; (b) posterior regions.

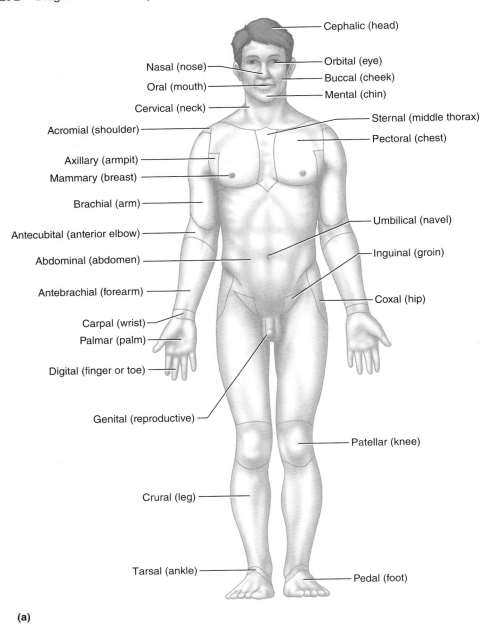

Cephalic (head)

Nasal (nose)

Orbital (eye)

Buccal (cheek)

Oral (mouth)

Mental (chin)

Cervical (neck)

Sternal (middle thorax)

Acromial (shoulder)

Pectoral (chest)

Axillary (armpit)

Mammary (breast)

Brachial (arm)

Umbilical (navel)

Antecubital (anterior elbow)

Inguinal (groin)

Abdominal (abdomen)

Antebrachial (forearm)

Coxal (hip)

Carpal (wrist)

Palmar (palm)

Digital (finger or toe)

Genital (reproductive)

Patellar (knee)

Crural (leg)

Tarsal (ankle)

Pedal (foot)

(a)

FIGURE 2.8 *Continued.*

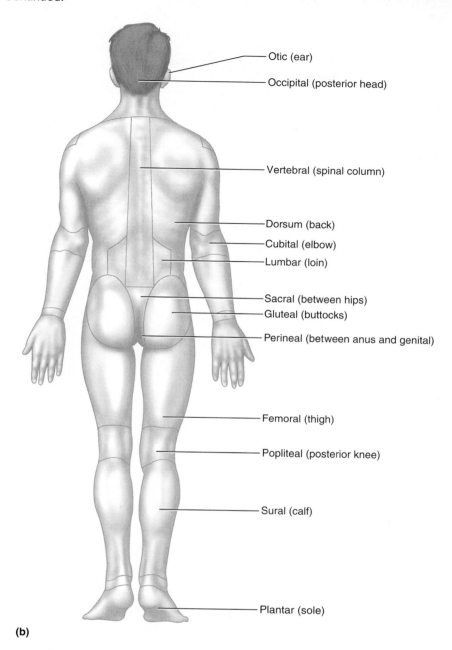

Otic (ear)

Occipital (posterior head)

Vertebral (spinal column)

Dorsum (back)

Cubital (elbow)

Lumbar (loin)

Sacral (between hips)

Gluteal (buttocks)

Perineal (between anus and genital)

Femoral (thigh)

Popliteal (posterior knee)

Sural (calf)

Plantar (sole)

(b)

Laboratory Assessment

2

Name _____

Date _____

Section _____

The ⬡ corresponds to the indicated outcome(s) found at the beginning of the laboratory exercise.

Body Organization, Membranes, and Terminology

Part A   Assessments

1. Match the body cavities in column A with the organs contained in the cavities in column B. Place the letter of your choice in the space provided. ⬡ ⬡

Column A		Column B
a. Abdominal cavity	_____	**1.** Liver
b. Cranial cavity	_____	**2.** Lungs
c. Pelvic cavity	_____	**3.** Spleen
d. Thoracic cavity	_____	**4.** Stomach
e. Vertebral canal (spinal cavity)	_____	**5.** Brain
	_____	**6.** Internal reproductive organs
	_____	**7.** Urinary bladder
	_____	**8.** Spinal cord
	_____	**9.** Heart
	_____	**10.** Small intestine

2. Label the body cavities in figure 2.9. Include an organ that would be located in each of the cavities, using the terms provided. ⬡ ⬡

FIGURE 2.9   Label body cavities 1–5. Add a representative organ located in the cavity, using the terms provided.

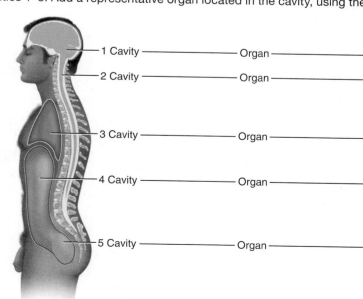

Terms:
Brain
Gallbladder
Lung
Spinal cord
Urethra

1 Cavity ——————— Organ —————————

2 Cavity ——————— Organ —————————

3 Cavity ——————— Organ —————————

4 Cavity ——————— Organ —————————

5 Cavity ——————— Organ —————————

Part B Assessments

Match the organ systems in column A with the principal functions in column B. Place the letter of your choice in the space provided. 🔼

Column A		Column B

Column A

a. Cardiovascular system
b. Digestive system
c. Endocrine system
d. Integumentary system
e. Lymphatic system
f. Muscular system
g. Nervous system
h. Reproductive system
i. Respiratory system
j. Skeletal system
k. Urinary system

Column B

_____ 1. The main system that secretes hormones

_____ 2. Provides an outer covering of the body for protection

_____ 3. Produces gametes (eggs and sperm)

_____ 4. Stimulates muscles to contract and interprets information from sensory organs

_____ 5. Provides a framework and support for soft tissues and produces blood cells in red marrow

_____ 6. Exchanges gases between air and blood

_____ 7. Transports excess fluid from tissues to blood and produces immune cells

_____ 8. Movement via contractions and creates most body heat

_____ 9. Removes liquid and wastes from blood and transports them to the outside of the body

_____ 10. Converts food molecules into forms that are absorbable

_____ 11. Transports nutrients, wastes, and gases throughout the body

Part C Assessments

Indicate whether each of the following sentences makes correct or incorrect usage of the word in boldface type (assume that the body is in the anatomical position). If the sentence is incorrect, in the space provided supply a term that will make it correct. 🔼

1. The mouth is **superior** to the nose.   _____

2. The stomach is **inferior** to the diaphragm.   _____

3. The trachea is **anterior** to the spinal cord.   _____

4. The larynx is **posterior** to the esophagus.   _____

5. The heart is **medial** to the lungs.   _____

6. The kidneys are **inferior** to the lungs.   _____

7. The hand is **proximal** to the elbow.   _____

8. The knee is **proximal** to the ankle.   _____

9. The thumb is the **lateral** digit of a hand.   _____

10. The popliteal surface region is **anterior** to the patellar region.   _____

Part D Assessments

Critical Thinking Assessment

State the quadrant of the abdominopelvic cavity where the pain or sound would be located for each of the six conditions listed. In some cases, there may be more than one correct answer, and pain is sometimes referred to another region. This phenomenon, called *referred pain,* occurs when pain is interpreted as originating from some area other than the parts being stimulated. When referred pain is involved in the patient's interpretation of the pain location, the proper diagnosis of the ailment is more challenging. For the purpose of this exercise, assume the pain is interpreted as originating from the organ involved. ⚠

1. Stomach ulcer _____

2. Appendicitis _____

3. Bowel sounds _____

4. Gallbladder attack _____

5. Kidney stone in left ureter _____

6. Ruptured spleen _____

Part E Assessments

Label figures 2.10 and 2.11.

FIGURE 2.10   Label the planes of the sectioned (cut) body part. ⚠

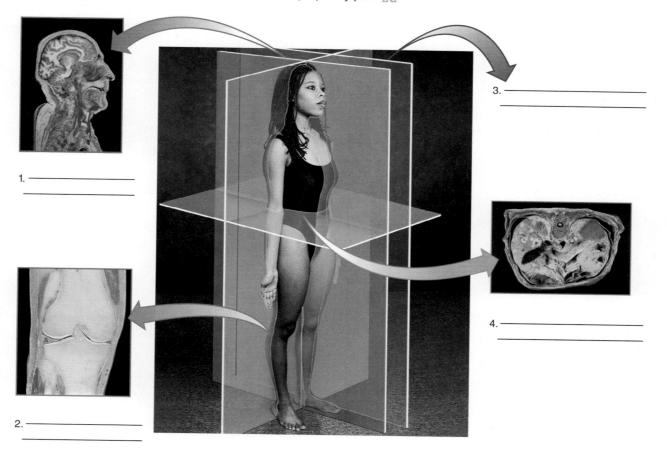

1. _____

2. _____

3. _____

4. _____

FIGURE 2.11 Label the indicated body surface regions. 5

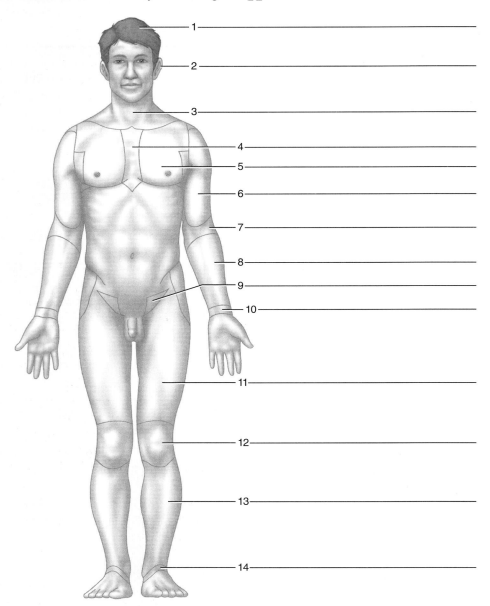

1 ————————————————

2 ————————————————

3 ————————————————

4 ————————————————

5 ————————————————

6 ————————————————

7 ————————————————

8 ————————————————

9

10 ————————————————

11 ————————————————

12 ————————————————

13 ————————————————

14 ————————————————

Laboratory Exercise 3

Chemistry of Life

Purpose of the Exercise

To review pH and basic categories of organic compounds, and to differentiate between types of organic compounds.

Materials Needed

For pH Tests:
Chopped fresh red cabbage
Beaker (250 mL)
Distilled water
Tap water
Vinegar
Baking soda
Pipets for measuring

For Organic Tests:
Test tubes
Test-tube rack
Test-tube clamps
China marker
Hot plate
Beaker for hot water bath (500 mL)
Pipets for measuring
Benedict's solution
Sudan IV dye
Biuret reagent (or 10% NaOH and 1% CuSO₄)

Laboratory scoop for measuring
Full-range pH test papers
7 assorted common liquids clearly labeled in closed bottles on a tray or in a tub
Droppers labeled for each liquid
Iodine-potassium-iodide (IKI) solution
Egg albumin
10% glucose solution
Clear carbonated soft drink
10% starch solution
Potatoes for potato water
Distilled water
Vegetable oil
Brown paper
Numbered unknown organic samples

Safety

▶ Review all safety guidelines inside the front cover of your laboratory manual.
▶ Clean laboratory surfaces before and after laboratory procedures using soap and water.
▶ Use extreme caution when working with chemicals.
▶ Safety goggles must be worn at all times.
▶ Wear disposable gloves while working with the chemicals.
▶ Precautions should be taken to prevent chemicals from contacting your skin.
▶ Do not mix any of the chemicals together unless instructed to do so.
▶ Clean up any spills immediately and notify the instructor at once.
▶ Wash your hands before leaving the laboratory.

Learning Outcomes

After completing this exercise, you should be able to

1. Demonstrate pH values of various substances through testing methods.
2. Determine categories of organic compounds with basic colorimetric tests.
3. Discover the organic composition of an unknown solution.

Pre-Lab

Carefully read the introductory material and examine the entire lab. Be familiar with pH and organic molecules from lecture or the textbook. Visit www.mhhe.com/martinseries2 for LabCam videos. Answer the pre-lab questions.

Pre-Lab Questions: Select the correct answer for each of the following questions:

1. The most basic unit of matter is
   a. energy.   **b.** an element.
   c. a molecule.   **d.** a cell.

2. Which number would represent a neutral pH?
   a. 100   **b.** 0
   c. 7   **d.** 14

3. The building block of a protein is
   a. an amino acid.   **b.** a monosaccharide.
   c. a lipid.   **d.** starch.

4. The building block of a carbohydrate is
   a. glycerol.   **b.** an amino acid.
   c. fatty acids.   **d.** a monosaccharide.

5. The test for starch uses which substance?
   a. IKI (iodine solution) **b.** Benedict's solution
   c. Sudan IV dye   **d.** Biuret reagent

6. A positive test for lipids results in a
   a. pinkish color change.
   b. yellow color change.
   c. translucent stain.
   d. blue to red color change.

7. Which of the following pH numbers would represent the strongest acid?
   a. 1   **b.** 5
   c. 7   **d.** 14

The complexities of the human body arise from the organization and interactions of chemicals. Organisms are made of matter, and the most basic unit of matter is the chemical element. The smallest unit of an element is an atom, and two or more of those can unite to form a molecule. All processes that occur within the body involve chemical reactions—interactions between atoms and molecules. We breathe to supply oxygen to our cells for energy. We eat and drink to bring chemicals into our bodies that our cells need. Water fills and bathes all of our cells and allows an amazing array of reactions to occur, all of which are designed to keep us alive. Chemistry forms the very basis of life and thus forms the foundation of anatomy and physiology.

The pH scale is from 0 to 14 with the midpoint of 7.0 (pure water) representing a neutral solution. If the solution has a pH of lower than 7.0, it represents an acidic solution with more hydrogen ions (H^+) than hydroxide ions (OH^-). The solution is a base if the pH is higher than 7.0 with more OH^- ions than H^+ ions. The weaker acids and bases are closer to 7, with the strongest acids near 0 and the strongest bases near 14. A seemingly minor change in the pH actually represents a more significant change as there is a tenfold difference in hydrogen ion concentrations with each whole number represented on the pH scale. Our cells can only function within minimal pH fluctuations.

The organic molecules (biomolecules) that are tested in this laboratory exercise include carbohydrates, lipids, and proteins. Carbohydrates are used to supply energy to our cells. The polysaccharide starch is composed of simple sugar (monosaccharide) building blocks. Lipids include fats, phospholipids, and steroids that provide some cellular structure and energy for cells. The building blocks of lipids include fatty acids and glycerol. Proteins compose important cellular structures, antibodies, and enzymes. Amino acids are the building blocks of proteins.

Procedure A—The pH Scale

In this section, you will be able to determine the pH of various substances. The abbreviation pH is traditionally used to measure the power of H (hydrogen). The pH is a mathematical way to refer to the negative logarithm of the hydrogen ion (H^+) concentration in a solution. Each change of a number on the pH scale actually represents a tenfold increase or decrease of the hydrogen ion concentration. The pH scale has a range from 0 to 14, with a pH of 7 representing neutral for distilled water. The numbers below 7 represent the acidic part of the scale, while the numbers above 7 represent the basic (alkalinity) part of the scale.

The pH of various body fluids is very important in the homeostasis of a particular body system and the entire living organism. Blood, urine, saliva, and gastric secretions are examples of some body fluids with characteristic pH ranges for a healthy individual. Some body fluids like blood must fall within a very narrow pH range, while others like urine have a much wider normal range. In order to maintain normal ranges of the pH of body fluids, various buffering systems are involved to resist pH fluctuations. Some of these systems entail chemical buffers and others consist of physiological buffers to help regulate the pH levels. The acid-base balance is important since our bodies are composed predominantly of water. Various acid-base relationships are covered within the discussion of each body system.

1. Study figure 3.1, which shows the pH scale. Note the range of the scale and the pH value that is considered neutral.
2. **Cabbage water tests.** Many tests can determine a pH value, but among the more interesting are colorimetric tests in which an indicator chemical changes color when it reacts. Many plant pigments, especially anthocyanins (which give plants color, from blue to red), can be used

FIGURE 3.1 As the concentration of hydrogen ions (H^+) increases, a solution becomes more acidic and the pH value decreases. As the concentration of ions that combine with hydrogen ions (such as hydroxide ions) increases, a solution becomes more basic (alkaline) and the pH value increases. The pH values of some common substances are shown.

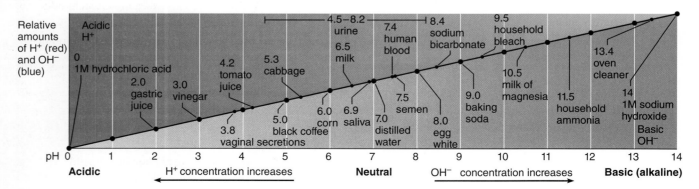

as colorimetric pH indicators. One that works well is the red pigment in red cabbage.

   a. Prepare cabbage water to be used as a general pH indicator. Fill a 250 mL beaker to the 100 mL level with chopped red cabbage. Add water to make 150 mL. Place the beaker on a hot plate and simmer the mixture until the pigments come out of the cabbage and the water turns deep purple. Allow the water to cool. (You may proceed to step 3 while you wait for this to finish.)

   b. Label three clean test tubes: one for water, one for vinegar, and one for baking soda.

   c. Place 2 mL of cabbage water into each test tube.

   d. To the first test tube, add 2 mL of distilled water and swirl the mixture. Note and record the color in Part A of Laboratory Assessment 3.

   e. Repeat this procedure for test tube 2, adding 2 mL of vinegar and swirling the mixture. Note and record the results.

   f. Repeat this procedure for test tube 3, adding one laboratory scoop of baking soda. Swirl the mixture. Note and record the results.

3. Testing with pH paper. Many commercial pH indicators are available. A very simple one to use is pH paper, which comes in small strips. Don gloves for this procedure so your skin secretions do not contaminate the paper and to protect you from the chemicals you are testing.

   a. Test the pH of distilled water by dropping one or two drops of distilled water onto a strip of pH paper, and note the color. Compare the color to the color guide on the container. Note the results of the test as soon as color appears, as drying of the paper will change the results. Record the pH value in Part A of the laboratory assessment.

   b. Repeat this procedure for tap water. Record the results.

   c. Individually test the various common substances found on your lab table. Record your results.

   d. Complete Part A of the laboratory assessment.

Procedure B—Organic Molecules

In this section, you will perform some simple tests to check for the presence of the following categories of organic molecules (biomolecules): protein, sugar, starch, and lipid. Specific color changes will occur if the target compound is present. Please note the original color of the indicator being added so you can tell if the color really changed. For example, Biuret reagent and Benedict's solution are both initially blue, so an end color of blue would indicate no change. Use caution when working with these chemicals and follow all directions. Carefully label the test tubes and avoid contaminating any of your samples. Only the Benedict's test requires heating and a time delay for accurate results. Do not heat any other tubes. To prepare for the Benedict's test, fill a 500 mL beaker half full with water and place it on the hot plate. Turn the hot plate on and bring the water to a boil. To save time, start the water bath before doing the Biuret test.

1. Biuret test for protein. In the presence of protein, Biuret reagent reacts with peptide bonds and changes to violet or purple. A pinkish color indicates that shorter polypeptides are present. The color intensity is proportional to the number of peptide bonds, thus the intensity reflects the length of polypeptides (amount of protein).

   a. Label six test tubes as follows: 1W, 1E, 1G, 1D, 1S, and 1P. The codes placed on the six test tubes used for the protein test include a number *1* referring to test 1 (biuret test for protein) followed by a letter. Letter *W* represents water (distilled); letter *E* represents egg (albumin); letter *G* represents glucose (10% solution); letter *D* represents drink (carbonated soft); letter *S* represents starch (solution); and letter *P* represents potato (juice). Similar codes are used for the tests for sugar (test 2) and starch (test 3) of Procedure B of this exercise.

   b. To each test tube, add 2 mL of one of the samples as follows:
     1W—2 mL distilled water
     1E—2 mL egg albumin
     1G—2 mL 10% glucose solution
     1D—2 mL carbonated soft drink
     1S—2 mL 10% starch solution
     1P—2 mL potato juice

   c. To each, add 2 mL of Biuret reagent, swirl the tube to mix it, and note the final color. [*Note:* If Biuret reagent is not available, add 2 mL of 10% NaOH (sodium hydroxide) and about 10 drops of 1% $CuSO_4$ (copper sulfate).] **Be careful—NaOH is very caustic.**

   d. Record your results in Part B of the laboratory assessment.

   e. Mark a "+" on any tubes with positive results and retain them for comparison while testing your unknown sample. Put remaining test tubes aside so they will not get confused with future trials.

2. Benedict's test for sugar (monosaccharides). In the presence of sugar, Benedict's solution changes from its initial blue to a green, yellow, orange, or reddish color, depending on the amount of sugar present. Orange and red indicate greater amounts of sugar.

   a. Label six test tubes as follows: 2W, 2E, 2G, 2D, 2S, and 2P.

   b. To each test tube, add 2 mL of one of the samples as follows:
     2W—2 mL distilled water
     2E—2 mL egg albumin
     2G—2 mL 10% glucose solution
     2D—2 mL carbonated soft drink
     2S—2 mL 10% starch solution
     2P—2 mL potato juice

c. To each, add 2 mL of Benedict's solution, swirl the tube to mix it, and place all tubes into the boiling water bath for 3 to 5 minutes.

d. Note the final color of each tube after heating and record the results in Part B of the laboratory assessment.

e. Mark a "+" on any tubes with positive results and retain them for comparison while testing your unknown sample. Put remaining test tubes aside so they will not get confused with future trials.

3. Iodine test for starch. In the presence of starch, iodine turns a dark purple or blue-black color. Starch is a long chain formed by many glucose units linked together side by side. This regular organization traps the iodine molecules and produces the dark color.

a. Label six test tubes as follows: 3W, 3E, 3G, 3D, 3S, and 3P.

b. To each test tube, add 2 mL of one of the samples as follows:

3W—2 mL distilled water
3E—2 mL egg albumin
3G—2 mL 10% glucose solution
3D—2 mL carbonated soft drink
3S—2 mL 10% starch solution
3P—2 mL potato juice

c. To each, add 0.5 mL of IKI (iodine solution) and swirl the tube to mix it.

d. Record the final color of each tube in Part B of the laboratory assessment.

e. Mark a "+" on any tubes with positive results and retain them for comparison while testing your unknown sample. Put remaining test tubes aside so they will not get confused with future trials.

4. Tests for lipids.

a. Label two separate areas of a piece of brown paper as "water" or "oil."

b. Place a drop of water on the area marked "water" and a drop of vegetable oil on the area marked "oil."

c. Let the spots dry several minutes, then record your observations in Part C of the laboratory assessment. Upon drying, oil leaves a translucent stain (grease spot) on brown paper; water does not leave such a spot.

d. In a test tube, add 2 mL of water and 2 mL of vegetable oil and observe. Shake the tube vigorously then let it sit for 5 minutes and observe again. Record your observations in Part B of the laboratory assessment.

e. Sudan IV is a dye that is lipid-soluble but not water-soluble. If lipids are present, the Sudan IV will stain them pink or red. Add a small amount of Sudan IV to the test tube that contains the oil and water. Swirl it; then let it sit for a few minutes. Record your observations in Part B of the laboratory assessment.

Procedure C—Identifying Unknown Compounds

Now you will apply the information you gained with the tests in the previous section. You will retrieve an unknown sample that contains none, one, or any combination of the following types of organic compounds: protein, sugar, starch, or lipid. You will test your sample using each test from the previous section and record your results in Part C of the laboratory assessment.

1. Label four test tubes as follows: (1, 2, 3, and 4).
2. Add 2 mL of your unknown sample to each test tube.
3. **Test for protein.** Add 2 mL of Biuret reagent to tube 1. Swirl the tube and observe the color. Record your observations in Part C of the laboratory assessment.
4. **Test for sugar.** Add 2 mL of Benedict's solution to tube 2. Swirl the tube and place it in a boiling water bath for 3 to 5 minutes. Record your observations in Part C of the laboratory assessment.
5. **Test for starch.** Add several drops of iodine solution to tube 3. Swirl the tube and observe the color. Record your observations in Part C of the laboratory assessment.
6. **Test for lipid.** Add 2 mL of water to tube 4. Swirl the tube and note if there is any separation.
7. Add a small amount of Sudan IV to test tube 4, swirl the tube and record your observations in Part C of the laboratory assessment.
8. Based on the results of these tests, determine if your unknown sample contains any organic compounds and, if so, what are they? Record and explain your identification in Part C of the laboratory assessment.

Name _____

Date _____

Section _____

The A corresponds to the indicated outcome(s) found at the beginning of the laboratory exercise.

Chemistry of Life

Part A   The pH Scale Assessments

1. Results from cabbage water test: A

Substance	Cabbage Water	Distilled Water	Vinegar	Baking Soda
Color				
Acid, base, or neutral?				

2. Results from pH paper tests: A

Substance Tested	Distilled Water	Tap Water	Sample 1:	Sample 2:	Sample 3:	Sample 4:	Sample 5:	Sample 6:	Sample 7:
pH Value									

3. Are the pH values the same for distilled water and tap water? _____

4. If not, what might explain this difference? _____

5. Draw the pH scale below, and indicate the following values: 0, 7 (neutral), and 14. Now label the scale with the names and indicate the location of the pH values for each substance you tested.

Part B Testing for Organic Molecules Assessments

1. **Biuret test results for protein.** Enter your results from the Biuret test for protein. A color change to purple indicates that protein is present; pink indicates that short polypeptides are present. **2**

Tube	Contents	Color	Protein Present (+) or Absent (−)
1W	Distilled water		
1E	Egg albumin		
1G	Glucose solution		
1D	Soft drink		
1S	Starch solution		
1P	Potato juice		

2. **Benedict's test results for most sugars (monosaccharides).** Enter your results from the Benedict's test for sugars. A color change to green, yellow, orange, or red indicates that sugar is present. Note the color after the mixture has been heated for 3 to 5 minutes. **2**

Tube	Contents	Color	Sugar Present (+) or Absent (−)
2W	Distilled water		
2E	Egg albumin		
2G	Glucose solution		
2D	Soft drink		
2S	Starch solution		
2P	Potato juice		

3. **Iodine test results for starch.** Enter your results from the iodine test for starch. A color change to dark blue or black indicates that starch is present. **2**

Tube	Contents	Color	Sugar Present (+) or Absent (−)
3W	Distilled water		
3E	Egg albumin		
3G	Glucose solution		
3D	Soft drink		
3S	Starch solution		
3P	Potato juice		

4. **Lipid test results.** What did you observe when you allowed the drops of oil and water to dry on the brown paper? **2**

_____

What did you observe when you mixed the oil and water together? _____

_____

What did you observe when you added the Sudan IV dye? **2** _____

If a person were on a low-carbohydrate, high-protein diet, which of the six substances tested would the person want to increase? _____

Explain your answer.

_____

_____

_____

_____

Part C   Testing for Unknown Organic Compounds Assessments

1. What is the number of your sample? _____

2. Record the results of your tests on the unknown here: **3**

Test Performed	Results
Biuret test for protein	
Benedict's test for sugars	
Iodine test for starch	
Water test for lipid	
Sudan IV test for lipid	

3. Based on these results, what organic compound(s) does your unknown sample contain? **3** _____

_____

4. Explain your answer. **3** _____

_____

Care and Use of the Microscope

Purpose of the Exercise

To become familiar with the major parts of a compound light microscope and their functions and to use the microscope to observe small objects.

Materials Needed

Compound light microscope
Lens paper
Microscope slides
Coverslips
Transparent plastic millimeter ruler
Prepared slide of letter *e*
Slide of three colored threads
Medicine dropper
Dissecting needle (needle probe)
Specimen examples for wet mounts
Methylene blue (dilute) or iodine-potassium-iodide
  stain
For Demonstration Activities:
Micrometer scale
Stereomicroscope (dissecting microscope)

Learning Outcomes

After completing this exercise, you should be able to

1. Differentiate the major functional parts of a compound light microscope.

2. Calculate the total magnification produced by various combinations of eyepiece and objective lenses.

3. Demonstrate proper use of the microscope to observe and measure small objects.

4. Prepare a simple microscope slide and sketch the objects you observed.

Pre-Lab

Carefully read the introductory material and examine the entire lab. Visit www.mhhe.com/martinseries2 for LabCam videos. Answer the pre-lab questions.

Pre-Lab Questions: Select the correct answer for each of the following questions:

1. The human eye cannot perceive objects less than
   a. one inch.
   b. one millimeter.
   c. 0.1 millimeter.
   d. one centimeter.

2. The objective lenses of the compound light microscope are attached to the
   a. stage.
   b. base.
   c. body tube.
   d. rotating nosepiece.

3. Which objective lens provides the least total magnification?
   a. scan
   b. low power
   c. high power
   d. oil immersion

4. The _____ increases or decreases the light intensity of the compound light microscope.
   a. eyepiece
   b. stage
   c. adjustment knob
   d. iris diaphragm

5. Basic lens cleaning is accomplished using
   a. water.
   b. lens paper.
   c. a paper towel.
   d. xylene.

6. When preparing a wet mount specimen for viewing, it should be covered with _____.
   a. clear paper
   b. another glass slide
   c. a coverslip
   d. transparent tape

7. The total magnification achieved using a 10× objective lens with a 10× eyepiece lens is 20×.
   True _____   False _____

8. A stereomicroscope is a good option to use for viewing rather large, opaque specimens.
   True _____   False _____

The human eye cannot perceive objects less than 0.1 mm in diameter, so a microscope is an essential tool for the study of small structures such as cells. The microscope usually used for this purpose is the *compound light microscope*. It is called compound because it uses two sets of lenses: an eyepiece or ocular lens and an objective lens system. The eyepiece lens system magnifies, or compounds, the image reaching it after the image is magnified by the objective lens system. Such an instrument can magnify images of small objects up to about one thousand times.

Procedure A—Microscope Basics

This procedure includes the terminology for the parts of the compound light microscope, the purpose of the parts, and how to manipulate the components for suitable image viewing. The proper storage of the microscope, lens-cleaning techniques, and magnification determination are also included.

1. Familiarize yourself with the following list of rules for care of the microscope:
   a. Keep the microscope under its *dustcover* and in a cabinet when it is not being used.
   b. Handle the microscope with great care. It is an expensive and delicate instrument. To move it or carry it, hold it by its *arm* with one hand and support its *base* with the other hand (fig. 4.1).
   c. Always store the microscope with the scanning or lowest power objective in place. Always start with this objective when using the microscope.
   d. To clean the lenses, rub them gently with *lens paper* or a high-quality cotton swab. If the lenses need additional cleaning, follow the directions in the "Lens-Cleaning Technique" section that follows.
   e. If the microscope has a substage lamp, be sure the electric cord does not hang off the laboratory table where someone might trip over it. The bulb life can be extended if the lamp is cool before the microscope is moved.
   f. Never drag the microscope across the laboratory table after you have placed it.
   g. Never remove parts of the microscope or try to disassemble the eyepiece or objective lenses.
   h. If your microscope is not functioning properly, report the problem to your laboratory instructor immediately.
2. Observe a compound light microscope and study figure 4.1 to learn the names of its major parts. Your microscope could be equipped with a binocular body and two eyepieces (fig. 4.2). The lens system of a compound

FIGURE 4.1 Major parts of a compound light microscope with a monocular body and a mechanical stage. Some microscopes are equipped with a binocular body.

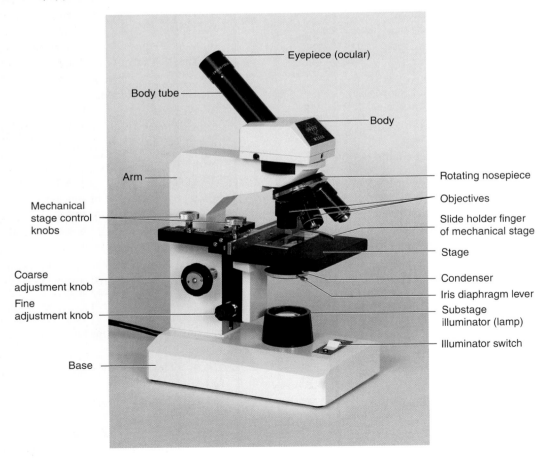

FIGURE 4.2 Scientist using a compound light microscope equipped with a binocular body and two eyepieces.

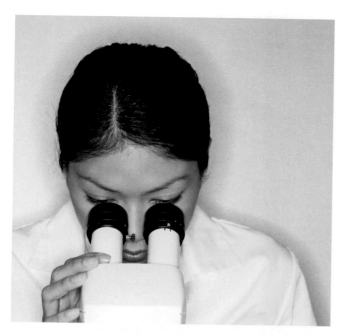

FIGURE 4.3 Microscope showing the path of light through it.

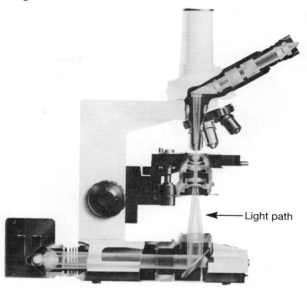

← Light path

Lens-Cleaning Technique

1. Moisten one end of a high-quality cotton swab with one drop of Kodak lens cleaner. Keep the other end dry.
2. Clean the optical surface with the wet end. Dry it with the other end, using a circular motion.
3. Use a hand aspirator to remove lingering dust particles.
4. Start with the scanning objective and work upward in magnification, using a new cotton swab for each objective.
5. When cleaning the eyepiece, do not open the lens unless it is absolutely necessary.
6. Use alcohol for difficult cleaning and only as a last resort use xylene. Regular use of xylene will destroy lens coatings.

microscope includes three parts—the condenser, objective lens, and eyepiece or ocular.

Light enters this system from a *substage illuminator (lamp)* or *mirror* and is concentrated and focused by a *condenser* onto a microscope slide or specimen placed on the *stage* (fig. 4.3). The condenser, which contains a set of lenses, usually is kept in its highest position possible.

The *iris diaphragm,* located between the light source and the condenser, can be used to increase or decrease the intensity of the light entering the condenser. Locate the lever that operates the iris diaphragm beneath the stage and move it back and forth. Note how this movement causes the size of the opening in the diaphragm to change. (Some microscopes have a revolving plate called a disc diaphragm beneath the stage instead of an iris diaphragm. Disc diaphragms have different-sized holes to admit varying amounts of light.) Which way do you move the diaphragm to increase the light intensity? _____ Which way to decrease it? _____

After light passes through a specimen mounted on a microscope slide, it enters an *objective lens system.* This lens projects the light upward into the *body tube,* where it produces a magnified image of the object being viewed.

The *eyepiece (ocular) lens system* then magnifies this image to produce another image seen by the eye. Typically, the eyepiece lens magnifies the image ten times (10×). Look for the number in the metal of the eyepiece that indicates its power (fig. 4.4). What is the eyepiece power of your microscope? _____

The objective lenses are mounted in a *rotating nosepiece* so that different magnifications can be achieved by rotating any one of several objective lenses into position above the specimen. Commonly, this set of lenses includes a scanning objective (4×), a low-power objective (10×), and a high-power objective, also called a high-dry-power objective (about 40×). Sometimes an oil immersion objective (about 100×) is present. Look for the number printed on each objective that indicates its power. What are the objective lens powers of your microscope?_____

To calculate the *total magnification* achieved when using a particular objective, multiply the power of the

FIGURE 4.4 The powers of this 10× eyepiece (a) and this 40× objective (b) are marked in the metal. DIN is an international optical standard on quality optics. The 0.65 on the 40× objective is the numerical aperture, a measure of the light-gathering capabilities.

(a)

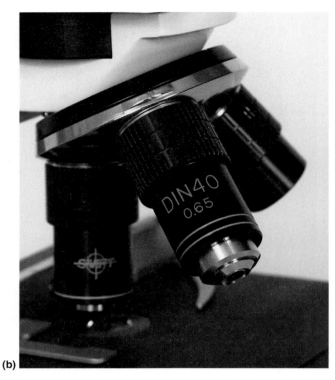

(b)

eyepiece by the power of the objective used. Thus, the 10× eyepiece and the 40× objective produce a total magnification of 10 × 40, or 400×. See table 4.1.
3. Complete Part A of Laboratory Assessment 4.
4. Turn on the substage illuminator and look through the eyepiece. You will see a lighted circular area called the *field of view.*

TABLE 4.1 Microscope Lenses

Objective Lens Name	Common Objective Lens Magnification	Common Eyepiece Lens Magnification	Total Magnification
Scan	4×	10×	40×
Low power (LP)	10×	10×	100×
High power (HP)	40×	10×	400×
Oil immersion	100×	10×	1,000×

Note: If you wish to observe an object under LP, HP, or oil immersion, locate and then center and focus the object first under scan magnification.

You can measure the diameter of this field of view by focusing the lenses on the millimeter scale of a transparent plastic ruler. To do this, follow these steps:
a. Place the ruler on the microscope stage in the spring clamp of a slide holder finger on a mechanical stage or under the stage (slide) clips. (*Note:* If your microscope is equipped with a mechanical stage, it may be necessary to use a short section cut from a transparent plastic ruler. The section should be several millimeters long and can be mounted on a microscope slide for viewing.)
b. Center the millimeter scale in the beam of light coming up through the condenser, and rotate the scanning objective into position.
c. While you watch from the side to prevent the lens from touching anything, raise the stage until the objective is as close to the ruler as possible, using the *coarse adjustment knob* (fig. 4.5). (*Note:* The adjustment knobs on some microscopes move the body and objectives downward and upward for focusing.)
d. Look into the eyepiece and use the *fine adjustment knob* to raise the stage until the lines of the millimeter scale come into sharp focus.
e. Adjust the light intensity by moving the *iris diaphragm lever* so that the field of view is brightly illuminated but comfortable to your eye. At the same time, take care not to over illuminate the field because transparent objects tend to disappear in bright light.

FIGURE 4.5 When you focus using a particular objective, you can prevent it from touching the specimen by watching from the side.

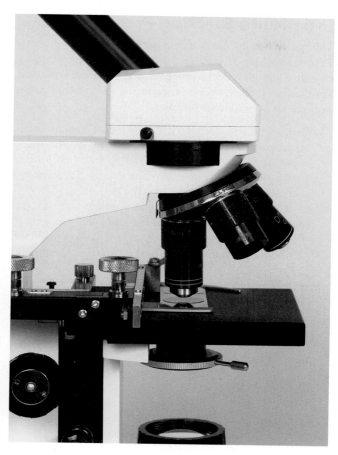

f. Position the millimeter ruler so that its scale crosses the greatest diameter of the field of view. Also, move the ruler so that one of the millimeter marks is against the edge of the field of view.

g. In millimeters, measure the distance across the field of view.

5. Complete Part B of the laboratory assessment.

6. Most microscopes are designed to be *parfocal*. This means that when a specimen is in focus with a lower-power objective, it will be in focus (or nearly so) when a higher-power objective is rotated into position. Always center the specimen in the field of view before changing to higher objectives.

   Rotate the low-power objective into position, and then look at the millimeter scale of the transparent plastic ruler. If you need to move the low-power objective to sharpen the focus, use the fine adjustment knob.

   Adjust the iris diaphragm so that the field of view is properly illuminated. Once again, adjust the millimeter ruler so that the scale crosses the field of view through its greatest diameter, and position the ruler so that a millimeter mark is against one edge of the field. Try to measure the distance across the field of view in millimeters.

7. Rotate the high-power objective into position, while you watch from the side, and then observe the millimeter scale on the plastic ruler. All focusing using high-power magnification should be done only with the fine adjustment knob. If you use the coarse adjustment knob with the high-power objective, you can accidently force the objective into the coverslip and break the slide. This is because the *working distance* (the distance from the objective lens to the slide on the stage) is much shorter when using higher magnifications.

   Adjust the iris diaphragm for proper illumination. Usually more illumination when using higher magnifications will help you to view the objects more clearly. Try to measure the distance across the field of view in millimeters.

8. Locate the numeral 4 (or 9) on the plastic ruler and focus on it using the scanning objective. Note how the number appears in the field of view. Move the plastic ruler to the right and note which way the image moves. Slide the ruler away from you and again note how the image moves.

9. Observe a letter *e* slide. Note the orientation of the letter *e* using the scan, LP, and HP objectives. As you increase magnifications, note that the amount of the letter *e* shown is decreased. Move the slide to the left, and then away from you, and note the direction in which the observed image moves. If a *pointer* is visible in your field of view, you can manipulate the pointer to a location (structure) within the field of view by moving the slide or rotating the ocular lens.

10. Examine the slide of the three colored threads using the low-power objective and then the high-power objective. Focus on the location where the three threads cross. By using the fine adjustment knob, determine the order from top to bottom by noting which color is in focus at different depths. The other colored threads will still be visible, but they will be blurred. Be sure to notice whether the stage or the body tube moves up and down with the adjustment knobs of the microscope being used for this depth determination. The vertical depth of the specimen clearly in focus is called the *depth of field (focus)*. Whenever specimens are examined, continue to use the fine adjustment focusing knob to determine relative depths of structures clearly in focus within cells,

Critical Thinking Activity

What was the sequence of the three colored threads from top to bottom? Explain how you came to that conclusion.

_____

_____

_____

Demonstration Activity

A compound light microscope is sometimes equipped with a micrometer scale mounted in the eyepiece. Such a scale is subdivided into fifty to one hundred equal divisions (fig. 4.6). These arbitrary divisions can be calibrated against the known divisions of a micrometer slide placed on the microscope stage. Once the values of the divisions are known, the length and width of a microscopic object can be measured by superimposing the scale over the magnified image of the object.

Observe the micrometer scale in the eyepiece of the demonstration microscope. Focus the low-power objective on the millimeter scale of a micrometer slide (or a plastic ruler) and measure the distance between the divisions on the micrometer scale in the eyepiece. What is the distance between the finest divisions of the scale in micrometers? _____

giving a three-dimensional perspective. The depth of field is less at higher magnifications.

11. Complete Parts C and D of the laboratory assessment.

Procedure B—Slide Preparation

This procedure includes techniques for making your own temporary slides. If your microscope has an oil immersion objective lens, the special procedures and care are described. For viewing thicker, opaque specimens, a stereomicroscope demonstration activity is included.

1. Prepare several temporary *wet mounts,* using any small, transparent objects of interest, and examine the specimens using the low-power objective and then a high-power objective to observe their details. To prepare a wet mount, follow these steps (fig. 4.7):

   a. Obtain a precleaned microscope slide.

   b. Place a tiny, thin piece of the specimen you want to observe in the center of the slide, and use a medicine dropper to put a drop of water over it. Consult with

---

FIGURE 4.6 The divisions of a micrometer scale in an eyepiece can be calibrated against the known divisions of a micrometer slide.

Source: Courtesy of Swift Instruments, Inc., San Jose, California

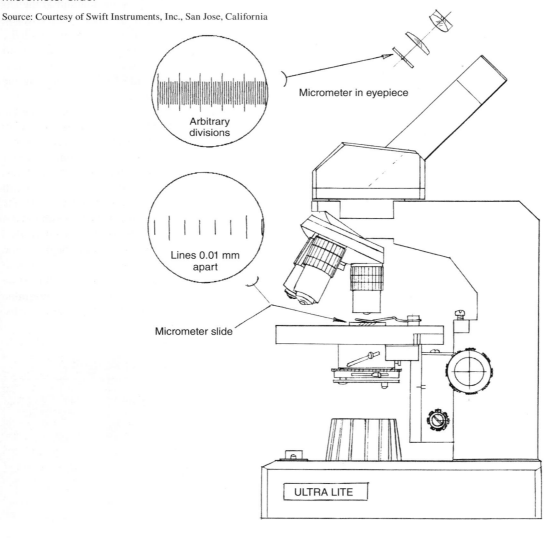

FIGURE 4.7 Steps in the preparation of a wet mount.

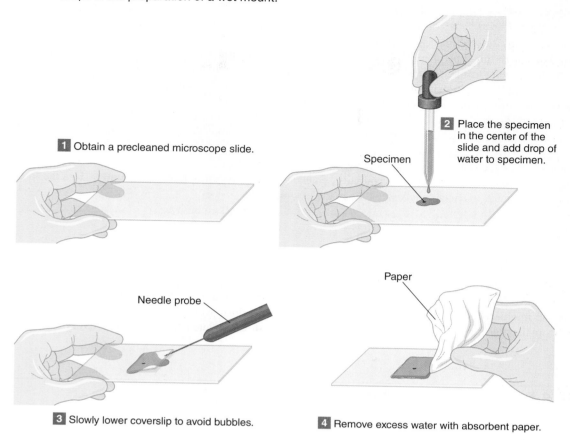

1 Obtain a precleaned microscope slide.

2 Place the specimen in the center of the slide and add drop of water to specimen.

Specimen

Needle probe

Paper

3 Slowly lower coverslip to avoid bubbles.

4 Remove excess water with absorbent paper.

your instructor if a drop of stain might enhance the image of any cellular structures of your specimen. If the specimen is solid, you might want to tease some of it apart with dissecting needles. In any case, the specimen must be thin enough so that light can pass through it. Why is it necessary for the specimen to be so thin?

_____

_____

c. Cover the specimen with a coverslip. Try to avoid trapping bubbles of air beneath the coverslip by slowly lowering it at an angle into the drop of water.

d. Remove any excess water from the edge of the coverslip with absorbent paper. If your microscope has an inclination joint, do not tilt the microscope while observing wet mounts because the fluid will flow.

e. Place the slide under the stage (slide) clips or in the slide holder on a mechanical stage, and position the slide so that the specimen is centered in the light beam passing up through the condenser.

f. Focus on the specimen using the scanning objective first. Next focus using the low-power objective, and then examine it with the high- power objective.

2. If an oil immersion objective is available, use it to examine the specimen. To use the oil immersion objective, follow these steps:

a. Center the object you want to study under the high-power field of view.

b. Rotate the high-power objective away from the microscope slide, place a small drop of immersion oil on the coverslip, and swing the oil immersion objective into position. To achieve sharp focus, use the fine adjustment knob only.

c. You will need to open the iris diaphragm more fully for proper illumination. More light is needed because the oil immersion objective covers a very small lighted area of the microscope slide.

d. The oil immersion objective must be very close to the coverslip to achieve sharp focus, so care must be taken to avoid breaking the coverslip or damaging the objective lens. For this reason, never lower the objective when you are looking into the eyepiece. Instead, always raise the objective to achieve focus, or prevent the objective from touching the coverslip by watching the microscope slide and coverslip from the side if the objective needs to be lowered. Usually

Demonstration Activity

A stereomicroscope (dissecting microscope) (fig. 4.8) is useful for observing the details of relatively large, opaque specimens. Although this type of microscope achieves less magnification than a compound light microscope, it has the advantage of producing a three-dimensional image rather than the flat, two-dimensional image of the compound light microscope. In addition, the image produced by the stereomicroscope is positioned in the same manner as the specimen, rather than being reversed and inverted as it is by the compound light microscope.

Observe the stereomicroscope. The eyepieces can be pushed apart or together to fit the distance between your eyes. Focus the microscope on the end of your finger. Which way does the image move when you move your finger to the right? _____

When you move it away? _____

If the instrument has more than one objective, change the magnification to a higher power. Use the instrument to examine various small, opaque objects available in the laboratory.

when using the oil immersion objective, only the fine adjustment knob needs to be used for focusing.

3. When you have finished working with the microscope, remove the microscope slide from the stage and wipe any oil from the objective lens with lens paper or a high-quality cotton swab. Swing the scanning objective or the low-power objective into position. Wrap the electric cord around the base of the microscope and replace the dustcover.

4. Complete Part E of the laboratory assessment.

FIGURE 4.8   A stereomicroscope, also called a dissecting microscope.

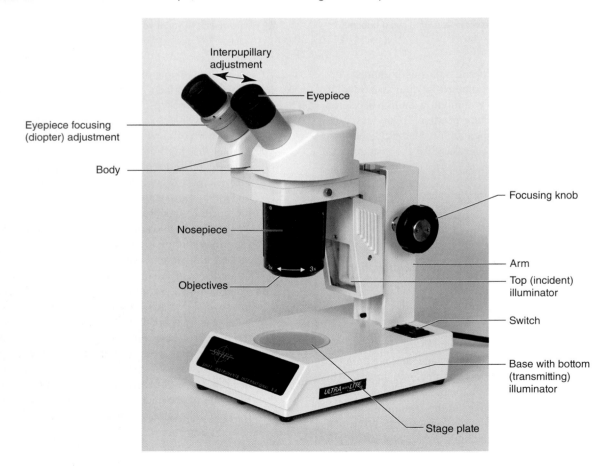

Name _____

Date _____

Section _____

The ⚠ corresponds to the indicated outcome(s) found at the beginning of the laboratory exercise.

Care and Use of the Microscope

Part A   Assessments

Revisit Procedure A, number 2, then complete the following: ⚠

1. What total magnification will be achieved if the 10× eyepiece and the 10× objective are used? _____

2. What total magnification will be achieved if the 10× eyepiece and the 100× objective are used? _____

Part B   Assessments

Revisit Procedure A, number 2, then complete the following:

1. Sketch the millimeter scale as it appears under the scanning objective magnification. (The circle represents the field of view through the microscope.)

2. In millimeters, what is the diameter of the scanning field of view? ⚠ _____

3. Microscopic objects often are measured in *micrometers*. A micrometer equals 1/1,000 of a millimeter and is symbolized by μm. In micrometers, what is the diameter of the scanning power field of view? ⚠ _____

4. If a circular object or specimen extends halfway across the scanning field, what is its diameter in millimeters? ⚠ _____

5. In micrometers, what is its diameter? ⚠ _____

Part C   Assessments

Complete the following:

1. Sketch the millimeter scale as it appears using the low-power objective.

2. What do you estimate the diameter of this field of view to be in millimeters? ⚠

_____

3. How does the diameter of the scanning power field of view compare with that of the low-power field?

_____

4. Why is it more difficult to measure the diameter of the high-power field of view than the low-power field?

_____

_____

5. What change occurred in the light intensity of the field of view when you exchanged the low-power objective for the high-power objective? _____

6. Sketch the numeral 4 (or 9) as it appears through the scanning objective of the compound microscope.

7. What has the lens system done to the image of the numeral? (Is it right side up, upside down, or what?) _____

8. When you moved the ruler to the right, which way did the image move? _____

9. When you moved the ruler away from you, which way did the image move? _____

Part D   Assessments

1. Label the microscope parts in figure 4.9. 🔺

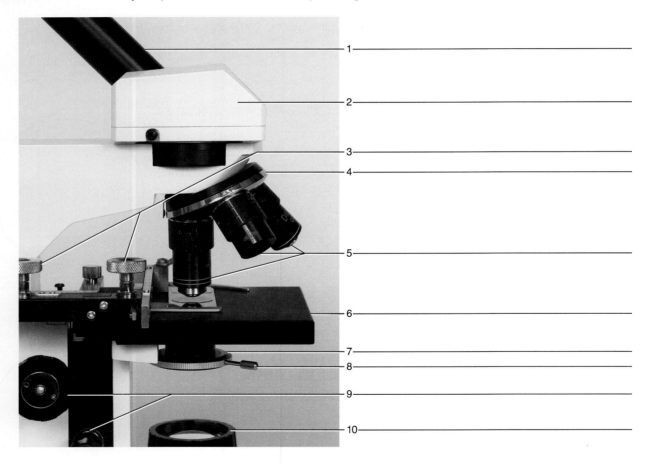

FIGURE 4.9   Identify the parts indicated on this compound light microscope.

2. Match the names of the microscope parts in column A with the descriptions in column B. Place the letter of your choice in the space provided. **1**

Column A

a. Adjustment knob (coarse)
b. Arm
c. Condenser
d. Eyepiece (ocular)
e. Field of view
f. Iris diaphragm
g. Nosepiece
h. Objective lens system
i. Stage
j. Stage (slide) clip

Column B

_____ 1. Increases or decreases the light intensity

_____ 2. Platform that supports a microscope slide

_____ 3. Concentrates light onto the specimen

_____ 4. Causes stage (or objective lens) to move upward or downward

_____ 5. After light passes through the specimen, it next enters this lens system

_____ 6. Holds a microscope slide in position

_____ 7. Contains a lens at the top of the body tube

_____ 8. Serves as a handle for carrying the microscope

_____ 9. Part to which the objective lenses are attached

_____ 10. Circular area seen through the eyepiece

Part E Assessments

Prepare sketches of the objects you observed using the microscope. For each sketch, include the name of the object, the magnification you used to observe it, and its estimated dimensions in millimeters and micrometers. **4**

5

Cell Structure and Function

Purpose of the Exercise

To review the structure and functions of major cellular components and to observe examples of human cells. To measure and compare the average cell's metabolic rate in individuals of different sizes (weight).

Materials Needed

Animal cell model
Clean microscope slides
Coverslips
Flat toothpicks
Medicine dropper
Methylene blue (dilute) or iodine-potassium-iodide
   stain
Prepared microscope slides of human tissues
Compound light microscope
Ph.I.L.S. 4.0

For Learning Extension Activities:
Single-edged razor blade
Plant materials such as leaves, soft stems, fruits, onion
   peel, and vegetables
Cultures of *Amoeba* and *Paramecium*

Safety

▶ Review all the safety guidelines inside the front cover.
▶ Clean laboratory surfaces before and after laboratory procedures.
▶ Wear disposable gloves for the wet-mount procedures of the cells lining the inside of the cheek.
▶ Work only with your own materials when preparing the slide of the cheek cells. Observe the same precautions as with all body fluids.
▶ Dispose of laboratory gloves, slides, coverslips, and toothpicks as instructed.
▶ Precautions should be taken to prevent cellular stains from contacting your clothes and skin.
▶ Wash your hands before leaving the laboratory.

Learning Outcomes

After completing this exercise, you should be able to

1. Name and locate the components of a cell.
2. Differentiate the functions of cellular components.
3. Prepare a wet mount of cells lining the inside of the cheek; stain the cells; and identify the plasma (cell) membrane, nucleus, and cytoplasm.
4. Examine cells on prepared slides of human tissues and identify their major components.
5. Explain the relationship of oxygen consumption with an animal's total weight.
6. Explain the relationship of oxygen consumption per gram of weight.
7. Explain the relationship of oxygen consumption and metabolic rate.
8. Calculate surface area–to–volume ratio and explain its significance to temperature regulation.
9. Integrate the concepts of weight, oxygen consumption, and metabolism and explain their application to the human body.

Pre-Lab

Carefully read the introductory material and examine the entire lab. Be familiar with the basic structures and functions of a cell from lecture or the textbook. Answer the pre-lab questions.

Pre-Lab Questions: Select the correct answer for each of the following questions:

1. Which of the following cellular structures is *not* easily visible with the compound light microscope?
   a. nucleus   **b.** DNA
   c. cytoplasm   **d.** plasma membrane
2. Which of the following cellular structures is located in the nucleus?
   a. nucleolus   **b.** ribosomes
   c. mitochondria   **d.** endoplasmic reticulum
3. The outer boundary of a cell is the
   a. mitochondrial   **b.** nuclear envelope.
       membrane.
   c. Golgi apparatus.   **d.** plasma membrane.

4. Microtubules, intermediate filaments, and microfilaments are components of
   a. vesicles.
   b. the Golgi apparatus.
   c. the cytoskeleton.
   d. ribosomes.
5. Easily attainable living cells observed in the lab are from
   a. inside the cheek.
   b. blood.
   c. hair.
   d. finger surface.
6. Cellular energy is called
   a. ER.
   b. ATP.
   c. DNA.
   d. RNA.
7. The smooth ER possesses ribosomes.
   True _____ False _____
8. The nuclear envelope contains nuclear pores.
   True _____ False _____

Cells are the "building blocks" from which all parts of the human body are formed. Their arrangement and interactions result in the shape, organization, and construction of the body and are responsible for carrying on its life processes. Using a compound light microscope and the proper stain, one can easily see the **plasma (cell) membrane,** the **cytoplasm,** and the **nucleus.** The cytoplasm is composed of a clear fluid, the *cytosol,* and numerous *cytoplasmic organelles* that are suspended in the cytosol.

The plasma membrane, composed of phospholipids, glycolipids, and glycoproteins, represents the cell boundary and functions in various methods of membrane transport. The nucleus is surrounded by a nuclear envelope, and many of the cytoplasmic organelles have membrane boundaries similar to the plasma membrane. Movements of substances across these membranes can be by passive processes, involving kinetic energy or hydrostatic pressure, or active processes, using the cellular energy of ATP (adenosine triphosphate).

FIGURE 5.1 The structures of a composite cell. The structures are not drawn to scale.

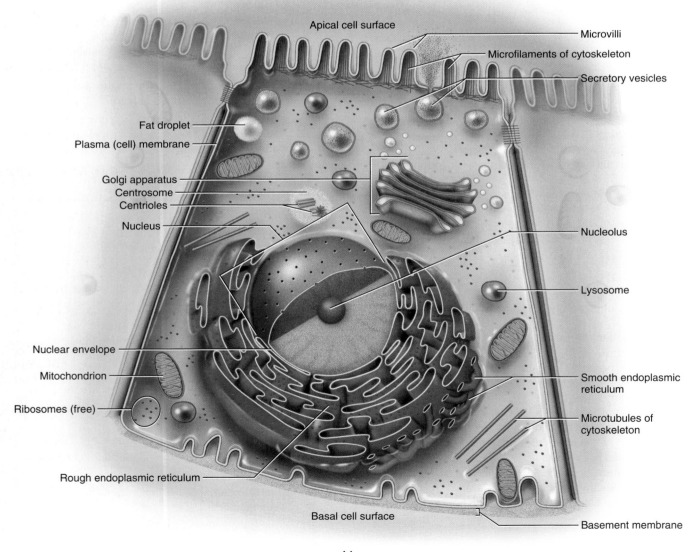

FIGURE 5.2 Structure of the plasma (cell) membrane.

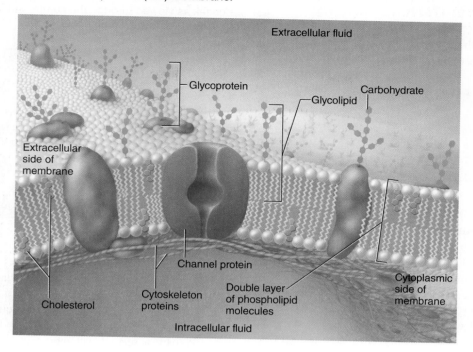

Metabolism refers to all the chemical reactions within cells that use or release energy. Is the metabolic rate of cells the same in a man who is six feet tall as in one who is five foot nine? Surprisingly, an understanding of temperature regulation is required to answer this question. *Temperature regulation* is accomplished through a balance of heat production and heat loss. Heat is produced by cells through metabolism (second law of thermodynamics: For every chemical reaction, some energy is always lost as heat). The greater the number of cells or the more metabolically active the cells, the more heat that is produced. Heat is lost through the surface of the skin. The larger the surface area of the skin, the more that is lost. For a given volume of cells, the metabolic rate of cells is established and maintained to offset the heat that is lost by the surface of the skin (within limits). The more heat that is lost at the skin, the more metabolically active the cells must be to produce the heat required to keep the body warm.

Procedure A—Cell Structure and Function

The boundary of the nucleus is a double-layered *nuclear envelope,* which has nuclear pores allowing the passage of genetic information. The nucleus contains fine strands of DNA (deoxyribonucleic acid) and structural and regulatory proteins called *chromatin,* which condenses to form chromosomes during cell divisions. The *nucleolus* is a nonmembranous structure composed of RNA (ribonucleic acid) and protein and is a site of ribosomal formation.

Cytoplasmic organelles provide for specialized metabolic functions. The *endoplasmic reticulum (ER)* has numerous canals that serve in transporting molecules as proteins

throughout the cytoplasm. *Ribosomes* synthesize proteins and are located free in the cytoplasm or on the surface of the endoplasmic reticulum (rough ER). If the ER lacks the ribosomes on the surface, it is called smooth ER and does not serve as a region of protein synthesis. The *Golgi apparatus (complex),* composed of flattened membranous sacs, is the site for packaging glycoproteins for transport and secretion. *Mitochondria* possess a double membrane and provide the main location for the cellular energy production of ATP. Mitochondria are often referred to as the "powerhouse" of a cell because of the energy production. *Lysosomes* are membranous sacs that contain intracellular digestive enzymes for destroying debris and worn-out organelles. *Vesicles* are membranous sacs produced by the cell that contain substances for storage or transport, or form from pinching off pieces of the plasma membrane. A *cytoskeleton* contains microtubules, intermediate filaments, and microfilaments that support cellular structures within the cytoplasm and are involved in cellular movements.

1. Study figures 5.1 and 5.2.
2. Observe the animal cell model and identify its major structures.
3. Complete Part A of Laboratory Assessment 5.
4. Prepare a wet mount of cells lining the inside of the cheek. To do this, follow these steps:
    a. Gently scrape (force is not necessary and should be avoided) the inner lining of your cheek with the broad end of a flat toothpick.
    b. Stir the toothpick in a drop of water on a clean microscope slide and dispose of the toothpick as directed by your instructor.

Critical Thinking Activity

The cells lining the inside of the cheek are frequently removed for making observations of basic cell structure. The cells are from stratified squamous epithelium. Explain why these cells are used instead of outer body surface tissue.

_____

_____

_____

Learning Extension Activity

Investigate the microscopic structure of various plant materials. To do this, prepare tiny, thin slices of plant specimens, using a single-edged razor blade. *(Take care not to injure yourself with the blade.)* Keep the slices in a container of water until you are ready to observe them. To observe a specimen, place it into a drop of water on a clean microscope slide and cover it with a coverslip. Use the microscope and view the specimen using low- and high-power magnifications. Observe near the edges where your section of tissue is most likely to be one cell thick. Add a drop of dilute methylene blue or iodine-potassium-iodide stain, and note if any additional structures become visible. How are the microscopic structures of the plant specimens similar to the human tissues you observed? _____

_____

How are they different? _____

_____

_____

Learning Extension Activity

Prepare a wet mount of the *Amoeba* and *Paramecium* by putting a drop of culture on a clean glass slide. Gently cover with a clean coverslip. Observe the movements of the *Amoeba* with pseudopodia and the *Paramecium* with cilia. Try to locate cellular components such as the plasma (cell) membrane, nuclear envelope, nucleus, mitochondria, and contractile vacuoles. Describe the movement of the *Amoeba*.

_____

_____

Describe the movement of the *Paramecium*.

_____

_____

_____

FIGURE 5.3 Stained cell lining the inside of the cheek as viewed through the compound light microscope using the high-power objective (400×).

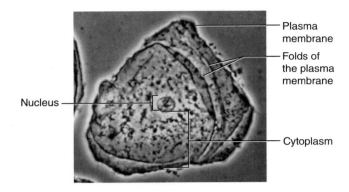

c. Cover the drop with a coverslip.

d. Observe the cheek cells by using the microscope. Compare your image with figure 5.3. To report what you observe, sketch a single cell in the space provided in Part B of the laboratory assessment.

5. Prepare a second wet mount of cheek cells, but this time, add a drop of dilute methylene blue or iodine-potassium-iodide stain to the cells. Cover the liquid with a coverslip and observe the cells with the microscope. Add to your sketch any additional structures you observe in the stained cells.

6. Answer the questions in Part B of the laboratory assessment.

7. Using the microscope, observe each of the prepared slides of human tissues. To report what you observe, sketch a single cell of each type in the space provided in Part C of the laboratory assessment.

8. Complete Parts C and D of the laboratory assessment.

Procedure B—Ph.I.L.S. Lesson 2 Metabolism: Size and Basal Metabolic Rate

1. Open Exercise 2: Metabolism: Size and Basal Metabolic Rate.

2. Read the objectives and introduction and take the pre-lab quiz.

3. After completing the pre-lab quiz, read through the wet lab. *Be sure to click open and view the videos that are indicated in red.*

4. The lab exercise will open when you click Continue after completing the wet lab (fig. 5.4).

Weighing the Mouse

5. Click the power switch to turn on the scale.

6. Click tare to set the scale to zero.

7. Weigh one of the white mice by clicking on one mouse and dragging it to the scale.

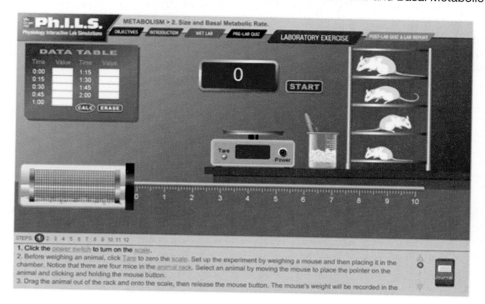

Setting up the Chamber

8. After weighing the mouse, place it in the chamber by clicking and dragging on the mouse. (To "click and drag," move the computer mouse to position the arrow on the mouse; left click on mouse and hold; drag the mouse to the chamber; and when positioned release the left click).

9. Place bubbles on the end of the calibrated tube by clicking on the pipette and dragging the pipette to the calibration tube. Be sure to **touch** *pipette to the end of the calibration tube.* If a bubble does not appear try again.

Measuring the Bubbles

10. Measure the initial position of the bubbles (10 mm) and record at 0:00 time in the data table. Notice that the data table includes a change in time every 15 seconds between 0:00 and 2:00 minutes.

11. After clicking the Start button, measure the position of the soap bubble within the calibration tube at 15-second intervals (you will hear a beep) and record in the data table. (The timer will begin when you hit Start). You may find it more manageable to click Pause after each 15-second interval, measure, record, and then click on Start to begin again.

12. When finished click Pause.

13. Click "calc" to see graph (linear regression). Journal will open.

Examine the linear regression graph that relates time and oxygen consumption. Does the amount of oxygen consumed for this mouse increase or decrease over time (i.e., Does the amount of oxygen consumed for this mouse increase or decrease the longer the mouse is in the chamber)?_____
Why? _____
_____
_____
_____

Repeating the Experiment

14. To complete with the same animal repeat numbers 9–13 (Ph.I.L.S. steps 4–11). *Perform at least three trials per mouse.*

15. To complete with a different animal, place the animal back in its cage by clicking and dragging the computer mouse. Repeat numbers 5–13 (Ph.I.L.S. steps 5–11).

16. You can print your graphs by hitting Control (Ctrl) on your keyboard while pushing *p*.

Interpreting Results

17. *With the graph still on the screen,* answer the questions in Part E of the laboratory assessment in the laboratory manual. If you accidentally close the graph, click on the Journal panel (red rectangle at bottom right of screen).

18. Click on Post-Lab Quiz and Lab Report.

19. Complete the Post-Lab Quiz by answering the ten questions on the computer screen.

20. Read the conclusion on the computer screen.

21. You may print the Lab Report for Metabolism: Size and Basal Metabolic Rate.

NOTES

Laboratory Assessment

5

Name _____

Date _____

Section _____

The ▲ corresponds to the indicated outcome(s) found at the beginning of the laboratory exercise.

Cell Structure and Function

Part A Assessments

1. Label the cellular structures in figure 5.5. ▲

FIGURE 5.5 Label the indicated cellular structures of this composite cell.

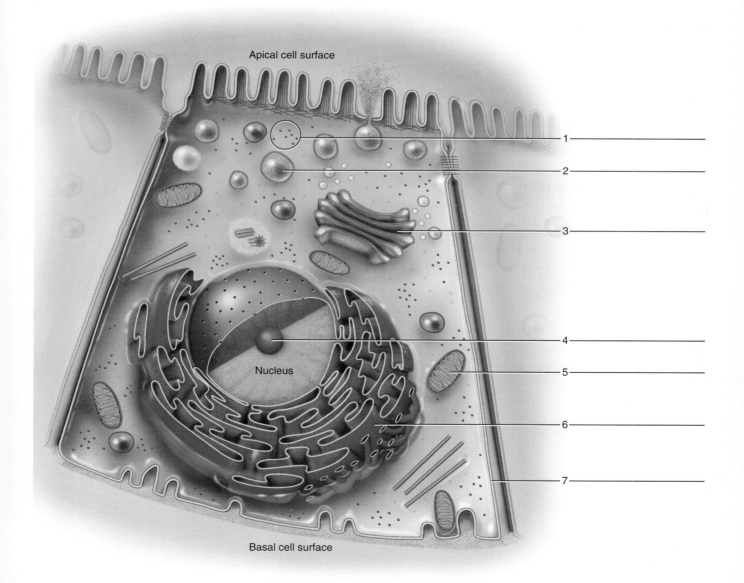

2. Match the cellular components in column A with the descriptions in column B. Place the letter of your choice in the space provided.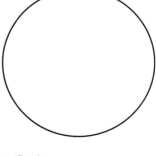

Column A	Column B

Column A

a. Chromatin
b. Cytoplasm
c. Endoplasmic reticulum
d. Golgi apparatus (complex)
e. Lysosome
f. Microtubule
g. Mitochondrion
h. Nuclear envelope
i. Nucleolus
j. Nucleus
k. Ribosome
l. Vesicle

Column B

_____ **1.** Loosely coiled fibers containing protein and DNA within nucleus

_____ **2.** Location of ATP production for cellular energy

_____ **3.** Small RNA-containing particles for the synthesis of proteins

_____ **4.** Membranous sac formed by the pinching off of pieces of plasma membrane

_____ **5.** Dense body of RNA and protein within the nucleus

_____ **6.** Part of the cytoskeleton involved in cellular movement

_____ **7.** Composed of membrane-bound canals for tubular transport throughout the cytoplasm

_____ **8.** Occupies space between plasma membrane and nucleus

_____ **9.** Flattened membranous sacs that package a secretion

_____ **10.** Membranous sac that contains digestive enzymes

_____ **11.** Separates nuclear contents from cytoplasm

_____ **12.** Spherical organelle that contains chromatin and nucleolus

Part B Assessments

Complete the following:

1. Sketch a single cheek cell. Label the cellular components you recognize. Add any additional structures observed to your sketch after staining was completed. (The circle represents the field of view through the microscope.) 3

2. After comparing the wet mount and the stained cheek cells, describe the advantage gained by staining cells.

Magnification _____ ×

_____

_____

Part C Assessments

Complete the following:

1. Sketch a single cell of each type you observed in the prepared slides of human tissues. Name the tissue, indicate the magnification used, and label the cellular components you recognize.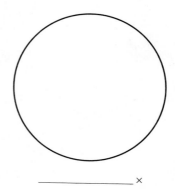

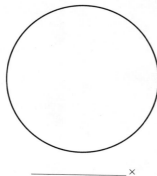

_____ ×

Tissue _____

_____ ×

Tissue _____

2. What do the various types of cells in these tissues have in common? _____

_____

3. What are the main differences you observed among these cells? _____

_____

_____

Part D Assessments

Electron micrographs represent extremely thin slices of cells. Each micrograph in figure 5.6 contains a section of a nucleus and some cytoplasm. Compare the organelles shown in these micrographs with organelles of the animal cell model and figure 5.1.

Identify the structures indicated by the arrows in figure 5.6. ⓵

1. _____

2. _____

3. _____

4. _____

5. _____

6. _____

7. _____

8. _____

9. _____

10. _____

FIGURE 5.6 Transmission electron micrographs of cellular components. The views are only portions of a cell. Magnifications: (a) 26,000×; (b) 10,000×. Identify the numbered cellular structures, using the terms provided. 1

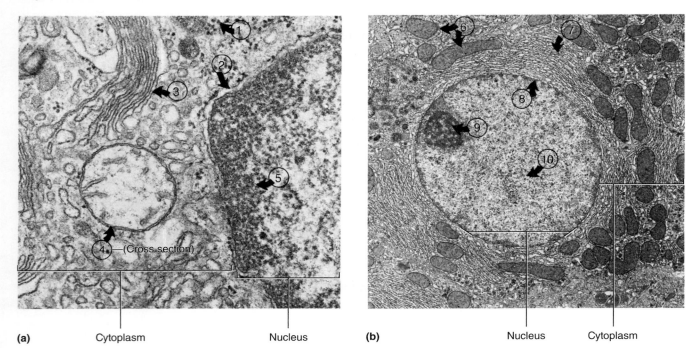

(a) Cytoplasm Nucleus (b) Nucleus Cytoplasm

Terms:
Chromatin (use 2 times)
Endoplasmic reticulum
Golgi apparatus
Mitochondria
Mitochondrion (cross section)
Nuclear envelope (use 2 times)
Nucleolus
Ribosomes

Answer the following questions after observing the transmission electron micrographs in figure 5.6.

11. What cellular structures were visible in the transmission electron micrographs that were not apparent in the cells you observed using the microscope? _____

_____

_____

Part E Ph.I.L.S. Lesson 2, Metabolism: Size and Basal Metabolic Rate Assessments

Interpreting Results

1. Examine the graph on the left of the final graph. On the x-axis is weight (grams) and on the y-axis is average oxygen consumed per minute (av 0_2/m). As the weight of the mouse decreases, does the average oxygen consumed per minute decrease or increase? _____ Is the relationship between weight and oxygen consumed direct or inverse? _____ **5**

2. Examine the graph on the right of the final graph. Weight (grams) is on the x-axis and average oxygen consumed per hour *per gram (of body weight)* (av 0_2/h/g) is on the y-axis. As the weight of the mouse decreases, does the average oxygen consumed per hour per gram decrease or increase? _____ Is the relationship between weight and oxygen consumed per gram of body weight direct or inverse? _____ **6**

3. Based on the results of the graph on the right *and* the fact that oxygen consumed is an indirect measure of metabolic rate, as the weight of the mouse decreases, does the metabolic rate decrease or increase? _____ Is the relationship between weight and metabolic rate direct or inverse? _____ **7**

Understanding the Results

4. **Surface Area–to–Volume Ratio:** As a (spherical) structure increases in radius, the surface area increases at a rate of r^2 and the volume increases at a rate of r^3. If the radius of the sphere is 2, the calculated surface area $2^2 = 2 \times 2 = 4$ and the calculated volume $2^3 = 2 \times 2 \times 2 = 8$ and the ratio of surface area/volume is $4/8 = 0.5$. If r is 4, calculate the surface area, volume, and surface area/volume ratio. Place your answers in the following box. **8**

   $r = 4$

   Surface area = _____   Volume = _____

   Surface area/volume ratio = _____
   ---

5. As the (spherical) structure becomes larger, is the *ratio* becoming larger or smaller? (Hint: Compare the surface area/volume ratio between two spheres—one with a radius of 2 and the other with a radius of 4.) _____ Is the volume increasing at a faster rate than the surface area? _____ **8**

Relating to the Body

Use this information to answer question 6. If the surface area–to–volume ratio is applied to the human body, volume is the size (weight) of the body and surface area is the area of skin surface. **8 9**

6. As a person increases in size:

   a. What happens to the ratio of surface area/volume? _____

   b. Is volume (size) increasing at a faster rate than surface area (skin)? _____

   c. Is this a more favorable ratio for keeping the body warm? _____

   d. Will each individual cell have to be as metabolically active as in a larger individual to keep the body warm? _____

   e. Is the metabolic rate of individual cells faster or slower? _____

Critical Thinking Assessment

Walt, a six-foot man, walks with his friend Albert, who is five–foot–nine.

a. Would you predict that Walt, with the larger body, is consuming more oxygen than Albert? _____

b. Would you predict that Walt is consuming more oxygen *per gram of weight* than Albert? _____

c. Which man would have the higher *metabolic rate per gram of weight*? _____

NOTES

Movements Through Membranes

Purpose of the Exercise

To demonstrate some of the physical processes by which substances move through membranes.

Safety

▶ Clean laboratory surfaces before and after laboratory procedures.
▶ Wear disposable gloves when handling chemicals and animal blood.
▶ Wear safety glasses when using chemicals.
▶ Dispose of laboratory gloves and blood-contaminated items as instructed.
▶ Wash your hands before leaving the laboratory.

Materials Needed

For Procedure A—
Diffusion:
Petri dish
White paper
Forceps

Potassium permanganate crystals
Millimeter ruler (thin and transparent)

For Procedure B—
Osmosis:
Thistle tube
Molasses (or Karo dark corn syrup)
Ring stand and clamp
Beaker
Rubber band

Millimeter ruler
Selectively permeable (semipermeable) membrane (presoaked dialysis tubing of 1 5/16" or greater diameter)

For Procedure C—
Hypertonic,
Hypotonic, and
Isotonic Solutions:
Test tubes
Marking pen
Test-tube rack
10 mL graduated cylinder
Medicine dropper

Uncoagulated animal blood
Distilled water
0.9% NaCl (aqueous solution)
3.0% NaCl (aqueous solution)
Clean microscope slides
Coverslips
Compound light microscope

For Procedure D—
Filtration:
Glass funnel
Filter paper
Ring stand and ring
Beaker
Powdered charcoal or ground black pepper
1% glucose (aqueous solution)

1% starch (aqueous solution)
Test tubes
10 mL graduated cylinder
Water bath (boiling water)
Benedict's solution
Iodine-potassium-iodide solution
Medicine dropper

For Procedure E—Ph.I.L.S. Lesson 1 Osmosis and
Diffusion: Varying Extracellular Concentration
Ph.I.L.S. 4.0

For Alternative Osmosis
Activity:
Fresh chicken egg
Beaker

Laboratory balance
Spoon
Vinegar
Corn syrup (Karo)

Learning Outcomes

After completing this exercise, you should be able to

1. Show diffusion and identify examples of diffusion.
2. Interpret diffusion by preparing and explaining a graph.
3. Show osmosis and identify examples of osmosis.
4. Distinguish among hypertonic, hypotonic, and isotonic solutions and examine the effects of these solutions on animal cells.
5. Show filtration and identify examples of filtration.
6. Identify the percent transmittance of light through a blood sample placed in varying concentrations of NaCl.
7. Interpret the results to explain the relationship of percent transmittance and condition of the red blood cells (normal, shriveled or ruptured) in the blood sample.
8. Interpret the results to explain the condition of the red blood cells (normal, shriveled, or ruptured) to the relative concentration of NaCl solutions (isotonic, hypertonic, or hypotonic).
9. Relate plasma concentration and red blood cell integrity with events that occur in the human body.

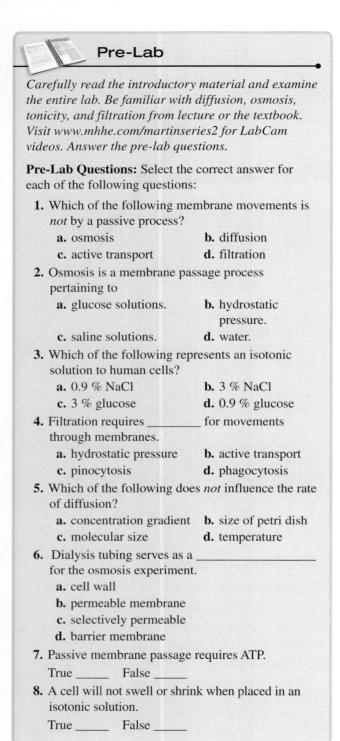

Pre-Lab

Carefully read the introductory material and examine the entire lab. Be familiar with diffusion, osmosis, tonicity, and filtration from lecture or the textbook. Visit www.mhhe.com/martinseries2 for LabCam videos. Answer the pre-lab questions.

Pre-Lab Questions: Select the correct answer for each of the following questions:

1. Which of the following membrane movements is *not* by a passive process?
   - **a.** osmosis
   - **b.** diffusion
   - **c.** active transport
   - **d.** filtration

2. Osmosis is a membrane passage process pertaining to
   - **a.** glucose solutions.
   - **b.** hydrostatic pressure.
   - **c.** saline solutions.
   - **d.** water.

3. Which of the following represents an isotonic solution to human cells?
   - **a.** 0.9 % NaCl
   - **b.** 3 % NaCl
   - **c.** 3 % glucose
   - **d.** 0.9 % glucose

4. Filtration requires _____ for movements through membranes.
   - **a.** hydrostatic pressure
   - **b.** active transport
   - **c.** pinocytosis
   - **d.** phagocytosis

5. Which of the following does *not* influence the rate of diffusion?
   - **a.** concentration gradient
   - **b.** size of petri dish
   - **c.** molecular size
   - **d.** temperature

6. Dialysis tubing serves as a _____ for the osmosis experiment.
   - **a.** cell wall
   - **b.** permeable membrane
   - **c.** selectively permeable
   - **d.** barrier membrane

7. Passive membrane passage requires ATP.
   - True _____ False _____

8. A cell will not swell or shrink when placed in an isotonic solution.
   - True _____ False _____

A plasma membrane functions as a gateway through which chemical substances and small particles may enter or leave a cell. These substances move through the membrane by passive (physical) processes such as diffusion, osmosis, and filtration, or by active (physiological) processes such as active transport, phagocytosis, or pinocytosis. Passive processes do not require the cellular energy of ATP, but occur due to molecular motion or hydrostatic pressure; active processes utilize ATP.

This laboratory exercise contains examples of passive membrane transport through either living plasma membranes or artificial membranes. *Diffusion* is the random motion of molecules from an area of higher concentration toward an area of lower concentration. The rate of diffusion depends upon factors such as the concentration gradient, the molecular size, and the temperature. Facilitated diffusion is a carrier-mediated transport, which is not represented in this laboratory exercise.

A special case of passive membrane passage, called *osmosis,* occurs when water molecules diffuse through a selectively permeable membrane from an area of higher water concentration toward an area of lower water concentration. *Tonicity* refers to the osmotic pressure of a solution in relation to a cell. A solution is isotonic if the osmotic pressure is the same as the cell. A 5% glucose solution and a 0.9% normal saline solution serve as isotonic solutions to human cells. A hypertonic solution has a higher concentration of solutes than the cell; a hypotonic solution has a lower concentration of solutes than a cell. The osmosis of water will occur out of a cell when immersed in a hypertonic solution, and the cell will shrink (crenate). Conversely, the osmosis of water will occur into a cell when immersed in a hypotonic solution, and the cell will swell, and may burst (lyse).

Filtration is the movement of water and solutes through a selectively permeable membrane by hydrostatic pressure upon the membrane. The pressure gradient forces the water and solutes (filtrate) from the higher hydrostatic pressure area to a lower hydrostatic pressure area. For example, blood pressure provides the hydrostatic pressure for water and dissolved substances to move through capillary walls.

Procedure A—Diffusion

Simple diffusion can occur in living organisms as well as in nonliving systems involving gases, liquids, and solids. In this procedure, diffusion will be examined in water without cells.

1. To demonstrate diffusion, refer to figure 6.1 as you follow these steps:
   - **a.** Place a petri dish, half filled with water, on a piece of white paper that has a millimeter ruler positioned on the paper. Wait until the water surface is still. Allow approximately 3 minutes.
     Note: The petri dish should remain level. A second millimeter ruler may be needed under the petri dish as a shim to obtain a level amount of the water inside the petri dish.
   - **b.** Using forceps, place one crystal of potassium permanganate near the center of the petri dish and near the millimeter ruler (fig. 6.1).
   - **c.** Measure the radius of the purple circle at 1-minute intervals for 10 minutes and record the results in Part A of Laboratory Assessment 6.

2. Complete Part A of the assessment.

FIGURE 6.1 To demonstrate diffusion, place one crystal of potassium permanganate in the center of a petri dish containing water. Place the crystal near the millimeter ruler (positioned under the petri dish).

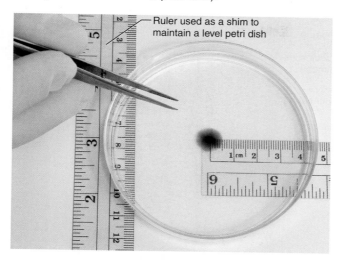

Ruler used as a shim to maintain a level petri dish

Learning Extension Activity

Repeat the demonstration of diffusion using a petri dish filled with ice-cold water and a second dish filled with very hot water. At the same moment, add a crystal of potassium permanganate to each dish and observe the circle as before. What difference do you note in the rate of diffusion in the two dishes? How do you explain this difference? _____

_____

_____

_____

_____

Procedure B—Osmosis

For the osmosis experiment, an artificial membrane will serve as a model for an actual plasma membrane. A layer of dialysis tubing that has been soaked for 30 minutes can easily be cut open because it becomes pliable. This membrane model works as a selectively (semipermeable; differentially) permeable membrane.

When water moves across a selectively permeable membrane down its concentration gradient, it is termed osmosis. Solute particles will also move down their concentration gradients across a membrane if the solute will pass through the membrane. However, if all the solute particles are nonpenetrating because they are too large to pass across a selectively permeable membrane, only water will move across the selectively permeable membrane. This will result in changes in the volume of water on a certain side of the membrane. This will create a volume change in relation to a cell or models used as simulated cells.

1. To demonstrate osmosis, refer to figure 6.2 as you follow these steps:
   a. One person plugs the tube end of a thistle tube with a finger.
   b. Another person then fills the bulb with molasses until it is about to overflow at the top of the bulb. Air remains trapped in the stem.
   c. Cover the bulb opening with a single-thickness piece of moist selectively permeable membrane.
   d. Tightly secure the membrane in place with several wrappings of a rubber band.
   e. Immerse the bulb end of the tube in a beaker of water. If leaks are noted, repeat the procedures.
   f. Support the upright portion of the tube with a clamp on a ring stand. Folded paper under the clamp will protect the thistle tube stem from breakage.
   g. Mark the meniscus level of the molasses in the tube. *Note:* The best results will occur if the mark of the molasses is a short distance up the stem of the thistle tube when the experiment starts.
   h. Measure the level changes after 10 minutes and 30 minutes and record the results in Part B of the laboratory assessment.
2. Complete Part B of the laboratory assessment.

Alternative Activity

Eggshell membranes possess selectively permeable properties. To demonstrate osmosis using a natural membrane, soak a fresh chicken egg in vinegar for about 24 hours to remove the shell. Use a spoon to carefully handle the delicate egg. Place the egg in a hypertonic solution (corn syrup) for about 24 hours. Remove the egg, rinse it, and using a laboratory balance, weigh the egg to establish a baseline weight. Place the egg in a hypotonic solution (distilled water). Remove the egg and weigh it every 15 minutes for an elapsed time of 75 minutes. Explain any weight changes that were noted during this experiment.

_____

_____

_____

_____

_____

FIGURE 6.2 (a) Fill the bulb of the thistle tube with molasses; (b) tightly secure a piece of selectively permeable (semipermeable) membrane over the bulb opening; and (c) immerse the bulb in a beaker of water. Note: These procedures require the participation of two people.

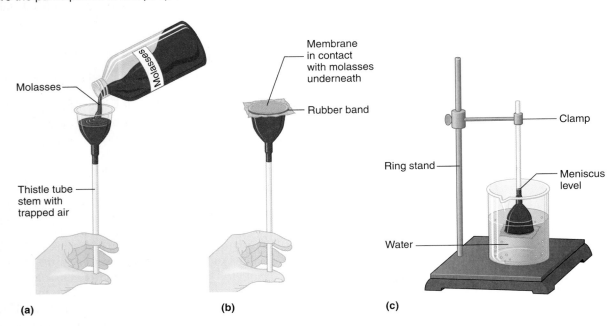

(a) (b) (c)

Procedure C—Hypertonic, Hypotonic, and Isotonic Solutions

Applications of volume changes to cells from osmosis are addressed in this procedure. If a cell is placed in an isotonic solution with the same water and solute concentration, there will be no net osmosis and the cell will retain its normal size. If a cell is placed in a hypertonic solution, the cell will lose water and shrink; if the cell is in a hypotonic solution it will swell and burst.

1. To demonstrate the effect of hypertonic, hypotonic, and isotonic solutions on animal cells, follow these steps:
   a. Place three test tubes in a rack and mark them *tube 1, tube 2,* and *tube 3.* (*Note:* One set of tubes can be used to supply test samples for the entire class.)
   b. Using 10 mL graduated cylinders, add 3 mL of distilled water to tube 1; add 3 mL of 0.9% NaCl to tube 2; and add 3 mL of 3.0% NaCl to tube 3.
   c. Place three drops of fresh, uncoagulated animal blood into each of the tubes, and gently mix the blood with the solutions. Wait 5 minutes.
   d. Using three separate medicine droppers, remove a drop from each tube and place the drops on three separate microscope slides marked *1, 2,* and *3.*
   e. Cover the drops with coverslips and observe the blood cells, using the high power of the microscope.
2. Complete Part C of the laboratory assessment.

Alternative Activity

Various substitutes for blood can be used for Procedure C. Onion, cucumber, or cells lining the inside of the cheek represent three possible options.

Procedure D—Filtration

Filtration is another example of passive membrane passage. Unlike the previous experiments, hydrostatic pressure provides the mechanism for movements across a membrane instead of molecular motion. Water and solutes that pass through the membrane move simultaneously in the same direction. Solutes that are too large to cross the membrane remain on their original side of the membrane. Filtration occurs faster when the hydrostatic pressures increase.

1. To demonstrate filtration, follow these steps:
   a. Place a glass funnel in the ring of a ring stand over an empty beaker. Fold a piece of filter paper in half and then in half again. Open one thickness of the filter paper to form a cone. Wet the cone, and place it in the funnel. The filter paper is used to demonstrate how movement across membranes is limited by the size of the molecules, but it does not represent a working model of biological membranes.
   b. Prepare a mixture of 5 cc (approximately 1 teaspoon) powdered charcoal (or ground black pepper) and equal amounts of 1% glucose solution and 1% starch solution in a beaker. Pour some of the mixture into the funnel until it nearly reaches the top of the filter-paper

FIGURE 6.3 Apparatus used to illustrate filtration.

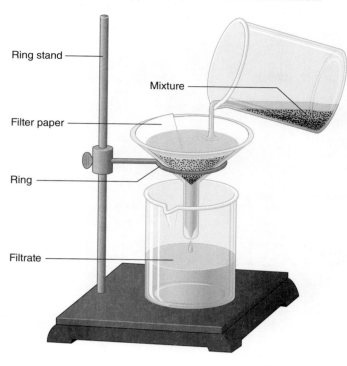

Ring stand

Mixture

Filter paper

Ring

Filtrate

cone. Care should be taken to prevent the mixture from spilling over the top of the filter paper. Collect the filtrate in the beaker below the funnel (fig. 6.3).

c. Test some of the filtrate in the beaker for the presence of glucose. To do this, place 1 mL of filtrate in a clean test tube and add 1 mL of Benedict's solution. Place the test tube in a water bath of boiling water for 2 minutes and then allow the liquid to cool slowly. If the color of the solution changes to green, yellow, or red, glucose is present (fig. 6.4).

FIGURE 6.4 Heat the filtrate and Benedict's solution in a boiling water bath for 2 minutes.

Water bath

Filtrate and Benedict's solution

Hot plate

d. Test some of the filtrate in the beaker for the presence of starch. To do this, place a few drops of filtrate in a test tube and add one drop of iodine-potassium-iodide solution. If the color of the solution changes to blue-black, starch is present.

e. Observe any charcoal in the filtrate.

2. Complete Part D of the laboratory assessment.

Procedure E—Ph.I.L.S. Lesson 1 Osmosis and Diffusion: Varying Extracellular Concentration

1. Open Exercise 1: Osmosis and Diffusion: Varying Extracellular Concentration.
2. Read the objectives and introduction and take the pre-lab quiz.
3. After completing the pre-lab quiz, read through the wet lab. *Be sure to click open and view the videos that are indicated in red in the wet lab.*
4. The lab exercise will open when you click Continue after completing the wet lab (fig. 6.5).

Setup

5. Click the power switch to turn on the spectrophotometer.
6. Set the wavelength at 510 nm (wavelength at which cell membranes absorb light if intact).
7. Add 1 mL of blood to each of the tubes using the pipette. Lift the pipette (green) and drag over to the flask of blood. Using the up arrow, add 1 mL of blood to the 1 mL mark of the pipette.
8. Drag the pipette to an empty tube and release the blood by pressing the down arrow.
9. Repeat numbers 7 and 8 (Ph.I.L.S. steps 4–6) until all of the test tubes are filled.

Calibrate Spectrophotometer

10. Set the transmittance value at 0 using the arrows at zero.
11. Open the lid of the "holder" of the spectrophotometer (If unsure, click on the word *spectrophotometer* in the instructions at the bottom of the screen) and click and drag the "blank" (tube on the far left of the rack) into the spectrophotometer. (To click and drag, move the mouse to position the arrow on the tube; left-click on mouse and hold; drag the tube to the holder of the spectrophotometer; and when positioned release the left click). Close the lid and set the transmittance at 100 using the arrows at Calibrate.
12. Click the lid of the spectrophotometer and then the tube (top of tube visible in the spectrophotometer) to remove it. The tube will come straight up out of the spectrophotometer. Then click on the tube to drag it back to the rack.
13. The Journal will open automatically.
14. After viewing, close the Journal by clicking on the X in the upper right corner of the graph.

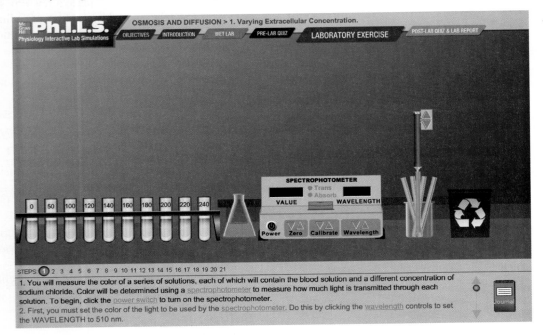

Measuring Transmittance

15. Click and drag the next tube to the spectrophotometer and close the lid. (A reading is taken when the lid is closed.) Open the lid. The Journal will open automatically and you can view the data.

16. Click on the test tube to remove it. (The tube will come up out of the spectrophotometer and the Journal will close automatically.) Then drag it back to the rack.

17. Repeat numbers 15 and 16 (Ph.I.L.S. steps 16–21) until all tubes have been tested.

Interpreting Results

18. *With the graph still on the screen,* answer the questions in Part E of the laboratory assessment in the lab manual. If you accidentally closed the graph, click on the Journal panel (red rectangle at bottom right of screen). To help you to remember the relationship of % transmittance and the state of the red blood cells, use the following:

 a. *More* light is transmitted in samples where the red blood cells have taken on water and *ruptured open* (hemolysis). (Red blood cells are destroyed and no longer block the light as effectively, so more light is transmitted through the sample.)

 b. *Less* light is transmitted in samples where the red blood cells have lost water and *shriveled* (crenated). (Red blood cells are losing water and becoming more "compact," and they block the light more effectively, so less light is transmitted through the sample.)

> Swelling cells: high transmittance
> Shrinking cells: low transmittance

19. Complete the Post-Lab Quiz (click open Post-Lab Quiz and Lab Report) by answering the ten questions on the computer screen.

20. Read the conclusion on the computer screen.

21. You may print the Lab Report for Osmosis and Diffusion: Varying Extracellular Concentration.

Name _____

Date _____

Section _____

The ⒶA corresponds to the indicated outcome(s) found at the beginning of the laboratory exercise.

Movements Through Membranes

Part A   Assessments

Complete the following:

1. Enter data for changes in the movement of the potassium permanganate. ⒶⒶ

Elapsed Time	Radius of Purple Circle in Millimeters
Initial	_____
1 minute	_____
2 minutes	_____
3 minutes	_____
4 minutes	_____
5 minutes	_____
6 minutes	_____
7 minutes	_____
8 minutes	_____
9 minutes	_____
10 minutes	_____

2. Prepare a graph that illustrates the diffusion distance of potassium permanganate in 10 minutes. ⒶⒶ

3. Explain your graph. ⒶⒶ _____

_____

_____

_____

_____

_____

_____

_____

_____

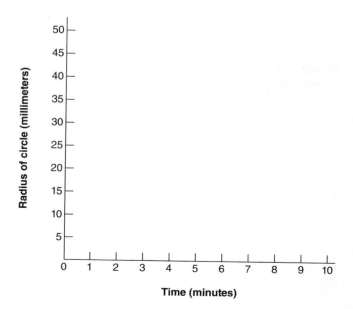

4. Define *diffusion.* _____

_____

_____

_____

Critical Thinking Assessment

By answering yes or no, indicate which of the following provides an example of diffusion. ⚠️

1. A perfume bottle is opened, and soon the odor can be sensed in all parts of the room. _____

2. A sugar cube is dropped into a cup of hot water, and, without being stirred, all of the liquid becomes sweet tasting. _____

3. Water molecules move from a faucet through a garden hose when the faucet is turned on. _____

4. A person blows air molecules into a balloon by forcefully exhaling. _____

5. A crystal of blue copper sulfate is placed in a test tube of water. The next day, the solid is gone, but the water is evenly colored. _____

Part B Assessments

Complete the following:

1. What was the change in the level of molasses in 10 minutes?_____

2. What was the change in the level of molasses in 30 minutes?_____

3. How do you explain this change? ⚠️ _____

_____

4. Define *osmosis.* _____

_____

Critical Thinking Assessment

By answering yes or no, indicate which of the following involves osmosis. ⚠️

1. A fresh potato is peeled, weighed, and soaked in a strong salt solution. The next day, it is discovered that the potato has lost weight. _____

2. Garden grass wilts after being exposed to dry chemical fertilizer. _____

3. Air molecules escape from a punctured tire as a result of high pressure inside. _____

4. Plant seeds soaked in water swell and become several times as large as before soaking. _____

5. When the bulb of a thistle tube filled with water is sealed by a selectively permeable membrane and submerged in a beaker of molasses, the water level in the tube falls. _____

Part C Assessments

Complete the following:

1. In the spaces, sketch a few blood cells from each of the test tubes and indicate the magnification. Ⓐ

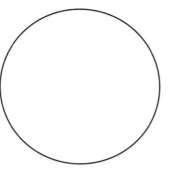 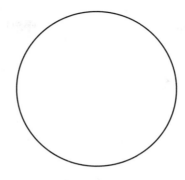 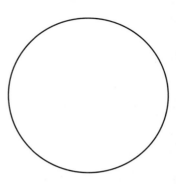

Tube 1
(distilled water)

_____×

Tube 2
(0.9% NaCl)

_____×

Tube 3
(3.0% NaCl)

_____×

2. Based on your results, which tube contained a solution hypertonic to the blood cells? Ⓐ _____

Give the reason for your answer. Ⓐ _____

_____

3. Which tube contained a solution hypotonic to the blood cells? Ⓐ _____

Give the reason for your answer. Ⓐ _____

_____

4. Which tube contained a solution isotonic to the blood cells? Ⓐ _____

Give the reason for your answer. Ⓐ _____

_____

5. Observe the RBCs shown in figure 6.6. Select the solutions in which the cells were placed to illustrate the effects of tonicity on cells.

FIGURE 6.6 Three blood cells placed in three different solutions: distilled water, 0.9% NaCl, and 3% NaCl. Select the solutions in which the cells were placed, using the terms provided. Ⓐ

Terms:
Hypertonic
Hypotonic
Isotonic

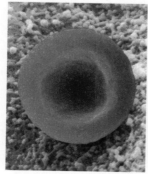  

(a) _____

(b) _____

(c) _____

Part D Assessments

Complete the following:

1. Which of the substances in the mixture you prepared passed through the filter paper into the filtrate? **5** _____

   _____

2. What evidence do you have for your answer to question 1? **5** _____

   _____

   _____

3. What force was responsible for the movement of substances through the filter paper? **5** _____

4. What substances did not pass through the filter paper? **5** _____

   _____

5. What factor prevented these substances from passing through? **5** _____

   _____

6. Define *filtration.* _____

   _____

Critical Thinking Assessment

By answering yes or no, indicate which of the following involves filtration. **5**

1. Oxygen molecules move into a cell and carbon dioxide molecules leave a cell because of differences in the concentrations of these substances on either side of the plasma membrane._____

2. Blood pressure forces water molecules from the blood outward through the thin wall of a blood capillary. _____

3. Urine is forced from the urinary bladder through the tubular urethra by muscular contractions._____

4. Air molecules enter the lungs through the airways when air pressure is greater outside these organs than inside._____

5. Coffee is made using a coffeemaker (not instant)._____

Part E Ph.I.L.S. Lesson 1, Osmosis and Diffusion: Varying Extracellular Concentration Assessments

1. Use the data from the journal on the computer; diagram a graph of an NaCl concentration and % transmittance. (Include title. Label each axis, including units. Plot the points and connect.) 6

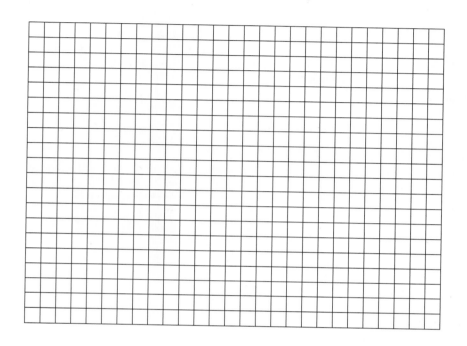

2. On the preceding graph, indicate the areas that correspond to the blood cells being in each of the following solutions: (A) isotonic solution; (B) hypertonic solution (C) hypotonic solution. 7 8

3. Physiological saline (0.9%) is isotonic. What concentration (mM) would be equivalent to physiological saline? 9

_____

4. Describe the net movement of water (into the cell, out of the cell, no net movement) when cells are placed in a(n): 9

   a. isotonic solution _____

   b. hypotonic solution _____

   c. hypertonic solution _____

Critical Thinking Assessment

A patient is given an IV (intravenous) of deionized water (water with no solutes). If a large enough amount of fluid is administered, predict the effects. Choose one of the words in parentheses for each blank to complete the sentence. Plasma becomes

_____ (hypotonic/hypertonic) to the red blood cells.

_____ Water moves from the (plasma or red blood cells)

_____ to the (plasma or red blood cells)

_____ causing the red blood cells to (shrivel or swell).

A person drinks ocean salt water (3.5% salt). If the person drinks a large amount of the salt water (the salt water is absorbed into the blood), the plasma becomes

_____ (hypotonic/hypertonic) to the red blood cells.

_____ Water moves from the (plasma or red blood cells)

_____ to the (plasma or red blood cells)

_____ causing the red blood cells to (shrivel or swell).

What would you predict if a patient was administered an IV containing a physiological saline solution (0.9% sodium

chloride)? _____

_____

Cell Cycle

Purpose of the Exercise

To review the stages in the cell cycle and to observe cells in various stages of their life cycles.

Materials Needed

Models of animal mitosis
Microscope slides of whitefish mitosis (blastula)
Compound light microscope

For Demonstration Activity:

Microscope slide of human chromosomes from leukocytes in mitosis
Oil immersion objective

Learning Outcomes

After completing this exercise, you should be able to

1. Describe the phases and structures of the cell cycle.
2. Identify and sketch the stages in the life cycle of a particular cell.
3. Arrange into a correct sequence a set of models or drawings of cells in various stages of their life cycles.

Pre-Lab

Carefully read the introductory material and examine the entire lab. Be familiar with the cell cycle from lecture or the textbook. Answer the pre-lab questions.

Pre-Lab Questions: Select the correct answer for each of the following questions:

1. Which of the following is *not* part of interphase?
   - **a.** cytokinesis
   - **b.** first gap phase
   - **c.** second gap phase
   - **d.** synthesis phase

2. A cell in the stage of G-zero
   - **a.** divides continuously.
   - **b.** ceases cell division.
   - **c.** exhibits cytokinesis.
   - **d.** exists in a gap phase.

3. Which is the correct sequence of the M (mitotic) phase?
   - **a.** telophase, prophase, metaphase, anaphase
   - **b.** prophase, anaphase, metaphase, telophase
   - **c.** prophase, telophase, metaphase, anaphase
   - **d.** prophase, metaphase, anaphase, telophase

4. Which of the following events occurs in prophase?
   - **a.** spindle disappears
   - **b.** nuclear envelopes reassemble
   - **c.** chromosomes condense
   - **d.** sister chromatids separate

5. Which of the following events occurs during telophase?
   - **a.** sister chromatids separate
   - **b.** chromosomes align along equator
   - **c.** cytokinesis
   - **d.** spindle apparatus forms

6. Sister chromatids separate to opposite poles during
   - **a.** telophase.
   - **b.** anaphase.
   - **c.** prophase.
   - **d.** metaphase.

7. A karyotype is used to count centrioles and spindle fibers.
   True _____ False _____

8. Daughter cells are formed from cell cycles.
   True _____ False _____

The cell cycle consists of the series of changes a cell undergoes from the time it is formed until it divides. Interphase, mitotic phase, and differentiation are stages of a cell cycle. Typically, a newly formed diploid cell with 46 chromosomes grows to a certain size and then divides to form two new cells (*daughter cells*), each with 46 chromosomes.

Before the cell divides, it must synthesize biochemicals and other contents. This period of preparation is called *interphase*. The extensive period of interphase is divided into three phases. The S phase, when DNA synthesis occurs, is between two gap phases (G_1 and G_2), when cell growth occurs and cytoplasmic organelles duplicate. Interphase is followed by the *M (mitotic) phase,* which has two major portions: (1) division of the nucleus, called *mitosis,* and (2) division of the cell's cytoplasm, called *cytokinesis.* Mitosis includes four recognized phases: *prophase, metaphase, anaphase,* and *telophase.* Eventually, some specialized cells, such as skeletal muscle cells and most nerve cells, cease further cell division, but remain alive. A mature living cell that ceases cell division is in a state called G_0 (G-zero) for the remainder of its life span. This state can last from hours to decades.

A special type of cell division, called *meiosis,* occurs in the reproductive system to produce haploid gametes with 23 chromosomes. Meiosis is not included in this laboratory exercise.

Procedure—Cell Cycle

A cell generally subsists for extended periods of time in interphase in which the metabolic processes within the nucleus and cytoplasm occur. Those processes were studied in Laboratory Exercise 5. The mitosis phase occurs for cellular division to enable growth and repair of cells. As you examine cells that have frequent cell division in this lab, you will find that cells in interphase are still the most frequently located. The slides selected for your study possess abundant mitosis phases so that you can more efficiently locate some of the interesting cellular structures that can be distinguished during mitosis. Keep in mind that the cell cycle is referred to by phases, not steps, as one phase or stage gradually becomes the next phase. You will probably observe some phases that are debatable as to which phase is represented, but this is common. The figures selected for this lab for comparison purposes represent phases that are close to the mid-portion of the phase.

1. Study the various stages of the cell's life cycle represented in figures 7.1, 7.2, 7.3, and 7.4.

FIGURE 7.1   The cell cycle: interphase (G_1, S, and G_2) and M (mitotic) phase (mitosis and cytokinesis).

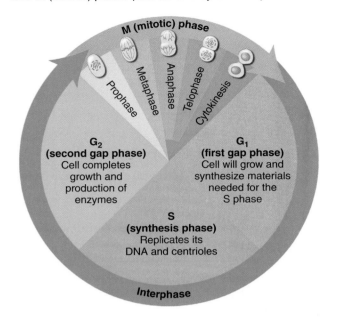

FIGURE 7.2   Cell in interphase (400×). The nucleoplasm contains a fine network of chromatin.

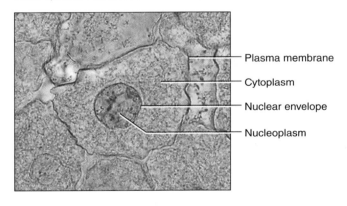

FIGURE 7.3   Cell in prophase (400×).

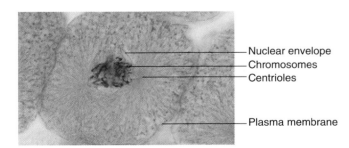

68

FIGURE 7.4 Structures in the dividing cell during anaphase.

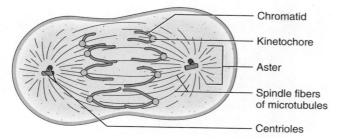

- Chromatid
- Kinetochore
- Aster
- Spindle fibers of microtubules
- Centrioles

2. Using figure 7.5 as a guide, observe the animal mitosis models and review the major events in a cell's life cycle represented by each of them. Be sure you can arrange these models in correct sequence if their positions are changed. The acronym IPMAT can help you arrange the correct order of phases in the cell cycle. This includes interphase followed by the four phases of mitosis. Cytokinesis overlaps anaphase and telophase.

3. Complete Part A of Laboratory Assessment 7.

4. Obtain a slide of the whitefish mitosis (blastula).
   a. Examine the slide using the high-power objective of a microscope. The tissue on this slide was obtained from a developing embryo (blastula) of a fish, and many of the embryonic cells are undergoing mitosis.

The chromosomes of these dividing cells are darkly stained.

b. Search the tissue for cells in various stages of cell division. There are several sections on the slide. If you cannot locate different stages in one section, examine the cells of another section because the stages occur randomly on the slide. Frequently refer to figures 7.2, 7.3, and 7.5 as references for identifications.

c. Each time you locate a cell in a different stage, sketch it in an appropriate circle in Part B of the laboratory assessment.

Critical Thinking Activity

Which stage (phase) of the cell cycle was the most numerous in the blastula? _____

Explain your answer. _____

_____

_____

_____

---

5. Complete Parts C and D of the laboratory assessment.

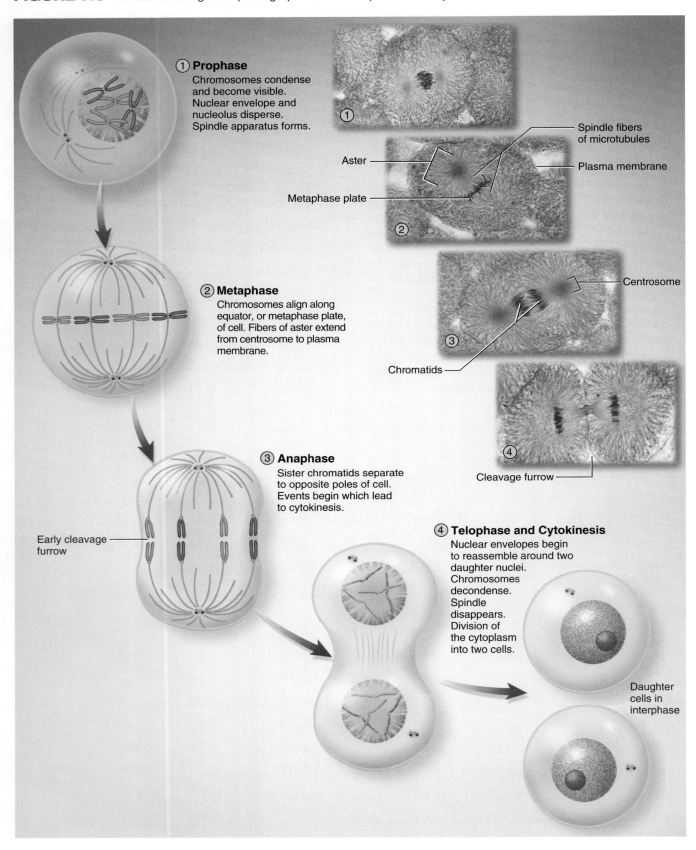

① **Prophase**
Chromosomes condense and become visible. Nuclear envelope and nucleolus disperse. Spindle apparatus forms.

Spindle fibers of microtubules

Aster

Plasma membrane

Metaphase plate

Centrosome

② **Metaphase**
Chromosomes align along equator, or metaphase plate, of cell. Fibers of aster extend from centrosome to plasma membrane.

Chromatids

Cleavage furrow

③ **Anaphase**
Sister chromatids separate to opposite poles of cell. Events begin which lead to cytokinesis.

Early cleavage furrow

④ **Telophase and Cytokinesis**
Nuclear envelopes begin to reassemble around two daughter nuclei. Chromosomes decondense. Spindle disappears. Division of the cytoplasm into two cells.

Daughter cells in interphase

Using the oil immersion objective of a microscope, see if you can locate some human chromosomes by examining a prepared slide of human chromosomes from leukocytes. The cells on this slide were cultured in a special medium and were stimulated to undergo mitosis. The mitotic process was arrested in metaphase by exposing the cells to a chemical called colchicine, and the cells were caused to swell osmotically. As a result of this treatment, the chromosomes were spread apart. A complement of human chromosomes should be visible when they are magnified about 1,000×. Each chromosome is double-stranded and consists of two chromatids joined by a common centromere (fig. 7.6).

FIGURE 7.6 (a) A complement of human chromosomes of a female (1,000×). (b) A karyotype can be constructed by arranging the homologous chromosome pairs together in a chart. A karyotype can aid in diagnosis of genetic conditions and abnormalities. The completed karyotype indicates a normal male.

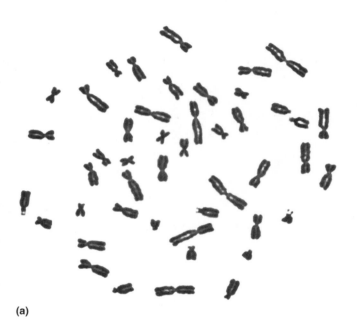

(a)

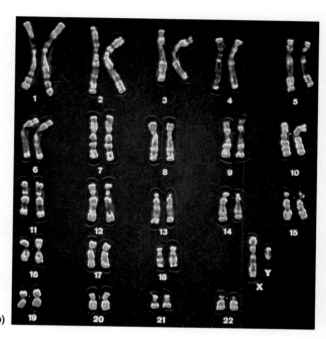

(b)

NOTES

Laboratory Assessment

7

Name _____

Date _____

Section _____

The ⟨A⟩ corresponds to the indicated outcome(s) found at the beginning of the laboratory exercise.

Cell Cycle

Part A Assessments

Complete the table. ⟨1⟩

Stage	Major Events Occurring
Interphase — G_1, S, and G_2	
M (Mitotic) Phase — **Mitosis** Prophase	
Metaphase	
Anaphase	
Telophase	
Cytokinesis	

Part B   Assessments

Sketch an interphase cell and cells in different stages of mitosis to illustrate the whitefish cell's life cycle. Label the major cellular structures represented in the sketches and indicate cytokinesis locations. (The circles represent fields of view through the microscope.) 2

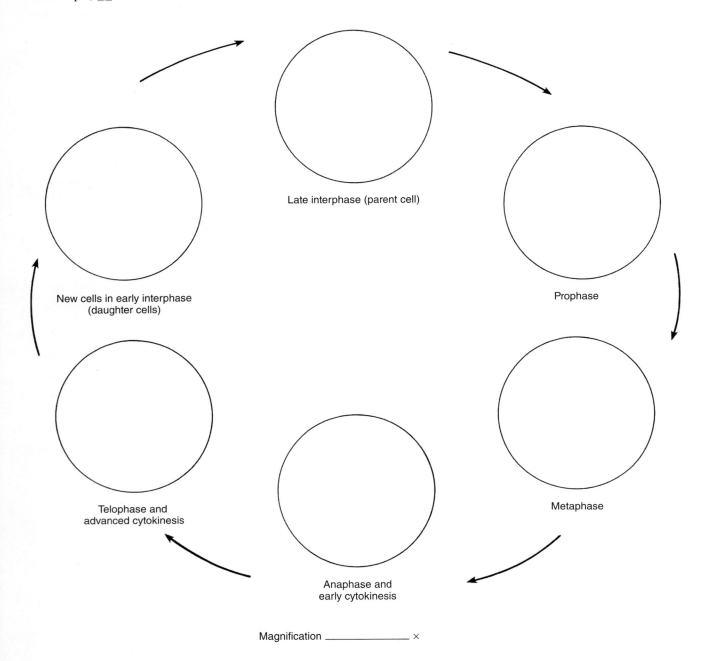

Late interphase (parent cell)

New cells in early interphase
(daughter cells)

Prophase

Telophase and
advanced cytokinesis

Metaphase

Anaphase and
early cytokinesis

Magnification _____ ×

Part C   Assessments

Complete the following:

1. In what ways are the new cells (daughter cells), which result from a cell cycle, similar? _____

_____

2. Distinguish between mitosis and cytokinesis. 1 _____

_____

Part D Assessments

1. Identify the mitotic stage represented by each of the micrographs in figure 7.7 (a–d). **3**

   a. _____

   b. _____

   c. _____

   d. _____

2. Identify the structures indicated by numbers in figure 7.7 by placing the correct numbers in the spaces provided.

   _____ Aster

   _____ Chromatid

   _____ Cleavage furrow

   _____ Metaphase plate

   _____ Nuclear envelope

   _____ Plasma membrane

   _____ Spindle fibers

FIGURE 7.7   Identify the mitotic stage and structures of the cell in each of these micrographs of the whitefish blastula (250×).

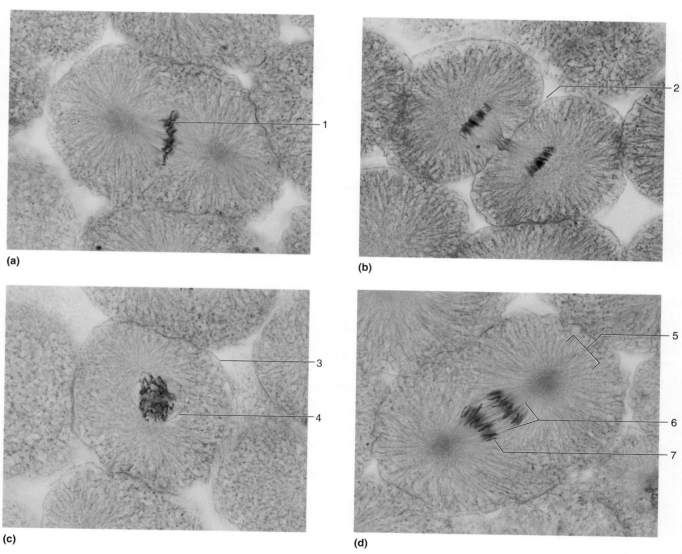

(a)

(b)

(c)

(d)

Epithelial Tissues

Purpose of the Exercise

To review the characteristics of epithelial tissues and to observe examples.

Materials Needed

Compound light microscope
Prepared slides of the following epithelial tissues:
 Simple squamous epithelium (lung)
 Simple cuboidal epithelium (kidney)
 Simple columnar epithelium (small intestine)
 Pseudostratified (ciliated) columnar epithelium
  (trachea)
 Stratified squamous epithelium (esophagus)
 Stratified cuboidal epithelium (salivary gland)
 Stratified columnar epithelium (urethra)
 Transitional epithelium (urinary bladder)

For Learning Extension Activity:
Colored pencils

Learning Outcomes

After completing this exercise, you should be able to

1. Sketch and label the characteristics of epithelial tissues that you were able to observe.
2. Differentiate the special characteristics of each type of epithelial tissue.
3. Indicate a location and function of each type of epithelial tissue.
4. Inventory the general characteristics of epithelial tissues that you were able to observe.

Pre-Lab

Carefully read the introductory material and examine the entire lab. Be familiar with epithelial tissues from lecture or the textbook. Answer the pre-lab questions.

Pre-Lab Questions: Select the correct answer for each of the following questions:

1. The tissues that cover external and internal body surfaces are
    a. epithelial.   **b.** connective.
    c. muscle.   **d.** nervous.
2. The cells with a flat shape are named
    a. stratified.   **b.** cuboidal.
    c. columnar.   **d.** squamous.
3. One edge of an epithelial cell attaches to the
    a. cilia.   **b.** basement membrane.
    c. nucleus.   **d.** cytoplasm.
4. If a structure cut has a circular appearance when observed, the sectional cut was a(n)
    a. oblique section.   **b.** cross section.
    c. longitudinal section. **d.** lengthwise section.
5. Which of the following is *not* a function of epithelial tissues?
    a. protection   **b.** secretion
    c. bind and support   **d.** absorption
6. Which of the following epithelial tissues is more than one cell layer thick?
    a. transitional   **b.** simple squamous
    c. pseudostratified   **d.** simple cuboidal
      columnar
7. The study of tissues is called histology.
    True _____   False _____
8. The basement membrane is composed of cilia.
    True _____   False _____

A tissue is composed of a layer or group of cells similar in size, shape, and function. The study of tissues is called **histology.** Within the human body, there are four major types of tissues: (1) *epithelial,* which cover the body's external and internal surfaces and most glands; (2) *connective,* which bind and support parts; (3) *muscle,* which make movement possible; and (4) *nervous,* which conduct impulses from one part of the body to another and help to control and coordinate body activities.

Epithelial tissues are tightly packed single (simple) to multiple (stratified) layers composed of cells that provide protective barriers. The underside of this tissue layer contains an acellular basement membrane layer of adhesives and collagen with which the epithelial cells anchor to an underlying connective tissue. A unique type of simple epithelium, pseudostratified columnar, appears to be multiple cells thick. However, this is a false appearance because all cells bind to the basement membrane. The cells readily divide and lack blood vessels (are avascular).

Epithelial tissues always have a free (apical) surface exposed to the outside or to an open space internally and a basal surface that attaches to the basement membrane. Blood vessels near the basement membrane allow epithelial cells to obtain their oxygen and nutrients. Many shapes of the cells, like squamous (flat), cuboidal, and columnar, exist that are used to name and identify the variations. Epithelial cell functions include protection, filtration, secretion, and absorption. Many of the prepared slides contain more than the tissue to be studied, so be certain that your view matches the correct tissue. Also be aware that stained colors of all tissues might vary.

The slides you observe might have representative tissue from more than one site, but the basic tissue structure will be similar to those represented in this laboratory exercise. Simple squamous epithelium is located in the air sacs (alveoli) of the lungs and inner linings of the heart and blood vessels. Simple cuboidal epithelium is situated in kidney tubules, thyroid gland, liver, and ducts of salivary glands. A nonciliated type of simple columnar epithelium is located in the linings of the uterus, stomach, and intestines, and a ciliated type lines the uterine tubes. A ciliated type of pseudostratified columnar epithelium is positioned in the linings of the upper respiratory tubes. A keratinized type of stratified squamous epithelium is found in the epidermis of the skin, and a nonkeratinized type in the linings of the oral cavity, esophagus, vagina, and anal canal. Stratified cuboidal epithelium is represented in larger ducts of glands while stratified columnar epithelium lines the urethra of males. Transitional epithelium lines the urinary bladder, ureters, and part of the urethra. Various glands of the body are composed of primarily glandular epithelium. Study of glands will be found in Laboratory Exercises 11, 39, and 40.

Procedure—Epithelial Tissues

This procedure represents the first of three laboratory exercises in a study of histology. So that you may clearly observe the details of tissue sections, the sectioned tissue has been stained, and the colors you observe will depend upon the particular stains used for the preparation. The slice of the histological section is usually only one or two cells thick, which will allow light to be transmitted through the entire section. Use the fine adjustment knob on the microscope to view the depth of the structures within the tissue section. Refer to the figures and the table in the laboratory exercise often as you examine each tissue.

1. Use the microscope to observe the prepared slides of types of epithelial tissues. As you observe each tissue, look for its special distinguishing features such as cell size, shape, and arrangement. Compare your prepared slides of epithelial tissues to the micrographs in figure 8.1 and the characteristics of each specific tissue in table 8.1 on p. 81.
2. As you observe the tissues in figure 8.1 and the prepared slides, note characteristics epithelial tissues have in common.
3. As you observe each type of epithelial tissue, prepare a labeled sketch of a representative portion of the tissue in Part A of Laboratory Assessment 8.
4. Test your ability to recognize each type of epithelial tissue. To do this, have a laboratory partner select one of the prepared slides, cover its label, and focus the microscope on the tissue. Then see if you can correctly identify the tissue.
5. Review the introductory material and table 8.1.
6. Complete Parts B and C of the laboratory assessment.

Learning Extension Activity

As you observe histology slides, be aware that tissues and some organs may have been sectioned in longitudinal (lengthwise cut), cross section (cut across), or oblique (angular cut) ways. Observe figure 8.2 for how this can be demonstrated on cuts of a banana. The direction in which the tissue or organ was cut will result in a certain perspective when it is sectioned, just as the banana was cut three different ways. Often a tissue slide has more than one tissue represented and has blood vessels or other structures cut in various ways. Locate an example of a longitudinal section, cross section, and oblique section of some structure on one of your tissue slides.

FIGURE 8.1 Micrographs of epithelial tissues. *Note:* A bracket to the right of a micrograph indicates the tissue layer.

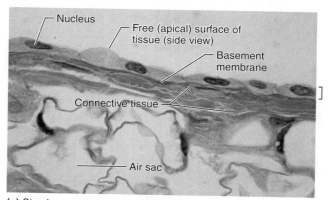

Nucleus — Free (apical) surface of tissue (side view) — Basement membrane — Connective tissue — Air sac

(a) Simple squamous epithelium (side view) (from lung) (1,100×)

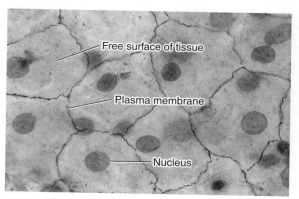

Free surface of tissue — Plasma membrane — Nucleus

(b) Simple squamous epithelium (surface view) (670×)

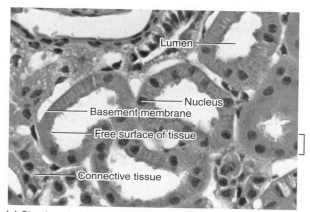

Lumen — Nucleus — Basement membrane — Free surface of tissue — Connective tissue

(c) Simple cuboidal epithelium (from kidney) (630×)

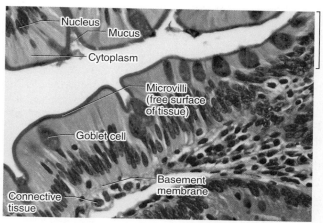

Nucleus — Mucus — Cytoplasm — Microvilli (free surface of tissue) — Goblet cell — Basement membrane — Connective tissue

(d) Simple columnar epithelium (nonciliated) (from intestine) (400×)

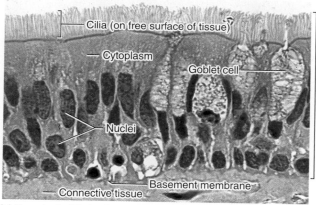

Cilia (on free surface of tissue) — Cytoplasm — Goblet cell — Nuclei — Basement membrane — Connective tissue

(e) Pseudostratified columnar epithelium with cilia (from trachea) (1,000×)

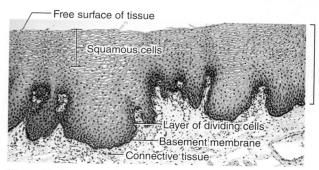

Free surface of tissue — Squamous cells — Layer of dividing cells — Basement membrane — Connective tissue

(f) Stratified squamous epithelium (nonkeratinized) (from esophagus) (400×)

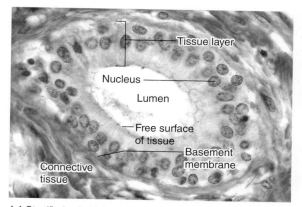

Tissue layer — Nucleus — Lumen — Free surface of tissue — Basement membrane — Connective tissue

(g) Stratified cuboidal epithelium (from salivary gland) (600×)

FIGURE 8.1 *Continued.*

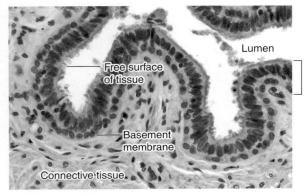

(h) Stratified columnar epithelium (from male urethra) (230×)

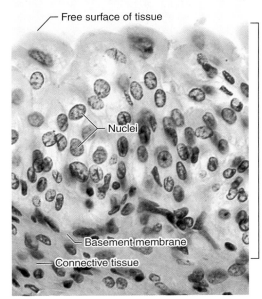

(i) Transitional epithelium (unstretched) (from urinary bladder) (400×)

FIGURE 8.2 Three possible cuts of a banana: (a) cross section; (b) oblique section; (c) longitudinal section. Sections through an organ, as a body tube, frequently produce views similar to the cut banana.

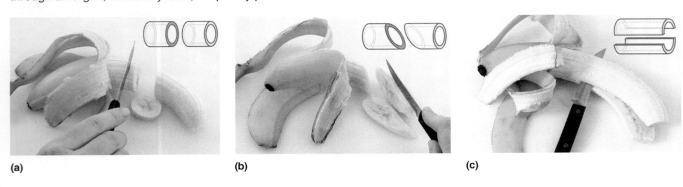

(a)   (b)   (c)

TABLE 8.1 Epithelial Tissues, Descriptions, Functions, and Representative Locations

Tissue Type	Descriptions	Functions	Representative Locations
Simple squamous epithelium	Single thin layer; flattened cells	Filtration; diffusion; osmosis; secretion	Air sacs (alveoli) of lungs; walls of capillaries; linings of blood vessels and ventral body cavity
Simple cuboidal epithelium	Single layer; cube-shaped cells	Secretion; absorption	Surface of ovaries; linings of kidney tubules; linings of ducts of certain glands
Simple columnar epithelium	Single layer; elongated narrow cells; some ciliated	Protection; secretion; absorption	Linings of uterus, stomach, gallbladder, and intestines
Pseudostratified columnar epithelium	Single layer; elongated cells; some cells do not reach free surface; often ciliated	Protection; secretion; movement of mucus and substances	Linings of respiratory passages
Stratified squamous epithelium	Many layers; top cells flattened; keratinized surface cells of epidermis	Protection; resists abrasion	Epidermis of skin; linings of oral cavity, esophagus, vagina, and anal canal
Stratified cuboidal epithelium	2 to 3 layers; cube-shaped cells	Protection; secretion	Linings of larger ducts of mammary glands, sweat glands, salivary glands, and pancreas
Stratified columnar epithelium	Superficial layer of elongated cells; basal layers of cube-shaped cells	Protection; secretion	Part of the male urethra; parts of the pharynx
Transitional epithelium	Many layers; cube-shaped and elongated cells; thinner layers when stretched	Distensibility; protection	Linings of urinary bladder and ureters and part of urethra
Glandular epithelium	Unicellular or multicellular	Secretion	Salivary glands; sweat glands; endocrine glands

Name _____

Date _____

Section _____

The ⚠ corresponds to the indicated outcome(s) found at the beginning of the laboratory exercise.

Epithelial Tissues

Part A   Assessments

In the space that follows, sketch a few cells of each type of epithelium you observed. For each sketch, label the major characteristics, indicate the magnification used, write an example of a location in the body, and provide a function. **1 2 3**

Simple squamous epithelium (____×)	Simple cuboidal epithelium (____×)
Location: _____	Location: _____
Function: _____	Function: _____
Simple columnar epithelium (____×)	Pseudostratified columnar epithelium with cilia (____×)
Location: _____	Location: _____
Function: _____	Function: _____

Stratified squamous epithelium (____×)

Location: _____

Function: _____

Stratified cuboidal epithelium (____×)

Location: _____

Function: _____

Stratified columnar epithelium (____×)

Location: _____

Function: _____

Transitional epithelium (____×)

Location: _____

Function: _____

Critical Thinking Assessment

As a result of your observations of epithelial tissues, which one(s) provide(s) the best protection? Explain your answer. _____

_____

_____

_____

Learning Extension Activity

Use colored pencils to differentiate various cellular structures in Part A. Select a different color for a nucleus, cytoplasm, plasma membrane, basement membrane, goblet cell, and cilia whenever visible.

Part B Assessments

Match the tissues in column A with the characteristics in column B. Place the letter of your choice in the space provided. (Some answers may be used more than once.) 2 3

Column A

a. Simple columnar epithelium
b. Simple cuboidal epithelium
c. Simple squamous epithelium
d. Pseudostratified columnar epithelium
e. Stratified squamous epithelium
f. Stratified columnar epithelium
g. Stratified cuboidal epithelium
h. Transitional epithelium

Column B

_____ 1. Consists of several layers of cube-shaped, elongated, and irregular cells allowing an expandable lining

_____ 2. Commonly possesses cilia that move dust and mucus out of the airways of the respiratory passages

_____ 3. Single layer of flattened cells

_____ 4. Single row of elongated cells, but some cells don't reach the free surface

_____ 5. Forms walls of capillaries and air sacs of lungs

_____ 6. Provides lining of urethra of males and parts of pharynx

_____ 7. Provides abrasion protection of skin epidermis and oral cavity

_____ 8. Forms inner lining of urinary bladder and ureters

_____ 9. Lines kidney tubules and ducts of salivary glands

_____ 10. Forms lining of stomach and intestines

_____ 11. Two or three layers of cube-shaped cells

_____ 12. Forms lining of oral cavity, anal canal, and vagina

Part C Assessments

As you examined each specific epithelial tissue, you should have noted some of the general characteristics that they possess as described in the laboratory exercise. List any of these you were able to observe. 4

_____

_____

_____

_____

NOTES

Connective Tissues

Purpose of the Exercise

To review the characteristics of connective tissues and to observe examples of the major types.

Materials Needed

Compound light microscope
Prepared slides of the following:
  Areolar connective tissue
  Adipose tissue
  Reticular connective tissue
  Dense regular connective tissue
  Dense irregular connective tissue
  Elastic connective tissue
  Hyaline cartilage
  Fibrocartilage
  Elastic cartilage
  Bone (compact, ground, cross section)
  Blood (human smear)
For Learning Extension Activity:
Colored pencils

Learning Outcomes

After completing this exercise, you should be able to

① Sketch and label the characteristics of connective tissues that you were able to observe.

② Differentiate the special characteristics of each of the major types of connective tissue.

③ Indicate a location and function of each type of connective tissue.

④ Inventory the general characteristics of connective tissues that you were able to observe.

Pre-Lab

Carefully read the introductory material and examine the entire lab. Be familiar with connective tissues from lecture or the textbook. Answer the pre-lab questions.

Pre-Lab Questions: Select the correct answer for each of the following questions:

1. Which of the following is *not* a function of connective tissues?
   a. fill spaces
   b. provide support
   c. bind structures together
   d. movement

2. The most abundant fibers of connective tissues are
   a. reticular.
   b. elastic.
   c. collagen.
   d. glycoprotein.

3. Which of the following is a loose connective tissue?
   a. fibrocartilage
   b. areolar
   c. spongy bone
   d. elastic connective

4. Which connective tissue is liquid?
   a. blood
   b. adipose
   c. reticular connective
   d. hyaline cartilage

5. The connective tissue that composes tendons and ligaments is
   a. dense regular connective.
   b. elastic connective.
   c. elastic cartilage.
   d. dense irregular connective.

6. The connective tissue _____ contains abundant fibers of collagen and cells in a solid matrix.
   a. fibrocartilage
   b. adipose
   c. bone
   d. areolar connective

7. Adipose, areolar, and reticular connective tissues are considered loose connective tissue types.
   True _____   False _____

8. Blood and lymph are considered loose connective tissue types.
   True _____   False _____

9. Cells secrete the fibers and a ground substance of a connective tissue.
   True _____   False _____

Connective tissues contain a variety of cell types and occur in all regions of the body. They bind structures together, provide support and protection, fill spaces, store fat, and transport blood cells.

Connective tissue cells are not as close together as epithelial cells and are often widely scattered in an abundance of a noncellular **extracellular matrix.** This extracellular matrix varies in quantity depending upon the specific tissue type and is like a filler material between the cells. The cells produce and secrete two components of the extracellular matrix known as the *ground substance* and *fibers.* The ground substance varies from a liquid to semisolid to solid, and has functions in support and as a medium for substances to move between cells and blood vessels. Fibers, the other component of the extracellular matrix, consist of fibrous proteins including *collagen, reticular,* and *elastic* fibers. Collagen fibers are thick threads, the most abundant of the three types, and are white in unstained tissues. Reticular fibers are fine threads of highly branched collagen with glycoprotein. Elastic fibers form complex networks of a highly branched protein called elastin, which is springlike and appears yellowish in unstained tissues. The fibers provide the binding properties within the tissue and between organs. You might compare connective tissue to making gelatin: the gelatin of various densities represents the ground substance, added fruit represents cells, and added strands represent the fibers.

Many of the prepared slides contain more than the tissue to be studied, so be certain that your view matches the correct tissue. Additional study of bone and blood will be found in Laboratory Exercises 12 and 41.

Procedure—Connective Tissues

1. Use a microscope to observe the prepared slides of various connective tissues. As you observe each tissue, look for its special distinguishing features. Compare your prepared slides of connective tissues to the micrographs in figure 9.1 and the characteristics of each specific tissue in tables 9.1 and 9.2 on page 90.
2. As you observe the tissues in figure 9.1 and the prepared slides, note the characteristics connective tissues have in common.
3. As you observe each type of connective tissue, prepare a labeled sketch of a representative portion of the tissue in Part A of Laboratory Assessment 9.
4. Test your ability to recognize each of these connective tissues by having a laboratory partner select a slide, cover its label, and focus the microscope on this tissue. Then see if you correctly identify the tissue.
5. Complete Parts B and C of the laboratory assessment.

FIGURE 9.1 Micrographs of connective tissues.

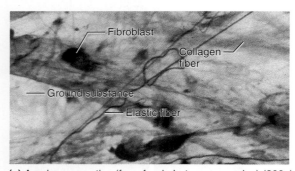

(a) Areolar connective (from fascia between muscles) (800×)

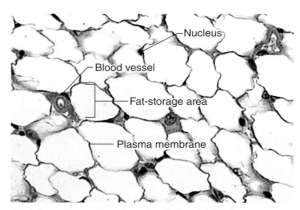

(b) Adipose tissue (from hypodermis layer) (400×)

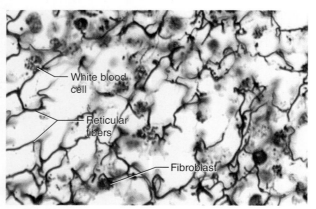

(c) Reticular connective (from spleen) (1,000×)

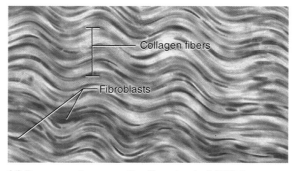

(d) Dense regular connective (from tendon) (500×)

FIGURE 9.1 *Continued.*

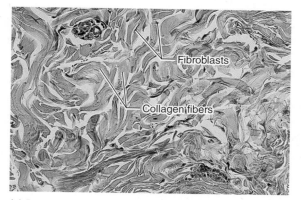

(e) Dense irregular connective (from dermis) (400×)

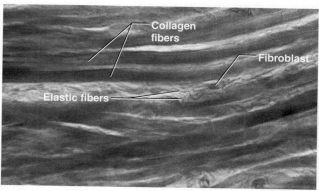

(f) Elastic connective (from artery wall) (600×)

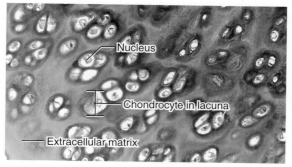

(g) Hyaline cartilage (from trachea) (400×)

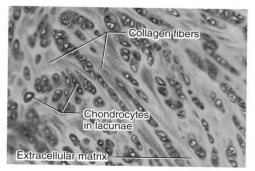

(h) Fibrocartilage (from intervertebral disc) (400×)

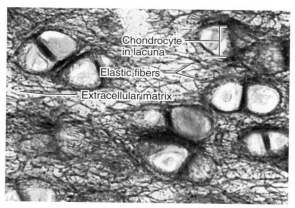

(i) Elastic cartilage (from ear) (1,000×)

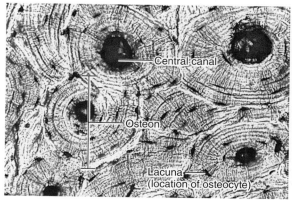

(j) Compact bone (200×)

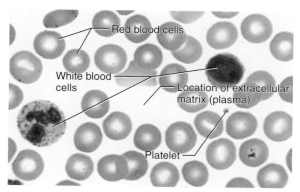

(k) Blood (400×)

TABLE 9.1 Connective Tissues, Descriptions, and Functions

Tissue Type	Descriptions	Functions
Areolar connective	Cells in abundant fluid-gel matrix; loose arrangement of collagen and elastic fibers	Loosely binds organs; holds tissue fluids
Adipose	Cells in sparse fluid-gel matrix; closely packed cells	Protects; insulates; stores fat
Reticular connective	Cells in fluid-gel matrix; reticular fibers	Supportive framework of organs
Dense regular connective	Cells in fluid-gel matrix; parallel, wavy collagen fibers	Tightly binds body parts
Dense irregular connective	Cells in fluid-gel matrix; random collagen fibers	Sustains tissue tension; durable
Elastic connective	Cells in fluid-gel matrix; collagen fibers densely packed; branched elastic fibers	Provides elastic quality
Hyaline cartilage	Cells in firm solid-gel matrix; fine network of collagen fibers; appears glassy	Supports; protects; provides framework
Fibrocartilage	Cells in firm solid-gel matrix; abundant collagen fibers	Supports; protects; absorbs shock
Elastic cartilage	Cells in firm solid-gel matrix; weblike elastic fibers	Supports; protects; provides flexible framework
Bone	Cells in solid matrix; many collagen fibers	Supports; protects; provides framework
Blood	Cells and platelets in fluid matrix called plasma	Transports nutrients, wastes, and gases; defends against disease; clotting

TABLE 9.2 Types of Mature Connective Tissues and Representative Locations*

Loose Connective	Dense Connective	Cartilage	Bone	Liquid Connective
Areolar Connective Locations: around body organs; binds skin to deeper organs	*Dense Regular Connective* Locations: tendons; ligaments	*Hyaline Cartilage* Locations: nasal septum; larynx; costal cartilage; ends of long bones; fetal skeleton	*Compact Bone* Locations: bone shafts; beneath periosteum	*Blood* Locations: lumens of blood vessels; heart chambers
Adipose Locations: hypodermis (subcutaneous layer); around kidneys and heart; yellow bone marrow; breasts	*Dense Irregular Connective* Locations: dermis; heart valves; periosteum on bone	*Fibrocartilage* Locations: between vertebrae; between pubic bones; pads (meniscus) in knee	*Spongy (Cancellous) Bone* Locations: ends of long bones; inside flat and irregular bones	*Lymph* Locations: lumens of lymphatic vessels
Reticular Connective Locations: spleen; thymus; lymph nodes; red bone marrow	*Elastic Connective* Locations: larger artery walls; vocal cords; some ligaments between vertebrae	*Elastic Cartilage* Locations: outer ear; epiglottis		

*This table represents a scheme to organize connective tissue relationships. Spongy bone and lymph tissues are not examined microscopically in this laboratory exercise.

Name _____

Date _____

Section _____

The 🅐 corresponds to the indicated outcome(s) found at the beginning of the laboratory exercise.

Connective Tissues

Part A   Assessments

In the space that follows, sketch a small section of each of the types of connective tissues you observed. For each sketch, label the major characteristics, indicate the magnification used, write an example of a location in the body, and provide a function. 🔺1 🔺2 🔺3

Areolar connective (_____×)
Location: _____
Function: _____

Adipose (_____×)
Location: _____
Function: _____

Reticular connective (_____×)
Location: _____
Function: _____

Dense regular connective (_____×)
Location: _____
Function: _____

Dense irregular connective (_____×)

Location: _____

Function: _____

Elastic connective (_____×)

Location: _____

Function: _____

Hyaline cartilage (_____×)

Location: _____

Function: _____

Fibrocartilage (_____×)

Location: _____

Function: _____

Elastic cartilage (_____×)

Location: _____

Function: _____

Compact bone (_____×)

Location: _____

Function: _____

```
                    Blood (_____×)
Location: _____
Function: _____
```

 Learning Extension Activity

Use colored pencils to differentiate various cellular structures in Part A. Select a different color for the cells, fibers, and ground substance whenever visible.

Part B   Assessments

Match the tissues in column A with the characteristics in column B. Place the letter of your choice in the space provided. (Some answers may be used more than once.) 2 3

Column A	Column B
a. Adipose	_____ **1.** Forms framework of outer ear
b. Areolar connective	_____ **2.** Functions as heat insulator beneath skin
c. Blood	_____ **3.** Contains large amounts of fluid and lacks fibers
d. Bone (compact)	_____ **4.** Cells in a solid matrix arranged around central canal
e. Dense irregular connective	_____ **5.** Binds skin to underlying organs
f. Dense regular connective	_____ **6.** Main tissue of tendons and ligaments
g. Elastic cartilage	
h. Elastic connective	_____ **7.** Forms the flexible part of the nasal septum
i. Fibrocartilage	_____ **8.** Pads between vertebrae that are shock absorbers
j. Hyaline cartilage	_____ **9.** Main tissue of dermis
k. Reticular connective	

_____ **10.** Occurs in some ligament attachments between vertebrae and larger artery walls

_____ **11.** Forms supporting tissue in walls of thymus and spleen

_____ **12.** Cells in a fluid-gel matrix with parallel collagen fibers

_____ **13.** Contains loose arrangement of elastic and collagen fibers

_____ **14.** Transports nutrients and defends against disease

Part C Assessments

As you examined each specific connective tissue, you should have noted some general characteristics that they possess as described in the laboratory exercise. List any of these you were able to observe. 4

_____

_____

_____

_____

_____

Critical Thinking Assessment

Abdominal impact injuries often involve the spleen. Explain the structural tissue characteristics that make the spleen so vulnerable to serious injury. 3

_____

_____

_____

_____

_____

10

Muscle and Nervous Tissues

Purpose of the Exercise

To review the characteristics of muscle and nervous tissues and to observe examples of these tissues.

Materials Needed

Compound light microscope
Prepared slides of the following:
    Skeletal muscle tissue
    Smooth muscle tissue
    Cardiac muscle tissue
    Nervous tissue (spinal cord smear and/or
        cerebellum)
For Learning Extension Activity:
Colored pencils

Learning Outcomes

After completing this exercise, you should be able to

1. Sketch and label the characteristics of muscle tissues and nervous tissues that you were able to observe.

2. Differentiate the special characteristics of each type of muscle tissue and nervous tissue.

3. Indicate a location and function of each type of muscle tissue and nervous tissue.

Pre-Lab

Carefully read the introductory material and examine the entire lab. Be familiar with muscle tissues and nervous tissue from lecture or the textbook. Answer the pre-lab questions.

Pre-Lab Questions: Select the correct answer for each of the following questions:

1. Which muscle tissue is under conscious control?
    a. skeletal   **b.** smooth
    c. cardiac   **d.** stomach

2. Which of the following organs lacks smooth muscle?
    a. blood vessels   **b.** stomach
    c. iris   **d.** heart

3. Which muscle tissue lacks striations?
    a. cardiac   **b.** skeletal
    c. smooth   **d.** voluntary

4. Which of the following conduct action potentials?
    a. smooth muscle cells   **b.** neurons
    c. neuroglia   **d.** skeletal muscle cells

5. Muscles of facial expression are
    a. smooth muscles.   **b.** skeletal muscles.
    c. involuntary muscles.   **d.** cardiac muscles.

6. Which muscle tissue is multinucleated?
    a. skeletal   **b.** smooth
    c. stomach   **d.** cardiac

7. Intercalated discs represent the junction where heart muscle cells fit together.
    True _____   False _____

8. Both cardiac and skeletal muscle cells are voluntary.
    True _____   False _____

Muscle tissues are characterized by the presence of elongated cells, often called muscle fibers, that can contract to create movements. Many of our muscles are attached to the skeleton, but muscles are also components of many of our internal organs. During muscle contractions, considerable body heat is generated to help maintain our body temperature. Because more heat is generated than is needed to maintain our body temperature, much of the heat is dissipated from our body through the skin.

The three types of muscle tissues are *skeletal, smooth,* and *cardiac.* Skeletal muscles are under our conscious control and are considered voluntary. Although most skeletal muscles are attached to bones via tendons, the tongue, facial muscles, and voluntary sphincters do not attach directly to the bone. Functions include body movements, maintaining posture, breathing, speaking, controlling waste eliminations, and protection. Skeletal muscles also aid the movement of lymph and venous blood during muscle contractions. Smooth muscle is considered involuntary and is located in many visceral organs, the iris, blood vessels, respiratory tubes, and attached to hair follicles. Functions include the motions of visceral organs (peristalsis), controlling pupil size, blood flow, and airflow, and creating "goose bumps" if we are too cold or frightened. Cardiac muscle is located only in the heart wall. It is considered involuntary and functions to pump blood.

Nervous tissues occur in the brain, spinal cord, and peripheral nerves. The tissue consists of two cell types: *neurons* and *neuroglia.* Neurons, also called nerve cells, contain a cell body with the nucleus and most of the cytoplasm, and cellular processes that extend from the cell body. Cellular processes include one to many dendrites and a single axon (nerve fiber). Neurons are considered excitable cells because they can generate signals called action potentials (nerve impulses) along the neuron to another neuron or a muscle or gland. Neuroglia (glial cells) of various types are more abundant than neurons; they cannot conduct nerve impulses, but they have important supportive and protective functions for neurons. Additional study of nervous tissue will be found in Laboratory Exercise 27.

Procedure—Muscle and Nervous Tissues

1. Using the microscope, observe each of the types of muscle tissues on the prepared slides. Look for the special features of each type. Compare your prepared slides of muscle tissues to the micrographs in figure 10.1

FIGURE 10.1 Micrographs of muscle and nervous tissues.

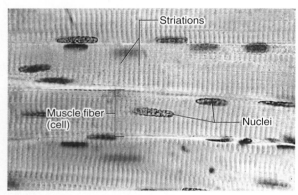

(a) Skeletal muscle (from leg) (400×)

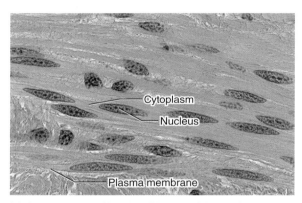

(b) Smooth muscle (from small intestine) (1,000×)

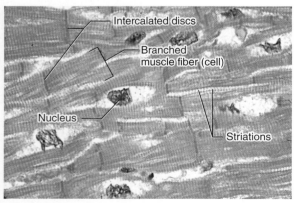

(c) Cardiac muscle (from heart) (400×)

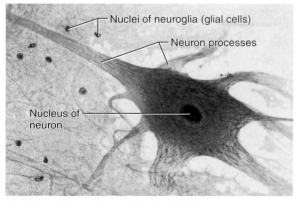

(d) Nervous tissue (from spinal cord) (400×)

96

and the characteristics of each specific tissue in tables 10.1 and 10.2.

2. As you observe each type of muscle tissue, prepare a labeled sketch of a representative portion of the tissue in Part A of Laboratory Assessment 10.

3. Observe the prepared slide of nervous tissue and identify neurons (nerve cells), neuron cellular processes, and neuroglia. Compare your prepared slide of nervous tissue to the micrograph in figure 10.1 and the characteristics in table 10.1.

4. Prepare a labeled sketch of nervous tissue in Part A of the laboratory assessment.

5. Test your ability to recognize each of these muscle and nervous tissues by having your laboratory partner select a slide, cover its label, and focus the microscope on this tissue. Then see if you correctly identify the tissue.

6. Complete Part B of the laboratory assessment.

TABLE 10.1 **Muscle and Nervous Tissues, Descriptions, Functions, and Representative Locations**

Tissue Type	Descriptions	Functions	Representative Locations
Skeletal muscle	Long, threadlike cells; striated; many nuclei near plasma membrane	Voluntary movements of skeletal parts; facial expressions	Muscles usually attached to bones
Smooth muscle	Shorter spindle-shaped cells; single central nucleus	Involuntary movements of internal organs	Walls of hollow internal organs
Cardiac muscle	Branched cells; striated; single nucleus (usually)	Heart contractions to pump blood; involuntary	Heart walls
Nervous tissue	Neurons with long cellular processes; neuroglia smaller and variable	Sensory reception and conduction of action potentials; neuroglia supportive	Brain, spinal cord, and peripheral nerves

TABLE 10.2 **Muscle Tissue Characteristics**

Characteristic	Skeletal Muscle	Smooth Muscle	Cardiac Muscle
Appearance of cells	Unbranched and relatively parallel	Spindle-shaped	Branched and connected in complex networks
Striations (pattern of alternating dark and light bands across cells)	Present and obvious	Absent	Present but faint
Nucleus	Multinucleated	Uninucleated	Uninucleated (usually)
Intercalated discs (junction where cells fit together)	Absent	Absent	Present
Control	Voluntary	Involuntary	Involuntary

NOTES

Name _____

Date _____

Section _____

The ⚠ corresponds to the indicated outcome(s) found at the beginning of the laboratory exercise.

Muscle and Nervous Tissues

Part A   Assessments

In the space that follows, sketch a few cells or fibers of each of the three types of muscle tissues and of nervous tissue as they appear through the microscope. For each sketch, label the major structures of the cells or fibers, indicate the magnification used, write an example of a location in the body, and provide a function. ⚠ ⚠ ⚠

Skeletal muscle tissue (_____×)
Location: _____
Function: _____

Smooth muscle tissue (_____×)
Location: _____
Function: _____

Cardiac muscle tissue (_____×)
Location: _____
Function: _____

Nervous tissue (_____×)
Location: _____
Function: _____

Learning Extension Activity

Use colored pencils to differentiate various cellular structures in Part A.

Part B Assessments

Match the tissues in column A with the characteristics in column B. Place the letter of your choice in the space provided. (Some answers may be used more than once.) 🄰 🄱

Column A	Column B
a. Cardiac muscle	_____ **1.** Coordinates, regulates, and integrates body functions
b. Nervous tissue	_____ **2.** Contains intercalated discs
c. Skeletal muscle	_____ **3.** Muscle that lacks striations
d. Smooth muscle	_____ **4.** Striated and involuntary
	_____ **5.** Striated and voluntary
	_____ **6.** Contains neurons and neuroglia
	_____ **7.** Muscle attached to bones
	_____ **8.** Muscle that composes heart
	_____ **9.** Moves food through the digestive tract
	_____ **10.** Transmits impulses along cellular processes
	_____ **11.** Muscle under conscious control
	_____ **12.** Muscle of blood vessels and iris

Integumentary System

Purpose of the Exercise

To observe the structures and tissues of the integumentary system and to review the functions of these parts.

Materials Needed

Skin model
Hand magnifier or dissecting microscope
Forceps
Microscope slide and coverslip
Compound light microscope
Prepared microscope slide of human scalp or axilla
Prepared slide of dark (heavily pigmented) human skin
Prepared slide of thick skin (plantar or palmar)

For Learning Extension Activity:
Tattoo slide
Stereomicroscope (dissecting microscope)

Learning Outcomes

After completing this exercise, you should be able to

1. Locate the structures of the integumentary system.
2. Describe the major functions of these structures.
3. Distinguish the locations and tissues among epidermis, dermis, and the hypodermis.
4. Sketch the layers of the skin and associated structures observed on the prepared slide.

Pre-Lab

Carefully read the introductory material and examine the entire lab. Be familiar with skin layers and accessory structures of the skin from lecture or the textbook. Answer the pre-lab questions.

Pre-Lab Questions: Select the correct answer for each of the following questions:

1. Which of the following is *not* a function of the integumentary system?
   a. protection
   b. excrete small amounts of waste
   c. movement
   d. aid in regulating body temperature

2. The two distinct skin layers are the
   a. epidermis and dermis.
   b. hypodermis and dermis.
   c. hypodermis and epidermis.
   d. dermis and hypodermis.

3. Apocrine sweat glands are located in _____ regions of the body.
   a. forehead   b. axillary and genital
   c. palmar   d. plantar

4. The hypodermis is composed of _____ tissues.
   a. adipose and stratified squamous epithelial
   b. areolar and dense irregular connective
   c. stratified squamous epithelial and adipose
   d. areolar and adipose connective

5. The _____ layer of the epidermis is only present in thick skin.
   a. stratum corneum   b. stratum lucidum
   c. stratum spinosum   d. stratum granulosum

6. Frequent cell division occurs in the _____ of the epidermis.
   a. stratum corneum   b. stratum spinosum
   c. stratum granulosum d. stratum basale

7. The greatest concentration of melanin is in the dermis.
   True _____   False _____

8. Thick skin of the palms and soles contains five strata of the epidermis.
   True _____   False _____

The integumentary system includes the skin, hair, nails, sebaceous (oil) glands, and sweat (sudoriferous) glands. These structures provide a protective covering for deeper tissues, aid in regulating body temperature, retard water loss, house sensory receptors, synthesize various chemicals, and excrete small quantities of wastes.

The skin consists of two distinct layers. The outer layer, the *epidermis,* consists of stratified squamous epithelium. The inner layer, the *dermis,* consists of a superficial papillary region of areolar connective tissue and a thicker and deeper reticular region of dense irregular connective tissue. Beneath the dermis is the hypodermis (subcutaneous layer; superficial fascia) composed of adipose and areolar connective tissues. The hypodermis is not considered a true layer of the skin.

Accessory structures of the skin include nails, hair follicles, and skin glands. The hair, which grows through a depression from the epidermis, possesses a hair papilla at the base of the hair which contains a network of capillaries that supply the nutrients for cell divisions for hair growth within the hair bulb. As the cells of the hair are forced toward the surface of the body, they become keratinized and pigmented and die. Attached to the follicle is the arrector pili muscle that can pull the hair to a more upright position, causing goose bumps when experiencing cold temperatures or fear. A sebaceous gland secretes an oily sebum into the hair follicles, which keeps the hair and epidermal surface pliable and somewhat waterproof.

Sweat glands (sudoriferous glands) are distributed over most regions of the body and consist of two types of glands. The widespread eccrine sweat glands are most numerous on the palms, soles of the feet, and the forehead. Their ducts open to the surface at a sweat pore. Their secretions increase during hot days, physical exercise, and stress; they serve an excretory function and can help prevent our body temperature from overheating. The apocrine sweat glands are most abundant in the axillary and genital regions. Apocrine sweat ducts open into the hair follicles and become active at puberty. Their secretions increase during stress and pain and have little influence on thermoregulation.

Procedure—Integumentary System

In this procedure you will use a skin model and make comparisons to the figures in the lab manual to locate the layers and accessory structures of the skin. Hair structures will be observed from your own body with different magnifications and then compared to additional detail using prepared slides. Several micrographs of different magnifications are provided of vertical sections of the skin layers and accessory structures. Use a combination of all of the micrographs as you observe the skin slides available in your laboratory.

1. Use figures 11.1 and 11.2 and locate as many of these structures as possible on a skin model.
2. Use figure 11.3 as a guide to locate the specific epidermal layers (strata) on a skin model. Note the locations and descriptions from table 11.1.

FIGURE 11.1 Layers of skin and deeper hypodermis indicated on (a) an illustration and (b) cadaver skin.

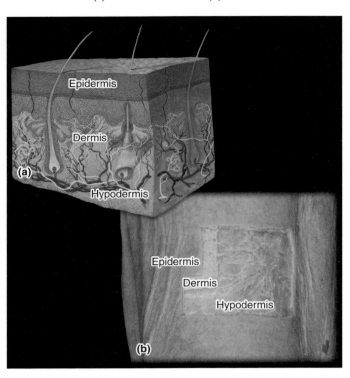

3. Use the hand magnifier or dissecting microscope and proceed as follows:
   a. Observe the skin, hair, and nails on your hand.
   b. Compare the type and distribution of hairs on the front and back of your forearm.
4. Use low-power magnification of the compound light microscope and proceed as follows:
   a. Pull out a single hair with forceps and mount it on a microscope slide under a coverslip.
   b. Observe the root and shaft of the hair and note the scalelike parts that make up the shaft.
5. Complete Parts A and B of Laboratory Assessment 11.
6. As vertical sections of human skin are observed, remember that the lenses of the microscope invert and reverse images. It is important to orient the position of the epidermis, dermis, and hypodermis layers using scan magnification before continuing with additional observations. Compare all of your skin observations to the various micrographs in figure 11.4. Use low-power magnification of the compound light microscope and proceed as follows:
   a. Observe the prepared slide of human scalp or axilla.
   b. Locate the epidermis, dermis, and hypodermis; a hair follicle; an arrector pili muscle; a sebaceous gland; and a sweat gland.
   c. Focus on the epidermis with high power and locate the stratum corneum, stratum granulosum, stratum spinosum, and stratum basale. Note how the shapes of the cells in these layers differ as described in table 11.1.

FIGURE 11.2 Vertical section of the skin and hypodermis (subcutaneous layer).

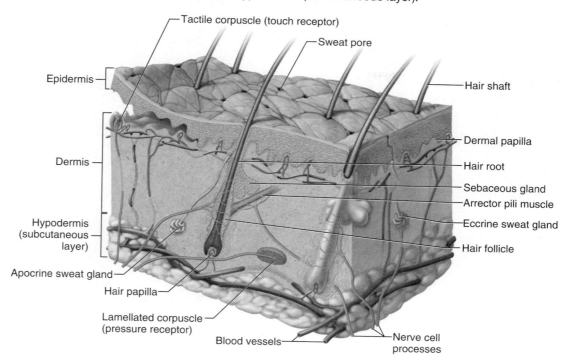

- Tactile corpuscle (touch receptor)
- Sweat pore
- Epidermis
- Hair shaft
- Dermal papilla
- Dermis
- Hair root
- Sebaceous gland
- Arrector pili muscle
- Hypodermis (subcutaneous layer)
- Eccrine sweat gland
- Hair follicle
- Apocrine sweat gland
- Hair papilla
- Lamellated corpuscle (pressure receptor)
- Blood vessels
- Nerve cell processes

FIGURE 11.3 Epidermal layers in this section of thick skin from the fingertip (400×).

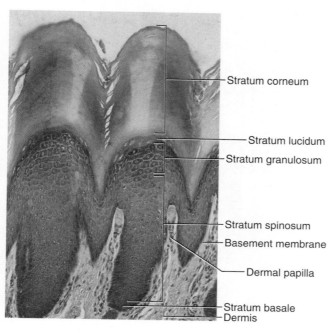

- Stratum corneum
- Stratum lucidum
- Stratum granulosum
- Stratum spinosum
- Basement membrane
- Dermal papilla
- Stratum basale
- Dermis

TABLE 11.1 Layers of the Epidermis

Layer	Location	Descriptions
Stratum corneum	Most superficial layer	Many layers of keratinized, dead epithelial cells; appear scaly and flattened; resists water loss, absorption, and abrasion
Stratum lucidum	Between stratum corneum and stratum granulosum on soles and palms of thick skin	Cells appear clear; nuclei, organelles, and plasma membranes no longer visible
Stratum granulosum	Beneath the stratum corneum (or stratum lucidum of thick skin)	Three to five layers of flattened granular cells; contain shrunken fibers of keratin and shriveled nuclei
Stratum spinosum	Beneath the stratum granulosum	Many layers of cells with centrally located, large, oval nuclei; develop fibers of keratin; cells becoming flattened in superficial portion
Stratum basale	Deepest layer	A single row of cuboidal or columnar cells; layer also includes melanocytes; frequent cell division; some cells become parts of more superficial layers

FIGURE 11.4 Features of human skin are indicated in these micrographs: (a) and (b) various structures of epidermis and dermis; (c) epidermis of dark skin; (d) base of hair structures.

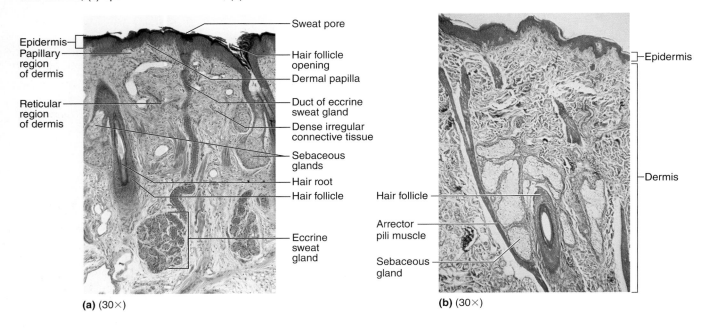

Epidermis
Papillary region of dermis
Reticular region of dermis

Sweat pore
Hair follicle opening
Dermal papilla
Duct of eccrine sweat gland
Dense irregular connective tissue
Sebaceous glands
Hair root
Hair follicle
Eccrine sweat gland

(a) (30×)

Epidermis
Dermis

Hair follicle
Arrector pili muscle
Sebaceous gland

(b) (30×)

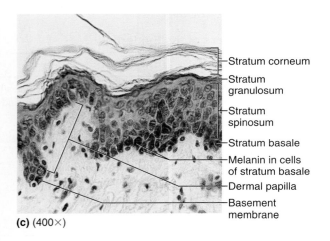

Stratum corneum
Stratum granulosum
Stratum spinosum
Stratum basale
Melanin in cells of stratum basale
Dermal papilla
Basement membrane

(c) (400×)

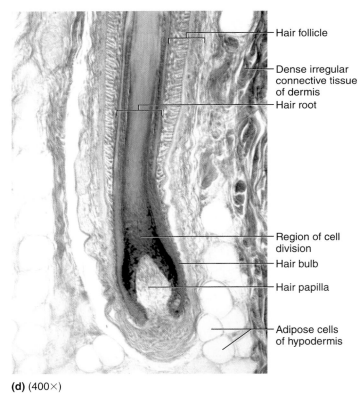

Hair follicle
Dense irregular connective tissue of dermis
Hair root
Region of cell division
Hair bulb
Hair papilla
Adipose cells of hypodermis

(d) (400×)

Critical Thinking Activity

Explain the advantage for melanin granules being located in the deep layers of the epidermis, and not the dermis or deeper hypodermis.

_____

_____

_____

_____

_____

_____

  d. Observe the dense irregular connective tissue that makes up the bulk of the dermis (reticular region).
  e. Observe the adipose tissue that composes most of the hypodermis (subcutaneous layer).
7. Observe the prepared slide of dark (heavily pigmented) human skin with low-power magnification. Note that the pigment is most abundant in the deepest layers of the epidermis. Focus on this region with the high-power objective. The pigment-producing cells, or melanocytes, are located among the stratum basale cells. Some melanin is retained within some cells of the stratum spinosum as cells are forced closer to the surface of the skin. Differences in skin color are primarily due to the quantity of the pigment _melanin_ produced by these cells. Exposure to ultraviolet (UV) rays of sunlight can increase the amount of melanin produced, causing a suntan. Melanin absorbs the UV radiation, which helps to protect the nuclei of cells.

8. Observe the prepared slide of thick skin from the palm of a hand or the sole of a foot. Locate the stratum lucidum, which is present only in thick skin. Locate the four other strata of thick skin and note the very thick stratum corneum (fig. 11.3).
9. Complete Part C of the laboratory assessment.
10. Using low-power magnification, locate a hair follicle sectioned longitudinally through its bulblike base. Also locate a sebaceous gland close to the follicle and find a sweat gland (fig. 11.4). Observe the detailed structure of these parts with high-power magnification.
11. Complete Parts D and E of the laboratory assessment.

Learning Extension Activity

Observe a vertical section of human skin through a tattoo, using low-power magnification. Note the location of the dispersed ink granules within the upper portion of the dermis. From a thin vertical section of a tattoo, it is not possible to determine the figure or word of the entire tattoo as seen on the surface of the skin. Compare this to the location of melanin granules found in dark (heavily pigmented) skin. Suggest reasons why a tattoo is permanent and a suntan is not.

_____

_____

_____

_____

_____

_____

NOTES

Name _____

Date _____

Section _____

The ⚠ corresponds to the indicated outcome(s) found at the beginning of the laboratory exercise.

Integumentary System

Part A   Assessments

1. Label the structures indicated in figure 11.5. ⚠

FIGURE 11.5   Label the features of the skin.

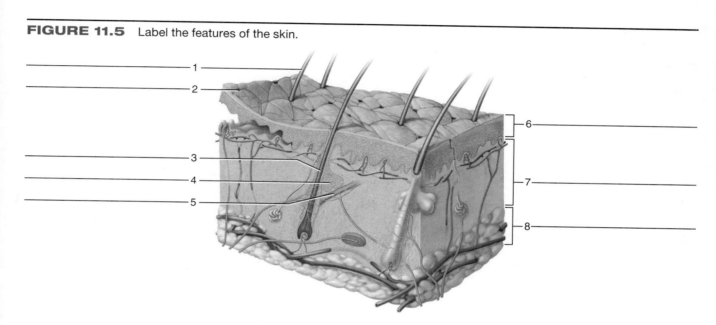

2. Match the structures in column A with the description and functions in column B. Place the letter of your choice in the space provided. ⚠ ⚠

Column A	Column B
a. Apocrine sweat gland	_____ **1.** An oily secretion that helps to waterproof body surface
b. Arrector pili muscle	_____ **2.** Outermost layer of epidermis
c. Dermis	_____ **3.** Become active at puberty
d. Eccrine (merocrine) sweat gland	_____ **4.** Epidermal pigment
e. Epidermis	_____ **5.** Inner layer of skin
f. Hair follicle	_____ **6.** Responds to elevated body temperature
g. Keratin	_____ **7.** General name of entire superficial layer of the skin
h. Melanin	_____ **8.** Gland that secretes an oily substance
i. Sebaceous gland	_____ **9.** Hard protein of nails and hair
j. Sebum	_____ **10.** Cell division and deepest layer of epidermis
k. Stratum basale	_____ **11.** Tubelike part that contains the root of the hair
l. Stratum corneum	_____ **12.** Causes hair to stand on end and goose bumps to appear

Part B Assessments

Complete the following:

1. How does the skin of your palm differ from that on the back (posterior) of your hand? _____

_____

_____

2. Describe the differences you observed in the type and distribution of hair on the front (anterior) and back (posterior) of your

forearm. _____

_____

3. Explain how a hair is formed. ⚠2 _____

_____

Part C Assessments

Complete the following:

1. Distinguish the locations and tissues among epidermis, dermis, and hypodermis layers. ⚠3 _____

_____

_____

_____

2. How do the cells of stratum corneum and stratum basale differ? ⚠3 _____

_____

_____

3. State the specific location of melanin observed in dark skin. ⚠3 _____

_____

4. What special qualities does the connective tissue of the dermis have? ⚠3 _____

_____

Part D Assessments

Complete the following:

1. What part of the hair extends from the hair papilla to the body surface? ⚠ _____

2. In which layer of skin are sebaceous glands found? ⚠ _____

3. How are sebaceous glands associated with hair follicles? ⚠ _____

_____

4. In which layer of skin are sweat glands usually located? ⚠ _____

Part E Assessments

Sketch a vertical section of human skin, using the scanning objective. Label the skin layers and a hair follicle, a sebaceous gland, and a sweat gland. ⚠

Bone Structure and Classification

Purpose of the Exercise

To review the way bones are classified and to examine the structure of a long bone and bone tissue.

Materials Needed

Prepared microscope slide of ground compact bone
Human bone specimens, including long, short, flat, irregular, and sesamoid types
Human long bone, sectioned longitudinally
Fresh animal bones, sectioned longitudinally and transversely
Compound light microscope
Dissecting microscope

For Demonstration Activity:

Fresh chicken bones (radius and ulna from wings)
Vinegar or dilute hydrochloric acid

Safety

▶ Wear disposable gloves for handling fresh bones and for the demonstration of a bone soaked in vinegar or dilute hydrochloric acid.
▶ Wash your hands before leaving the laboratory.

Learning Outcomes

After completing this exercise, you should be able to

1. Arrange five groups of bones based on their shapes and identify an example for each group.
2. Locate the major structures of a long bone.
3. Distinguish between compact and spongy bone.
4. Differentiate the structural and functional characteristics of bone tissue.

Pre-Lab

Carefully read the introductory material and the entire lab. Be familiar with bone structure and bone tissue from lecture or the textbook.
Visit www.mhhe.com/martinseries2 for LabCam videos. Answer the pre-lab questions.

Pre-Lab Questions: Select the correct answer for each of the following questions:

1. Which of the following tissues is *not* part of a bone as an organ?
   a. dense connective   **b.** cartilage
   c. muscle   **d.** blood

2. The organic matter of living bone includes
   a. calcium phosphate.
   b. collagen fibers and cells.
   c. calcium carbonate.
   d. magnesium and fluoride.

3. The _____ is an example of a sesamoid bone.
   a. vertebra   **b.** femur
   c. carpal   **d.** patella

4. The epiphyseal plate represents the
   a. ends of the epiphyses.
   b. shaft between the epiphyses.
   c. growth zone of hyaline cartilage.
   d. membrane around a bone.

5. The central canal of bone tissue contains
   a. blood vessels and nerves.   **b.** osteocytes.
   c. red bone marrow.   **d.** yellow bone marrow.

6. A _____ is an example of an irregular bone.
   a. femur   **b.** carpal
   c. rib   **d.** vertebra

7. A femur includes both compact and spongy bone tissues.
   True _____ False _____

8. Chicken bones, with both organic and inorganic components, possess the quality of tensile strength.
   True _____ False _____

9. Trabeculae are structural characteristics of compact bone.
   True _____ False _____

A bone represents an organ of the skeletal system. As such, it is composed of a variety of tissues including bone tissue, cartilage, dense connective tissue, blood, and nervous tissue. Bones are not only alive, but also multifunctional. They support and protect softer tissues, provide points of attachment for muscles, house blood-producing cells, and store inorganic salts.

Living bone is a combination of about one-third organic matter and two-thirds inorganic matter. The organic matter consists mostly of embedded cells and collagen fibers. The inorganic matter is mostly complex salt crystals, hydroxyapatite, consisting of calcium phosphate. Lesser amounts of calcium carbonate and ions of magnesium, fluoride, and sodium become incorporated into the inorganic crystals.

Bones are classified according to their shapes as long, short, flat, irregular, or sesamoid (round). Long bones are much longer than they are wide and have expanded ends. Short bones are somewhat cube shaped, with similar lengths and widths. Flat bones have wide surfaces, but they are sometimes curved, such as those of the cranium. Irregular bones have numerous shapes and often have articulations with more than one other bone. Sesamoid bones are small and embedded within a tendon near joints where compression often occurs. The patella is a sesamoid bone that is included in the total skeletal number of 206. Any additional sesamoid bones that may develop in compression areas of the hand or foot are not considered among the 206 bones of the adult skeleton. Although bones of the skeleton vary greatly in size and shape, they have much in common structurally and functionally.

Procedure—Bone Structure and Classification

During embryonic and fetal development, much of the supportive tissue is cartilage. Cartilage is retained in certain regions of the adult skeleton. Articular cartilage remains on the articulating surfaces of movable joints, costal cartilage connects the ribs to the sternum, and intervertebral discs are between the vertebrae. Hyaline cartilage composes the articular cartilage and the costal cartilage, while fibrocartilage is represented between the vertebrae.

Many bones contain compact and spongy bone components. The compact bone structure contains cylinder-shaped units called osteons. Each osteon contains a central canal that includes blood vessels and nerves in living bone. The cells or osteocytes are located in concentric circles within small spaces known as lacunae. The cellular processes of the osteocytes pass through microscopic canaliculi, which allow for the transport of nutrient and waste substances between cells and the central canal. The extracellular matrix occupies most of the area of an osteon.

Spongy bone does not possess the typical osteon units of compact bone. The osteocytes of spongy bone are located within a lattice of bony plates known as trabeculae. Canaliculi allow for diffusion of substances between the cells and the marrow that is positioned between the trabeculae.

1. Observe the individual bone specimens and arrange them into groups, according to the following shapes and examples (figs. 12.1 and 12.2).

    long—femur; humerus; phalanges

    short—carpals; tarsals

    flat—ribs; scapula; most cranial bones

    irregular—vertebra; some facial bones as sphenoid

    sesamoid (round)—patella

Critical Thinking Activity

Explain how bone cells embedded in a solid ground substance obtain nutrients and eliminate wastes.

_____

_____

_____

_____

FIGURE 12.1 Representative examples of bones classified by shape with a skeleton location icon for the bones illustrated.

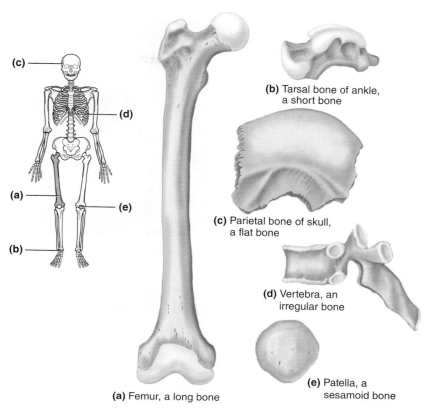

(b) Tarsal bone of ankle, a short bone

(c) Parietal bone of skull, a flat bone

(d) Vertebra, an irregular bone

(e) Patella, a sesamoid bone

(a) Femur, a long bone

FIGURE 12.2 Radiograph (X ray) of a child's hand showing numerous epiphyseal plates. Only single epiphyseal plates develop in the hand and fingers.

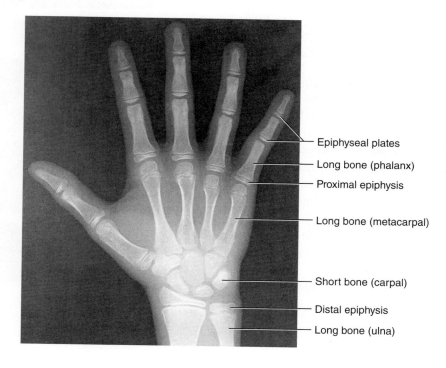

Epiphyseal plates

Long bone (phalanx)

Proximal epiphysis

Long bone (metacarpal)

Short bone (carpal)

Distal epiphysis

Long bone (ulna)

FIGURE 12.3 The major structures of a long bone (living femur).

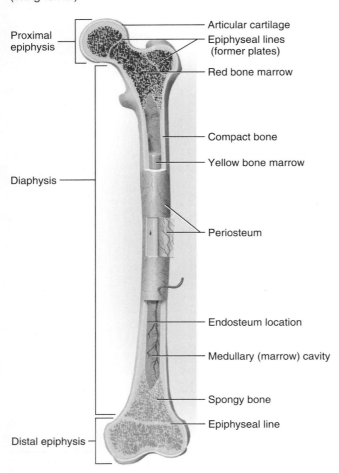

- Proximal epiphysis
- Articular cartilage
- Epiphyseal lines (former plates)
- Red bone marrow
- Compact bone
- Yellow bone marrow
- Diaphysis
- Periosteum
- Endosteum location
- Medullary (marrow) cavity
- Spongy bone
- Epiphyseal line
- Distal epiphysis

FIGURE 12.4 Representative locations of compact and spongy bone.

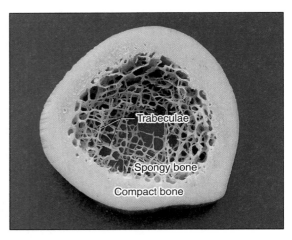

Trabeculae
Spongy bone
Compact bone

(a) Femur, cross section

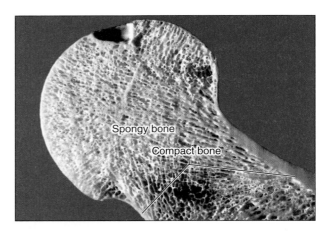

Spongy bone
Compact bone

(b) Femur, longitudinal section

2. Become familiar with the features of a long bone (fig. 12.3).
3. Note representative compact and spongy bone locations (fig. 12.4).
4. Examine the microscopic structure of bone tissue by observing a prepared microscope slide of ground compact bone. Use figures 12.5 and 12.6 of bone tissue to locate the following features:

   osteon (Haversian system)—cylinder-shaped unit

   central canal (Haversian canal; osteonic canal)—contains blood vessels and nerves

   lamella—concentric ring of matrix around central canal

   lacuna—small chamber for an osteocyte

   bone extracellular matrix—collagen and calcium phosphate

   canaliculus—minute tube containing cellular process

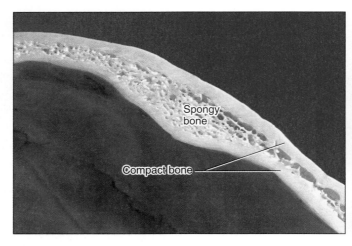

Spongy bone
Compact bone

(c) Skull bone, sectioned

FIGURE 12.5 The structures associated with a bone.

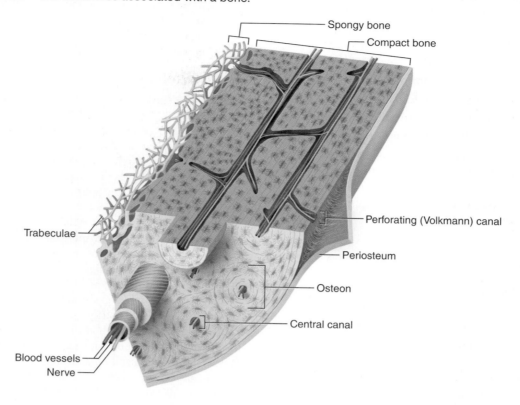

FIGURE 12.6 Micrograph of ground compact bone tissue (160×).

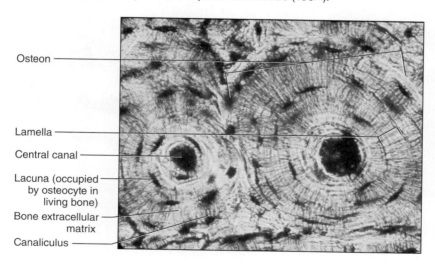

5. Complete Part A of Laboratory Assessment 12.
6. Examine the sectioned long bones and locate the following (figs. 12.2, 12.3, and 12.4):

epiphysis
   proximal—nearest torso
   distal—farthest from torso

epiphyseal plate—growth zone of hyaline cartilage

articular cartilage—on ends of epiphyses

diaphysis—shaft between epiphyses

periosteum—membrane around bone (except articular cartilage) of dense irregular connective tissue

compact (dense) bone—forms diaphysis and epiphyseal surfaces

spongy (cancellous) bone—within epiphyses
   trabeculae—a structural lattice of plates in spongy bone

medullary (marrow) cavity—hollow chamber

endosteum—thin membrane lining medullary cavity of reticular connective tissue

yellow bone marrow—occupies medullary cavity and stores adipose tissue

red bone marrow—occupies spongy bone in some epiphyses and flat bones and produces blood cells

7. Use the dissecting microscope to observe the compact bone and spongy bone of the sectioned specimens. Also examine the marrow in the medullary cavity and the spaces within the spongy bone of the fresh specimen.
8. Complete Parts B and C of the laboratory assessment.

 Demonstration Activity

Examine a fresh chicken bone and a chicken bone that has been soaked for several days in vinegar or overnight in dilute hydrochloric acid. Wear disposable gloves for handling these bones. This acid treatment removes the inorganic salts from the bone extracellular matrix. Rinse the bones in water and note the texture and flexibility of each (fig. 12.7a). The bone becomes soft and flexible without the support of the inorganic salts with calcium.

Examine the specimen of chicken bone that has been exposed to high temperature (baked at 121°C/250°F for 2 hours). This treatment removes the protein and other organic substances from the bone extracellular matrix (fig. 12.7b). The bone comes brittle and fragile without the benefit of the collagen fibers. A living bone with a combination of the qualities of inorganic and organic substances provides tensile strength. View the LabCam video on the chicken bone mentioned in the Pre-Lab.

FIGURE 12.7 Results of fresh chicken bone demonstration: (a) soaked in vinegar; (b) baked in oven.

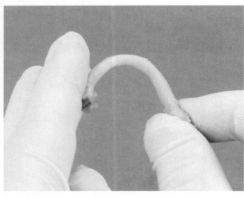

(a)

(b)

Name _____

Date _____

Section _____

The ⬛ corresponds to the indicated outcome(s) found at the beginning of the laboratory exercise.

Bone Structure and Classification

Part A Assessments

Complete the following statements: (*Note:* Questions 1–6 pertain to bone classification by shape.)

1. A bone that has a wide surface is classified as a(an) _____ bone. ⬛

2. The bones of the wrist are examples of _____ bones. ⬛

3. The bone of the thigh is an example of a(an) _____ bone. ⬛

4. Vertebrae are examples of _____ bones. ⬛

5. The patella (kneecap) is an example of a large _____ bone. ⬛

6. The bones of the skull that form a protective covering for the brain are examples of _____ bones. ⬛

7. Distinguish between the epiphysis and the diaphysis of a long bone. ⬛ _____

_____

8. Describe where cartilage is found on the surface of a long bone. ⬛ _____

_____

9. Describe where the periosteum is found on the surface of a long bone. ⬛ _____

_____

Part B Assessments

Complete the following:

1. Distinguish the locations and tissues between the periosteum and the endosteum. ⬛ _____

_____

_____

2. What structural differences did you note between compact bone and spongy bone? ⬛ _____

_____

_____

_____

3. How are these structural differences related to the locations and functions of these two types of bone? ▲3 _____

_____

_____

4. From your observations, how does the marrow in the medullary cavity compare with the marrow in the spaces of the spongy bone? ▲4 _____

_____

_____

Part C   Assessments

Identify the structures indicated in figures 12.8, 12.9, and 12.10.

FIGURE 12.8   Label the structures of this long bone, using the terms provided. ▲2 ▲3

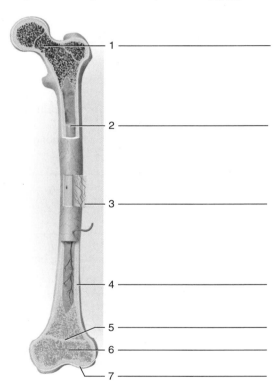

Terms:
Articular cartilage
Compact bone
Epiphyseal line
Periosteum
Red bone marrow
Spongy bone
Yellow bone marrow

1 — _____

2 — _____

3 — _____

4 — _____

5 — _____

6 — _____

7 — _____

FIGURE 12.9 Identify the structures indicated in (a) the unsectioned long bone (fifth metatarsal) and (b) the partially sectioned long bone, using the terms provided. ⚁ ⚂

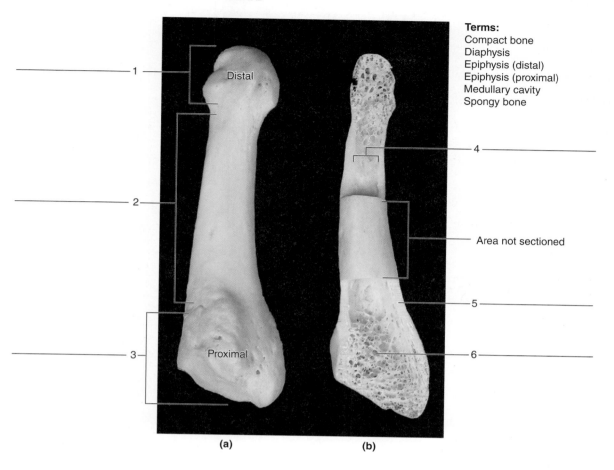

Terms:
Compact bone
Diaphysis
Epiphysis (distal)
Epiphysis (proximal)
Medullary cavity
Spongy bone

Distal

Proximal

Area not sectioned

(a)   (b)

FIGURE 12.10 Label the structures associated with bone, using the terms provided. ⚂ ⚃

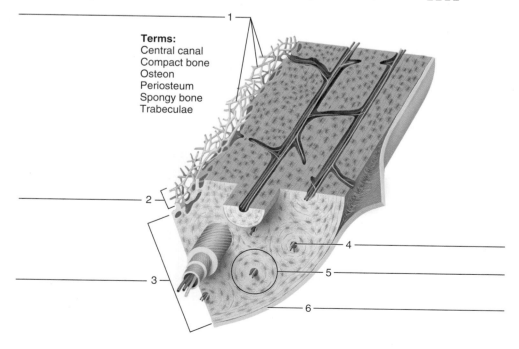

Terms:
Central canal
Compact bone
Osteon
Periosteum
Spongy bone
Trabeculae

NOTES

Organization of the Skeleton

Purpose of the Exercise

To review the organization of the skeleton, the major bones of the skeleton, and the terms used to describe skeletal structures.

Materials Needed

Human skeleton, articulated
Human skeleton, disarticulated

For Learning Extension Activity:
Colored pencils

For Demonstration Activity:
Radiographs (X rays) of skeletal structures
Stereomicroscope (dissecting microscope)

Learning Outcomes

After completing this exercise, you should be able to

1. Locate and label the major bones of the human skeleton.
2. Distinguish between the axial skeleton and the appendicular skeleton.
3. Associate the terms used to describe skeletal structures and locate examples of such structures on the human skeleton.

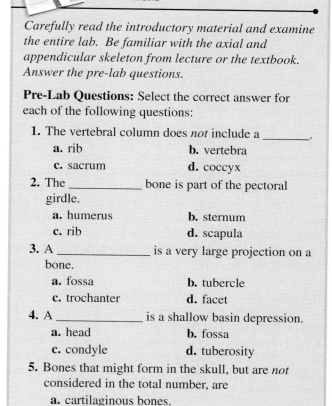

Pre-Lab

Carefully read the introductory material and examine the entire lab. Be familiar with the axial and appendicular skeleton from lecture or the textbook. Answer the pre-lab questions.

Pre-Lab Questions: Select the correct answer for each of the following questions:

1. The vertebral column does *not* include a _____.
   a. rib
   b. vertebra
   c. sacrum
   d. coccyx

2. The _____ bone is part of the pectoral girdle.
   a. humerus
   b. sternum
   c. rib
   d. scapula

3. A _____ is a very large projection on a bone.
   a. fossa
   b. tubercle
   c. trochanter
   d. facet

4. A _____ is a shallow basin depression.
   a. head
   b. fossa
   c. condyle
   d. tuberosity

5. Bones that might form in the skull, but are *not* considered in the total number, are
   a. cartilaginous bones.
   b. sutural bones.
   c. middle ear bones.
   d. sesamoid bones.

6. The _____ is a bone in the upper limb.
   a. ulna
   b. scapula
   c. clavicle
   d. fibula

7. A _____ is a depression type of bone feature (bone marking).
   a. foramen
   b. crest
   c. sulcus
   d. tuberosity

The skeleton can be divided into two major portions: (1) the *axial skeleton*, which consists of the bones and cartilages of the head, neck, and trunk, and (2) the *appendicular skeleton*, which consists of the bones of the limbs and those that anchor the limbs to the axial skeleton. The bones that anchor the limbs include the pectoral and pelvic girdles.

Ossification is the bone formation process. This process occurs during fetal development and continues through childhood growth. Bone forms either from intramembranous origins or endochondral origins. During the growth of *intramembranous bones*, membrane-like connective tissue layers similar to the dermis develop in an area destined to become the flat bones of the skull. Eventually, bone-forming cells (osteoblasts) alter the membrane areas into bone tissue. In contrast, *endochondral bones* develop from hyaline cartilage and through the ossification process develop into most of the bones of the skeleton. In either type of ossification, both compact and spongy bone develops.

The number of bones in the adult skeleton is often reported to be 206. Men and women, although variations can exist, possess the same total bone number of 206. However, at birth the number of bones is closer to 275 as many ossification centers are still composed of cartilage, and some bones form during childhood. For example, each hip bone (coxal bone) includes an ilium, ischium, and pubis, and many long bones have three ossification centers separated by epiphyseal plates. The sternum, composed of a manubrium, body, and xiphoid process, becomes a single bone much later than when we reach our full height. Some people have additional bones not considered in the total number. Sesamoid bones other than the the patellae may develop in the hand or the foot. Also, extra bones sometimes form in the skull within the sutures; these are called sutural (wormian) bones.

Special terminology is used to describe the features of a bone. The term used depends on whether the feature is a type of projection, articulation, depression, or opening. Many of these features can be noted when viewing radiographs. Some of the features can be palpated (touched) if they are located near the surface of the body.

Procedure—Organization of the Skeleton

1. Study figure 13.1 and use it as a reference to examine the bones in the human skeleton. As you locate the following bones, note the number of each in the skeleton. Palpate as many of the corresponding bones in your skeleton as possible.

   axial skeleton
   - skull
     - cranium (8)
     - face (14)
   - middle ear bone (6)
   - hyoid bone—supports the tongue (1)
   - vertebral column
     - vertebra (24)
     - sacrum (1)
     - coccyx (1)
   - thoracic cage
     - rib (24)
     - sternum (1)

   appendicular skeleton
   - pectoral girdle
     - scapula (2)
     - clavicle (2)
   - upper limbs
     - humerus (2)
     - radius (2)
     - ulna (2)
     - carpal (16)
     - metacarpal (10)
     - phalanx (28)
   - pelvic girdle
     - hip bone (coxal bone; pelvic bone; innominate bone) (2)
   - lower limbs
     - femur (2)
     - tibia (2)
     - fibula (2)
     - patella (2)
     - tarsal (14)
     - metatarsal (10)
     - phalanx (28)

Total	**206 bones**

2. Complete Part A of Laboratory Assessment 13.
3. Bone features (bone markings) can be grouped together in a category of projections, articulations, depressions, or openings. Within each category more specific examples occur. The bones illustrated in figure 13.2 represent specific examples of locations of specific features in the human body. Locate each of the following features on the example bone from a disarticulated skeleton, noting the size, shape, and location in the human skeleton.

   Projections: sites for tendon and ligament attachment
   crest—ridgelike
   epicondyle—superior to condyle
   line (linea)—slightly raised ridge
   process—prominent
   protuberance—outgrowth
   ramus—extension
   spine—thornlike
   trochanter—large

FIGURE 13.1 Major bones of the skeleton: (a) anterior view; (b) posterior view. The axial portion is orange, and the appendicular portion is yellow.

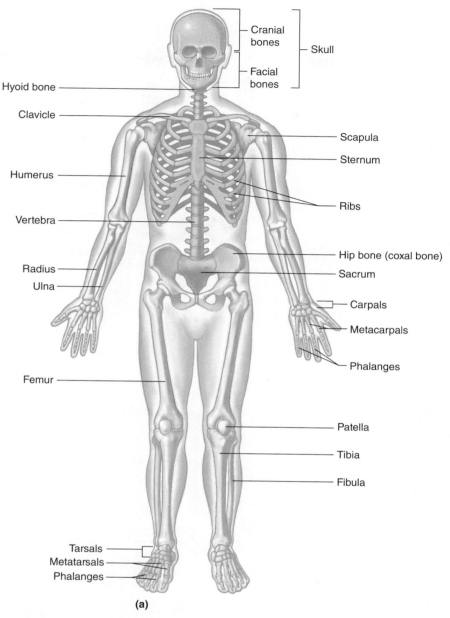

(a)

tubercle—small knoblike

tuberosity—rough elevation

<u>Articulations:</u> **where bones connect at a joint or articulate with each other**

condyle—rounded process

facet—nearly flat

head—expanded end

<u>Depressions:</u> **recessed areas in bones**

alveolus—socket

fossa—shallow basin

fovea—tiny pit

notch—indentation on edge

sulcus—narrow groove

<u>Openings:</u> **open spaces in bones**

canal—tubular passage

fissure—slit

foramen—hole

meatus—tubelike opening

sinus—cavity

FIGURE 13.1 *Continued.*

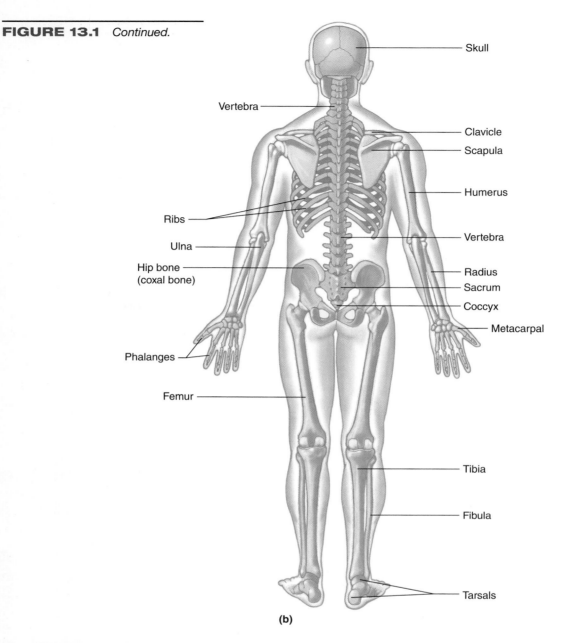

Skull

Vertebra

Clavicle

Scapula

Humerus

Ribs

Vertebra

Ulna

Radius

Hip bone
(coxal bone)

Sacrum

Coccyx

Metacarpal

Phalanges

Femur

Tibia

Fibula

Tarsals

(b)

Critical Thinking Activity

Locate and name the largest foramen in the skull.

_____

_____

Locate and name the largest foramen in the skeleton.

_____

_____

_____

4. Complete Parts B, C, and D of the laboratory assessment.

Demonstration Activity

Images on radiographs (X rays) are produced by allowing X rays from an X-ray tube to pass through a body part and to expose photographic film positioned on the opposite side of the part. The image that appears on the film after it is developed reveals the presence of parts with different densities. Bone, for example, is very dense tissue and is a good absorber of X rays. Thus, bone generally appears light on the film. Air-filled spaces, on the other hand, absorb almost no X rays and appear as dark areas on the film. Liquids and soft tissues absorb intermediate quantities of X rays, so they usually appear in various shades of gray.

Examine the available radiographs of skeletal structures by holding each film in front of a light source. Identify as many of the bones and features as you can.

FIGURE 13.2 Representative examples of bone features (bone markings) on bones of the skeleton (a–h). The two largest foramina of the skeleton have complete labels.

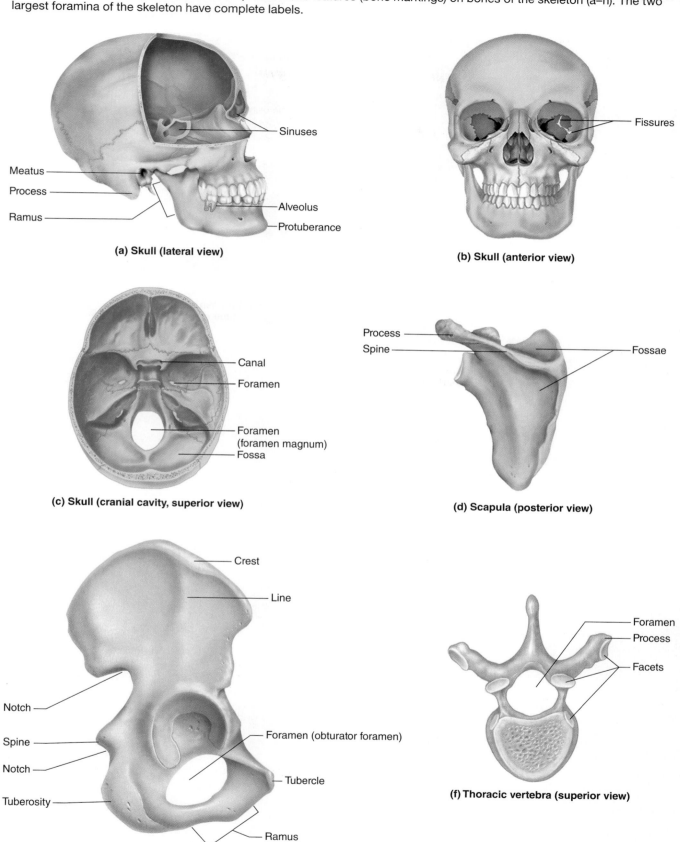

(a) Skull (lateral view)

Sinuses
Meatus
Process
Ramus
Alveolus
Protuberance

(b) Skull (anterior view)

Fissures

(c) Skull (cranial cavity, superior view)

Canal
Foramen
Foramen (foramen magnum)
Fossa

(d) Scapula (posterior view)

Process
Spine
Fossae

(e) Hip bone (lateral view)

Crest
Line
Notch
Spine
Notch
Tuberosity
Foramen (obturator foramen)
Tubercle
Ramus

(f) Thoracic vertebra (superior view)

Foramen
Process
Facets

FIGURE 13.2 *Continued.*

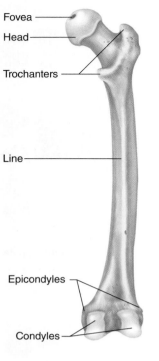

Fovea

Head

Trochanters

Line

Epicondyles

Condyles

**(g) Femur
(posterior view)**

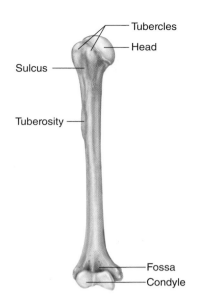

Tubercles

Head

Sulcus

Tuberosity

Fossa

Condyle

**(h) Humerus
(anterior view)**

Name _____

Date _____

Section _____

The ⒜ corresponds to the indicated outcome(s) found at the beginning of the laboratory exercise.

Organization of the Skeleton

Part A   Assessments

Label the bones indicated in figure 13.3. ⒈

FIGURE 13.3   Label the major bones of the skeleton: (a) anterior view; (b) posterior view, using the terms provided. ⒈ ⒉

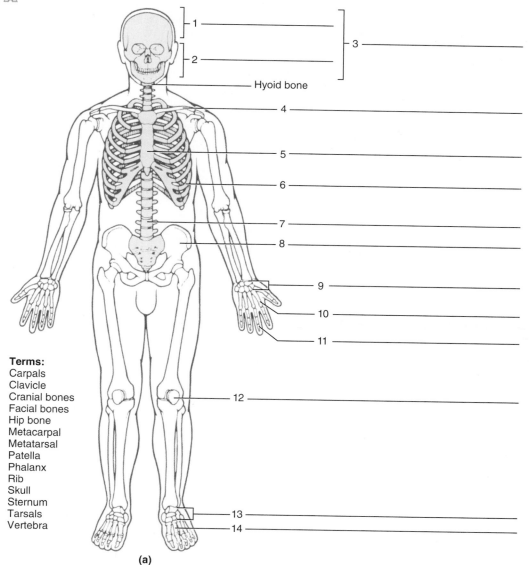

1
2
3
Hyoid bone
4
5
6
7
8
9
10
11
12
13
14

Terms:
Carpals
Clavicle
Cranial bones
Facial bones
Hip bone
Metacarpal
Metatarsal
Patella
Phalanx
Rib
Skull
Sternum
Tarsals
Vertebra

(a)

FIGURE 13.3 *Continued.*

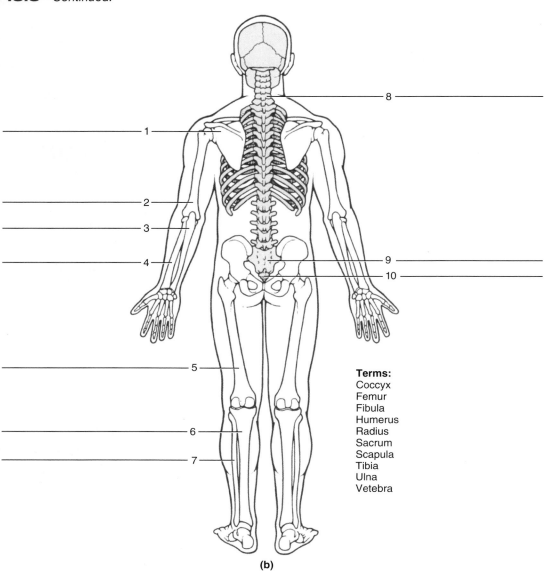

1

2

3

4

5

6

7

8

9

10

(b)

Terms:
Coccyx
Femur
Fibula
Humerus
Radius
Sacrum
Scapula
Tibia
Ulna
Vetebra

Learning Extension Activity

Use colored pencils to distinguish the individual bones in figure 13.3.

Part B   Assessments

1. Identify the bones indicated in figure 13.4.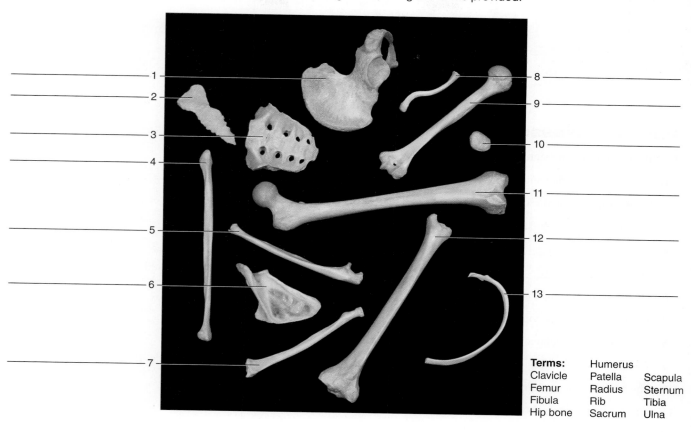

FIGURE 13.4   Identify the bones in this random arrangement, using the terms provided.

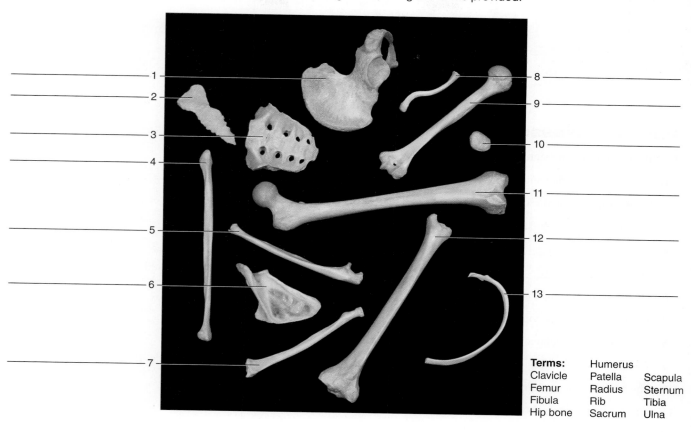

Terms: Humerus
Clavicle   Patella   Scapula
Femur   Radius   Sternum
Fibula   Rib   Tibia
Hip bone   Sacrum   Ulna

2. List any of the bones shown in figure 13.4 that are included as part of the axial skeleton. _____

Part C   Assessments

1. Match the terms in column A with the definitions in column B. Place the letter of your choice in the space provided.

Column A	Column B
a. Condyle	_____ **1.** Small, nearly flat articular surface
b. Crest	_____ **2.** Deep depression or shallow basin
c. Facet	_____ **3.** Rounded process
d. Foramen	_____ **4.** Opening or hole
e. Fossa	_____ **5.** Projection extension
f. Line	_____ **6.** Ridgelike projection
g. Ramus	_____ **7.** Slightly raised ridge

2. Match the terms in column A with the definitions in column B. Place the letter of your choice in the space provided. 3

Column A	Column B
a. Fovea	_____ **1.** Tubelike opening
b. Head	_____ **2.** Tiny pit or depression
c. Meatus	_____ **3.** Small, knoblike projection
d. Sinus	_____ **4.** Thornlike projection
e. Spine	_____ **5.** Rounded enlargement at end of bone
f. Trochanter	_____ **6.** Air-filled cavity within bone
g. Tubercle	_____ **7.** Relatively large process

Part D Assessments

Complete the following statements:

1. The extra bones that sometimes develop between the flat bones of the skull are called_____ bones. 1

2. Small bones occurring in some tendons are called _____ bones. 1

3. The cranium and facial bones compose the _____ . 1

4. The _____ bone supports the tongue. 1

5. The _____ at the inferior end of the sacrum is composed of several fused vertebrae. 1

6. Most ribs are attached anteriorly to the _____ . 1

7. The thoracic cage is composed of _____ pairs of ribs. 1

8. The scapulae and clavicles together form the _____ . 1

9. Which of the following bones is *not* part of the appendicular skeleton: clavicle, femur, scapula, sternum? _____ 2

10. The wrist is composed of eight bones called _____ . 1

11. The hip bones (coxal bones) are attached posteriorly to the _____ . 1

12. The _____ bone covers the anterior surface of the knee. 1

13. The bones that articulate with the distal ends of the tibia and fibula are called _____ . 1

14. All finger and toe bones are called _____ . 1

Skull

Purpose of the Exercise

To examine the structure of the human skull and to identify the bones and major features of the skull.

Materials Needed

Human skull, articulated
Human skull, disarticulated
Human skull, sagittal section

Learning Outcomes

After completing this exercise, you should be able to

1. Locate and label the bones of the skull and their major features.

2. Locate and label the major sutures of the cranium.

3. Locate the sinuses of the skull.

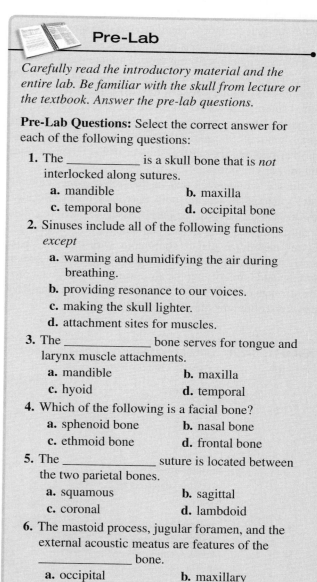

Pre-Lab

Carefully read the introductory material and the entire lab. Be familiar with the skull from lecture or the textbook. Answer the pre-lab questions.

Pre-Lab Questions: Select the correct answer for each of the following questions:

1. The _____ is a skull bone that is *not* interlocked along sutures.
   - **a.** mandible
   - **b.** maxilla
   - **c.** temporal bone
   - **d.** occipital bone

2. Sinuses include all of the following functions *except*
   - **a.** warming and humidifying the air during breathing.
   - **b.** providing resonance to our voices.
   - **c.** making the skull lighter.
   - **d.** attachment sites for muscles.

3. The _____ bone serves for tongue and larynx muscle attachments.
   - **a.** mandible
   - **b.** maxilla
   - **c.** hyoid
   - **d.** temporal

4. Which of the following is a facial bone?
   - **a.** sphenoid bone
   - **b.** nasal bone
   - **c.** ethmoid bone
   - **d.** frontal bone

5. The _____ suture is located between the two parietal bones.
   - **a.** squamous
   - **b.** sagittal
   - **c.** coronal
   - **d.** lambdoid

6. The mastoid process, jugular foramen, and the external acoustic meatus are features of the _____ bone.
   - **a.** occipital
   - **b.** maxillary
   - **c.** temporal
   - **d.** ethmoid

7. Foramina, canals, and fissures serve as passageways for blood vessels and nerves.
   True _____ False _____

8. The jugular foramen is larger than the foramen magnum.
   True _____ False _____

A human skull consists of twenty-two bones that, except for the lower jaw, are firmly interlocked along sutures. Eight of these immovable bones make up the braincase, or cranium, and thirteen more immovable bones and the mandible form the facial skeleton.

Bones located within the temporal bone are the auditory ossicles (middle ear bones). They include the malleus, incus, and stapes that are examined in Laboratory Exercise 37. The stapes is considered the smallest bone in the human body among the total number of 206. The hyoid bone is suspended from the temporal bones and serves for tongue and larynx muscle attachments. A fractured hyoid bone is sometimes used as an indication of strangulation as the cause of death.

After intramembranous ossification of the flat bones of the skull is complete, a fibrous type of joint remains between them. These joints are considered immovable and are known as sutures. The suture between the two parietal bones is the sagittal suture. The lambdoid suture is between the occipital and parietal bones. The coronal suture is bordered by the frontal bone and the parietal bones. The squamous suture is primarily along the superior border of the temporal bone with the parietal bone.

Sinus cavities are found in the frontal bone, ethmoid bone, sphenoid bone, and both maxillary bones. Because they are associated with the nasal passages, they are called paranasal sinuses. These air-filled sinuses are lined with a mucous membrane that is continuous with the nasal passages. The sinuses function to lighten the skull, assist in warming and humidifying the air during breathing, and provide resonance to our voices. If infected, the large maxillary sinuses are often misinterpreted to be a toothache.

Procedure—Skull

1. Examine human articulated, disarticulated, and sectioned skulls available in the laboratory. Identify the bones and features illustrated in figures 14.1, 14.2, 14.3, 14.4, 14.5, and 14.6 on the actual skulls. Note that the figures are in full color, with the bones labeled in **boldface** and feature labels not in boldface. It is best to work with a partner or in small groups.

2. Examine the **cranial bones** of the articulated human skull and the sectioned skull. Also observe the corresponding disarticulated bones. Locate the following bones and features in the laboratory specimens and, at the same time, palpate as many of these bones and features in your skull as possible.

   frontal bone (1)
   - supraorbital foramen (or notch)—transmits blood vessels and nerves
   - frontal sinus—two present

   parietal bone (2)
   - sagittal suture—where parietal bones meet
   - coronal suture—where parietal bones meet frontal bone

   occipital bone (1)
   - lambdoid suture—where parietal bones meet occipital bone
   - external occipital protuberance—midline projection on posterior surface

---

FIGURE 14.1 Bones and features of the skull, anterior view. (**Boldface** indicates a bone name; not boldface indicates a feature.)

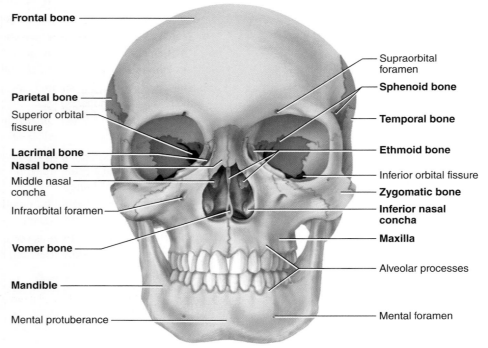

FIGURE 14.2 Bones and features of the skull, lateral view. (**Boldface** indicates a bone name; not boldface indicates a feature.)

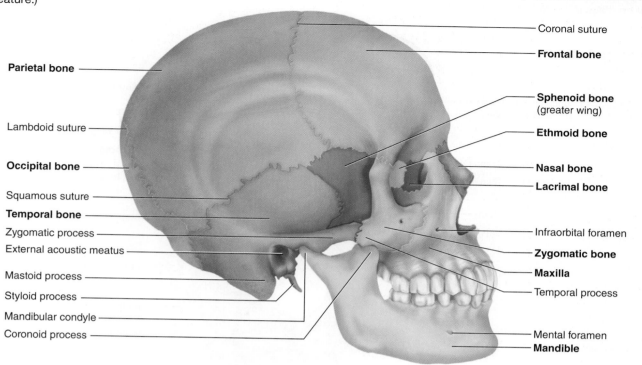

- Coronal suture
- **Frontal bone**
- **Parietal bone**
- **Sphenoid bone** (greater wing)
- Lambdoid suture
- **Ethmoid bone**
- **Occipital bone**
- **Nasal bone**
- **Lacrimal bone**
- Squamous suture
- **Temporal bone**
- Zygomatic process
- Infraorbital foramen
- External acoustic meatus
- **Zygomatic bone**
- Mastoid process
- **Maxilla**
- Styloid process
- Temporal process
- Mandibular condyle
- Coronoid process
- Mental foramen
- **Mandible**

FIGURE 14.3 Bones and features of the skull, inferior view. (**Boldface** indicates a bone name; not boldface indicates a feature.)

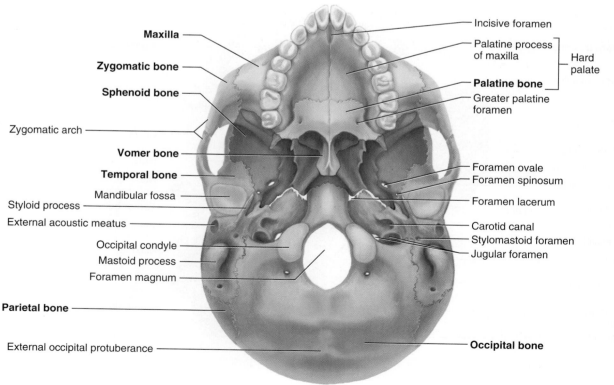

- **Maxilla**
- Incisive foramen
- Palatine process of maxilla
- Hard palate
- **Zygomatic bone**
- **Sphenoid bone**
- **Palatine bone**
- Greater palatine foramen
- Zygomatic arch
- **Vomer bone**
- Foramen ovale
- Foramen spinosum
- **Temporal bone**
- Mandibular fossa
- Foramen lacerum
- Styloid process
- External acoustic meatus
- Carotid canal
- Occipital condyle
- Stylomastoid foramen
- Mastoid process
- Jugular foramen
- Foramen magnum
- **Parietal bone**
- External occipital protuberance
- **Occipital bone**

FIGURE 14.4 Bones and features of the floor of the cranial cavity, superior view. (**Boldface** indicates a bone name; not boldface indicates a feature.)

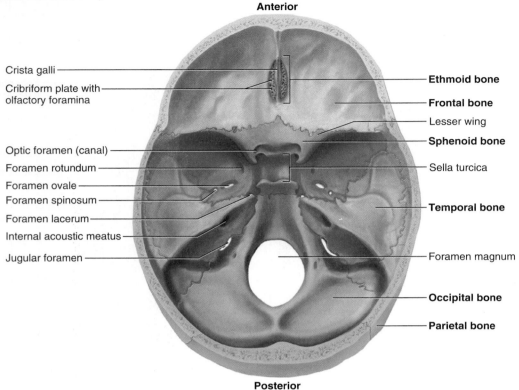

Anterior

Crista galli

Cribriform plate with olfactory foramina

Ethmoid bone

Frontal bone

Lesser wing

Sphenoid bone

Optic foramen (canal)

Foramen rotundum

Foramen ovale

Foramen spinosum

Foramen lacerum

Internal acoustic meatus

Jugular foramen

Sella turcica

Temporal bone

Foramen magnum

Occipital bone

Parietal bone

Posterior

FIGURE 14.5 Bones and features of the sagittal section of the skull. (**Boldface** indicates a bone name; not boldface indicates a feature.)

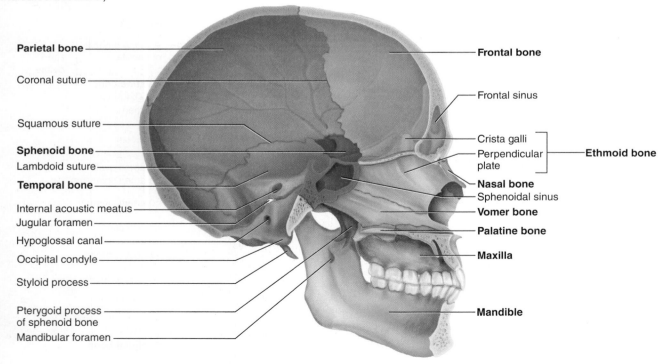

Parietal bone

Coronal suture

Squamous suture

Sphenoid bone

Lambdoid suture

Temporal bone

Internal acoustic meatus

Jugular foramen

Hypoglossal canal

Occipital condyle

Styloid process

Pterygoid process of sphenoid bone

Mandibular foramen

Frontal bone

Frontal sinus

Crista galli

Perpendicular plate

Ethmoid bone

Nasal bone

Sphenoidal sinus

Vomer bone

Palatine bone

Maxilla

Mandible

134

FIGURE 14.6   Bones of a disarticulated skull.

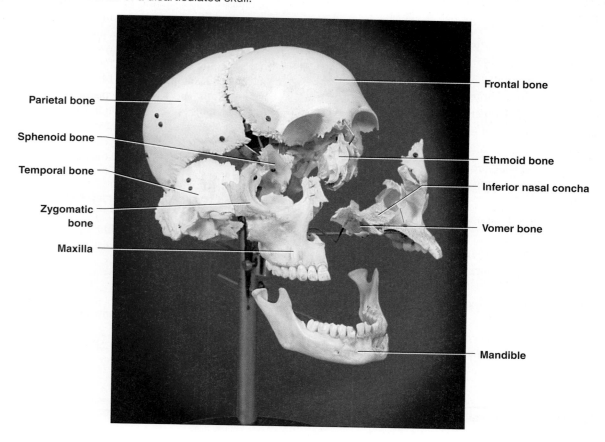

Parietal bone

Sphenoid bone

Temporal bone

Zygomatic bone

Maxilla

Frontal bone

Ethmoid bone

Inferior nasal concha

Vomer bone

Mandible

- foramen magnum—where brainstem and spinal cord meet
- occipital condyle—articulates with first cervical vertebra

temporal bone (2)

- squamous suture—where parietal bones meet temporal bone
- external acoustic meatus—opening of outer ear
- mandibular fossa—articulates with mandible
- mastoid process—attachment site for muscles
- styloid process—long, pointed projection
- carotid canal—transmits internal carotid artery
- jugular foramen—transmits internal jugular vein
- internal acoustic meatus—opening for cranial nerves
- zygomatic process—forms much of zygomatic arch

sphenoid bone (1)

- sella turcica—saddle-shaped depression for pituitary gland
- greater and lesser wings—lateral extensions
- sphenoidal sinus—two present

ethmoid bone (1)

- cribriform plate—contains many olfactory foramina for olfactory nerves
- perpendicular plate—forms most of nasal septum
- superior nasal concha—small extensions in nasal cavity
- middle nasal concha—medium extensions in nasal cavity
- ethmoidal sinus—many present
- crista galli—attachment site of some brain membranes

3. Examine the **facial bones** of the articulated and sectioned skulls and the corresponding disarticulated bones. Locate the following:

maxilla (2)

- maxillary sinus—largest sinuses
- palatine process—forms anterior hard palate
- alveolar process—supports the teeth
- alveolar arch—formed by alveolar processes

palatine bone—L-shaped bones (2)

zygomatic bone (2)

- temporal process—extension that joins zygomatic process
- zygomatic arch—formed by zygomatic and temporal processes

lacrimal bone—thin and scalelike (2)

nasal bone—forms bridge of nose (2)

vomer bone—forms part of nasal septum (1)

inferior nasal concha—largest of the conchae (2)

mandible (1)

- ramus—upward projection
- mandibular condyle—articulates with temporal bone
- coronoid process—attachment site for chewing muscles
- alveolar process—supports the teeth
- mandibular foramen—admits blood vessels and nerves for teeth
- mental foramen—passageway for blood vessels and nerves
- mental protuberance—chin projection

4. Complete Part A of Laboratory Assessment 14.

5. Examine any skulls that are sectioned to expose the paranasal sinuses. Sinuses are located in the frontal, ethmoid, sphenoid, and maxillary bones. Compare the skulls with figures 14.5, 14.6, and 14.7.

6. Some of the major passageways through the skull bones have specific names for the foramen, canal, fissure, or meatus. Reexamine figures 14.1 through 14.5 and the actual skulls and locate the passageways of the skull listed in table 14.1. The major structures transmitted through the specific passageway are included in the table.

7. Complete Parts B, C, and D of the laboratory assessment.

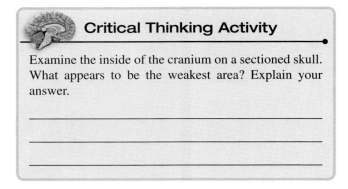

Critical Thinking Activity

Examine the inside of the cranium on a sectioned skull. What appears to be the weakest area? Explain your answer.

_____

_____

_____

FIGURE 14.7 Paranasal sinuses are located in the frontal bone, ethmoid bone, sphenoid bone, and both maxillary bones.

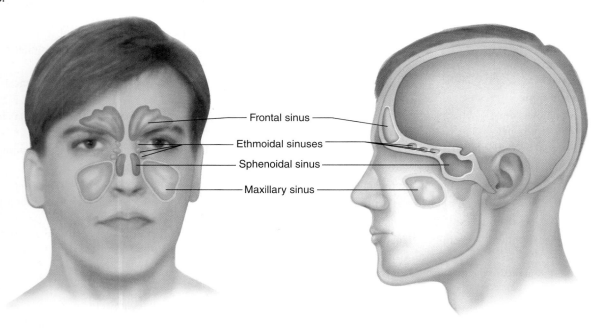

Frontal sinus

Ethmoidal sinuses

Sphenoidal sinus

Maxillary sinus

TABLE 14.1 Major Passageways of the Skull

Passageway	Location Bone	Major Stuctures Transmitted Through
Carotid canal	Temporal bone	Internal carotid artery
Foramen lacerum	Between temporal, sphenoid, and occipital bones	Branch of pharyngeal artery (opening becomes mostly closed by connective tissues)
Foramen magnum	Occipital bone	Inferior part of brainstem connecting to spinal cord, vertebral arteries, and accessory nerve
Foramen ovale	Sphenoid bone	Mandibular division of trigeminal nerve
Foramen rotundum	Sphenoid bone	Maxillary division of trigeminal nerve
Foramen spinosum	Sphenoid bone	Middle meningeal blood vessels
Greater palatine foramen	Palatine bone	Palatine blood vessels and nerves
Hypoglossal canal	Occipital bone	Hypoglossal nerve
Incisive foramen	Maxilla	Nasopalatine nerves
Inferior orbital fissure	Between maxilla, sphenoid, and zygomatic bones	Infraorbital nerve and blood vessels
Infraorbital foramen	Maxilla	Infraorbital blood vessels and nerves
Internal acoustic meatus	Temporal bone	Branches of facial and vestibulocochlear nerves
Jugular foramen	Between temporal and occipital bones	Glossopharyngeal, vagus, and accessory nerves, and internal jugular vein
Mandibular foramen	Mandible	Inferior alveolar blood vessels and nerves
Mental foramen	Mandible	Mental nerve and blood vessels
Optic formen (canal)	Sphenoid bone	Optic nerve and ophthalmic artery
Stylomastoid foramen	Temporal bone	Facial nerve and blood vessels
Superior orbital fissure	Sphenoid bone	Oculomotor, trochlear, and abducens nerves, and ophthalmic division of trigeminal nerve
Supraorbital foramen	Frontal bone	Supraorbital blood vessels and nerves

NOTES

Name _____

Date _____

Section _____

The A corresponds to the indicated outcome(s) found at the beginning of the laboratory exercise.

Skull

Part A   Assessments

Identify the numbered bones and features of the skulls indicated in figures 14.8, 14.9, 14.10, and 14.11.

FIGURE 14.8   Identify the bones and features indicated on this anterior view of the skull, using the terms provided. (If the line lacks the word *bone*, label the particular feature.) A

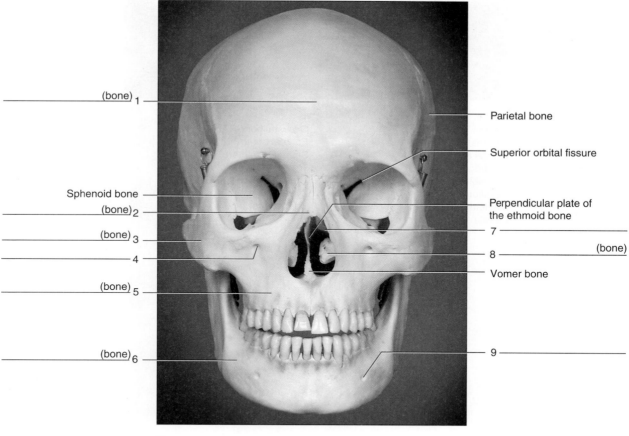

_____ (bone) 1

Sphenoid bone _____

_____ (bone) 2

_____ (bone) 3

_____ 4

_____ (bone) 5

_____ (bone) 6

Parietal bone

Superior orbital fissure

Perpendicular plate of
the ethmoid bone

7 _____

8 _____ (bone)

Vomer bone

9 _____

Terms:

Frontal bone
Inferior nasal concha
Infraorbital foramen
Mandible
Maxilla

Mental foramen
Middle nasal concha
  (of ethmoid bone)
Nasal bone
Zygomatic bone

FIGURE 14.9 Identify the bones and features indicated on this lateral view of the skull, using the terms provided. 1 2

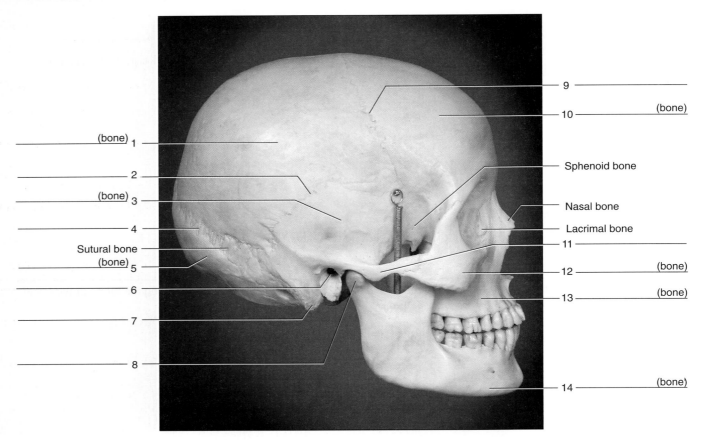

Terms:

Coronal suture
External acoustic meatus
Frontal bone
Lambdoid suture
Mandible
Mandibular condyle
Mastoid process
Maxilla

Occipital bone
Parietal bone
Squamous suture
Temporal bone
Zygomatic bone
Zygomatic process
  (of temporal bone)

FIGURE 14.10 Identify the bones and features indicated on this inferior view of the skull, using the terms provided. 1

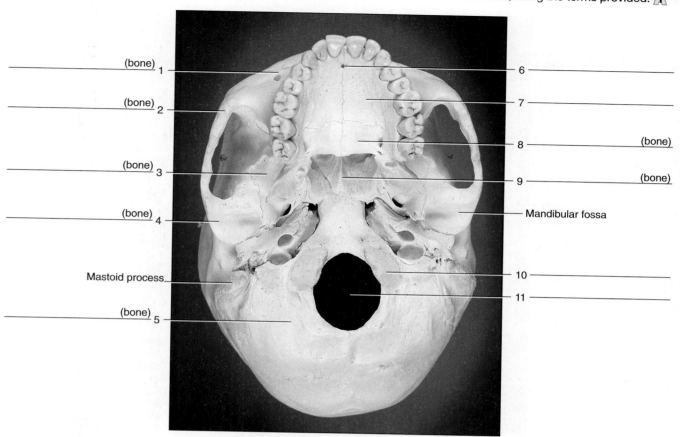

(bone) 1

(bone) 2

(bone) 3

(bone) 4

Mastoid process

(bone) 5

6

7

8 (bone)

9 (bone)

Mandibular fossa

10

11

Terms:

Foramen magnum
Incisive foramen
Maxilla
Occipital bone
Occipital condyle
Palatine bone

Palatine process of maxilla
Sphenoid bone
Temporal bone
Vomer bone
Zygomatic bone

FIGURE 14.11 Identify the bones and features on this floor of the cranial cavity of a skull, using the terms provided.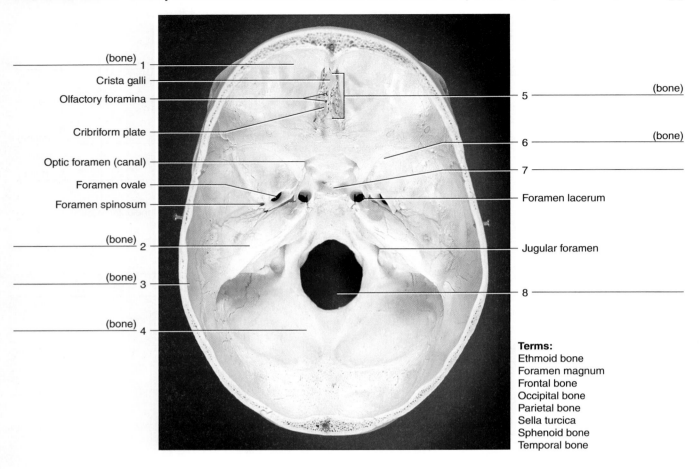

Terms:
Ethmoid bone
Foramen magnum
Frontal bone
Occipital bone
Parietal bone
Sella turcica
Sphenoid bone
Temporal bone

Part B   Assessments

Match the bones in column A with the features in column B. Place the letter of your choice in the space provided. (Some answers are used more than once.)

Column A
a. Ethmoid bone
b. Frontal bone
c. Occipital bone
d. Parietal bone
e. Sphenoid bone
f. Temporal bone

Column B
_____ 1. Forms sagittal, coronal, squamous, and lambdoid sutures

_____ 2. Cribriform plate

_____ 3. Crista galli

_____ 4. External acoustic meatus

_____ 5. Foramen magnum

_____ 6. Mandibular fossa

_____ 7. Mastoid process

_____ 8. Middle nasal concha

_____ 9. Occipital condyle

_____ 10. Sella turcica

_____ 11. Styloid process

_____ 12. Supraorbital foramen

Part C   Assessments

Complete the following statements:

1. The _____ suture joins the frontal bone to the parietal bones. **2**

2. The parietal bones are firmly interlocked along the midline by the _____ suture. **2**

3. The _____ suture joins the parietal bones to the occipital bone. **2**

4. The temporal bones are joined to the parietal bones along the _____ sutures. **2**

5. Name the three cranial bones that contain sinuses. **3** _____

_____

6. Name a facial bone that contains a sinus. **3** _____

Part D   Assessments

Match the bones in column A with the characteristics in column B. Place the letter of your choice in the space provided. **1**

Column A	Column B
a. Inferior nasal concha	_____ **1.** Forms bridge of nose
b. Lacrimal bone	_____ **2.** Only movable bone in the facial skeleton
c. Mandible	_____ **3.** Contains coronoid process
d. Maxilla	_____ **4.** Creates prominence of cheek inferior and lateral to the eye
e. Nasal bone	_____ **5.** Contains sockets of upper teeth
f. Palatine bone	_____ **6.** Forms inferior portion of nasal septum
g. Vomer bone	_____ **7.** Forms anterior portion of zygomatic arch
h. Zygomatic bone	_____ **8.** Scroll-shaped bone
	_____ **9.** Forms anterior roof of mouth
	_____ **10.** Contains mental foramen
	_____ **11.** Forms posterior roof of mouth
	_____ **12.** Scalelike part in medial wall of orbit

NOTES

15

Vertebral Column and Thoracic Cage

Purpose of the Exercise

To examine the vertebral column and the thoracic cage of the human skeleton and to identify the bones and major features of these parts.

Materials Needed

Human skeleton, articulated
Samples of cervical, thoracic, and lumbar vertebrae
Human skeleton, disarticulated

Learning Outcomes

After completing this exercise, you should be able to

1. Identify the structures and functions of the vertebral column.
2. Locate the features of a vertebra.
3. Distinguish the cervical, thoracic, and lumbar vertebrae and the sacrum and coccyx.
4. Identify the structures and functions of the thoracic cage.
5. Distinguish between true and false ribs.

Pre-Lab

Carefully read the introductory material and examine the entire lab. Be familiar with the vertebral column and the thoracic cage from lecture or the textbook. Answer the pre-lab questions.

Pre-Lab Questions: Select the correct answer for each of the following questions:

1. The most superior bone of the vertebral column is the
   a. coccyx.   b. vertebra prominens.
   c. axis.   d. atlas.

2. The vertebral column possesses
   a. four curvatures.   b. three curvatures.
   c. one curvature.   d. no curvatures as it is straight.

3. Humans have _____ pairs of true ribs.
   a. two   b. five
   c. seven   d. twelve

4. The _____ ribs do *not* have costal cartilage attachments to the sternum.
   a. false   b. floating
   c. true   d. superior

5. Humans possess _____ cervical vertebrae.
   a. twenty-six   b. twelve
   c. seven   d. five

6. The superior end of the sacrum articulates with the
   a. coccyx.   b. femur.
   c. twelfth thoracic   d. fifth lumbar
      vertebra.   vertebra.

7. The anterior (sternal) end of a rib articulates with a thoracic vertebra.
   True _____   False _____

8. All cervical, thoracic, and lumbar vertebrae possess a vertebral foramen.
   True _____   False _____

9. A feature of the second cervical vertebra is the dens.
   True _____   False _____

The vertebral column, consisting of twenty-six bones, extends from the skull to the pelvis and forms the vertical axis of the human skeleton. The *vertebral column* includes seven cervical vertebrae, twelve thoracic vertebrae, five lumbar vertebrae, one sacrum of five fused vertebrae, and one coccyx of usually four fused vertebrae. To help you to remember the number of cervical, thoracic, and lumbar vertebrae from superior to inferior, consider this saying: breakfast at 7, lunch at 12, and dinner at 5. These vertebrae are separated from one another by cartilaginous intervertebral discs and are held together by ligaments.

The *thoracic cage* surrounds the thoracic and upper abdominal cavities. It includes the ribs, the thoracic vertebrae, the sternum, and the costal cartilages. The thoracic cage provides protection for the heart and lungs.

Procedure A—Vertebral Column

The vertebral column extends from the first cervical vertebra next to the skull to the inferior tip of the coccyx. The first cervical vertebra (C1) is also known as the *atlas* and has a posterior tubercle instead of a more pronounced spinous process. The second cervical vertebra (C2), known as the *axis,* has a superior projection, the dens (odontoid process) that serves as a pivot point for some rotational movements. The seventh cervical vertebra (C7) is often referred to as the *vertebra prominens* because the spinous process is elongated and easily palpated as a surface feature. The seven cervical vertebrae have the distinctive feature of transverse foramina for passageways of blood vessels serving the brain.

The twelve thoracic vertebrae have facets for the articulation sites of the twelve pairs of ribs. They are larger than the cervical vertebrae and have spinous processes that are rather long and have an inferior angle. The five lumbar vertebrae have the largest bodies, allowing better support and resistance to twisting of the trunk, and spinous processes that are rather short and blunt.

The five sacral vertebrae of a child fuse into a single bone by the age of about 26. The posterior ridge known as the *medial sacral crest* represents fused spinous processes. The usual four coccyx vertebrae fuse into a single bone by the age of about 30.

Critical Thinking Activity

Note the four curvatures of the vertebral column. What functional advantages exist with curvatures for skeletal structure instead of a straight vertebral column?

_____

_____

_____

_____

_____

The four curvatures of the vertebral column develop either before or after birth. The thoracic and sacral curvatures (primary curvatures) form by the time of birth. The cervical curvature develops by the time a baby is able to hold the head erect and crawl, while the lumbar curvature forms by the time the child is able to walk. The cervical and lumbar curvatures represent the secondary curvatures. The four curvatures allow for flexibility and resiliency of the vertebral column and for it to function somewhat like a spring instead of a rigid rod.

1. Examine figure 15.1 and the vertebral column of the human skeleton. Locate the following bones and features. At the same time, locate as many of the cor-

FIGURE 15.1 Bones and features of the vertebral column (right lateral view).

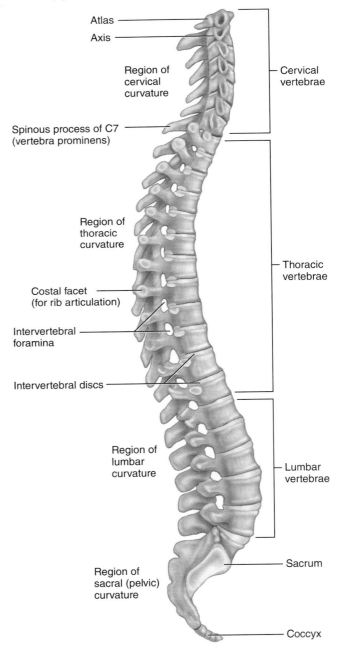

Atlas

Axis

Region of cervical curvature

Cervical vertebrae

Spinous process of C7 (vertebra prominens)

Region of thoracic curvature

Thoracic vertebrae

Costal facet (for rib articulation)

Intervertebral foramina

Intervertebral discs

Region of lumbar curvature

Lumbar vertebrae

Region of sacral (pelvic) curvature

Sacrum

Coccyx

responding bones and features in your skeleton as possible.

cervical vertebrae   **(7)**
- atlas (C1)
- axis (C2)
- vertebra prominens (C7)

thoracic vertebrae   **(12)**

lumbar vertebrae   **(5)**

sacrum   **(1)**

coccyx   **(1)**

intervertebral discs—fibrocartilage

vertebral canal—contains spinal cord

cervical curvature

thoracic curvature

lumbar curvature

sacral (pelvic) curvature

intervertebral foramina—passageway for spinal nerves

2. Compare the available samples of cervical, thoracic, and lumbar vertebrae along with figures 15.2 and 15.3. Note differences in size and shapes and locate the following features:

vertebral foramen—location of spinal cord

body—largest part of vertebra; main support portion

pedicles—form lateral area of vertebral foramen

laminae—thin plates forming posterior area of vertebral foramen

spinous process—posterior projection

transverse processes—lateral projections

facets—articulating surfaces

superior articular processes—superior projections

inferior articular processes—inferior projections

inferior vertebral notch—space for nerve passage

transverse foramina—only present on cervical vertebrae; passageway for blood vessels

dens (odontoid process) of axis—superior process of C2 and is a pivot location at atlas

---

FIGURE 15.2 The superior features of (a) the atlas, and the superior and right lateral features of (b) the axis. (The broken arrow indicates a transverse foramen.)

Number code:
1. Superior articular facet
2. Vertebral foramen
3. Dens (odontoid process)
4. Facet that articulates with occipital condyle
5. Transverse process
6. Transverse foramen
7. Spinous process
8. Body

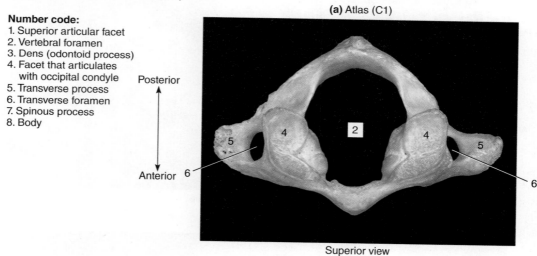

(a) Atlas (C1)

Superior view

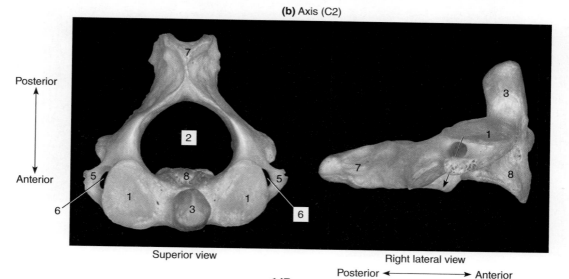

(b) Axis (C2)

Superior view   Right lateral view

147

FIGURE 15.3 The superior and right lateral features of the (a) cervical, (b) thoracic, and (c) lumbar vertebrae. (The broken arrow indicates a transverse foramen.)

Superior views   Right lateral views

Number code:
1. Superior articular process
2. Transverse process
3. Lamina
4. Spinous process
5. Pedicle
6. Body
7. Inferior vertebral notch
8. Vertebral foramen
9. Transverse foramen
10. Costal facets

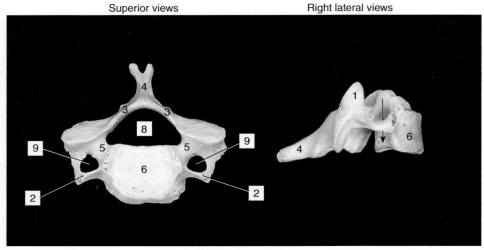

(a) Cervical vertebra

Posterior

Anterior

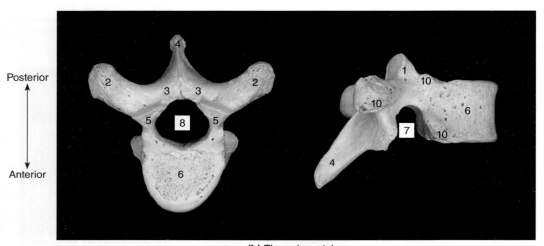

(b) Thoracic vertebra

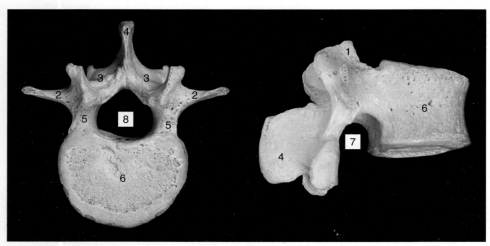

(c) Lumbar vertebra

Posterior ◄────────► Anterior

FIGURE 15.4 The sacrum and coccyx.

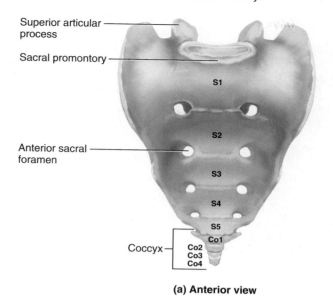

(a) Anterior view

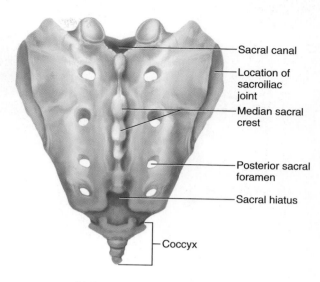

(b) Posterior view

3. Examine the sacrum and coccyx along with figure 15.4. Locate the following features:

   sacrum—formed by five fused vertebrae
   - superior articular process—superior projection with a facet articulation site
   - anterior sacral foramen— passageway for blood vessels and nerves
   - posterior sacral foramen—passageway for blood vessels and nerves
   - sacral promontory—anterior border of S1; important landmark for obstetricians
   - sacral canal—portion of vetrebral canal
   - median sacral crest—area of fused spinous processes
   - sacral hiatus—inferior opening of vertebral canal

   coccyx—formed by three to five fused vertebrae

4. Complete Parts A and B of Laboratory Assessment 15.

Procedure B—Thoracic Cage

The twelve thoracic vertebrae are associated with the twelve pairs of ribs. The superior seven pairs of ribs are connected directly to the sternum with costal cartilage and are known as true ribs. The five inferior pairs, known as false ribs, either connect indirectly to the sternum with costal cartilage or do not connect to the sternum. Pairs eleven and twelve are called the floating ribs because they only connect with the thoracic vertebrae and not with the sternum.

The manubrium, body, and xiphoid process represent three regions that eventually fuse into a single flat bone, the sternum. The heart is located mainly beneath the body portion of the sternum. Any chest compressions adminis-

tered during cardiopulmonary resuscitation should be over the body area of the sternum, not the xiphoid process region. Chest compressions over the xiphoid process region can force the xiphoid process deep into the liver or the inferior portion of the heart and cause a fatal hemorrhage to occur.

1. Examine figures 15.5 and 15.6 and the thoracic cage of the human skeleton. Locate the following bones and features:

   rib
   - head—expanded end near thoracic vertebra
   - tubercle—projection near thoracic vertebra
   - neck—narrow region between head and tubercle
   - shaft—main portion
   - anterior (sternal) end—costal cartilage location
   - facets—articulation surfaces
   - true ribs—pairs 1–7
   - false ribs—pairs 8–12; includes floating ribs
     - floating ribs—pairs 11–12

   costal cartilages—hyaline cartilage
   sternum
   - jugular (suprasternal) notch—superior concave border of manubrium
   - clavicular notch—articulation site of clavicle
   - manubrium—superior part
   - sternal angle—junction of manubrium and body at level of second rib pair
   - body—largest, middle part
   - xiphoid process—inferior part; remains cartilaginous until adulthood

2. Complete Parts C and D of the laboratory assessment.

FIGURE 15.5 Superior view of a typical rib with the articulation sites with a thoracic vertebra.

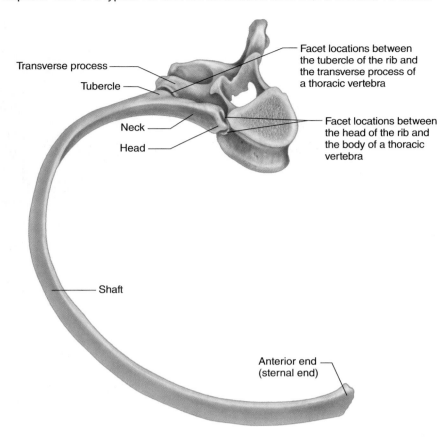

Transverse process

Tubercle

Neck

Head

Facet locations between the tubercle of the rib and the transverse process of a thoracic vertebra

Facet locations between the head of the rib and the body of a thoracic vertebra

Shaft

Anterior end (sternal end)

FIGURE 15.6 Bones and features of the thoracic cage (anterior view).

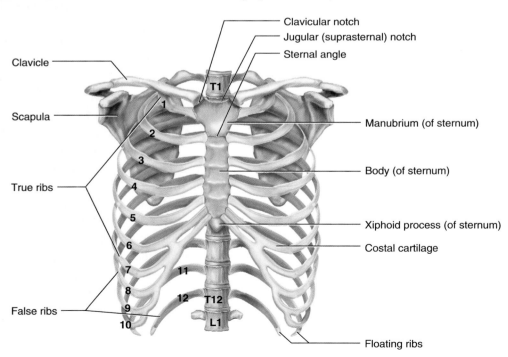

Clavicle

Scapula

True ribs

False ribs

Clavicular notch

Jugular (suprasternal) notch

Sternal angle

T1

Manubrium (of sternum)

Body (of sternum)

Xiphoid process (of sternum)

Costal cartilage

1
2
3
4
5
6
7
8
9
10
11
12

T12

L1

Floating ribs

Name _____

Date _____

Section _____

The ⚠ corresponds to the indicated outcome(s) found at the beginning of the laboratory exercise.

Vertebral Column and Thoracic Cage

Part A Assessments

Complete the following statements:

1. The vertebral column encloses and protects the _____. ⚠

2. The vertebral column extends from the skull to the _____. ⚠

3. The seventh cervical vertebra is called the _____ and has an obvious spinous process surface feature that can be palpated. ⚠

4. The _____ of the vertebrae support the weight of the head and trunk. ⚠

5. The _____ separate adjacent vertebrae, and they soften the forces created by walking. ⚠

6. The intervertebral foramina provide passageways for _____. ⚠

7. Transverse foramina of _____ vertebrae serve as passageways for blood vessels leading to the brain. ⚠

8. The first vertebra also is called the _____. ⚠

9. When the head is moved from side to side, the first vertebra pivots around the _____ of the second vertebra. ⚠

10. The _____ vertebrae have the largest and strongest bodies. ⚠

11. The number of vertebrae that fuse in the adult to form the sacrum is _____. ⚠

Part B Assessments

1. Based on your observations, compare typical cervical, thoracic, and lumbar vertebrae in relation to the characteristics indicated in the table. The table is partly completed. For your responses, consider characteristics such as size, shape, presence or absence, and unique features. ⚠ ⚠

Vertebra	Number	Size	Body	Spinous Process	Transverse Foramina
Cervical	7		smallest	C2 through C6 are forked	
Thoracic		intermediate			
Lumbar					absent

2. Identify the bones and features in figures 15.7 and 15.8.

FIGURE 15.7 Label the bones and features of a lateral view of a vertebral column by placing the correct numbers in the spaces provided. 🛆 🛆

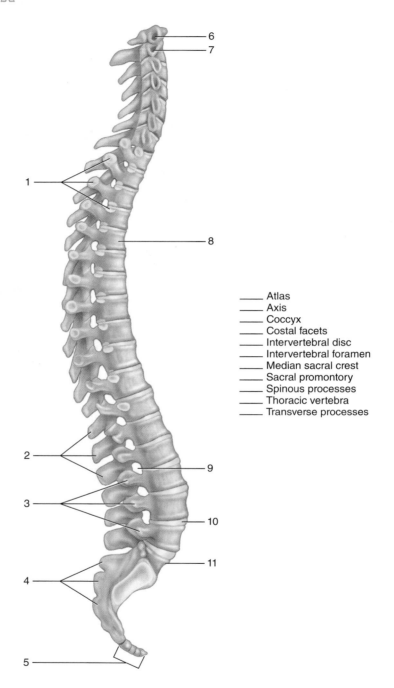

_____ Atlas
_____ Axis
_____ Coccyx
_____ Costal facets
_____ Intervertebral disc
_____ Intervertebral foramen
_____ Median sacral crest
_____ Sacral promontory
_____ Spinous processes
_____ Thoracic vertebra
_____ Transverse processes

FIGURE 15.8 Identify the bones and features indicated in this radiograph of the neck (lateral view), using the terms provided. ⚠1 ⚠2

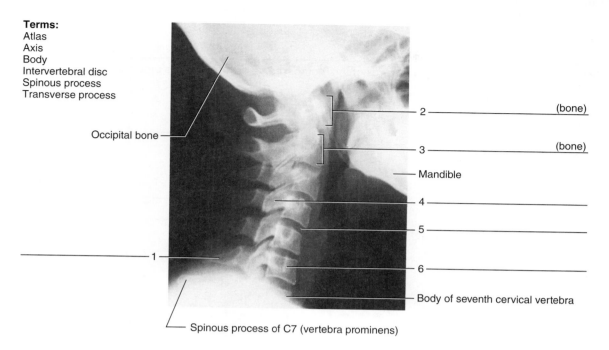

Terms:
Atlas
Axis
Body
Intervertebral disc
Spinous process
Transverse process

Occipital bone

2 _____ (bone)

3 _____ (bone)

Mandible

4 _____

5 _____

6 _____

Body of seventh cervical vertebra

1 _____

Spinous process of C7 (vertebra prominens)

Part C   Assessments

Complete the following statements:

1. The manubrium, body, and xiphoid process form a bone called the _____. ⚠4

2. The last two pairs of ribs that have no cartilaginous attachments to the sternum are sometimes called _____ ribs. ⚠5

3. There are _____ pairs of true ribs. ⚠5

4. Costal cartilages are composed of _____ tissue. ⚠4

5. The manubrium articulates with the _____ on its superior border. ⚠4

6. List three general functions of the thoracic cage. ⚠4 _____

_____

7. The sternal angle indicates the location of the _____ pair of ribs. ⚠4

Part D   Assessments

Identify the bones and features indicated in figure 15.9.

FIGURE 15.9   Label the bones and features of the thoracic cage, using the terms provided. **A** **5**

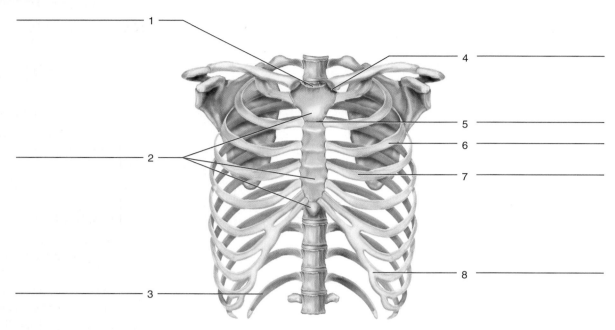

1

2

3

4

5

6

7

8

Terms:
Costal cartilage of false rib
Costal cartilage of true rib
Clavicular notch
Floating rib
Jugular notch
Sternal angle
Sternum
True rib

Pectoral Girdle and Upper Limb

Purpose of the Exercise

To examine the bones of the pectoral girdle and upper limb and to identify the major features of these bones.

Materials Needed

Human skeleton, articulated
Human skeleton, disarticulated

Learning Outcomes

After completing this exercise, you should be able to

(1) Locate and identify the bones of the pectoral girdle and their major features.

(2) Locate and identify the bones of the upper limb and their major features.

Pre-Lab

Carefully read the introductory material and examine the entire lab. Be familiar with the pectoral girdle and upper limb bones from lecture or the textbook. Answer the pre-lab questions.

Pre-Lab Questions: Select the correct answer for each of the following questions:

1. The clavicle and the scapula form the
   a. pectoral girdle.   **b.** pelvic girdle.
   c. upper limb.   **d.** axial skeleton.

2. Anatomically, *arm* represents
   a. shoulder to fingers.   **b.** shoulder to wrist.
   c. elbow to wrist.   **d.** shoulder to elbow.

3. Which of the following is *not* part of the scapula?
   a. spine   **b.** manubrium
   c. acromion   **d.** supraspinous fossa

4. Which carpal is included in the proximal row?
   a. hamate   **b.** capitate
   c. trapezium   **d.** lunate

5. Which of the following is the most proximal part of the upper limb?
   a. styloid process of radius
   b. styloid process of ulna
   c. head of humerus
   d. medial epicondyle of humerus

6. Which of the following is the most distal feature of the humerus?
   a. anatomical neck   **b.** deltoid tuberosity
   c. capitulum   **d.** head

7. The capitate is one of the eight carpals in a wrist.
   True _____ False _____

8. The clavicle articulates with the sternum and the humerus.
   True _____ False _____

A **pectoral girdle (shoulder girdle)** consists of an anterior clavicle and a posterior scapula. A *pectoral girdle* represents an incomplete ring (girdle) of bones as the posterior scapulae do not meet each other, but muscles extend from their medial borders to the vertebral column. The clavicles on their medial ends form a joint with the manubrium of the sternum. The pectoral girdle supports the upper limb and serves as attachments for various muscles that move the upper limb. This allows considerable flexibility of the shoulder. Relatively loose attachments of the pectoral girdle with the humerus allow a wide range of movements, but shoulder joint injuries are somewhat common. Additionally, the clavicle is a frequently broken bone when one reaches with an upper limb to break a fall.

Each upper limb includes a humerus in the arm, a radius and ulna in the forearm, and eight carpals, five metacarpals, and fourteen phalanges in the hand. (Anatomically, *arm* represents the region from shoulder to elbow, *forearm* is elbow to the wrist, and *hand* includes the wrist to the ends of digits.) These bones form the framework of the upper limb. They also function as parts of levers when muscles contract.

Procedure A—Pectoral Girdle

The clavicle and scapula of the pectoral girdle provide for attachments of neck and trunk muscles. The clavicle is not a straight bone, but rather has two curves making it slightly S-shaped. The clavicle serves as a brace bone to keep the upper limb to the side of the body. The bone is easily fractured because it is so close to the anterior surface and because the shoulder or upper limb commonly breaks a fall. Fortunately bones have tensile strength, and when a force is exerted upon the clavicle from the lateral side of the body, the bone will bend to some extent at both curves until the threshold is reached, causing a fracture to occur.

The scapula has many tendon attachment sites for muscles of the neck, trunk, and upper limb. Because the scapula does not connect directly to the axial skeleton and because it has so many muscle attachments, there is a great deal of flexibility for shoulder movements. Unfortunately, this much flexibility can result in dislocating the humerus from the scapula. This type of injury can occur during gymnastics maneuvers or when suddenly pulling or swinging a child by an upper limb.

1. Examine figures 16.1 and 16.2 along with the bones of the pectoral girdle. Locate the following features of the clavicle and the scapula. At the same time, locate as many of the corresponding surface bones and features of your own skeleton as possible.

   clavicle
   - sternal (medial) end—articulates with the manubrium of the sternum
   - acromial (lateral) end—articulates with the acromion of the scapula

---

FIGURE 16.1 Bones and features of the right shoulder and upper limb (anterior view).

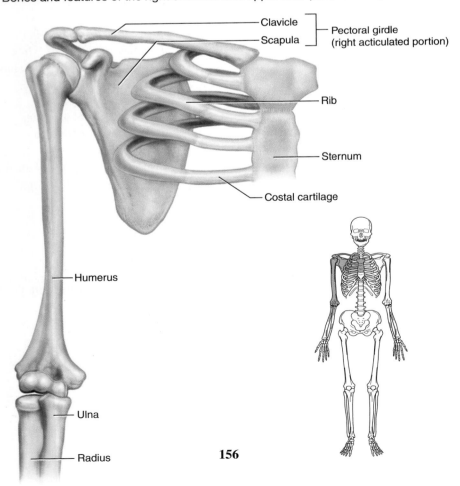

156

scapula
- spine
- acromion—lateral end of spine
- glenoid cavity—shallow socket; articulates with head of humerus
- coracoid process—beaklike projection
- borders
    - superior border
    - medial (vertebral) border
    - lateral (axillary) border
- fossae—shallow depressions
    - supraspinous fossa
    - infraspinous fossa
    - subscapular fossa
- angles
    - superior angle
    - inferior angle

- tubercles
    - supraglenoid tubercle
    - infraglenoid tubercle

2. Complete Parts A and B of Laboratory Assessment 16.

Critical Thinking Activity

Why is the clavicle a bone that can easily fracture?

_____
_____
_____
_____
_____
_____

FIGURE 16.2 Right scapula (a) posterior surface (b) lateral view (c) anterior surface.

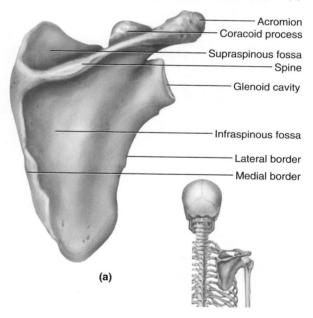

Acromion
Coracoid process
Supraspinous fossa
Spine
Glenoid cavity
Infraspinous fossa
Lateral border
Medial border

(a)

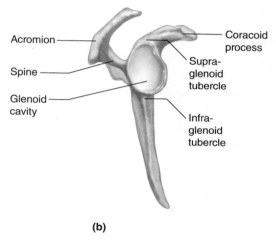

Acromion
Spine
Glenoid cavity
Coracoid process
Supra-glenoid tubercle
Infra-glenoid tubercle

(b)

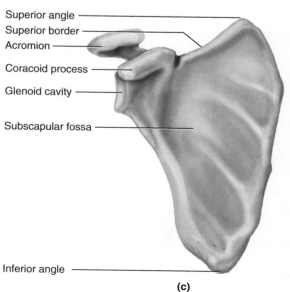

Superior angle
Superior border
Acromion
Coracoid process
Glenoid cavity
Subscapular fossa
Inferior angle

(c)

Procedure B—Upper Limb

The humerus has sometimes been called the "funny bone" because of a tingling sensation (temporary pain) if it is bumped on the medial epicondyle where the ulnar nerve passes. The two bones of the forearm are nearly parallel when in anatomical position, with the radius positioned on the lateral side and the ulna on the medial side. All of the bones of the hand, including those of the wrist (carpals), palm (metacarpals), and digits (phalanges), are visible from an anterior view.

1. Examine figures 16.3, 16.4, and 16.5. Locate the following bones and features of the upper limb:

   humerus
   - proximal features
     - head—articulates with glenoid cavity
     - greater tubercle—on lateral side
     - lesser tubercle—on anterior side
     - anatomical neck—tapered region near head
     - surgical neck—common fracture site
     - intertubercular sulcus—furrow for tendon of biceps muscle
   - shaft
     - deltoid tuberosity
   - distal features
     - capitulum—lateral condyle; articulates with radius
     - trochlea—medial condyle; articulates with ulna
     - medial epicondyle
     - lateral epicondyle
     - coronoid fossa—articulates with coronoid process of ulna
     - olecranon fossa—articulates with olecranon process of ulna

FIGURE 16.3 Right humerus (a) anterior features and (b) posterior features.

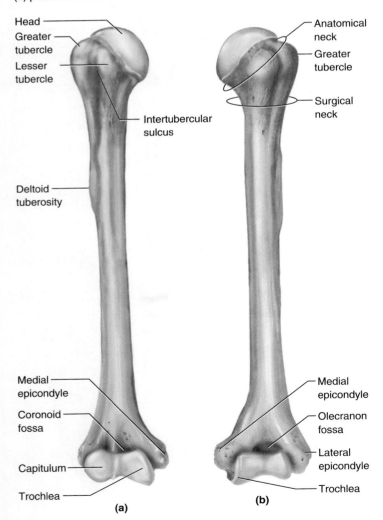

Head
Greater tubercle
Lesser tubercle
Intertubercular sulcus
Deltoid tuberosity
Medial epicondyle
Coronoid fossa
Capitulum
Trochlea
(a)

Anatomical neck
Greater tubercle
Surgical neck
Medial epicondyle
Olecranon fossa
Lateral epicondyle
Trochlea
(b)

FIGURE 16.4 Anterior features of the right radius and ulna.

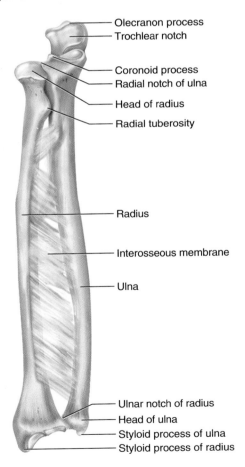

Olecranon process
Trochlear notch
Coronoid process
Radial notch of ulna
Head of radius
Radial tuberosity
Radius
Interosseous membrane
Ulna
Ulnar notch of radius
Head of ulna
Styloid process of ulna
Styloid process of radius

FIGURE 16.5 Anterior (palmar) view of the bones of the right hand. The proximal row of carpals are colored yellow and the distal row of carpals are colored green.

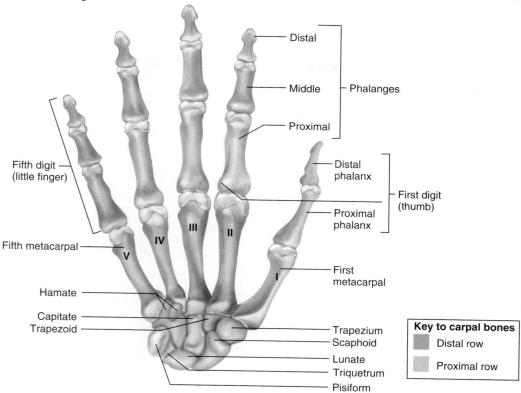

radius—lateral bone of forearm
- head of radius—allows rotation at elbow
- radial tuberosity
- styloid process of radius
- ulnar notch of radius—articulation site with ulna

ulna—medial bone of forearm; longer than radius
- trochlear notch
- radial notch of ulna—articulation site with head of radius
- olecranon process
- coronoid process
- styloid process of ulna
- head of ulna—at distal end

carpal bones—positioned in two irregular rows
- proximal row (listed lateral to medial)
  - scaphoid
  - lunate
  - triquetrum
  - pisiform
- distal row (listed medial to lateral)
  - hamate
  - capitate
  - trapezoid
  - trapezium

The following mnemonic device will help you learn the eight carpals:

So **L**ong **T**op **P**art
Here **C**omes **T**he **T**humb

The first letter of each word corresponds to the first letter of a carpal. This device arranges the carpals in order for the proximal, transverse row of four bones from lateral to medial, followed by the distal, transverse row from medial to lateral, which ends nearest the thumb. This arrangement assumes the hand is in the anatomical position.

metacarpals (I–V)

phalanges—located in digits (fingers)
- proximal phalanx
- middle phalanx—not present in first digit
- distal phalanx

2. Complete Parts C, D, and E of the laboratory assessment.

NOTES

Name _____

Date _____

Section _____

The 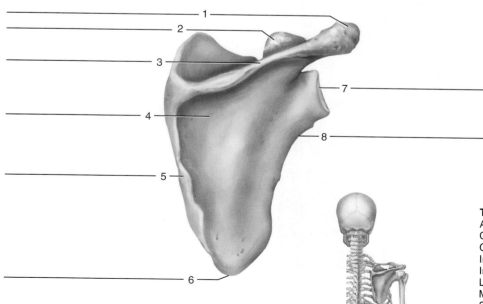 corresponds to the indicated outcome(s) found at the beginning of the
laboratory exercise.

Pectoral Girdle and Upper Limb

Part A Assessments

Complete the following statements:

1. The pectoral girdle is an incomplete ring because it is open in the back between the _____. ⚠

2. The medial end of a clavicle articulates with the _____ of the sternum. ⚠

3. The lateral end of a clavicle articulates with the _____ process of the scapula. ⚠

4. The _____ is a bone that serves as a brace between the sternum and the scapula. ⚠

5. The _____ divides the scapula into unequal portions. ⚠

6. The lateral tip of the shoulder is the _____ of the scapula. ⚠

7. Near the lateral end of the scapula, the _____ process of the scapula curves anteriorly and inferiorly from the clavicle. ⚠

8. The glenoid cavity of the scapula articulates with the _____ of the humerus. ⚠

Part B Assessments

Label the structures indicated in figure 16.6.

FIGURE 16.6   Label the posterior surface of the right scapula, using the terms provided. ⚠

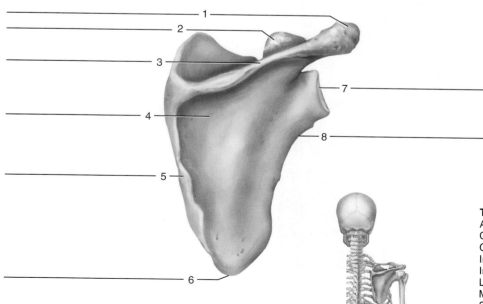

Terms:
Acromion
Coracoid process
Glenoid cavity
Inferior angle
Infraspinous fossa
Lateral border
Medial border
Spine

161

Part C Assessments

Match the bones in column A with the bones and features in column B. Place the letter of your choice in the space provided. 2

Column A

a. Carpals
b. Humerus
c. Metacarpals
d. Phalanges
e. Radius
f. Ulna

Column B

_____ 1. Capitate

_____ 2. Coronoid fossa

_____ 3. Deltoid tuberosity

_____ 4. Greater tubercle

_____ 5. Five palmar bones

_____ 6. Fourteen bones in digits

_____ 7. Intertubercular sulcus

_____ 8. Lunate

_____ 9. Olecranon fossa

_____ 10. Radial tuberosity

_____ 11. Trapezium

_____ 12. Trochlear notch

Part D Assessments

Identify the bones and features indicated in the radiographs of figures 16.7, 16.8, and 16.9.

FIGURE 16.7 Identify the bones and features indicated on this radiograph of the right elbow (anterior view), using the terms provided. 2

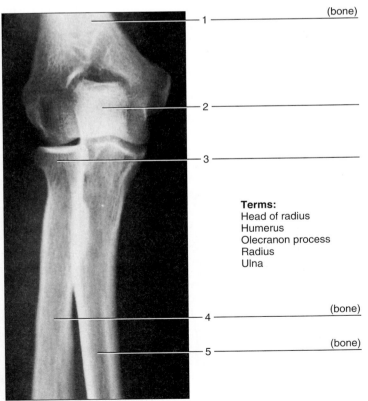

1 ——————————— (bone)

2 ———————————

3 ———————————

Terms:
Head of radius
Humerus
Olecranon process
Radius
Ulna

4 ——————————— (bone)

5 ——————————— (bone)

FIGURE 16.8 Identify the bones and features indicated on this radiograph of the anterior view of the right shoulder, using the terms provided. △ △

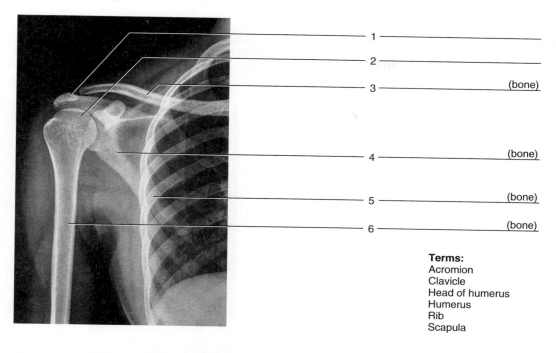

1 _____

2 _____

3 _____ (bone)

4 _____ (bone)

5 _____ (bone)

6 _____ (bone)

Terms:
Acromion
Clavicle
Head of humerus
Humerus
Rib
Scapula

FIGURE 16.9 Identify the bones indicated on this radiograph of the right hand (anterior view), using the terms provided. △

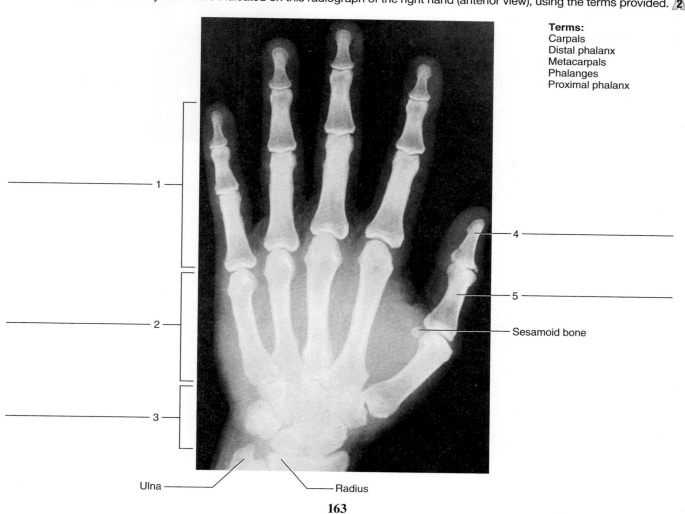

Terms:
Carpals
Distal phalanx
Metacarpals
Phalanges
Proximal phalanx

1 _____

2 _____

3 _____

4 _____

5 _____

Sesamoid bone

Ulna _____

Radius

Part E   Assessments

Identify the features of a humerus in figure 16.10 and the bones of the hand in figure 16.11.

FIGURE 16.10   Label the anterior features of a right humerus. **2**

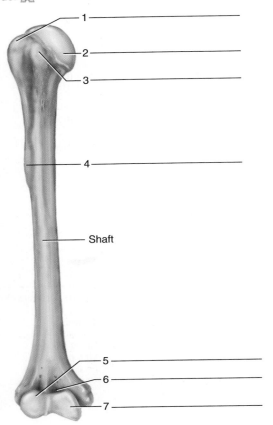

Shaft

FIGURE 16.11   Complete the labeling of the bones numbered on this anterior view of the right hand by placing the correct numbers in the spaces provided. **2**

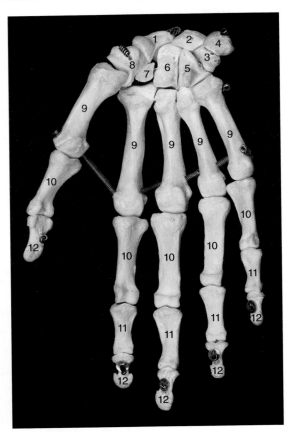

	Capitate	4	Pisiform
	Distal phalanges		Proximal phalanges
	Hamate		Scaphoid
	Lunate		Trapezium
	Metacarpals		Trapezoid
	Middle phalanges	3	Triquetrum

Pelvic Girdle and Lower Limb

Purpose of the Exercise

To examine the bones of the pelvic girdle and lower limb, and to identify the major features of these bones.

Materials Needed

Human skeleton, articulated
Human skeleton, disarticulated
Male and female pelves

Learning Outcomes

After completing this exercise, you should be able to

1. Locate and identify the bones of the pelvic girdle and their major features.
2. Differentiate a male and female pelvis.
3. Locate and identify the bones of the lower limb and their major features.

Pre-Lab

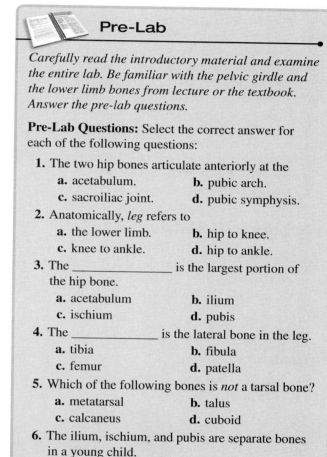

Carefully read the introductory material and examine the entire lab. Be familiar with the pelvic girdle and the lower limb bones from lecture or the textbook. Answer the pre-lab questions.

Pre-Lab Questions: Select the correct answer for each of the following questions:

1. The two hip bones articulate anteriorly at the
   a. acetabulum.   **b.** pubic arch.
   c. sacroiliac joint.   **d.** pubic symphysis.

2. Anatomically, *leg* refers to
   a. the lower limb.   **b.** hip to knee.
   c. knee to ankle.   **d.** hip to ankle.

3. The _____ is the largest portion of the hip bone.
   a. acetabulum   **b.** ilium
   c. ischium   **d.** pubis

4. The _____ is the lateral bone in the leg.
   a. tibia   **b.** fibula
   c. femur   **d.** patella

5. Which of the following bones is *not* a tarsal bone?
   a. metatarsal   **b.** talus
   c. calcaneus   **d.** cuboid

6. The ilium, ischium, and pubis are separate bones in a young child.
   True _____   False _____

7. Ischial spines, ischial tuberosities, and iliac crests are closer together in a pelvis of a female than in a pelvis of a male.
   True _____   False _____

8. Each digit of a foot has three phalanges.
   True _____   False _____

The pelvic girdle includes two hip bones that articulate with each other anteriorly at the pubic symphysis. Posteriorly, each hip bone articulates with a sacrum at a sacroiliac joint. Together, the pelvic girdle, sacrum, and coccyx comprise the pelvis. The pelvis, in turn, provides support for the trunk of the body and provides attachments for the lower limbs. The pelvis supports and protects the viscera in the pelvic region of the abdominopelvic cavity. The pelvic outlet, with boundaries of the coccyx, inferior border of the pubic symphysis, and between the ischial tuberosities, is clinically important in females. The pelvic outlet must be large enough to successfully accommodate the fetal head during a vaginal delivery. Each acetabulum of a hip bone articulates with the head of the femur of a lower limb. The hip joint structures provide a more stable joint compared to a shoulder joint.

The bones of the lower limb form the framework of the thigh, leg, and foot. (Anatomically, *thigh* represents the region from hip to knee, *leg* is from knee to ankle, and *foot* includes the ankle to the end of the toes.) Each limb includes a femur in the thigh, a patella in the knee, a tibia and fibula in the leg, and seven tarsals, five metatarsals, and fourteen phalanges in the foot. These bones and large muscles are for weight-bearing support and locomotion and thus are considerably larger and possess more stable joints than those of an upper limb.

Procedure A—Pelvic Girdle

The pelvic girdle supports the majority of the weight of the head, neck, and trunk. Each hip bone (coxal bone) originates in three separate ossification areas known as the ilium, ischium, and pubis. These are separate bones of a child but fuse into an individual hip bone. All three parts of the hip bone fuse within the acetabulum, and the ischium and pubis also fuse along the inferior portion of the obturator foramen. The acetabulum is a well-formed deep socket for the head of the femur. The obturator foramen is the largest foramen in the skeleton; it serves as a passageway for blood vessels and nerves between the pelvic cavity and the thigh.

Several of the bone features (bone markings) of the pelvis can be palpated because they are located near the surface of the body. The medial sacral crest is near the middle of the posterior surface of the pelvis, somewhat superior to the coccyx. The ilium is the largest portion of the hip bone, and its iliac crest can be palpated along the anterior and lateral portions. By following the anterior portion of the iliac crest, the anterior superior iliac spine can be felt very close to the surface.

1. Examine figures 17.1 and 17.2.
2. Observe the bones of the pelvic girdle and locate the following:

hip bone (coxal bone; pelvic bone; innominate bone)
- ilium
  - iliac crest
  - anterior superior iliac spine
  - anterior inferior iliac spine
  - posterior superior iliac spine
  - posterior inferior iliac spine
  - greater sciatic notch—portion in ischium
  - iliac fossa

FIGURE 17.1 Bones of the pelvis (anterosuperior view).

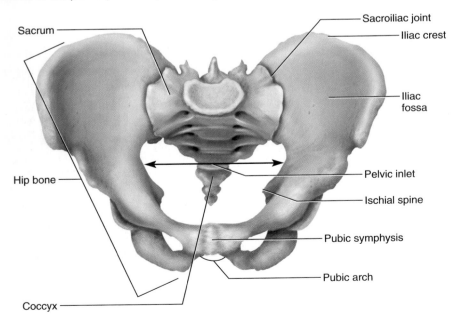

FIGURE 17.2 Right hip bone (a) lateral view (b) medial view. The three colors used enable the viewing of fusion locations of the ilium, ischium, and pubis of the adult skeleton.

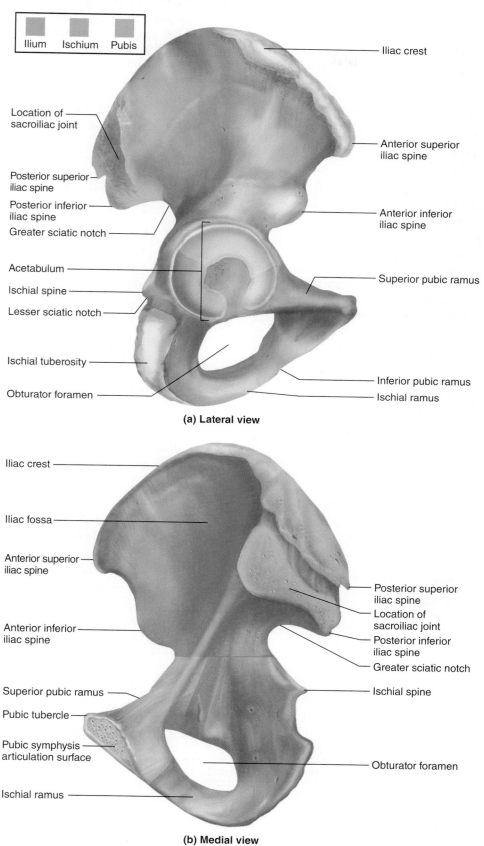

| Ilium | Ischium | Pubis |

(a) Lateral view

- Iliac crest
- Location of sacroiliac joint
- Anterior superior iliac spine
- Posterior superior iliac spine
- Posterior inferior iliac spine
- Greater sciatic notch
- Anterior inferior iliac spine
- Acetabulum
- Ischial spine
- Superior pubic ramus
- Lesser sciatic notch
- Ischial tuberosity
- Obturator foramen
- Inferior pubic ramus
- Ischial ramus

(b) Medial view

- Iliac crest
- Iliac fossa
- Anterior superior iliac spine
- Posterior superior iliac spine
- Location of sacroiliac joint
- Anterior inferior iliac spine
- Posterior inferior iliac spine
- Greater sciatic notch
- Superior pubic ramus
- Ischial spine
- Pubic tubercle
- Pubic symphysis articulation surface
- Obturator foramen
- Ischial ramus

- ischium
    - ischial tuberosity—supports weight of body when seated
    - ischial spine
    - ischial ramus
    - lesser sciatic notch
- pubis
    - pubic symphysis—cartilaginous joint between pubic bones
    - pubic tubercle
    - superior pubic ramus
    - inferior pubic ramus
    - pubic arch—between pubic bones of pelvis
- acetabulum—formed by portions of ilium, ischium, and pubis
- obturator foramen—formed by portions of ischium and pubis

3. Observe the male pelvis and the female pelvis. Use table 17.1 as a guide as comparisons are made between the pelves of males and females.
4. Complete Part A of Laboratory Assessment 17.

Procedure B—Lower Limb

The femur of the lower limb is the longest and strongest bone of the skeleton. The neck of the femur has somewhat of a lateral angle and is the weakest part of the bone and a common fracture site, especially if the person has some degree of osteoporosis. This fracture site is usually called a broken hip; however it is actually a broken femur in the region of the hip joint. Often this fracture is attributed to a fall causing the fracture, when many times the fracture occurs first, followed by the fall. The greater trochanter can be palpated along the proximal and lateral region of the thigh about one hand length below the iliac crest. The medial and lateral epicondyles can be palpated near the knee.

Large muscles are positioned on the anterior and posterior surfaces as well as the lateral and medial surfaces of the femur and the hip joint region. The thigh can be pulled in any direction depending upon which muscle group is contracting at that time. The large anterior muscles of the thigh (quadriceps femoris) possess a common patellar tendon with an enclosed sesamoid bone, the patella. The muscle attachment continues distally to the tibial tuberosity.

1. Examine figures 17.3, 17.4, and 17.5.
2. Observe the bones of the lower limb and locate each of the following:

femur
- proximal features
    - head
    - fovea capitis
    - neck
    - greater trochanter
    - lesser trochanter
- shaft
    - gluteal tuberosity
    - linea aspera
- distal features
    - lateral epicondyle
    - medial epicondyle
    - lateral condyle
    - medial condyle

patella
tibia
- medial condyle
- lateral condyle
- tibial tuberosity
- anterior border (crest; margin)
- medial malleolus

TABLE 17.1 Differences Between Male and Female Pelves

Structure of Comparison	Male Pelvis	Female Pelvis
General structure	Heavier thicker bones and processes	Lighter thinner bones and processes
Sacrum	Narrower and longer	Wider and shorter
Coccyx	Less movable	More movable
Pelvic outlet	Smaller	Larger
Greater sciatic notch	Narrower	Wider
Obturator foramen	Round	Triangular to oval
Acetabula	Larger; closer together	Smaller; farther apart
Pubic arch	Usually 90° or less; more V-shaped	Usually greater than 90°
Ischial spines	Longer; closer together	Shorter; farther apart
Ischial tuberosities	Rougher; closer together	Smoother; farther apart
Iliac crests	Less flared; closer together	More flared; farther apart

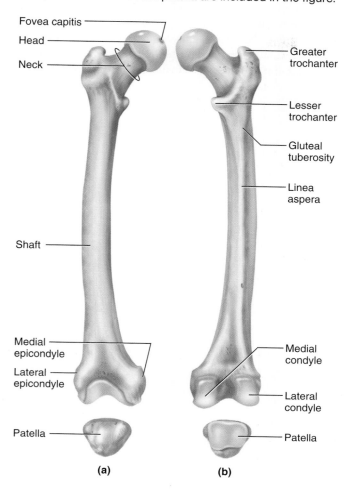

FIGURE 17.3 The features of (a) the anterior surface and (b) the posterior surface of the right femur. The anterior and posterior views of the patella are included in the figure.

Fovea capitis
Head
Neck
Greater trochanter
Lesser trochanter
Gluteal tuberosity
Linea aspera
Shaft
Medial epicondyle
Lateral epicondyle
Medial condyle
Lateral condyle
Patella
Patella

(a)   (b)

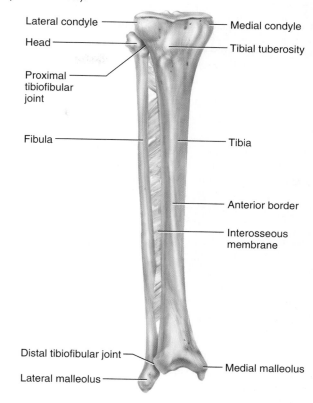

FIGURE 17.4 Features of the right tibia and fibula (anterior view).

Lateral condyle
Head
Proximal tibiofibular joint
Fibula
Medial condyle
Tibial tuberosity
Tibia
Anterior border
Interosseous membrane
Distal tibiofibular joint
Lateral malleolus
Medial malleolus

fibula
- head
- lateral malleolus

tarsal bones
- talus
- calcaneus
- navicular
- cuboid
- lateral cuneiform
- intermediate (middle) cuneiform
- medial cuneiform

metatarsal bones

phalanges
- proximal phalanx
- middle phalanx—absent in first digit
- distal phalanx

3. Complete Parts B, C, and D of the laboratory assessment.

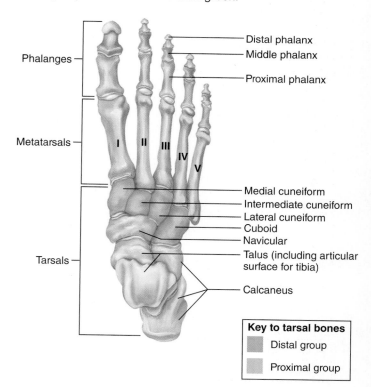

FIGURE 17.5 Superior surface view of right foot. The proximal group of tarsals are colored yellow and the distal group of tarsals are colored green.

Phalanges
Distal phalanx
Middle phalanx
Proximal phalanx
Metatarsals
I   II   III   IV   V
Medial cuneiform
Intermediate cuneiform
Lateral cuneiform
Cuboid
Navicular
Talus (including articular surface for tibia)
Tarsals
Calcaneus

Key to tarsal bones
Distal group
Proximal group

Compare the size and depth of an acetabulum of a hip bone to the glenoid cavity of the scapula._____

_____

_____

_____

_____

How do these socket differences between the hip joint and the shoulder joint relate to strength and mobility?

_____

_____

_____

_____

_____

Critical Thinking Activity

As a review of the entire skeleton, use the disarticulated skeleton and arrange all of the bones in relative position to re-create their normal positions. Work with a partner or in a small group and place the bones on the surface of a laboratory table. Check your results with the articulated skeletons.

Name _____

Date _____

Section _____

The ⚠️ corresponds to the indicated outcome(s) found at the beginning of the laboratory exercise.

Pelvic Girdle and Lower Limb

Part A   Assessments

Complete the following statements:

1. The pelvic girdle consists of two _____. ⚠️

2. The head of the femur articulates with the _____ of the hip bone. ⚠️

3. The _____ is the largest portion of the hip bone. ⚠️

4. The distance between the _____ represents the shortest diameter of the pelvic outlet. ⚠️

5. The pubic bones come together anteriorly to form a cartilaginous joint called the _____. ⚠️

6. The _____ is the superior margin of the ilium that causes the prominence of the hip. ⚠️

7. When a person sits, the _____ of the ischium supports the weight of the body. ⚠️

8. The angle formed by the pubic bones below the pubic symphysis is called the _____. ⚠️

9. The _____ is the largest foramen in the skeleton. ⚠️

10. The ilium joins the sacrum at the _____ joint. ⚠️

Critical Thinking Assessment

Examine the male and female pelves. Look for major differences between them. Note especially the flare of the iliac bones, the angle of the pubic arch, the distance between the ischial spines and ischial tuberosities, and the curve and width of the sacrum. In what ways are the differences you observed related to the function of the female pelvis as a birth canal? ⚠️

_____

_____

_____

Part B   Assessments

Match the bones in column A with the features in column B. Place the letter of your choice in the space provided. ⚠️

Column A	Column B	
a. Femur	_____ 1. Middle phalanx	_____ 7. Tibial tuberosity
b. Fibula	_____ 2. Lesser trochanter	_____ 8. Talus
c. Metatarsals	_____ 3. Medial malleolus	_____ 9. Linea aspera
d. Patella	_____ 4. Fovea capitis	_____ 10. Lateral malleolus
e. Phalanges	_____ 5. Calcaneus	_____ 11. Sesamoid bone
f. Tarsals	_____ 6. Lateral cuneiform	_____ 12. Five bones that form the instep
g. Tibia		

Part C Assessments

Identify the bones and features indicated in the radiographs of figures 17.6, 17.7, and 17.8.

FIGURE 17.6 Identify the bones and features indicated on this radiograph of the anterior view of the pelvic region, using the terms provided. △1 △3

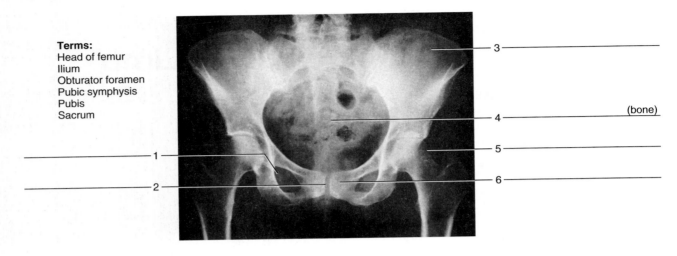

Terms:
Head of femur
Ilium
Obturator foramen
Pubic symphysis
Pubis
Sacrum

FIGURE 17.7 Identify the bones and features indicated in this radiograph of the right knee (anterior view), using the terms provided. △3

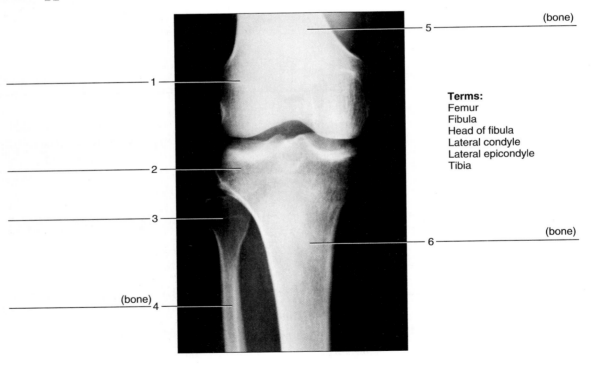

Terms:
Femur
Fibula
Head of fibula
Lateral condyle
Lateral epicondyle
Tibia

FIGURE 17.8 Identify the bones indicated in this radiograph of the right foot (medial side), using the terms provided. 🔺3

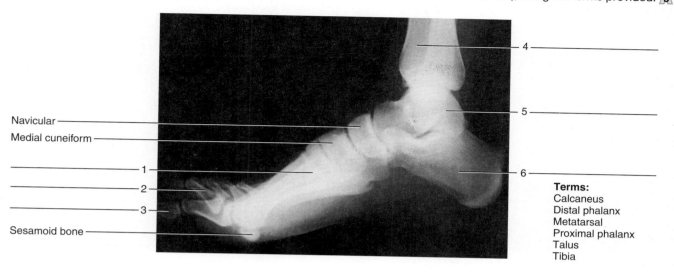

Navicular

Medial cuneiform

1

2

3

Sesamoid bone

4

5

6

Terms:
Calcaneus
Distal phalanx
Metatarsal
Proximal phalanx
Talus
Tibia

Part D   Assessments

Identify the bones of the foot in figure 17.9 and the features of a femur in figure 17.10.

FIGURE 17.9 Identify the bones indicated on this superior view of the right foot, using the terms provided. 🔺3

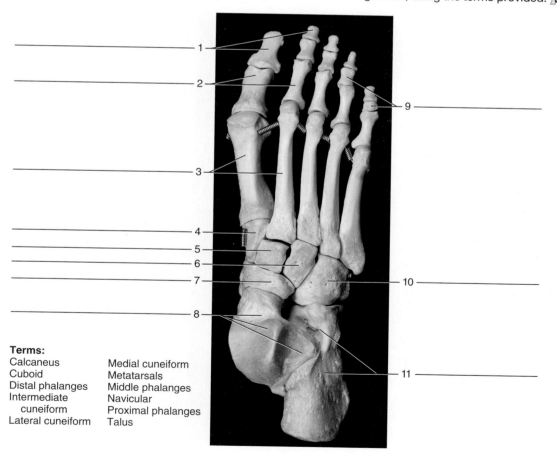

1

2

3

4

5

6

7

8

9

10

11

Terms:
Calcaneus
Cuboid
Distal phalanges
Intermediate
  cuneiform
Lateral cuneiform

Medial cuneiform
Metatarsals
Middle phalanges
Navicular
Proximal phalanges
Talus

FIGURE 17.10   Label the anterior features of a right femur. 3

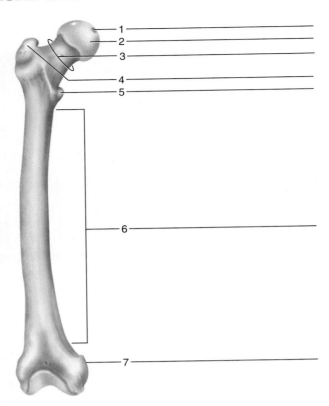

1 _____

2 _____

3 _____

4 _____

5 _____

6 _____

7 _____

Fetal Skeleton

Purpose of the Exercise

To examine and make comparisons between fetal skeletons and adult skeletons.

Materials Needed

Human skull
Human adult skeleton
Human fetal skeleton
Meterstick
Metric ruler
Note: If a fetal skeleton is not available, use the figures represented in this laboratory exercise with the metersticks included in the photographs.

Learning Outcomes

After completing this exercise, you should be able to

1. Locate and describe the function of the fontanels of a fetal skull.

2. Measure and compare the size of the cranial bones and the facial bones between a fetal skull and the adult skull.

3. Compare the fetal frontal and temporal bones to adult examples.

4. Measure and compare the total height and the upper and lower limb lengths for a fetal skeleton and an adult skeleton.

5. Locate and describe six fetal bones or body regions that are not completely ossified but become single bones or groups of bones in the adult skeleton.

6. Estimate the gestational age of a fetus from measured lengths.

7. Distinguish intramembranous ossification bones from endochondral ossification bones.

 Pre-Lab

Carefully read the introductory material and examine the entire lab. Be familiar with bone development and a fetal skull from lecture or the textbook. Visit www.mhhe.com/martinseries2 for LabCam videos. Answer the pre-lab questions.

Pre-Lab Questions: Select the correct answer for each of the following questions:

1. Development of the embryo and fetus progresses from
   a. medial toward lateral.
   b. cephalic toward caudal.
   c. anterior toward posterior.
   d. inferior toward superior.

2. The last fontanel to close is the
   a. posterior.   b. sphenoidal.
   c. mastoid.   d. anterior.

3. Fontanels function for all of the following *except*
   a. formation of the vertebrae.
   b. compression of the cranium during vaginal delivery.
   c. additional cranial skull growth after birth.
   d. additional brain development after birth.

4. Measurements of height after 20 weeks of development are from
   a. crown-to-heel.   b. crown-to-rump.
   c. head-to-toes.   d. sitting height.

5. The anterior fontanel involves all of the following bones *except* the
   a. left parietal bone.   b. right parietal bone.
   c. occipital bone.   d. frontal bone.

6. The occipital bone forms from intramembranous ossification.
   True _____ False _____

7. The upper and lower limb bones develop from intramembranous ossification.
   True _____ False _____

8. The skeleton of a newborn baby consists of fewer bones than 206.
   True _____ False _____

A human adult skeleton consists of 206 bones. However, the skeleton of a newborn child may have nearly 275 bones due to the numerous ossification centers and epiphyseal plates present for many bones. Development of the embryo and the fetus progresses from cephalic toward caudal regions of the body. As a result, there is a noticeable difference between the proportions of the head, the entire body, and the upper and lower limbs between a fetus and an adult. Additionally, the bones most important for protection are among the early ossification sites.

The ossification process occurs during fetal development and continues through childhood growth. Intramembranous ossification forms the flat bones of the cranium, some facial bones, and portions of the mandible and clavicles. Endochondral ossification from hyaline cartilage forms most of the bones of the skeleton.

Procedure A—Skulls

1. Examine the fetal skull from three views (figures 18.1, 18.2, and 18.3). Fontanels (sometimes called "soft spots") are fibrous membranes that have not completed ossification. Fontanels allow for compression of the cranium during a vaginal delivery and for some additional brain development after birth. The posterior and lateral fontanels close during the first year after birth, but the anterior fontanel does not close until nearly two years of age. Locate the following fontanels:

Mastoid fontanel	**(2)**
Sphenoidal fontanel	**(2)**
Posterior fontanel	**(1)**
Anterior fontanel	**(1)**

2. Complete Part A of Laboratory Assessment 18.
3. Examine the anterior views of a fetal skull next to an adult skull (fig. 18.4). Use a meterstick and measure the height of the entire cranium (braincase) and the height of the face of the fetal skull (fig. 18.4*a*). The brain, which forms in early development, occupies the large cranial cavity. This gives a perspective of a large head. However, the small face is related to undersized maxillae, maxillary sinuses, mandible, lack of teeth eruptions, and nasal cavities. The facial portion of a fetal skull at birth represents about one-eighth of the entire skull, whereas the adult facial portion represents nearly one-half of the complete skull.
4. As growth of the fetus advances, the relative size of the cranium becomes smaller as the size of the face increases. Use a meterstick and measure the height of the cranium and face of an adult skull (fig. 18.4*b*).
5. Record your measurements in Part B of the laboratory assessment.
6. Make careful observations of the ossification of the frontal bone and the temporal bone.
7. Complete Part B of the laboratory assessment.

FIGURE 18.1 Superior view of the fetal skull.

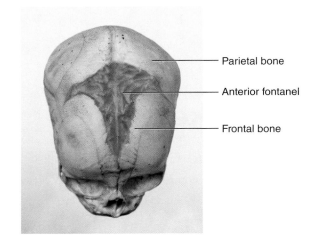

Parietal bone

Anterior fontanel

Frontal bone

FIGURE 18.2 Lateral view of the fetal skull.

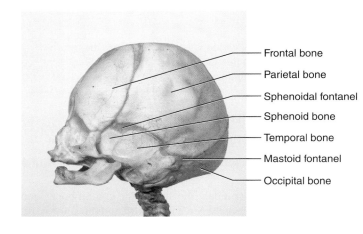

Frontal bone

Parietal bone

Sphenoidal fontanel

Sphenoid bone

Temporal bone

Mastoid fontanel

Occipital bone

FIGURE 18.3 Posterior view of fetal skull.

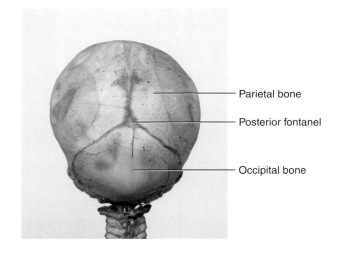

Parietal bone

Posterior fontanel

Occipital bone

FIGURE 18.4 Anterior views of skulls next to metric scales: (a) fetal skull; (b) adult skull. The dotted lines indicate the division between the cranium and the face.

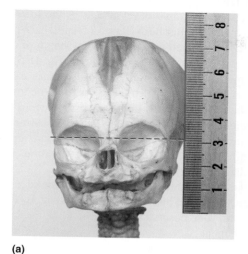

(a)

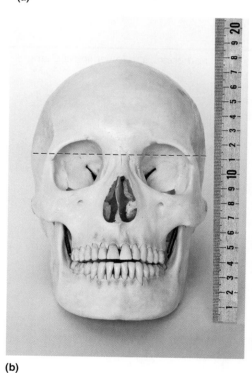

(b)

FIGURE 18.5 Anterior view of a fetal skeleton next to a metric scale. To estimate the gestational age of this fetus, measure the total height (crown-to-heel).

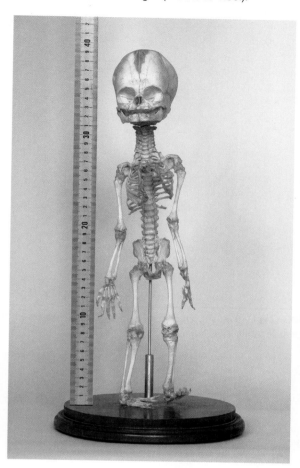

Procedure B—Entire Skeleton

1. Observe and then measure the total height (crown-to-heel) of the entire fetal skeleton (fig. 18.5) and an adult skeleton. Compare the height of the fetal skeleton with the data in table 18.1. Observe and then measure the total height (crown-to-heel) and the sitting height (crown-to-rump) of the human fetus shown in figure 18.6. Figure 18.7 illustrates the proportionate changes of bodies during development from an embryo to the

TABLE 18.1   Embryonic and Fetal Height (Length)

Gestational Age (Months)	Sitting Height (crown-to-rump)*	Total Height (crown-to-heel)*
1	0.6 cm	
2	3 cm	
3	7–9 cm	9–10 cm
4	12–14 cm	16–20 cm
5	18–19 cm	25–27 cm
6		30–32 cm
7		35–37 cm
8		40–45 cm
9		50–53 cm

*Measurements of height until about 20 weeks are more often from crown-to-rump; measurements after 20 weeks are more often from crown-to-heel.

FIGURE 18.6 Human fetus next to a metric scale. To estimate the gestational age of this young fetus, measure the total height (crown-to-heel) and the sitting height (crown-to-rump).

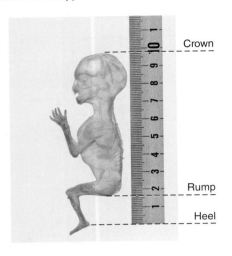

Crown

Rump

Heel

adult. Record the measurements in Part C of the laboratory assessment.

2. Much of the development of the upper limb occurs during fetal development, especially during the second trimester. The rapid development of the lower limb is much greater after birth (figs. 18.5 and 18.7).

3. Observe and then measure the total length of the upper limb and the lower limb of the fetus (fig. 18.5). Record the measurements in Part C of the laboratory assessment.

4. Observe the adult skeleton and measure the total length of the upper limb and the lower limb. Record the measurements in Part C of the laboratory assessment.

5. Observe very carefully a hip bone, sternum, and sacrum of a fetal skeleton. Record your observations in Part C of the laboratory assessment.

6. Make careful observations of the anterior knee, thoracic cage, wrist, and ankles of the fetal skeleton (fig. 18.5). Record your observations in Part C of the laboratory assessment.

7. Complete Parts C and D of the laboratory assessment.

FIGURE 18.7 During development, proportions of the head, entire body, and upper and lower limbs change considerably.

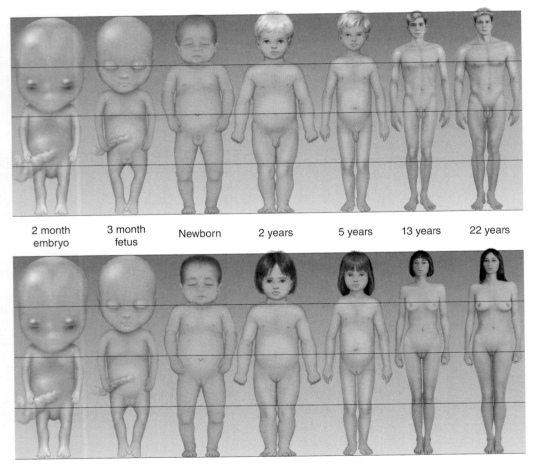

2 month embryo   3 month fetus   Newborn   2 years   5 years   13 years   22 years

Name _____

Date _____

Section _____

The ⬭ corresponds to the indicated outcome(s) found at the beginning of the laboratory exercise.

Fetal Skeleton

Part A Assessments

Complete the following:

1. List all of the bones that form part of the border with the mastoid fontanel. ⬭
   _____

2. List all of the bones that form part of the border with the sphenoidal fontanel. ⬭
   _____

3. List all of the bones that form part of the border with the posterior fontanel. ⬭
   _____

4. List all of the bones that form part of the border with the anterior fontanel. ⬭
   _____

5. Describe the function of fontanels. ⬭
   _____

6. What is the final outcome of the fontanels? ⬭
   _____
   _____

Part B Assessments

1. Enter the measurements of the fetal and adult skulls in the table. ⬭

Skull	Total Height of Skull (cm)	Cranial Height (cm)	Facial Height (cm)
Fetus			
Adult			

2. Compare the frontal bone of the fetal skull with the adult skull. Describe some unusual aspects of the fetal frontal bone as compared to the adult structure. ⬭
   _____
   _____

179

3. Compare the temporal bone of the fetal skull with the adult skull. Describe some unusual aspects of the fetal temporal bone as compared to the adult structure. **3**

_____

_____

4. Summarize the measured and observed relationships between a fetal skull and an adult skull. **2**

_____

_____

_____

_____

Critical Thinking Assessment

Why is it difficult to match the identifications of baby pictures with those of adults? **2**

_____

_____

_____

_____

Part C Assessments

1. Enter the measurements of the fetal and adult skeletons in the table. **4**

Skeleton	Total Height (cm)	Upper Limb Length (cm)	Lower Limb Length (cm)
Fetus			
Adult			

2. Summarize the relationship of total body height to limb lengths for the fetal skeleton and the adult skeleton. **4**

_____

_____

_____

_____

3. Locate and describe the discernible ossification of the following fetal bones and body regions and compare them with the adult skeleton. **5**

    a. Hip bone:

_____

_____

b. Sternum:

_____

_____

c. Sacrum:

_____

_____

d. Anterior knee:

_____

_____

e. Thoracic cage:

_____

_____

f. Wrist (carpals) and ankle (tarsals):

_____

_____

4. Estimate the age of fetal development of the fetal skeleton in figure 18.5 and the one that is represented at your school. Estimate the age of the human fetus shown in figure 18.6. Use the data provided in table 18.1 for your determination of the estimates. **6**

    a. Fetal skeleton in figure 18.5: _____

    b. Fetal skeleton at your school: _____

    c. Fetus in figure 18.6: _____

Part D Assessments

Identify the fetal skeleton structures in figures 18.8 and 18.9.

FIGURE 18.8 Label the fetal bones and fontanels by placing the correct numbers in the spaces provided. **1 5**

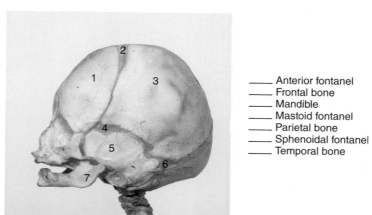

_____ Anterior fontanel
_____ Frontal bone
_____ Mandible
_____ Mastoid fontanel
_____ Parietal bone
_____ Sphenoidal fontanel
_____ Temporal bone

FIGURE 18.9 Distinguish the type of ossification formation by selecting the correct choice from the terms provided. 7

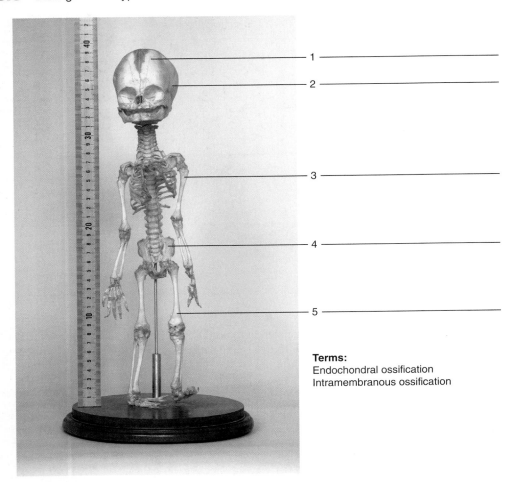

1 _____

2 _____

3 _____

4 _____

5 _____

Terms:
Endochondral ossification
Intramembranous ossification

Joint Structure and Movements

Purpose of the Exercise

To examine examples of the three types of structural and functional joints, to identify the major features of these joints, and to review the types of movements produced at synovial joints.

Materials Needed

Human skull
Human skeleton, articulated
Models of synovial joints (shoulder, elbow, hip, and knee)
For Demonstration Activities:
Fresh animal joint (knee joint preferred)
Radiographs of major joints

Safety

▶ Wear disposable gloves when handling the fresh animal joint.
▶ Wash your hands before leaving the laboratory.

Learning Outcomes

After completing this exercise, you should be able to

① Distinguish structural features among fibrous, cartilaginous, and synovial joints.

② Distinguish functional characteristics of synarthroses, amphiarthroses, and diarthroses.

③ Locate examples of each type of structural and functional joint.

④ Examine the structure and types of movements of the shoulder, elbow, hip, and knee joints.

⑤ Identify the types of movements that occur at synovial joints.

Pre-Lab

Carefully read the introductory material and examine the entire lab. Be familiar with joint structures and movements from lecture or the textbook. Visit www.mhhe.com/martinseries2 for LabCam videos. Answer the pre-lab questions.

Pre-Lab Questions: Select the correct answer for each of the following questions:

1. Fibrous joints are structural types containing
   a. cartilage.
   b. dense fibrous connective tissue.
   c. synovial membranes.
   d. synovial fluid.

2. Which synovial joint allows movements in all planes?
   a. saddle   b. pivot
   c. ball-and-socket d. plane

3. Synarthosis pertains to functional joints that are
   a. immovable.   b. slightly movable.
   c. freely movable. d. synovial types.

4. Movement away from the midline is called
   a. rotation.   b. adduction.
   c. extension.   d. abduction.

5. Movements that combine flexion, abduction, extension, and adduction are called
   a. rotations.   b. excursions.
   c. circumductions. d. hyperextensions.

6. The synovial membrane and the fibrous capsule combine and form the
   a. articular cartilage. b. periosteum.
   c. spongy bone.   d. joint (articular) capsule.

7. Rotation occurs at a pivot type of synovial joint.
   True _____   False _____

8. Dorsiflexion and plantar flexion are characteristic movements of the hand.
   True _____   False _____

Joints are junctions between bones. Although they vary considerably in structure, they can be classified according to the type of tissue that binds the bones together. Thus, the three groups of structural joints can be identified as *fibrous joints, cartilaginous joints,* and *synovial joints.* Fibrous joints are filled with dense fibrous connective tissue, cartilaginous joints are filled with a type of cartilage, and synovial joints contain synovial fluid inside a joint cavity.

Joints can also be classified by the degree of functional movement allowed: *synarthroses* are immovable, *amphiarthroses* allow slight movement, and *diarthroses* allow free movement. Movements occurring at freely movable synovial joints are due to the contractions of skeletal muscles. In each case, the type of movement depends on the type of joint involved and the way in which the muscles are attached to the bones on either side of the joint.

Procedure A—Types of Joints

Joints of the body are areas where various types of articulations occur. Because there are variations in the types of connective tissues involved in the joint structure and variations in the degree of movement possible, two types of classifications are used. Structural joints are fibrous, cartilaginous, or synovial; functional joints are synarthrosis, amphiarthrosis, or diarthrosis. The joints are areas where bones are held together, but allowing various degrees of body movements. Some joints are only temporary; when complete ossification occurs, the entire joint becomes bone. Examples of these temporary joints include a right and left frontal bone becoming a single frontal bone, and the eventual complete ossification of the epiphyseal plates in the growth areas of the long bones.

1. Use table 19.1 as a reference and examine the human skull and articulated skeleton. Locate examples of the following types of structural and functional joints:

   fibrous joints
   - suture—synarthrosis
   - gomphosis—synarthrosis
   - syndesmosis—amphiarthrosis

   cartilaginous joints
   - synchondrosis—synarthrosis
   - symphysis—amphiarthrosis

   synovial joints—diarthrosis

2. Complete Part A of Laboratory Assessment 19.

3. Using table 19.1 and figure 19.1 as references, locate examples of the following types of synovial joints in the skeleton. At the same time, examine the corresponding joints in the models and in your body. Experiment with each joint to experience its range of movements.

ball-and-socket joint	**hinge joint**
condylar (ellipsoid) joint	**pivot joint**
plane (gliding) joint	**saddle joint**

4. Complete Parts B and C of the laboratory assessment.

TABLE 19.1 Classification of Joints

Structural Classification	Structural Features	Functional Classification	Location Examples
Fibrous			
Suture	Dense fibrous connective tissue	Synarthrosis—immovable	Sutures of skull
Gomphosis	Periodontal ligament	Synarthrosis—immovable	Tooth sockets
Syndesmosis	Dense fibrous connective tissue	Amphiarthrosis—slightly movable	Tibiofibular joint
Cartilaginous			
Synchondrosis	Hyaline cartilage	Synarthrosis—immovable	Epiphyseal plates
Symphysis	Fibrocartilage	Amphiarthrosis—slightly movable	Pubic symphysis; intervertebral discs
Synovial			
Ball-and-socket	All synovial joints contain articular cartilage, a synovial membrane, and a joint cavity filled with synovial fluid	Diarthrosis—movements in all planes	Hip; shoulder
Hinge		Diarthrosis—flexion and extension	Elbow; knee; interphalangeal
Condylar (ellipsoid)		Diarthrosis—flexion, extension, abduction, and adduction	Radiocarpal; metacarpophalangeal
Pivot		Diarthrosis—rotation	Radioulnar at elbow; dens at atlas
Saddle		Diarthrosis—variety of movements mainly in two planes	Base of thumb with trapezium
Plane (gliding)		Diarthrosis—sliding or twisting	Intercarpal; intertarsal

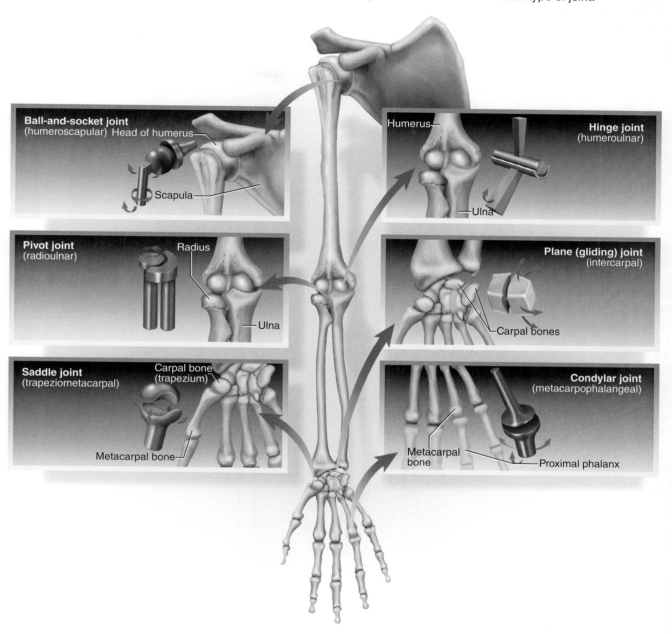

Ball-and-socket joint
(humeroscapular) Head of humerus
Scapula

Pivot joint
(radioulnar) Radius
Ulna

Saddle joint
(trapeziometacarpal) Carpal bone (trapezium)
Metacarpal bone

Humerus **Hinge joint**
(humeroulnar)
Ulna

Plane (gliding) joint
(intercarpal)
Carpal bones

Condylar joint
(metacarpophalangeal)
Metacarpal bone
Proximal phalanx

Procedure B—Examples of Synovial Joints

Synovial joints have some characteristic structural features and functional characteristics. The structure of the synovial joint includes a joint (articular) capsule composed of two layers: an outer fibrous capsule and an inner synovial membrane. The fibrous capsule is composed of a dense connective tissue that is continuous with the periosteum of the bone, while the synovial membrane secretes a lubricating synovial fluid that fills the enclosed joint cavity. Articular (hyaline) cartilage covers the surface of the bone within the joint. Various ligaments and tendons also cross synovial joints.

In the knee, two crescent-shaped pieces of cartilage called menisci extend partway across the joint from the medial and lateral sides. The meniscus will help stabilize the knee joint, where a great amount of pressure occurs, and it also helps to absorb shock. Although the knee is basically a hinge joint, there are some condylar and plane components of the joint. The condylar component of the joint allows some limited rotation between the articulations of the rounded condyles of the femur and the tibia when the knee is partly in the flexed position. The plane joint between the patellar surface of the femur and the patella allows some sliding movements. The knee joint also consists of two cruciate ligaments that cross each other deep within the joint. The posterior cruciate ligament (PCL) helps prevent excessive flexion; the anterior cruciate ligament (ACL) helps prevent hyperextension of the knee joint. A torn ACL is a common type of knee injury.

FIGURE 19.2 Basic structure of a synovial joint.

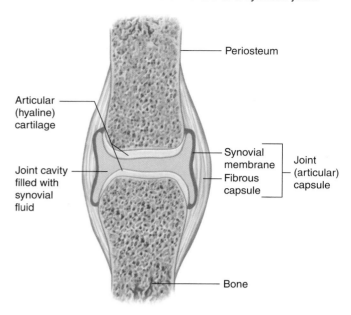

1. Study the major features of a synovial joint in figure 19.2.
2. Examine models of the shoulder, elbow, hip, and knee joints.
3. Locate the major knee joint structures of figure 19.3 that are visible on the knee joint model.
4. Complete Part D of the laboratory assessment.

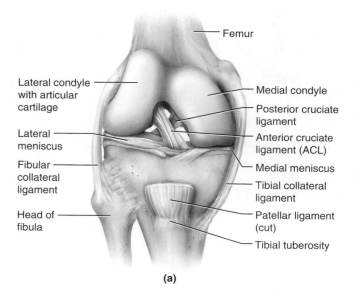

Femur

Lateral condyle with articular cartilage

Medial condyle

Posterior cruciate ligament

Lateral meniscus

Anterior cruciate ligament (ACL)

Fibular collateral ligament

Medial meniscus

Tibial collateral ligament

Head of fibula

Patellar ligament (cut)

Tibial tuberosity

(a)

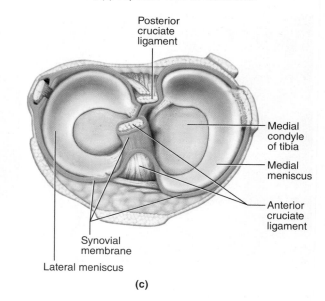

Posterior cruciate ligament

Medial condyle of tibia

Medial meniscus

Anterior cruciate ligament

Synovial membrane

Lateral meniscus

(c)

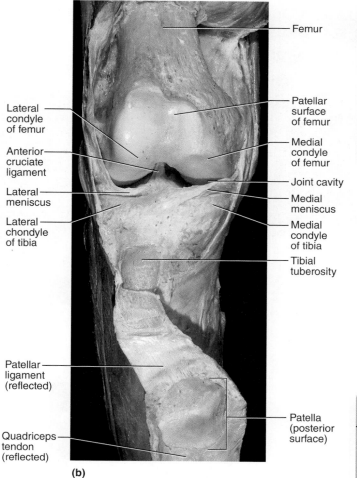

Femur

Lateral condyle of femur

Patellar surface of femur

Anterior cruciate ligament

Medial condyle of femur

Joint cavity

Lateral meniscus

Medial meniscus

Lateral chondyle of tibia

Medial condyle of tibia

Tibial tuberosity

Patellar ligament (reflected)

Patella (posterior surface)

Quadriceps tendon (reflected)

(b)

Demonstration Activity

Examine a longitudinal section of a fresh synovial animal joint. Locate the dense connective tissue that forms the fibrous capsule and the hyaline cartilage that forms the articular cartilage on the ends of the bones. Locate the synovial membrane on the inner surface of the joint capsule. Does the joint have any semilunar cartilages (menisci)? _____

Procedure C—Joint Movements

1. Examine figures 19.4 and 19.5.
2. Most joints are extended and/or adducted when the body is in anatomical position. Skeletal muscle action involves the movable end (*insertion*) being pulled toward the stationary end (*origin*). In the limbs, the origin is usually proximal to the insertion; in the trunk, the origin is usually medial to the insertion. Use these concepts as reference points as you move joints. Move various parts of your body to demonstrate the joint movements described in tables 19.2 and 19.3.
3. Complete Part E of the laboratory assessment.

Demonstration Activity

Study the available radiographs of joints by holding the films in front of a light source. Identify the type of joint and the bones incorporated in the joint. Also identify other major visible features.

FIGURE 19.4 Examples of angular, rotational, and gliding movements of synovial joints.

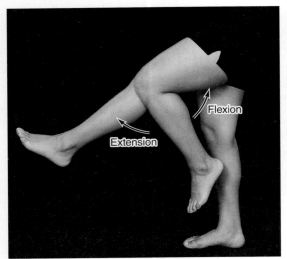

(a) Flexion and extension of the knee joint

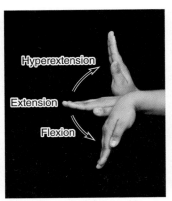

(b) Flexion, extension, and hyper-extension of the wrist joint

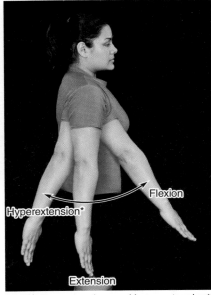

(c) Flexion, extension, and hyperextension* of the shoulder joint (*Hyperextension of the shoulder and hip joints is considered normal extension in some health professions.)

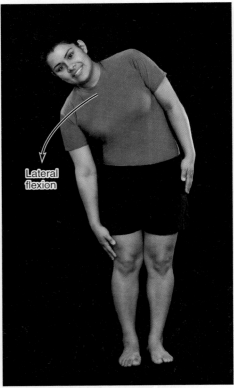

(d) Lateral flexion of the trunk (vertebral column)

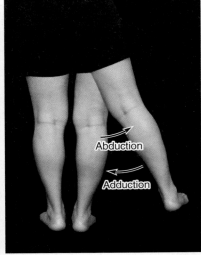

(e) Abduction and adduction of the hip joint

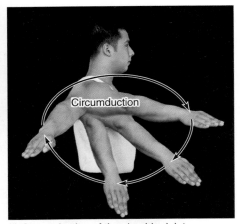

(f) Circumduction of the shoulder joint

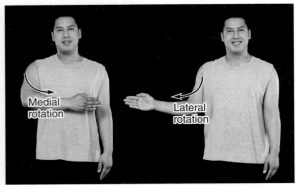

(g) Medial and lateral rotation of the arm at the shoulder joint

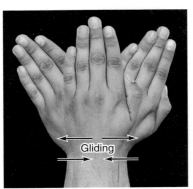

(h) Gliding movements among the carpals of the wrist

FIGURE 19.5 Special movements of synovial joints.

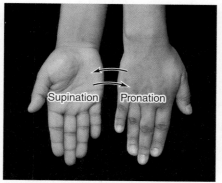

(a) Supination and pronation of the hand involving movement at the radioulnar joint

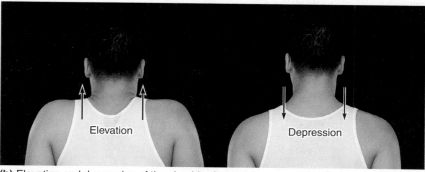

(b) Elevation and depression of the shoulder (scapula)

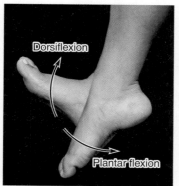

(c) Dorsiflexion and plantar flexion of the foot at the ankle joint

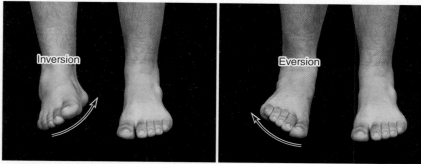

(d) Inversion and eversion of the right foot at the ankle joint (The left foot is unchanged for comparison.)

(e) Protraction and retraction of the head

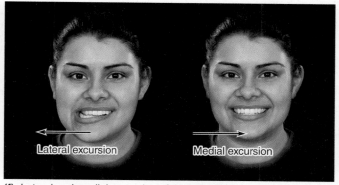

(f) Lateral and medial excursion of the mandible

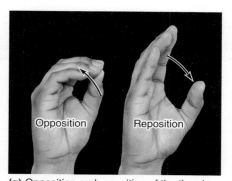

(g) Opposition and reposition of the thumb at the carpometacarpal (saddle) joint

TABLE 19.2   Angular, Rotational, and Gliding Movements

Movement	Description
Flexion	Decrease of an angle (usually in the sagittal plane)
Lateral flexion	Bending the trunk (vertebral column) to the side
Extension	Increase of an angle (usually in the sagittal plane)
Hyperextension	Extension beyond anatomical position
Abduction	Movement away from the midline (usually in the frontal plane)
Adduction	Movement toward the midline (usually in the frontal plane)
Circumduction	Circular movement (combines flexion, abduction, extension, and adduction)
Rotation	Movement of part around its long axis
Medial (internal)	Inward rotation
Lateral (external)	Outward rotation
Gliding	Back-and-forth and side-to-side sliding movements of plane joints

TABLE 19.3   Special Movements (pertain to specific joints)

Movement	Description
Supination	Movement of palm of hand anteriorly or upward
Pronation	Movement of palm of hand posteriorly or downward
Elevation	Movement of body part upward
Depression	Movement of body part downward
Dorsiflexion	Movement of ankle joint so dorsum (superior) of foot becomes closer to anterior surface of leg (as standing on heels)
Plantar flexion	Movement of ankle joint so the plantar surface of foot becomes closer to the posterior surface of leg (as standing on toes)
Inversion (called supination in some health professions)	Medial movement of sole of foot at ankle joint
Eversion (called pronation in some health professions)	Lateral movement of sole of foot at ankle joint
Protraction	Anterior movement in the transverse plane
Retraction	Posterior movement in the transverse plane
Excursion	Special movements of mandible when grinding food
Lateral	Movement of mandible laterally
Medial	Movement of mandible medially
Opposition	Movement of the thumb to touch another finger of the same hand
Reposition	Return of thumb to anatomical position

Critical Thinking Activity

Have your laboratory partner do some of the preceding movements and see if you can correctly identify the movements made.

Critical Thinking Activity

Describe a body position that can exist when all major body parts are flexed._____

_____

_____

_____

Name _____

Date _____

Section _____

The A corresponds to the indicated outcome(s) found at the beginning of the laboratory exercise.

Joint Structure and Movements

Part A   Assessments

Match the terms in column A with the descriptions in column B. Place the letter of your choice in the space provided. A 2

Column A

a. Gomphosis
b. Suture
c. Symphysis
d. Synchondrosis
e. Syndesmosis

Column B

_____ 1. Immovable joint between flat bones of the skull united by a thin layer of dense connective tissue

_____ 2. Fibrocartilage fills the slightly movable joint

_____ 3. Temporary joint in which bones are united by bands of hyaline cartilage

_____ 4. Slightly movable joint in which bones are united by interosseous membrane

_____ 5. Joint formed by union of tooth root in bony socket

Part B   Assessments

Identify the types of structural and functional joints numbered in figure 19.6. 3

Structural Classification   **Functional Classification**

1. _____   _____

2. _____   _____

3. _____   _____

4. _____   _____

5. _____   _____

6. _____   _____

7. _____   _____

8. _____   _____

9. _____   _____

FIGURE 19.6 Identify the types of structural and functional joints numbered in these illustrations.

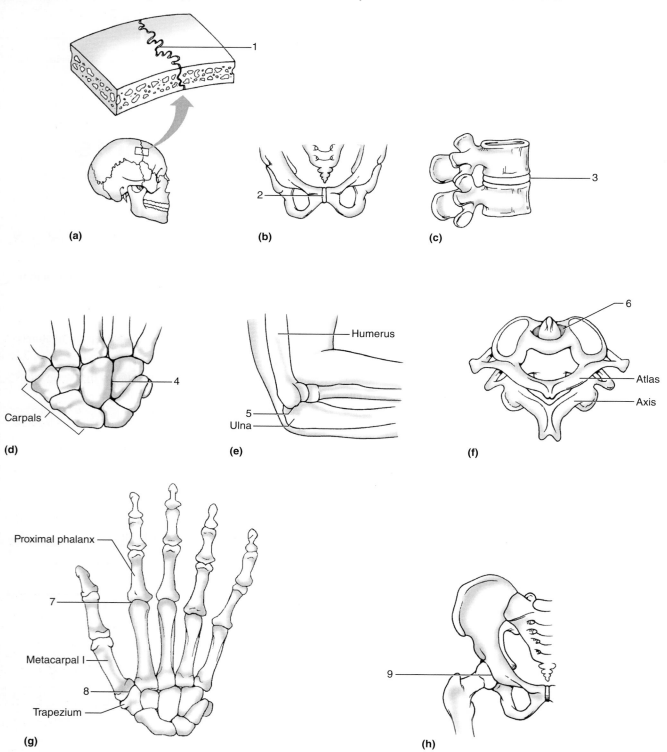

(a)

(b)

(c)

Humerus

Carpals

Ulna

Atlas

Axis

(d)

(e)

(f)

Proximal phalanx

Metacarpal I

Trapezium

(g)

(h)

Part C Assessments

Match the types of synovial joints in column A with the examples in column B. Place the letter of your choice in the space provided. **3**

Column A

a. Ball-and-socket
b. Condylar (ellipsoid)
c. Hinge
d. Pivot
e. Plane (gliding)
f. Saddle

Column B

_____ **1.** Hip joint

_____ **2.** Metacarpal–phalanx

_____ **3.** Proximal radius–ulna

_____ **4.** Humerus–ulna of the elbow joint

_____ **5.** Phalanx–phalanx

_____ **6.** Shoulder joint

_____ **7.** Knee joint between femur and tibia

_____ **8.** Carpal–metacarpal of the thumb

_____ **9.** Carpal–carpal

_____ **10.** Tarsal–tarsal

Part D Assessments

Complete the missing components of the following table: **4** **5**

Name of Joint	Type of Joint	Bones Included	Types of Movement Possible
Shoulder joint		Humerus, scapula	
Elbow joint	Hinge, plane, pivot		Flexion and extension between humerus and ulna; twisting between radius and humerus; rotation between head of radius and ulna
Hip joint			Movements in all planes and rotation
Knee joint	Hinge (modified), condylar, plane		

Part E Assessments

Identify the types of joint movements numbered in figure 19.7. **5**

1. _____ (of head)

2. _____ (of shoulder)

3. _____ (of shoulder)

4. _____
   _____ (of hand at radioulnar joint)

5. _____
   _____ (of hand at radioulnar joint)

6. _____ (of arm at shoulder)

7. _____ (of arm at shoulder)

8. _____ (of hand at wrist)

9. _____ (of hand at wrist)

10. _____ (of thigh at hip)

11. _____ (of thigh at hip)

12. _____ (of lower limb at hip)

13. _____ (of chin/mandible)

14. _____ (of chin/mandible)

15. _____ (of vertebral column)

16. _____ (of vertebral column)

17. _____ (of head and neck)

18. _____ (of head and neck)

19. _____ (of arm at shoulder)

20. _____ (of arm at shoulder)

21. _____ (of forearm at elbow)

22. _____ (of forearm at elbow)

23. _____ (of thigh at hip)

24. _____ (of thigh at hip)

25. _____ (of leg at knee)

26. _____ (of leg at knee)

27. _____ (of foot at ankle)

28. _____ (of foot at ankle)

FIGURE 19.7 Identify each of the types of movements numbered and illustrated: (a) anterior view; (b) lateral view of head; (c) lateral view. (Original illustration drawn by Ross Martin.)

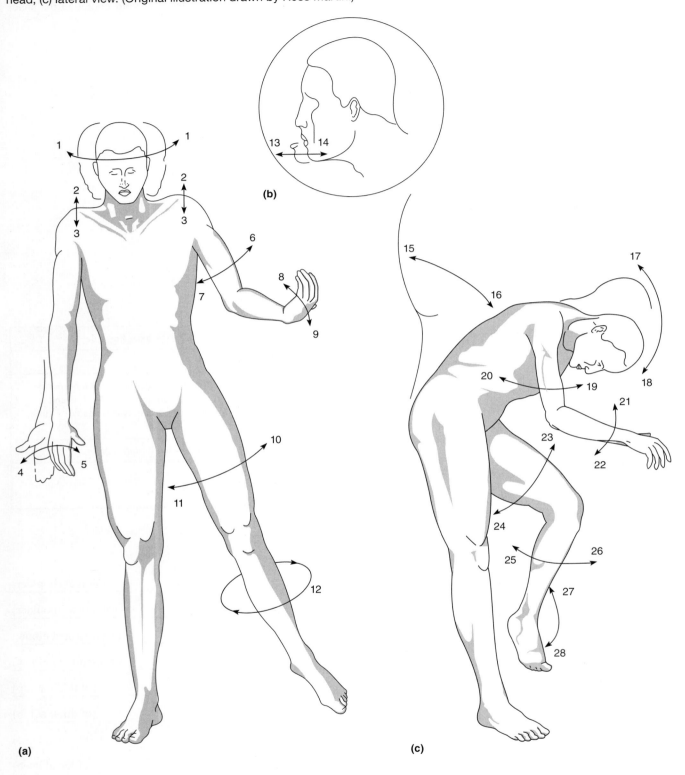

Skeletal Muscle Structure and Function

Purpose of the Exercise

To study the structure and function of skeletal muscles as cells and as organs.

Materials Needed

Compound light microscope
Prepared microscope slide of skeletal muscle tissue
  (longitudinal section and cross section)
Human torso model with musculature
Model of skeletal muscle fiber
Ph.I.L.S. 4.0

For Demonstration Activity:
Fresh round beefsteak

Learning Outcomes

After completing this exercise, you should be able to

1. Locate the structures of a skeletal muscle fiber (cell).

2. Describe how connective tissue is associated with muscle tissue within a skeletal muscle.

3. Distinguish between the origin and insertion of a muscle.

4. Describe and demonstrate the general actions of prime movers (agonists), synergists, fixators, and antagonists.

5. Diagram and label a muscle twitch.

6. Demonstrate *threshold* (the minimum voltage required to observe the appearance of muscle contraction).

7. Illustrate *recruitment* (the stimulation of additional motor units in the muscle) by observing the change in amplitude of the contraction.

8. Determine maximum contraction (the stimulation of all motor units in the muscle) by observing no further increase in the amplitude of the contraction.

9. Measure amplitude of muscle contraction when the muscle is stimulated with varying degrees of voltage.

10. Integrate the concepts of recruitment and maximum stimulation as applied to the muscles of the body.

Pre-Lab

Carefully read the introductory material and examine the entire lab. Be familiar with skeletal muscle tissue and muscle structure and function from lecture or the textbook. Answer the pre-lab questions.

Pre-Lab Questions: Select the correct answer for each of the following questions:

1. The outermost layer of connective tissue of a muscle is the
   a. fascicle.   **b.** endomysium.
   c. epimysium.   **d.** perimysium.

2. The thick myofibril filament of a sarcomere is composed of a protein
   a. myosin.   **b.** actin.
   c. titin.   **d.** sarcolemma.

3. The muscle primarily responsible for a movement is the
   a. synergist.   **b.** prime mover (agonist).
   c. antagonist.   **d.** origin.

4. The neuron and the collection of muscle fibers it innervates is called the
   a. neural impulse.   **b.** muscle fiber.
   c. motor neuron.   **d.** motor unit.

5. The functional contractile unit of a muscle fiber (cell) is a
   a. sarcoplasm.   **b.** sarcolemma.
   c. sarcomere.   **d.** sarcoplasmic reticulum.

6. The role of a particular muscle is always the same.
   True _____   False _____

7. A synergistic muscle contraction assists the prime mover.
   True _____   False _____

8. The plasma membrane of a muscle fiber (cell) is called the sarcolemma.
   True _____   False _____

A skeletal muscle represents an organ of the muscular system and is composed of several types of tissues. These tissues include skeletal muscle tissue, nervous tissue, and various connective tissues.

Each skeletal muscle is encased and permeated with connective tissue sheaths. The connective tissues surround and extend into the structure of a muscle and separate it into compartments. The entire muscle is encased by the *epimysium*. The *perimysium* covers bundles of cells (*fascicles*), within the muscle. The deepest connective tissue surrounds each individul muscle fiber (cell) as a thin *endomysium*. The connective tissues provide support and reinforcement during muscular contractions and allow portions of a muscle to contract somewhat independently. The connective tissue often extends beyond the end of a muscle, providing an attachment to other muscles or to bones. Some collagen fibers of the connective tissue are continuous with the tendon and the periosteum, making for a strong structural continuity.

Muscles are named according to their location, size, shape, action, attachments, number of origins, or the direction of the fibers. Examples of how muscles are named include: gluteus maximus (location and size); adductor longus (action and size); sternocleidomastoid (attachments); serratus anterior (shape and location); biceps (two origins); and orbicularis oculi (direction of fibers and location).

Skeletal muscles, such as the biceps brachii, are composed of many muscle fibers (cells). These muscle fibers are innervated by motor neurons of the nervous system. Anatomically, each motor neuron is arranged to extend to a specific number of muscle fibers. (Note that each muscle fiber is innervated by only one motor neuron.) The motor neuron and the collection of muscle fibers it innervates is called a *motor unit*. When the motor neuron relays a neural impulse to the muscle, all the muscle fibers that it innervates (i.e., the motor unit) will be stimulated to contract. If only a few motor units are stimulated, the muscle as a whole exerts little force. If a larger number of motor units are stimulated, the muscle will exert a greater force. Thus, to lift a heavy object like a suitcase requires the activation of more motor units in the biceps brachii than the lifting of a lighter object like a book.

Procedure A—Skeletal Muscle Structure

1. Examine the microscopic structure of a longitudinal section of skeletal muscle by observing a prepared microscope slide of this tissue. Use figure 20.1 of skeletal muscle tissue to locate the following features:

   skeletal muscle fiber (cell)

   nuclei

   striations (alternating light and dark)

2. Skeletal muscle cells are stimulated by nerve impulses over cellular processes (axons) of motor neurons. The *neuromuscular junctions* are the sites where the axons terminate at a muscle fiber (fig. 20.1).

3. Study figures 20.2 and 20.3. Note the arrangement of muscle fibers in relation to the connective tissues of the

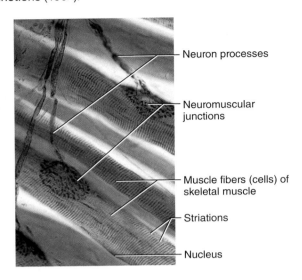

FIGURE 20.1 Micrograph of a longitudinal section of skeletal muscle fibers with associated neuromuscular junctions (400×).

Neuron processes

Neuromuscular junctions

Muscle fibers (cells) of skeletal muscle

Striations

Nucleus

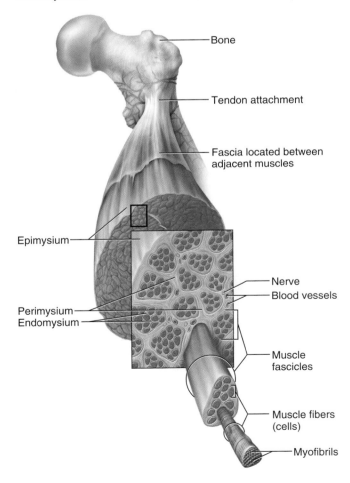

FIGURE 20.2 Skeletal muscle structure from the gross anatomy to the microscopic arrangement. Note the distribution pattern of the epimysium, perimysium, and endomysium.

Bone

Tendon attachment

Fascia located between adjacent muscles

Epimysium

Nerve

Blood vessels

Perimysium
Endomysium

Muscle fascicles

Muscle fibers (cells)

Myofibrils

FIGURE 20.3 Cross section of a fascicle and associated connective tissues (scanning electron micrograph, 320×).

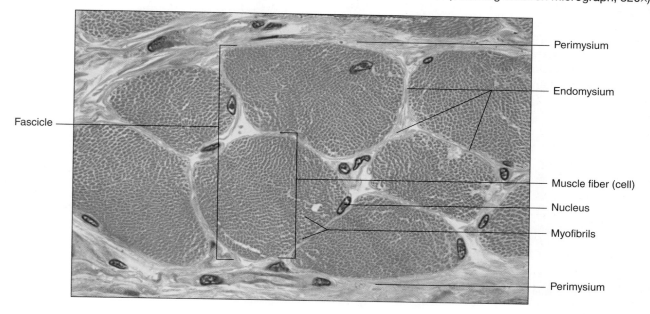

Fascicle

Perimysium

Endomysium

Muscle fiber (cell)

Nucleus

Myofibrils

Perimysium

epimysium, perimysium, and endomysium of an individual muscle. Examine the microscopic structure of a cross section of skeletal muscle tissue using figure 20.3 for identification of the structures.

4. Examine the human torso model and locate examples of tendons and aponeuroses. An *origin tendon* is attached to a fixed location, while the *insertion tendon* is attached to a more movable location. Sheets of connective tissue, called aponeuroses, also serve for some muscle attachments. An example of a large aponeurosis visible on the torso or a muscle chart is in the abdominal area. It appears as a broad white sheet of connective tissue for some of the abdominal muscle attachments to each other. Locate examples of cordlike tendons in your body. The large calcaneal tendon is easy to palpate in a posterior ankle just above the heel.

 Demonstration Activity

Examine the fresh round beefsteak. It represents a cross section through the beef thigh muscles. Note the white lines of connective tissue that separate the individual skeletal muscles. Also note how the connective tissue extends into the structure of a muscle and separates it into small compartments of muscle tissue. Locate the epimysium and the perimysium of an individual muscle.

⚠ **Safety**

▶ Wear disposable gloves when handling the fresh beefsteak.
▶ Wash your hands before leaving the laboratory.

FIGURE 20.4 Structures of a segment of a muscle fiber (cell).

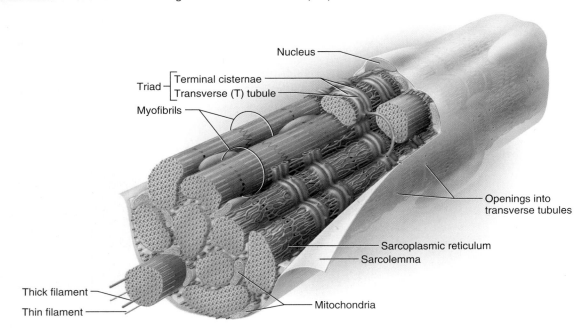

5. Using figures 20.4 and 20.5 as a guide, examine the model of the skeletal muscle fiber. Locate the following:

sarcolemma—plasma membrane of a muscle fiber

sarcoplasm—cytoplasm of a muscle fiber

myofibril—bundle of protein filaments

- thick filament—composed of contractile protein myosin
- thin filament—composed mostly of contractile protein actin
- elastic filament—composed of springy protein titin

sarcoplasmic reticulum (SR)—smooth ER of muscle fiber

- terminal cisternae—extended ends of SR next to transverse tubules

transverse (T) tubules—inward extensions of sarcolemma through the muscle fiber

sarcomere—functional contractile unit within muscle fiber (fig. 20.5)

- A (anisotropic) band—dark band of thick filaments and some overlap of thin filaments
- I (isotropic) band—light band of thin filaments
- H zone (band)—light band in middle of A band
- M line—dark line in middle of H zone
- Z disc (line)—thin and elastic filaments anchored at ends of sarcomere

FIGURE 20.5 Sarcomere contractile unit of a muscle fiber: (a) micrograph (16,000×); (b) illustration of the repeating pattern of striations (bands).

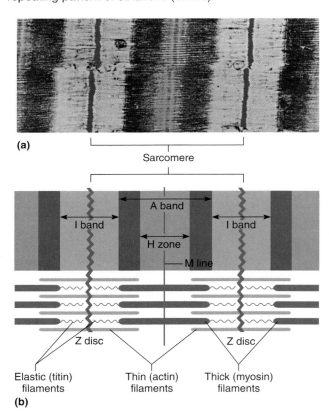

FIGURE 20.6 Antagonistic muscle pairs are located on opposite sides of the same body region as shown in an arm. The muscle acting as the prime mover depends upon the movement that occurs.

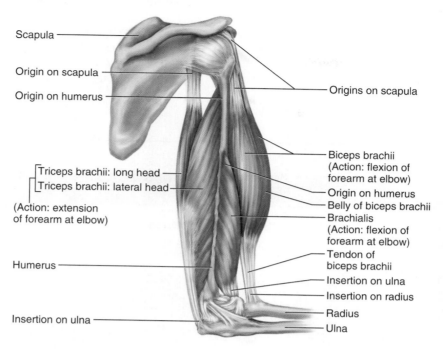

Scapula

Origin on scapula

Origin on humerus

Origins on scapula

Triceps brachii: long head
Triceps brachii: lateral head
(Action: extension of forearm at elbow)

Biceps brachii (Action: flexion of forearm at elbow)
Origin on humerus
Belly of biceps brachii
Brachialis (Action: flexion of forearm at elbow)

Humerus

Tendon of biceps brachii
Insertion on ulna
Insertion on radius
Radius
Ulna

Insertion on ulna

TABLE 20.1   Various Roles of Muscles

Functional Category	Description
Prime mover (agonist)	Muscle primarily responsible for the action (movement)
Antagonist	Muscle responsible for action in the opposite direction of a prime mover or for resistance to a prime mover
Synergist	Muscle contraction assists a prime mover
Fixator	Special type of synergist muscle; muscle contraction will stabilize a joint so another contracting muscle exerts a force on something else

6. Complete Part A and B of the Laboratory Assessment 20.
7. Study figure 20.6 and table 20.1.
8. Locate the biceps brachii, brachialis, and triceps brachii and their origins and insertions in the human torso model and in your body.
9. Make various movements with your upper limb at the shoulder and elbow. For each movement, determine the location of the muscles functioning as prime movers (agonists) and as antagonists. When a prime mover contracts (shortens) and a joint moves, its antagonist relaxes (lengthens). Antagonistic muscle pairs pull from opposite sides of the same body region. Synergistic muscles often supplement the contraction force of a prime mover, or by acting as fixators, they might also stabilize nearby joints. **The role of a muscle as a prime mover, antagonist, or synergist depends upon the movement under consideration, as their roles change.**
10. Complete Part C of the laboratory assessment.

Procedure B—Ph.I.L.S. Lesson 5 Skeletal Muscle Function: Stimulus-Dependent Force Generation

1. Open Exercise 5: Skeletal Muscle Function: Stimulus-Dependent Force Generation.
2. Read the objectives and introduction and take the pre-lab quiz. See figure 20.7.
3. After completing the pre-lab quiz, read through the wet lab. *Be sure to click open and view the videos that are indicated in red.*
4. The lab exercise will open when you have completed the wet lab (fig. 20.8).
5. Follow the instructions at the bottom of the screen to turn on the power of the data acquisition unit and the virtual computer screen and to connect the transducer and electrodes.

FIGURE 20.7   Motor unit.

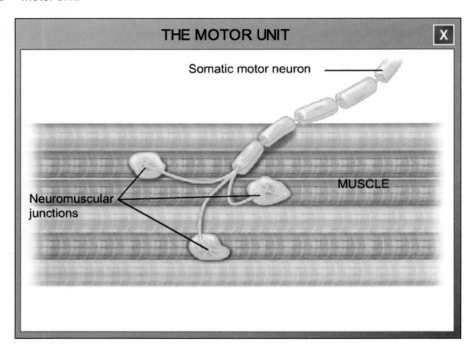

FIGURE 20.8   Opening screen for the laboratory exercise on Skeletal Muscle Function: Stimulus-Dependent Force Generation.

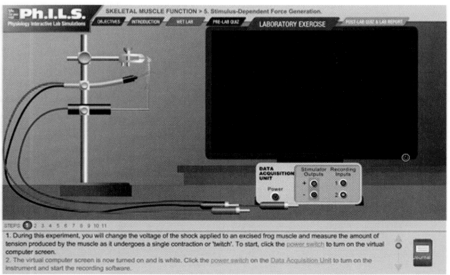

6. Set the stimulus voltage to at least 1.0 volts. (This voltage will elicit a muscle contraction.)
7. To stimulate the muscle, click the Shock button in the control panel. (A graph will appear with a red line and a blue line.)
8. Measure the tension (degree of contraction) by positioning the crosshairs (using the mouse) at the top of the wave. If not in the correct location, reposition by dragging to a new position. Now position the crosshairs at the bottom of the deflection (after the wave or 0 volts) and click. The result will display in the upper left of the virtual computer screen. Amplitude (amp) is the change in tension of the muscle representing the degree of movement of the transducer when the muscle contracted.
9. Click the Journal panel (red rectangle at bottom right of screen) to enter your value into the Journal. A table with volts and amplitude (amp) and a graph will appear. Note the range in values for volts from 0 to 1.6 volts.
10. Close the Journal window by clicking on the X in the right-hand corner.

11. To complete the table and produce the graph,
   a. Click Erase.
   b. Set voltage (1.1 volts).
   c. Click Shock.
   d. Measure.
   e. Click Journal.
   f. Repeat all steps except set voltage at new value. Complete all values in the table (range from 0 to 1.6 in increments of 0.1 volts).
12. After finishing the laboratory exercise, you can print the line tracing by clicking on the P in the top left of the virtual computer screen. An example of a line tracing is shown in figure 20.9.
13. Click on Post-Lab Quiz and Lab Report.
14. Complete the Post-Lab Quiz by answering the ten questions on the computer screen.
15. Read the conclusion on the computer screen.
16. You may print the Lab Report for the Skeletal Muscle Function: Stimulus-Dependent Force Generation computer simulation.
17. Complete Part D of the laboratory assessment in your laboratory manual.

FIGURE 20.9 Example: Line tracings showing the amount of tension produced by applying shocks of different voltages to the exposed frog gastrocnemius.

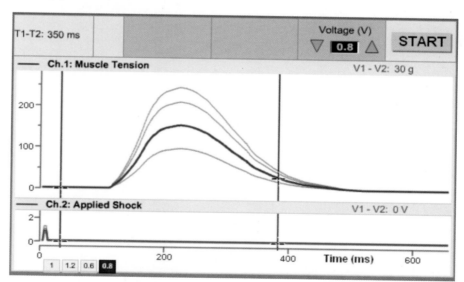

Laboratory Assessment

20

Name _____

Date _____

Section _____

The A corresponds to the indicated outcome(s) found at the beginning of the laboratory exercise.

Skeletal Muscle Structure and Function

Part A Assessments

Match the terms in column A with the definitions in column B. Place the letter of your choice in the space provided. A A

Column A	Column B
a. Endomysium	_____ **1.** Membranous channel extending inward from muscle fiber membrane
b. Epimysium	_____ **2.** Cytoplasm of a muscle fiber
c. Fascia	_____ **3.** Connective tissue located between adjacent muscles
d. Fascicle	_____ **4.** Layer of connective tissue that separates a muscle into small bundles called fascicles
e. Myosin	
f. Perimysium	_____ **5.** Plasma membrane of a muscle fiber
g. Sarcolemma	_____ **6.** Layer of connective tissue that surrounds a skeletal muscle
h. Sarcomere	_____ **7.** Unit of alternating light and dark striations between Z discs (lines)
i. Sarcoplasm	_____ **8.** Layer of connective tissue that surrounds an individual muscle fiber
j. Sarcoplasmic reticulum	
k. Tendon	_____ **9.** Cellular organelle in muscle fiber corresponding to the endoplasmic reticulum
l. Transverse (T) tubule	
	_____ **10.** Cordlike part that attaches a muscle to a bone
	_____ **11.** Protein found within thick filament
	_____ **12.** A small bundle of muscle fibers within a muscle

Part B Assessments

Provide the labels for the electron micrograph in figure 20.10.

FIGURE 20.10 Label this transmission electron micrograph (16,000×) of a relaxed sarcomere by placing the correct numbers in the spaces provided. A

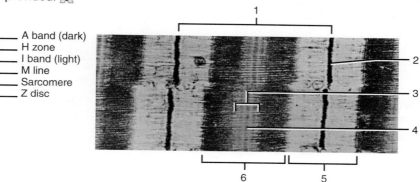

_____ A band (dark)
_____ H zone
_____ I band (light)
_____ M line
_____ Sarcomere
_____ Z disc

Part C   Assessments

Complete the following statements:

1. The _____ of a muscle is usually attached to a fixed location. **3**

2. The _____ of a muscle is usually attached to a movable location. **3**

3. A muscle responsible for most of a movement is called a(n) _____. **4**

4. Assisting muscles are called _____. **4**

5. Antagonists are muscles that resist the actions of _____ and cause movement in the opposite direction. **4**

6. When the forearm is extended at the elbow joint, the _____ muscle acts as the prime mover. **4**

7. When the biceps brachii acts as the prime mover, the _____ muscle assists as a synergist. **4**

Part D   Ph.I.L.S. Lesson 5, Skeletal Muscle Function: Stimulus-Dependent Force Generation Assessments

1. Diagram a graph of a twitch and label: latent period, contraction, and relaxation of a twitch. **5**

2. What was the threshold voltage for stimulation of the frog's gastrocnemius muscle? **6** **9** _____

3. What voltage produced maximal contraction? **8** **9** _____

4. What is occurring between the threshold and the maximum contraction? **7** **9** _____

5. Explain the relationship of increasing voltage stimulation to the recruitment of motor units. **6** **7** **8** _____

_____

_____

Critical Thinking Assessment

You are getting ready to leave the lab. You lift your pencil to put it in your book bag. Then you lift your lab book and place it in your book bag. Which scenario resulted from the stimulation of more motor units? **10**

_____

Which would require more neural signals being sent by the nervous system to the muscles of your arm? **10**

_____

What concept covered in this lab (threshold, recruitment, or maximum contraction) explains why there are certain objects that are too heavy for you to lift? **10**

_____

_____

Electromyography: BIOPAC© Exercise

Purpose of the Exercise

To measure and calculate fist clench strength from electromyograms obtained using the BIOPAC system and to listen to the EMG "sounds"; to measure and calculate the force generated during motor unit recruitment and to observe muscle fatigue.

Materials Needed

Computer system (Mac OS X 10.4–10.6, or PC running Windows XP, Vista, or 7)
BIOPAC Student Lab software v. 3.7.7 or above
BIOPAC Acquisition Unit MP36, MP35, MP30 (Windows only), or MP45 [Note: The MP30 is being phased out.] with (AC100A) transformer
BIOPAC serial cable (CBLSERA)
BIOPAC electrode lead set (SS2L)
BIOPAC disposable vinyl electrodes (EL503), 6 electrodes per subject
BIOPAC electrode gel (GEL1) and abrasive pad (ELPAD) or alcohol wipes
BIOPAC Headphones (OUT 1 for MP3X or 40 HP for MP45) optional
BIOPAC SS25 Hand Dynamometer (for Part 2)

Safety

► The BIOPAC electrode lead set is safe and easy to use and should be used only as described in the procedures section of the laboratory exercise.
► The electrode lead clips are color-coded. Make sure they are connected to the properly placed electrodes as demonstrated in figure 21.6.
► The vinyl electrodes are disposable and meant to be used only once. Each subject should use a new set.

Learning Outcomes

After completing this exercise, you should be able to

1. Record the maximum clench strength for dominant and nondominant hands.
2. Examine, record, and correlate motor unit recruitment with increased power of skeletal muscle contraction.
3. Listen to EMG "sounds" and correlate sound intensity with motor unit recruitment.
4. Examine, record, and calculate motor unit recruitment with increased power of skeletal muscle contraction.
5. Determine the maximum clench force between dominant and nondominant hands (and male and female, if possible).
6. Record the force produced by clenched muscles during fatigue.
7. Calculate the time it takes clenched muscles to fatigue.

Pre-Lab

Carefully read the introductory material and examine the entire lab. Be familiar with motor unit recruitment and fatigue of muscles from lecture or the textbook.

This is the first of a series of laboratory exercises that will utilize the BIOPAC computer-based data acquisition and analysis system. This is a complete system consisting of hardware and software, which provides data similar to that obtained by a physiograph. Once the software package has been loaded onto your computer, each lesson will use a different set of hardware. Connections and how to use the instrumentation will be explained in each lesson.

The hardware includes the MP3X or MP45 Acquisition Unit, connection cables, transformers, transducers, electrodes and electrode cables, and other accessories. The electrodes and other accessories often contain sensors that pick up a signal and send it to the MP3X or MP45 Acquisition Unit. The software picks up electrical signals coming into the MP3X or MP45 Acquisition Unit, which are then displayed on the screen as a waveform. The software then guides the user through the lesson and manages data saving and review.

The BIOPAC computer-based data acquisition and analysis system works in a manner similar to other recording equipment in sensing electrical signals in the body. See figure 21.1 for the basic setup of the BIOPAC Student Lab System. The major advantage of a computerized system such as the BIOPAC system is that it has dynamic calibration. With other equipment such as a physiograph, the instructor or student needs to manually adjust the dials and settings to get a good signal for the recording. With the BIOPAC Student Lab, the adjustments are made automatically for each subject during the calibration procedure in each lesson.

Before you start any lesson, the software should already be installed on the computer, and you should read through each lesson completely. The BIOPAC Student Laboratory Manual that comes with the system has a full description of the lessons and some troubleshooting information. The lessons as described in this manual are shortened versions of these. Figure 21.2 and table 21.1 have been included here as a quick reference guide to the Display Tools for Analysis as you proceed through the lessons.

In this laboratory exercise you will investigate the electrical properties of skeletal muscle using *electromyography*. This is a recording of skin-surface voltage that is produced by underlying skeletal muscle contraction. The voltage produced

FIGURE 21.1 The basic setup of the BIOPAC Student Lab System includes a computer, an MP3X Data Acquisition Unit, and various transducers.

Source: Courtesy of and © *BIOPAC* Systems, Inc.

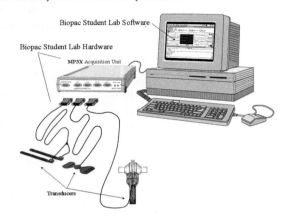

FIGURE 21.2 Display window icons.

Source: Courtesy of and © *BIOPAC* Systems, Inc.

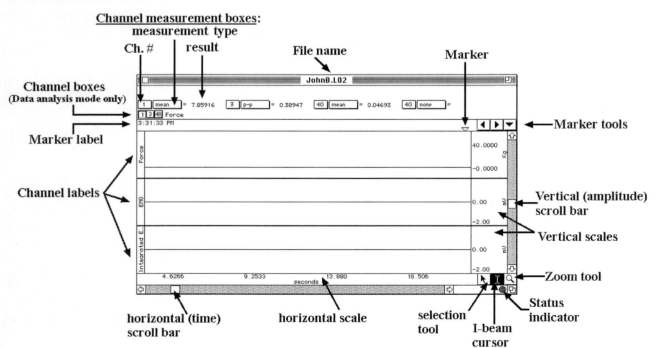

TABLE 21.1 Display Tools for Analysis

Viewing Tool	Overview
Selection Tool	The **Selection** icon is located in the lower right corner of the display window. The **Selection Tool** is a general-purpose cursor, used for selecting waveforms and scrolling through data.
I-Beam Tool	The **I-beam** icon is located in the lower right corner of the display window. The **I-beam Tool** is used to select an area for measurement. To activate it, click on it and move the mouse such that the cursor is positioned at the beginning of the region that you want to select. The cursor *must* be positioned within the data window. Hold the mouse button down and move (drag) the cursor to the second position. Release the mouse button to complete the selection. The selected area should remain darkened. Measurements will apply to the selected area.
Zoom Tool	The **Zoom** icon is in the lower right corner of the display window. Once selected, the **Zoom Tool** allows you to define an area of the waveform by click-hold-drag-releasing the mouse over the desired section. The horizontal (time) and vertical (amplitude) scales will expand to display just the selected section.
Zoom Previous	Click on the **Display** menu to access this option. After you have used the **Zoom Tool, Zoom Previous** will revert the display to the previous scale settings.
Horizontal (time) Scroll Bar	The **Horizontal Scroll Bar** is located below the time division scale. It only works when you are viewing only a portion of the wave form data. To move the display's time position, click on the scroll box arrows or hold the mouse button down on the scroll box and drag the box left (earlier) or right (later).
Vertical (amplitude) Scroll Bar	The **Vertical Scroll Bar** is located to the right of the amplitude (mV) division scale. To move the selected channel's vertical position, click on the scroll box or hold the mouse button down on the scroll box and drag the box up or down.
Autoscale waveforms	Click on the **Display** menu to access this option. Optimizes the vertical (amplitude) scale so that all the data will be shown on the screen.
Autoscale horizontal	Click on the **Display** menu to access this option. Compresses or expands the horizontal (time) scale so the entire recording will fit on the screen.
Grids	Click on the **File** menu and choose **Display Preferences** to access this option. Grids provide a quick visual guide to determine amplitude and duration. To turn the grids display on or off, change the **Preferences** setting to either **Show Grids** or **Hide Grids.**
Overlap button	Emulates an oscilloscope display.
Split button	Emulates a chart recorder display. This is the default display mode.
Adjust Baseline button	The **Adjust Baseline** button only appears in Lesson 3 ECG I. Allows you to position the waveform up or down in small increments so that the baseline can be set to exactly zero. This is not needed to get accurate amplitude measurements, but may be desired before making a printout or when using grids. When the **Adjust Baseline** button is pressed, **Up** and **Down** buttons will appear. Simply click on these to move the waveform up or down.

by contraction is weak, but it can be picked up by surface sensors to produce the recording, referred to as an *electromyogram (EMG)*.

Skeletal muscle is one of three types of muscle tissue in the body. Skeletal muscle is mainly attached to the various bones of the skeleton; it helps provide a framework and support for the body, as well as the ability to move in response to willing an action to occur. (Cardiac muscle is found only in the heart, where it originates and spreads the heartbeats that keep us alive. Smooth muscle is located mostly in the tubular organs of our internal systems; it accomplishes many types of internal movement, such as the churning action of the stomach and the transport of urine from the kidney to the bladder.)

Each skeletal muscle in our body is composed of thousands of skeletal muscle fibers (cells). Skeletal muscle fibers are unique among the three types of muscle tissue, in that they contract only in response to stimulation by a motor neuron. The response of a single skeletal muscle fiber to stimulation by its motor neuron is called a *twitch*. A twitch consists of a rapid period of tension development and contraction, followed by a decrease in tension and relaxation of the muscle.

Each motor neuron associated with our skeletal muscle fibers innervates (stimulates, controls) a certain number of fibers, from fewer than 10 to thousands. A single motor neuron plus all of the muscle fibers it controls is called a *motor unit*. In a motor unit, whenever the motor neuron is activated, all of its muscle fibers are commanded to contract simultaneously. Therefore, in real life, entire motor units, rather than single muscle fibers, are contracting to accomplish body movements.

When a single muscle fiber is stimulated by its motor neuron to contract, then relaxes completely, and then contracts again, the subsequent contractions all produce the same amount of muscle tension (fig. 21.3a). However, if a muscle fiber is stimulated again, before it has time to relax completely, then the second twitch will generate more tension than the first. This is called *temporal (wave) summation,* and it is the body's way of increasing the force generated by our muscles (fig. 21.3b). When a muscle fiber is stimulated to contract by several stimuli in rapid succession, and there is no opportunity for muscle relaxation between stimuli, the tension increases until it reaches the maximum possible force for that muscle fiber, and then remains at that level until the stimulation stops; this type of sustained contraction is called *tetanus*, or a *tetanic contraction* (fig. 21.3c).

FIGURE 21.3 Myograms of (a) isolated muscle twitches, (b) temporal summation, and (c) a tetanic contraction.

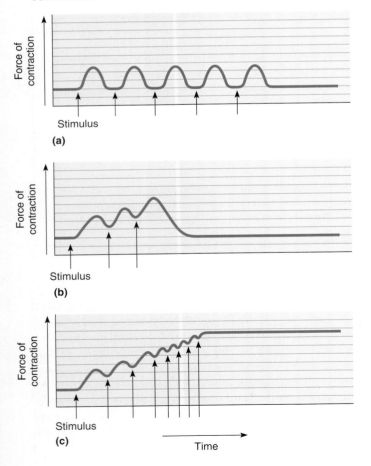

Although there are many motor units in a skeletal muscle, not all of them are stimulated during every muscle contraction. For example, if a person decides to pick up a 1 lb. weight, just a small number of motor units will be activated to contract. If, however, that person picks up a 10 lb. weight instead, many more motor units will be stimulated, in order to generate enough muscle tension to pick up the heavier object. An increase in the number of activated motor units in a skeletal muscle, in order to increase the tension in the muscle, is called *motor unit recruitment.* See figure 21.4 for an illustration of motor unit recruitment.

When skeletal muscles are at rest (not in the process of contracting), they maintain a slight amount of contraction, called muscle tone, or *tonus*. In resting skeletal muscles, motor units take turns contracting, so that the muscle is always ready to contract on short notice. This helps keep the motor units in good health, and in the case of the postural muscles in the back, it allows us to maintain good posture for many hours at a time.

Recruitment and tonus will be demonstrated in Part 1 of this laboratory exercise; as you increase the clench force of your first, more motor units will be recruited to generate increasing amounts of muscle tension. You will be able to observe tonus in the short periods of time between fist clenches. The periods of tonus will show up in the BIOPAC data window as very low mV readings.

Strenuous exercise or work, which requires repeated or sustained maximal muscle contraction, sometimes results in *muscle fatigue,* a state in which muscle contraction decreases, even when the muscle is still being stimulated. Several factors can induce muscle fatigue, but the main cause appears to be a buildup of lactic acid from the anaerobic respiration of glucose, the resulting decrease in blood pH in the vicinity, and the decreased response of the muscle at low pH. Part 2 of this lab exercise will demonstrate muscle fatigue. You will use a *hand dynamometer* to generate a recording called a *dynagram*. See figure 21.5 for the appearance and proper grip positions of two different types of dynamometers. Muscle fatigue will be induced by a sustained, maximum fist clench, and will appear on the dynagram as a steady decline in muscle tension.

Part 1—Electromyography: Standard and Integrated EMG

Procedure A—Setup

1. With your computer turned **ON** and the BIOPAC MP3X or MP45 unit turned **OFF,** plug the electrode lead set (SSL2) into Channel 3 and plug the headphones into the back of the unit if desired. Plug the BIOPAC SS25 Hand Dynamometer into Channel 1 only if you are going to do Part 2 of this lab exercise. The dynamometer is not used in Part 1.
2. Turn on the MP3X or MP45 Data Acquisition Unit.
3. Attach the electrodes to the dominant forearm (the right arm if the subject is right-handed, the left if he or she is left-handed) of the subject as shown in figure 21.6. This

FIGURE 21.4 Motor unit recruitment. (a) A small number of activated motor units only generates enough force to pick up a lightweight object. (b) Recruitment of more motor units has increased the force to the extent that a much heavier object can be lifted.

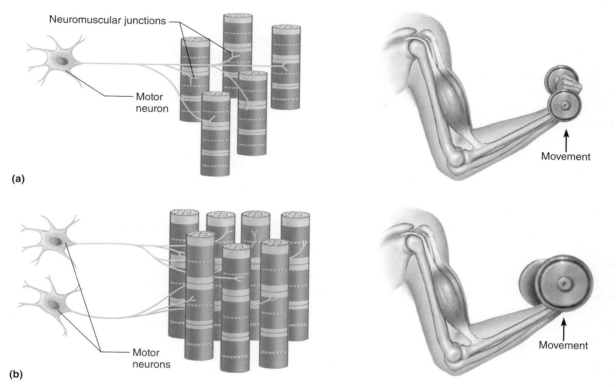

(a)

(b)

FIGURE 21.5 There are two types of hand dynamometers that you may have in your lab. Proper grip positions are also shown.

Source: Courtesy of and © *BIOPAC* Systems, Inc.

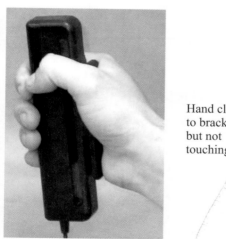

SS25LA grip position

SS25L grip position

FIGURE 21.6 Electrode lead attachment.

Source: Courtesy of and © *BIOPAC* Systems, Inc.

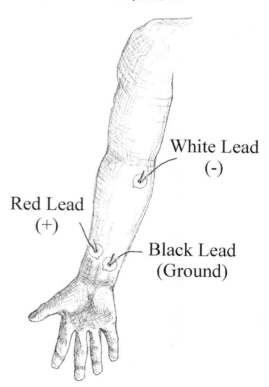

will be **Forearm I.** To help to ensure a good contact, make sure the area to be in contact with the electrodes is clean by wiping with the abrasive pad (ELPAD) or an alcohol wipe. A small amount of BIOPAC electrode gel (GEL1) may also be used to make better contact between the sensor in the electrode and the skin. For optimal adhesion, the electrodes should be placed on the skin at least 5 minutes before the calibration procedure.

4. Attach the electrode lead set by the pinch connectors as shown in figure 21.6. Make sure the proper color is attached to the appropriate electrode.

5. Start the BIOPAC Student Lab Program for Electromyography I, lesson L01-EMG-1.

6. Designate a unique filename to be used to save the subject's data.

Procedure B—Calibration

1. Click on **Calibrate.** A dialog box will direct you through the calibration procedure. After reading the box, click **OK.**

2. Wait about 2 seconds, clench the fist as hard as possible, then release. Wait for the calibration to stop (this will take about 8 seconds).

3. The calibration recording should look similar to figure 21.7. If it does not, click **Redo Calibration.** If the baseline is not at zero, the subject did not wait 2 seconds before clenching.

Procedure C—Recording

1. Fill out the subject profile on Laboratory Assessment 21. Click on **Record** and have the subject:
   a. Clench-Release-Wait, holding for 2 seconds during the clench and wait phases.
   b. Repeat this cycle a total of 4 times. During the first cycle the clench should be gentle; then increase the

strength with each successive cycle so that the fourth clench is the maximum force.

2. Click **Suspend.** The recording should be similar to figure 21.8. If it is not, click **Redo** and repeat step 1.

3. You may be able to remove the electrodes from the dominant forearm and place them and the lead set on the nondominant forearm of the same subject. If they do not adhere well, use new electrodes for the nondominant arm. Click **Resume** and repeat from step 1 of Procedure B—Calibration.

4. Click **Stop** when you have completed the procedure.

5. If you want to listen to the EMG signal, put on the headphones and click **Listen.** Have the subject experiment by changing the clench force as you watch the screen. When finished, click **Stop.** This data will not be saved for analysis.

Procedure D—Data Analysis

1. You can either analyze the current file now or save it and do the analysis later by using the **Review Saved Data** mode and selecting the correct file.

2. Note the channel number designations: **CH 1** displays **EMG** (the actual voltage recording) and **CH 40** the **Integrated EMG** (which reflects the absolute intensity of the voltage).

3. Notice the measurement box settings: **CH 40** should be set on "mean," and the others should be set on "none."

4. After clicking the I-beam icon (this will activate the select function), select the first EMG cluster as shown in figure 21.9.

5. The values in the measurement boxes should be recorded in the table in Part IA of Laboratory Assessment 21.

6. Repeat for the other three clusters and record the data in Part IA of the laboratory assessment.

7. Complete Part I of the laboratory assessment.

FIGURE 21.7 Calibration recording part 1 (single wave).

Source: Courtesy of and © *BIOPAC* Systems, Inc.

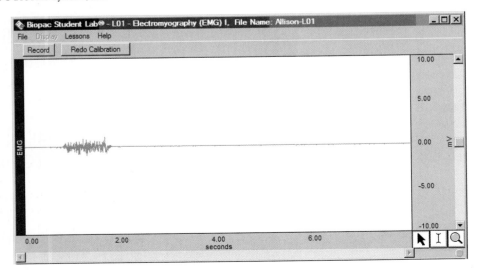

FIGURE 21.8 Clench-release recording.

Source: Courtesy of and © *BIOPAC* Systems, Inc.

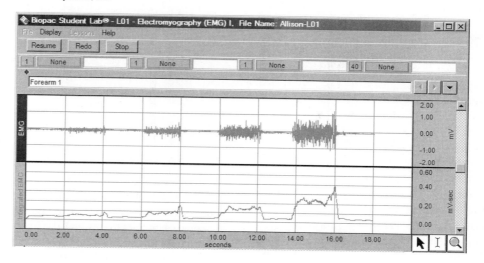

FIGURE 21.9 I-beam measurement.

Source: Courtesy of and © *BIOPAC* Systems, Inc.

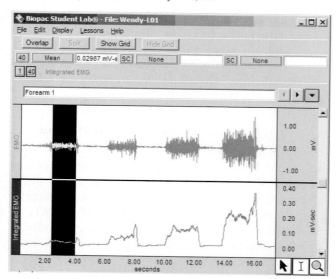

Part 2—Electromyography: Motor Unit Recruitment and Fatigue

Procedure A—Setup

If continuing with the same subject used in Part 1, move to setup step 5.

1. With your computer turned **ON** and the BIOPAC MP3X or MP45 unit turned **OFF,** plug the electrode lead set (SSL2) into Channel 1 and BIOPAC SS25 Hand Dynamometer into Channel 2. Plug the headphones into the back of the unit if desired.

2. Turn on the MP3X or MP45 Data Acquisition Unit.

3. Attach the electrodes to the dominant forearm (the right arm if the subject is right-handed, the left if he or she is left-handed) of the subject as shown in figure 21.6. This will be **Forearm 1.** To help to ensure a good contact, make sure the area to be in contact with the electrodes is clean by wiping with the abrasive pad (ELPAD) or alcohol wipe. A small amount of BIOPAC electrode gel (GEL1) may also be used to make better contact between the sensor in the electrode and the skin. For optimal adhesion, the electrodes should be placed on the skin at least 5 minutes before the calibration procedure.

4. Attach the electrode lead set by the pinch connectors as shown in figure 21.6. Make sure the proper color is attached to the appropriate electrode.

5. Start the BIOPAC Student Lab Program for Electromyography II, lesson L02-EMG-2.

6. Designate a unique filename to be used to save the subject's data, then click **OK.**

Procedure B—Calibration

1. Click **Calibrate.** Set the hand dynamometer on the table and click **OK.** This establishes a zero-force calibration before you continue the calibration sequence.

2. Grasp the hand dynamometer with the dominant hand (Forearm 1) as close to the dynagrip crossbar as possible without touching the crossbar. Note the position of your hand and try to replicate this during the recording phase.

3. Follow the instructions on the two successive pop-up windows and click **OK** when ready for each.

4. Wait 2 seconds, then clench the hand dynamometer as hard as possible, then release.

5. Wait for the calibration to stop. This will take about 8 seconds.

FIGURE 21.10 Calibration recording part 2 (single wave).

Source: Courtesy of and © *BIOPAC* Systems, Inc.

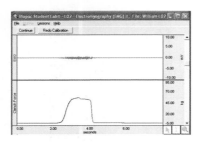

6. If the calibration recording does not resemble figure 21.10, click **Redo Calibration.**
7. Before proceeding to the recording procedure, you need to determine the force increments that you will use. Using the maximum force from the calibration, the increments should be as follows: 0–25 kg use 5 kg increments; 25–50 kg use 10 kg increments; > 50 kg use 20 kg increments.

Procedure C—Recording

1. Click on **Record** and have the subject:
   - Clench-Release-Wait and repeat with increasing clench force by the increments determined in step 7 above until you reach your maximum clench force.
   - There should be a 2-second interval between clenches.
2. Click **Suspend.** The recording should resemble figure 21.11 for Motor Unit Recruitment. If it does not, click **Redo** and repeat step 1.
3. Click **Resume.**
4. Clench the hand dynamometer with your maximum force. Note the force and try to maintain it.
5. When the maximum clench force has decreased by 50%, click **Suspend.** The recording should resemble figure 21.12 for Fatigue. If it does not, click **Redo** and repeat steps 3–5.
6. If correct, click **Stop.**
7. If you want to listen to the EMG signal, put on the headphones and click **Listen.** Have the subject experiment

FIGURE 21.11 Motor unit recruitment.

Source: Courtesy of and © *BIOPAC* Systems, Inc.

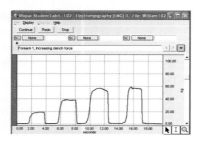

FIGURE 21.12 Fatigue.

Source: Courtesy of and © *BIOPAC* Systems, Inc.

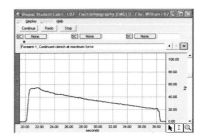

by changing the clench force as you watch the screen. When finished, click **Stop.**
8. Remove the electrodes from the dominant forearm and place new electrodes and the lead set on the nondominant forearm of the same subject. Click **Forearm 2.** This will return you to the Calibration sequence. Proceed as with the dominant arm for Calibration and Recording.
9. When finished, click **Done.**

Procedure D—Data Analysis

1. You can either analyze the current file or save it and do the analysis later using the **Review Saved Data** mode and selecting the correct file. The file with the extension "1-L02" is Forearm 1 and "2-L02" is Forearm 2.
2. Note the channel number designations: **CH 1** displays EMG (hidden); **CH 40** displays the **Integrated EMG** (mV), and **CH 41** displays clench force.
3. Note the channel/measurement boxes as follows:

> **CH 41: Mean; CH 40: Mean; CH 41: Value; and CH 40: DeltaT**

4. Use the I-beam cursor to select an area on the plateau phase of the first clench as shown in figure 21.13. Record the measurements in Laboratory Assessment 21 in the table for Motor Unit Recruitment (Part II A1).
5. Repeat step 4 on the plateau for each successive clench.
6. Scroll to the second recording segment (for Fatigue).

FIGURE 21.13 Proper selection of the plateau area of the first selected clench.

Source: Courtesy of and © *BIOPAC* Systems, Inc.

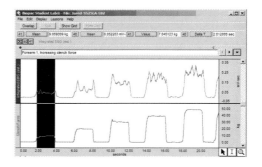

7. Use the I-beam cursor to select a point of maximal clench force at the start of Segment 2. Record the measurements in the laboratory assessment in the table for Fatigue (Part II A2).

8. Calculate 50% of the maximum clench force from step 7. Record the calculation in the laboratory assessment in the table for Fatigue (Part II A2).

9. Select the area from the point of 50% clench force to the maximum clench force by dragging the cursor. See figure 21.14 for proper selection of the area. Note the time to fatigue (CH 40/Δ T) and record in the laboratory assessment in the table for Fatigue (Part II A2).

10. Repeat from step 2 for the second forearm.

11. Complete Part II of the laboratory assessment.

12. Exit the program.

FIGURE 21.14 Proper selection of the area between 50% clench force and maximum clench force.

Source: Courtesy of and © *BIOPAC* Systems, Inc.

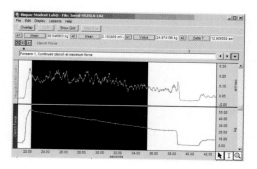

NOTES

Name _____

Date _____

Section _____

The ⚠ corresponds to the indicated outcome(s) found at the beginning of the laboratory exercise.

Electromyography: BIOPAC© Exercise

Subject Profile

Name _____   Height _____

Age _____   Weight _____

Gender: Male/Female   Dominant Forearm: right/left

I. Standard and Integrated EMG

Part A   Data and Calculations Assessments

1. EMG Measurements ⚠ ⚠

Cluster #	Dominant arm	Non-dominant arm
	40 Mean	40 Mean
1		
2		
3		
4		

Note: "Clusters" are the EMG bursts associated with each clench.

2. Use the mean measurement from the preceding table to compute the percentage increase in EMG activity recorded between the weakest clench and the strongest clench of Forearm 1.

Calculation:

Answer: _____ %

Part B   Assessments

Complete the following:

1. If you did both forearms on the same subject, are the mean measurements for the right and left maximum grip EMG cluster the same? Which one suggests the greater grip strength? Explain. ⚠

_____

_____

_____

2. What factors in addition to gender contribute to observed differences in clench strength? **2**

_____

_____

_____

3. Explain the source of signals detected by the EMG electrodes. **3**

_____

_____

_____

4. What is meant by the term *motor unit recruitment?*

_____

_____

_____

5. Define electromyography.

_____

_____

_____

II. Motor Unit Recruitment and Fatigue

Part A   Data and Calculations Assessments

1. Increasing Clench Force Data: **4** **5**

Peak #	Assigned Force Increment SS25L/LA = Kg SS56L = kgf/m²	Dominant forearm		Non-dominant forearm	
		Force at Peak	Integrated EMG (mV)	Force at Peak	Integrated EMG (mV)
		41 Mean	**40** Mean	**41** Mean	**40** Mean
1					
2					
3					
4					
5					
6					
7					
8					

2. Maximum Clench Force Data: 6 7

Dominant forearm			Non-dominant forearm		
Maximum Clench Force	50% of Max Clench Force	Time* to Fatigue	Maximum Clench Force	50% of Max Clench Force	Time* to Fatigue
41 Value	calculate	40 Delta T	41 Value	calculate	40 Delta T

*Note: You do not need to indicate the Delta T (time to fatigue) polarity. The polarity of the Delta T measurement reflects the direction the "I-beam" cursor was dragged to select the data. Data selected left to right will have a positive ("+") polarity, while data selected right to left will have a negative ("−") polarity.

Part B   Assessments

Complete the following:

1. Compare the strength of the dominant forearm to that of the nondominant forearm. 5

2. Is there a difference in the absolute values of force generated by males and females in your class? What might explain the difference? 5

3. Is there a difference in fatigue times between the two forearms? Why might this happen?

4. Define fatigue.

5. Define dynamometry.

Muscles of the Head and Neck

Purpose of the Exercise

To review the locations, actions, origins, and insertions of the muscles of the head and neck.

Materials Needed

Human torso model with musculature
Human skull
Human skeleton, articulated

For Learning Extension Activity:
Long rubber bands

Learning Outcomes

After completing this exercise, you should be able to

1. Locate and identify the muscles of facial expression, the muscles of mastication, the muscles that move the head and neck, and the muscles that move the hyoid bone and larynx.

2. Describe and demonstrate the action of each of these muscles.

3. Locate the origin and insertion of each of these muscles in a human skeleton and the musculature of the human torso model.

Pre-Lab

Carefully read the introductory material and examine the entire lab. Be familiar with the locations, actions, origins, and insertions of the muscles of the head and neck from lecture or the textbook. Answer the pre-lab questions.

Pre-Lab Questions: Select the correct answer for each of the following questions:

1. Muscles of mastication are all inserted on the
   - **a.** mandible.
   - **b.** maxilla.
   - **c.** tongue.
   - **d.** teeth.

2. Muscles of mastication include the following *except* the
   - **a.** lateral pterygoid.
   - **b.** masseter.
   - **c.** temporalis.
   - **d.** platysma.

3. Facial expressions are so variable partly because many facial muscles are inserted
   - **a.** on the mandible.
   - **b.** in the skin.
   - **c.** on the maxilla.
   - **d.** on facial bones.

4. Which of the following muscles does *not* move the hyoid bone?
   - **a.** sternocleidomastoid
   - **b.** mylohyoid
   - **c.** sternohyoid
   - **d.** thyrohyoid

5. Which of the following is *not* a facial expression muscle?
   - **a.** zygomaticus major
   - **b.** buccinator
   - **c.** orbicularis oris
   - **d.** scalenes

6. Flexion of the head and neck is an action of the _____ muscle.
   - **a.** semispinalis capitis
   - **b.** splenius capitis
   - **c.** sternocleidomastoid
   - **d.** trapezius

7. Which of the following muscles is the most visible from an anterior view of a person in anatomical position?
   - **a.** splenius capitis
   - **b.** nasalis
   - **c.** temporalis
   - **d.** medial pterygoid

The skeletal muscles of the head include the muscles of facial expression. Human facial expressions are more variable than those of other mammals because many of the muscles have insertions in the skin, rather than on bones, allowing great versatility of movements and emotions. The extrinsic muscles of eye movements will be covered in a later laboratory exercise.

The muscles of mastication (chewing) are all inserted on the mandible. Muscles for chewing allow the movements of depression and elevation of the mandible to open and close the jaw. However, jaw movements also involve protraction, retraction, lateral excursion, and medial excursion needed to bite off pieces of food and grind the food.

The muscles that move the head are located in the neck. The actions of flexion, extension, and hyperextension of the head and the cervical vertebral column occur when right and left paired muscles work simultaneously. Additional actions, such as lateral flexion and rotation, result from muscles on one side only contracting or alternating muscles contracting.

A group of neck muscles that have attachments on the hyoid bone and larynx assist in swallowing and speech. Some of the neck muscles that move the head and neck have their attachments in the thorax.

Procedure—Muscles of the Head and Neck

The muscle lists and tables in this laboratory exercise reflect muscle groupings using a combination of related locations and functions. Study and master one group at a time as a separate task. Frequently refer to the illustrations, tables, skeletons, and models. Consider working with a partner or a small group to study and review.

1. Study figure 22.1 and table 22.1.
2. Locate the following muscles in the human torso model and in your body whenever possible:

   muscles of facial expression
   - epicranius (occipitofrontalis)
     - frontal belly (frontalis)
     - occipital belly (occipitalis)
   - nasalis
   - orbicularis oculi
   - orbicularis oris
   - risorius
   - zygomaticus major
   - zygomaticus minor
   - buccinator
   - platysma

3. Demonstrate the actions of these muscles in your body.
4. Locate the origins and insertions of these muscles in the human skull and skeleton.
5. Study figures 22.1 and 22.2 and table 22.2.
6. Locate the following muscles in the human torso model and in your body whenever possible:

   muscles of mastication
   - masseter
   - temporalis
   - lateral pterygoid
   - medial pterygoid

FIGURE 22.1 Muscles of facial expression and mastication (a) anterior view and (b) lateral view.

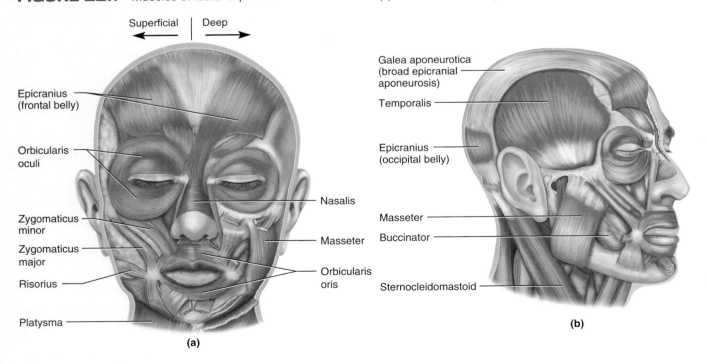

220

TABLE 22.1 Muscles of Facial Expression

Muscle	Origin	Insertion	Action
Epicranius	Occipital bone; galea aponeurotica	Skin of eyebrows; galea aponeurotica	Raises eyebrows; retracts scalp
Nasalis	Maxilla lateral to nose	Bridge of nose	Widens nostrils
Orbicularis oculi	Maxillary and frontal bones	Skin around eye	Closes eyes as in blinking
Orbicularis oris	Muscles near the mouth	Skin of lips	Closes lips; protrudes lips as for kissing
Risorius	Fascia near ear	Corner of mouth	Draws corner of mouth laterally
Zygomaticus major	Zygomatic bone	Corner of mouth	Raises corner of mouth as when smiling and laughing
Zygomaticus minor	Zygomatic bone	Corner of mouth	Raises corner of mouth as when smiling and laughing
Buccinator	Lateral surfaces of maxilla and mandible	Orbicularis oris	Compresses cheeks inward as when blowing air
Platysma	Fascia in upper chest	Lower border of mandible and skin at corner of mouth	Draws angle of mouth downward as when pouting or expressing horror

FIGURE 22.2   Deep muscles of mastication.

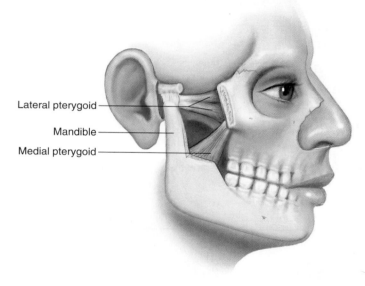

Lateral pterygoid
Mandible
Medial pterygoid

TABLE 22.2 Muscles of Mastication

Muscle	Origin	Insertion	Action
Masseter	Lower border of zygomatic arch	Lateral surface of mandible	Elevates mandible
Temporalis	Temporal bone	Coronoid process and anterior ramus of mandible	Elevates mandible
Lateral pterygoid	Sphenoid bone	Anterior surface of mandibular condyle	Depresses and protracts mandible and moves it from side to side as when grinding food
Medial pterygoid	Sphenoid, palatine, and maxillary bones	Medial surface of mandible	Elevates mandible and moves it from side to side as when grinding food

7. Demonstrate the actions of these muscles in your body.
8. Locate the origins and insertions of these muscles in the human skull and skeleton.
9. Study figures 22.3 and 22.4 and table 22.3.
10. Locate the following muscles in the human torso model and in your body whenever possible:

   muscles that move head and neck
   - sternocleidomastoid
   - trapezius (superior part)
   - scalenes (anterior, middle, and posterior)
   - splenius capitis
   - semispinalis capitis

11. Demonstrate the actions of these muscles in your body.
12. Locate the origins and insertions of these muscles in the human skull and skeleton.
13. Complete Part A of Laboratory Assessment 22.

14. Study figure 22.5 and table 22.4.
15. Locate the following muscles in the human torso model:

   muscles that move hyoid bone and larynx
   - suprahyoid muscles
     - digastric (2 parts)
     - stylohyoid
     - mylohyoid
   - infrahyoid muscles
     - sternohyoid
     - omohyoid (2 parts)
     - sternothyroid
     - thyrohyoid

16. Demonstrate the actions of these muscles in your body.
17. Locate the origins and insertions of these muscles in the human skull and skeleton.
18. Complete Parts B, C, and D of the laboratory assessment.

221

Learning Extension Activity

A long rubber band can be used to simulate muscle locations, origins, insertions, and actions on the human torso model, the skeleton, or a laboratory partner. Hold one end of the rubber band firmly on the origin location of a muscle, then slightly stretch the rubber band and hold the other end on the insertion site. Allow the insertion end to slowly move toward the origin end to simulate the contraction and action of the muscle.

FIGURE 22.3   Posterior muscles that move the head and neck.

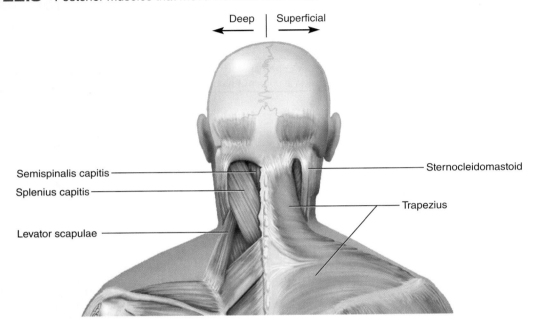

FIGURE 22.4   Muscles that move the head and neck, lateral view.

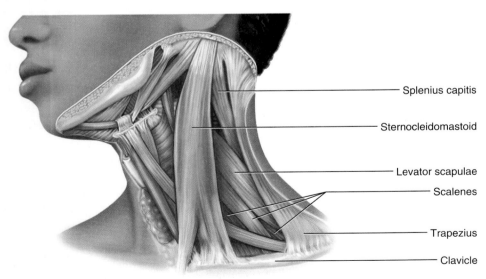

TABLE 22.3 Muscles That Move Head and Neck

Muscle	Origin	Insertion	Action
Sternocleidomastoid	Manubrium of sternum and medial clavicle	Mastoid process of temporal bone	Flexion of head and neck; rotation of head to left or right
Trapezius (superior part)	Occipital bone and spinous processes of C7 and several thoracic vertebrae	Clavicle and the spine and acromion of scapula	Extends head (as a synergist); (primary actions on scapula are covered in another lab)
Scalenes (anterior, middle, and posterior)	Transverse processes of all cervical vertebrae	Ribs 1–2	Elevates ribs 1–2; flexion and rotation of neck
Splenius capitis	Spinous processes of C7–T6	Mastoid process and occipital bone	Extends head; rotates head
Semispinalis capitis	Processes of inferior cervical and superior thoracic vertebrae	Occipital bone	Extends head; rotates head

FIGURE 22.5 Muscles that move the hyoid bone and larynx assist in swallowing and speech and are grouped into suprahyoid and infrahyoid muscles.

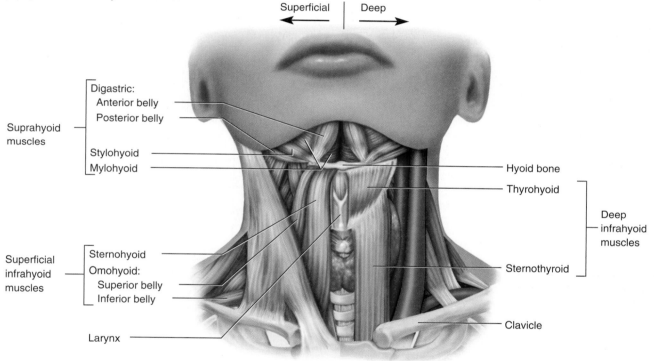

TABLE 22.4 Muscles That Move Hyoid Bone and Larynx

Muscle	Origin	Insertion	Action
Digastric (2 parts)	Inferior mandible (anterior belly) and mastoid process (posterior belly)	Hyoid bone	Opens mouth; depresses mandible; elevates hyoid bone
Stylohyoid	Styloid process of temporal bone	Hyoid bone	Retracts and elevates hyoid bone
Mylohyoid	Mandible	Hyoid bone	Elevates hyoid bone during swallowing
Sternohyoid	Manubrium and medial clavicle	Hyoid bone	Depresses hyoid bone
Omohyoid (2 parts)	Superior border of scapula	Hyoid bone	Depresses hyoid bone
Sternothyroid	Manubrium	Thyroid cartilage of larynx	Depresses larynx
Thyrohyoid	Thyroid cartilage of larynx	Hyoid bone	Depresses hyoid bone; elevates larynx

Name _____

Date _____

Section _____

The ⒜ corresponds to the indicated outcome(s) found at the beginning of the laboratory exercise.

Muscles of the Head and Neck

Part A   Assessments

Identify the muscles indicated in the head and neck in figures 22.6 and 22.7.

FIGURE 22.6   Label the anterior muscles of the head. ⒜

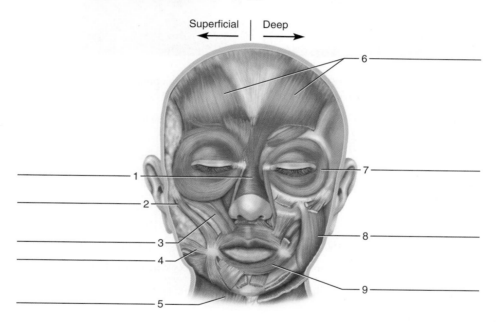

FIGURE 22.7 Identify the muscles of the head and neck in this lateral view of a cadaver, using the terms provided. 1

Anatomy & Physiology REVEALED® aprevealed.com

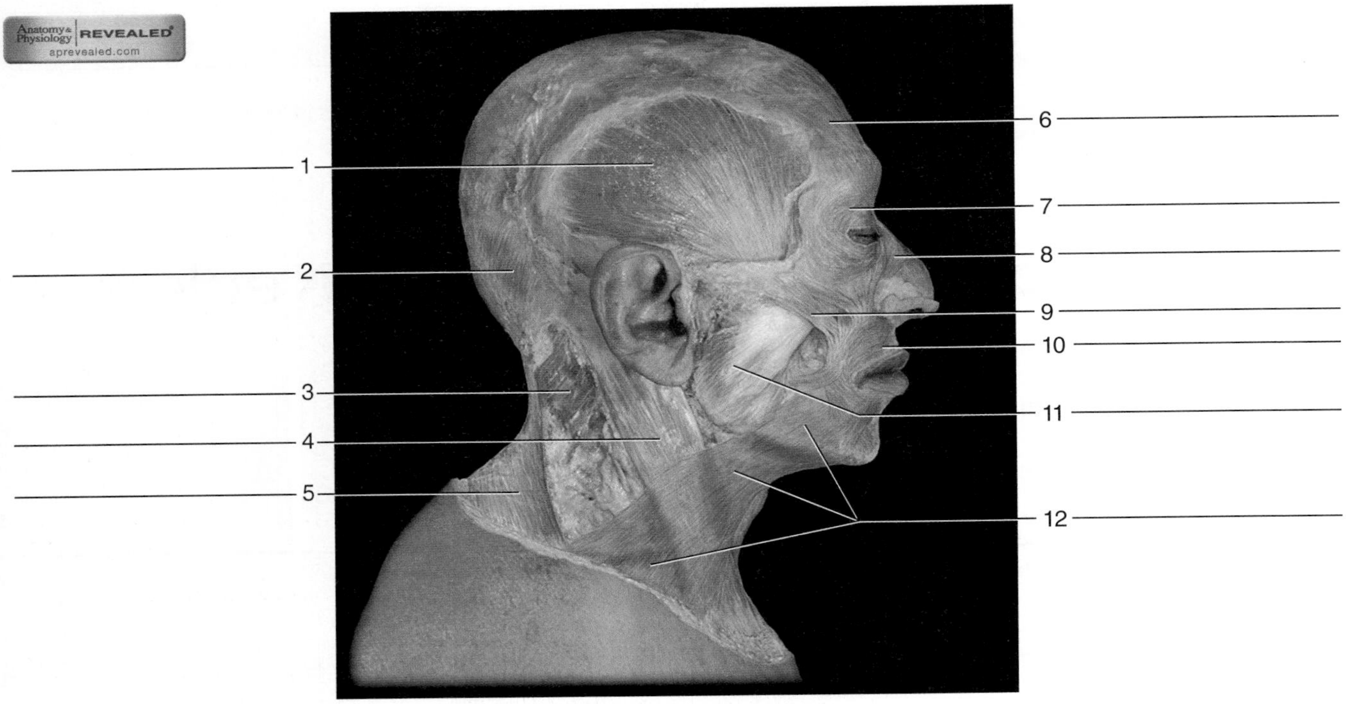

Terms:

Frontal belly (of epicranius)	Orbicularis oculi	Sternocleidomastoid
Masseter	Orbicularis oris	Temporalis
Nasalis	Platysma	Trapezius
Occipital belly (of epicranius)	Splenius capitis	Zygomaticus major

Part B Assessments

Complete the following statements:

1. When the ______________ contracts, the corner of the mouth is drawn upward and laterally when laughing. 2
2. The ______________ acts to compress the wall of the cheeks when air is blown out of the mouth. 2
3. The ______________ causes the lips to close and pucker during kissing, whistling, and speaking. 2
4. The temporalis acts to ______________. 2
5. The ______________ pterygoid can close the jaw and pull it sideways. 2
6. The ______________ pterygoid can protrude the jaw, pull the jaw sideways, and open the mouth. 2
7. The ______________ can close the eye, as in blinking. 2
8. The ______________ can pull the head toward the chest. 2
9. The muscle used for pouting and to express horror is the ______________. 2
10. The muscle used to widen the nostrils is the ______________. 2
11. The muscle used to elevate the hyoid bone during swallowing is the ______________. 2
12. The ______________ raises the eyebrows and moves the scalp. 2

Part C Assessments

Name the muscle indicated by the following combinations of origin and insertion.

Origin	Insertion	Muscle
1. Manubrium of sternum	Thyroid cartilage of larynx	_____
2. Zygomatic arch	Lateral surface of mandible	_____
3. Sphenoid bone	Anterior surface below mandibular condyle	_____
4. Manubrium of sternum and medial clavicle	Mastoid process of temporal bone	_____
5. Lateral surfaces of mandible and maxilla	Orbicularis oris	_____
6. Fascia in upper chest	Lower border of mandible and skin around corner of mouth	_____
7. Temporal bone	Coronoid process and anterior ramus of mandible	_____
8. Spinous processes of cervical and thoracic vertebrae (C7–T6)	Mastoid process of temporal bone and occipital bone	_____
9. Styloid process of temporal bone	Hyoid bone	_____
10. Transverse processes of cervical vertebrae	Ribs 1–2	_____
11. Frontal and maxillary bones	Skin around eye	_____
12. Medial clavicle and manubrium	Hyoid bone	_____

Part D Assessments

Critical Thinking Assessment

Identify the muscles of various facial expressions in the photographs of figure 22.8.

1. _____

2. _____

3. _____

4. _____

5. _____

FIGURE 22.8 Identify the muscles of facial expression being contracted in each of these photographs (a–c), using the terms provided. **1** **2**

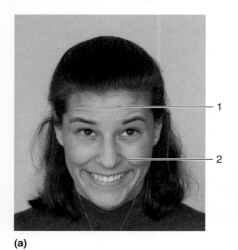

(a)

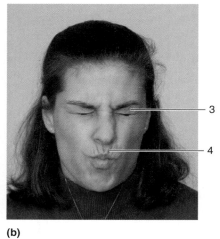

(b)

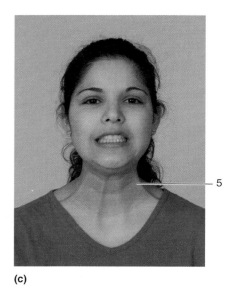

(c)

Terms:
Epicranius (frontal belly)
Orbicularis oculi
Orbicularis oris
Platysma
Zygomaticus major

Muscles of the Chest, Shoulder, and Upper Limb

Purpose of the Exercise

To review the locations, actions, origins, and insertions of the muscles in the chest, shoulder, and upper limb.

Materials Needed

Human torso model
Human skeleton, articulated
Muscular models of the upper limb

For Learning Extension Activity:
Long rubber bands

Learning Outcomes

After completing this exercise, you should be able to

1 Locate and identify the muscles of the chest, shoulder, and upper limb.

2 Describe and demonstrate the action of each of these muscles.

3 Locate the origin and insertion of each of these muscles in a human skeleton and on the muscular models.

Pre-Lab

Carefully read the introductory material and examine the entire lab. Be familiar with the locations, actions, origins, and insertions of the muscles of the chest, shoulder, and upper limb from lecture or the textbook. Visit www.mhhe.com/martinseries2 for LabCam videos. Answer the pre-lab questions.

Pre-Lab Questions: Select the correct answer for each of the following questions:

1. Chest and shoulder muscles that move the arm at the shoulder joint have insertions on the
   a. scapula.   **b.** clavicle and ribs.
   c. humerus.   **d.** radius or ulna.

2. Muscles located in the arm that move the forearm at the elbow joint have insertions on the
   a. scapula and clavicle.   **b.** humerus.
   c. radius or ulna.   **d.** carpals, metacarpals, and phalanges.

3. Rotator cuff muscles all have origins on the
   a. scapula.   **b.** clavicle.
   c. humerus.   **d.** radius.

4. Which of the following is *not* a rotator cuff muscle?
   a. supraspinatus   **b.** infraspinatus
   c. teres major   **d.** teres minor

5. The belly of the muscle is usually _____ the region pulled.
   a. located on   **b.** opposite
   c. distal to   **d.** proximal to

6. The belly of the flexor digitorum profundus muscle is located in the
   a. posterior superficial region of the forearm.
   b. anterior deep region of the forearm.
   c. anterior superficial region of the forearm.
   d. posterior superficial region of the arm.

7. There are more muscles that move the forearm than move the hand.
   True _____   False _____

8. The action of the brachialis muscle is flexion of the elbow joint.
   True _____   False _____

Many of the chest and shoulder muscles have their insertions on the scapula and provide various movements of the pectoral girdle. Some chest and shoulder muscles crossing the shoulder joint have insertions on the humerus and are responsible for moving the arm in various ways at the shoulder joint. Muscles located in the arm having insertions on the radius and ulna are responsible for forearm movements at the elbow joint. Muscles located in the forearm with insertions on carpals, metacarpals, or phalanges are responsible for various movements of the hand.

The muscles of the upper limb vary from large sizes that provide powerful actions to numerous smaller muscles in the forearm and hand that allow finer motor skills. The human hand, with its opposable thumb, enables us to grasp objects and provides great dexterity for tasks such as writing, drawing, and playing musical instruments.

As you demonstrate the actions of muscles in an upper limb, use the standard anatomical regions to describe them. The arm refers to shoulder to elbow; the forearm refers to elbow to wrist; and the hand includes the wrist, palm, and fingers. Also remember that the belly of a muscle responsible for a joint movement is usually proximal to the region being pulled. If a muscle crosses two joints, the actions allowed are even more diverse.

Procedure—Muscles of the Chest, Shoulder, and Upper Limb

The muscle lists and tables in this laboratory exercise reflect muscle groupings using a combination of related locations and functions. Study and master one group at a time as a separate task. Frequently refer to the illustrations, tables, skeletons, and models. Consider working with a partner or a small group to study and review.

1. Study figures 23.1 and 23.2 and table 23.1.
2. Locate the following muscles in the human torso model and models of the upper limb. Also locate in your body as many of the muscles as you can.

   muscles that move the pectoral girdle
   - anterior muscles
     - pectoralis minor
     - serratus anterior
   - posterior muscles
     - trapezius
     - rhomboid major
     - rhomboid minor
     - levator scapulae

FIGURE 23.1 Anterior muscles of the chest, shoulder, and arm. Some regional muscles are also labeled.

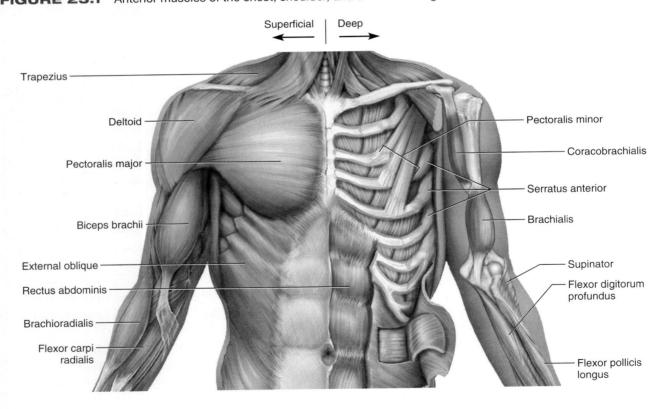

Superficial | Deep

Trapezius

Deltoid

Pectoralis major

Biceps brachii

External oblique

Rectus abdominis

Brachioradialis

Flexor carpi radialis

Pectoralis minor

Coracobrachialis

Serratus anterior

Brachialis

Supinator

Flexor digitorum profundus

Flexor pollicis longus

FIGURE 23.2 Posterior muscles of the shoulder, back, and arm.

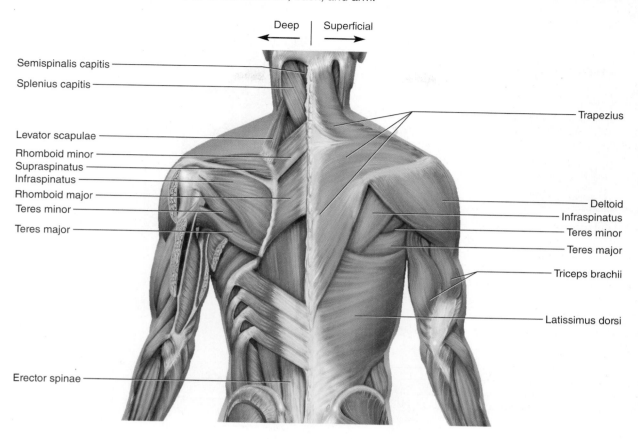

TABLE 23.1 Muscles That Move the Pectoral Girdle

Muscle	Origin	Insertion	Action
Pectoralis minor	Sternal ends of ribs 3–5	Coracoid process of scapula	Pulls scapula forward and downward; raises ribs
Serratus anterior	Lateral surfaces of most ribs	Medial border of scapula	Pulls scapula anteriorly (protracts scapula)
Trapezius	Occipital bone and spinous processes of C7 and upper thoracic vertebrae	Clavicle and the spine and acromion of scapula	Rotates scapula; various fibers raise scapula, pull scapula medially, or pull scapula and shoulder downward
Rhomboid major	Spinous processes of thoracic vertebrae (T2–T5)	Medial border of scapula	Retracts and elevates scapula
Rhomboid minor	Spinous processes of cervical vertebra C7 and thoracic vertebra T1	Medial border of scapula	Retracts and elevates scapula
Levator scapulae	Transverse processes of cervical vertebrae (C1–C4)	Medial border of scapula	Elevates scapula

FIGURE 23.3 Rotator cuff muscles: (a) anterior view; (b) posterior view; (c) lateral view. Some regional bones, features, and muscles are also labeled.

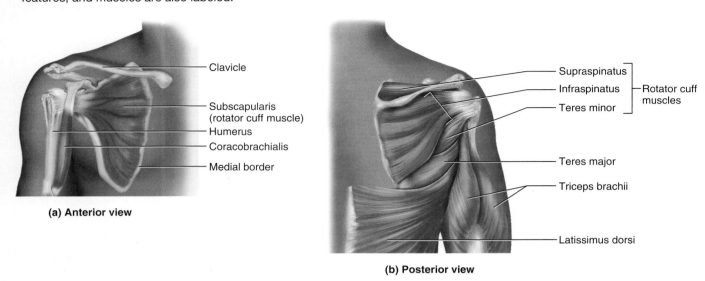

(a) Anterior view

Clavicle
Subscapularis (rotator cuff muscle)
Humerus
Coracobrachialis
Medial border

(b) Posterior view

Supraspinatus
Infraspinatus — Rotator cuff muscles
Teres minor
Teres major
Triceps brachii
Latissimus dorsi

(c) Lateral view

Anterior | Posterior

Acromion
Clavicle
Coracoid process
Glenoid cavity
Lateral border
Supraspinatus
Infraspinatus — Rotator cuff muscles
Teres minor
Subscapularis

TABLE 23.2 Muscles That Move the Arm

Muscle	Origin	Insertion	Action
Pectoralis major	Clavicle, sternum, and costal cartilages of upper ribs	Intertubercular sulcus of humerus	Flexes, adducts, and rotates arm medially
Latissimus dorsi	Spinous processes of lumbar and lower thoracic vertebrae, iliac crest, and lower ribs	Intertubercular sulcus of humerus	Extends, adducts, and rotates the arm medially
Deltoid	Acromion and spine of the scapula and the clavicle	Deltoid tuberosity of humerus	Abducts, extends, and flexes arm
Teres major	Lateral border of scapula	Intertubercular sulcus of humerus	Extends, adducts, and rotates arm medially
Coracobrachialis	Coracoid process of scapula	Shaft of humerus	Flexes and adducts the arm
Supraspinatus	Supraspinous fossa of scapula	Greater tubercle of humerus	Abducts the arm
Infraspinatus	Infraspinous fossa of scapula	Greater tubercle of humerus	Rotates arm laterally
Teres minor	Lateral border of scapula	Greater tubercle of humerus	Rotates arm laterally
Subscapularis	Subscapular fossa of scapula	Lesser tubercle of humerus	Rotates arm medially

3. Demonstrate the actions of these muscles in your body.
4. Locate the origins and insertions of these muscles in the human skeleton.
5. Study figures 23.1, 23.2, and 23.3, and table 23.2.
6. Locate the following muscles in the human torso model and models of the upper limb. Also locate in your body as many of the muscles as you can.

muscles that move the arm
- origins on axial skeleton
    - pectoralis major
    - latissimus dorsi
- origins on scapula
    - deltoid
    - teres major
    - coracobrachialis
- rotator cuff (SITS) muscles (origins also on scapula)
    - supraspinatus
    - infraspinatus
    - teres minor
    - subscapularis

7. Demonstrate the actions of these muscles in your body.
8. Locate the origins and insertions of these muscles in the human skeleton.
9. Study figures 23.1, 23.2, 23.4, and 23.5, and table 23.3.

FIGURE 23.4 The (a) superficial, (b) intermediate, and (c) deep anterior forearm muscles.

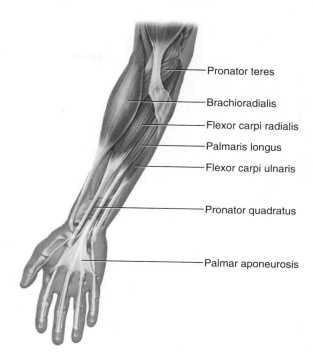

- Pronator teres
- Brachioradialis
- Flexor carpi radialis
- Palmaris longus
- Flexor carpi ulnaris
- Pronator quadratus
- Palmar aponeurosis

(a) Anterior muscles: superficial

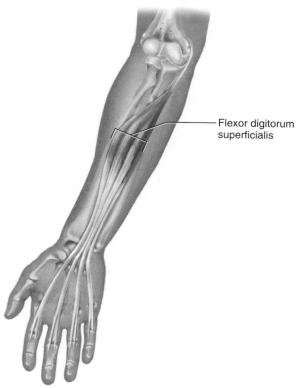

- Flexor digitorum superficialis

(b) Anterior muscles: intermediate

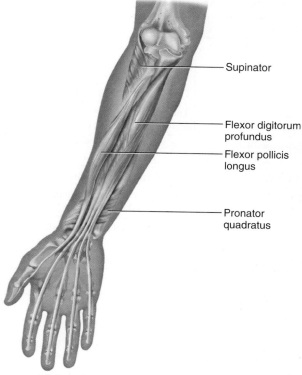

- Supinator
- Flexor digitorum profundus
- Flexor pollicis longus
- Pronator quadratus

(c) Anterior muscles: deep

233

FIGURE 23.5 The (a) superficial and (b) deep posterior forearm muscles.

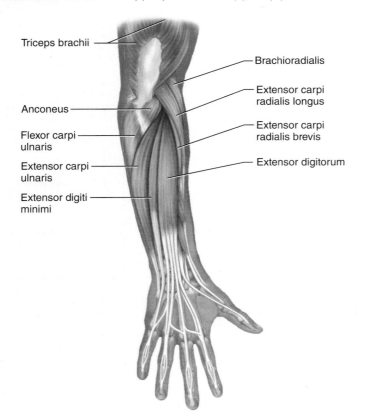

Triceps brachii

Brachioradialis

Extensor carpi radialis longus

Anconeus

Extensor carpi radialis brevis

Flexor carpi ulnaris

Extensor digitorum

Extensor carpi ulnaris

Extensor digiti minimi

(a) Posterior muscles: superficial

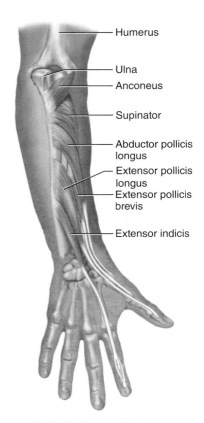

Humerus

Ulna

Anconeus

Supinator

Abductor pollicis longus

Extensor pollicis longus

Extensor pollicis brevis

Extensor indicis

(b) Posterior muscles: deep

TABLE 23.3 Muscles That Move the Forearm

Muscle	Origin	Insertion	Action
Biceps brachii	Coracoid process (short head) and tubercle above glenoid cavity of scapula (long head)	Radial tuberosity	Flexes elbow and supinates forearm and hand
Brachialis	Distal anterior shaft of humerus	Coronoid process of ulna	Flexes elbow
Triceps brachii	Tubercle below glenoid cavity and lateral and medial surfaces of humerus	Olecranon process of ulna	Extends elbow
Brachioradialis	Distal lateral end of humerus	Lateral surface of radius near styloid process	Flexes elbow
Anconeus	Lateral epicondyle of humerus	Olecranon process of ulna	Extends elbow
Supinator	Lateral epicondyle of humerus and ulna distal to radial notch	Lateral surface of proximal radius	Supinates forearm and hand
Pronator teres	Medial epicondyle of humerus and coronoid process of ulna	Lateral surface of radius	Pronates forearm and hand
Pronator quadratus	Anterior distal end of ulna	Anterior distal end of radius	Pronates forearm and hand

10. Locate the following muscles in the human torso model and models of the upper limb. Also locate in your body as many of the muscles as you can.

muscles that move the forearm
- muscle bellies in arm
  - biceps brachii
  - brachialis
  - triceps brachii
- muscle bellies in forearm
  - brachioradialis
  - anconeus
  - supinator
  - pronator teres
  - pronator quadratus

11. Demonstrate the actions of these muscles in your body.
12. Locate the origins and insertions of these muscles in the human skeleton.
13. Study figures 23.4 and 23.5 and table 23.4.

TABLE 23.4 Muscles That Move the Hand

Muscle	Origin	Insertion	Action
Palmaris longus (sometimes absent)	Medial epicondyle of humerus	Palmar aponeurosis	Flexes wrist
Flexor carpi radialis	Medial epicondyle of humerus	Base of metacarpals II–III	Flexes wrist and abducts hand
Flexor carpi ulnaris	Medial epicondyle of humerus and olecranon process of ulna	Carpals (pisiform and hamate) and metacarpal V	Flexes wrist and adducts hand
Flexor digitorum superficialis	Medial epicondyle of humerus, coronoid process of ulna, and radius	Middle phalanges of fingers 2–5	Flexes fingers and wrist
Flexor digitorum profundus	Anterior ulna and interosseous membrane	Distal phalanges of fingers 2–5	Flexes fingers 2–5
Flexor pollicis longus	Anterior radius and interosseous membrane	Distal phalanx of thumb	Flexes thumb
Extensor carpi radialis longus	Lateral supracondylar ridge of humerus	Base of metacarpal II	Extends wrist and abducts hand
Extensor carpi radialis brevis	Lateral epicondyle of humerus	Base of metacarpal III	Extends wrist and abducts hand
Extensor digitorum	Lateral epicondyle of humerus	Posterior phalanges of fingers 2–5	Extends fingers
Extensor digiti minimi	Lateral epicondyle of humerus	Proximal phalanx of finger 5	Extends finger 5 (little finger)
Extensor carpi ulnaris	Lateral epicondyle of humerus and posterior ulna	Base of metacarpal V	Extends wrist and adducts hand
Abductor pollicis longus	Posterior radius and ulna and interosseous membrane	Metacarpal I and trapezium	Abducts thumb
Extensor pollicis longus	Posterior ulna and interosseous membrane	Distal phalanx of thumb	Extends thumb
Extensor pollicis brevis	Posterior radius and interosseous membrane	Proximal phalanx of thumb	Extends thumb
Extensor indicis	Posterior ulna and interosseous membrane	Middle and distal phalanges of finger 2	Extends finger 2 (index finger)

14. Locate the following muscles in the human torso model and models of the upper limb. Also locate in your body as many of the muscles as you can.

muscles that move the hand
- anterior superficial flexor muscles
  - palmaris longus (absent sometimes)
  - flexor carpi radialis
  - flexor carpi ulnaris
  - flexor digitorum superficialis (intermediate depth)
- anterior deep flexor muscles
  - flexor digitorum profundus
  - flexor pollicis longus
- posterior superficial extensor muscles
  - extensor carpi radialis longus
  - extensor carpi radialis brevis
  - extensor digitorum
  - extensor digiti minimi
  - extensor carpi ulnaris
- posterior deep extensor muscles
  - abductor pollicis longus
  - extensor pollicis longus
  - extensor pollicis brevis
  - extensor indicis

15. Demonstrate the actions of these muscles in your body.
16. Locate the origins and insertions of these muscles in the human skeleton.
17. Complete Parts A, B, C, and D of Laboratory Assessment 23.

Learning Extension Activity

A long rubber band can be used to simulate muscle locations, origins, insertions, and actions on muscular models, the skeleton, or a laboratory partner. Hold one end of the rubber band firmly on the origin location of a muscle, then slightly stretch the rubber band and hold the other end on the insertion site. Allow the insertion end to slowly move toward the origin end to simulate the contraction and action of the muscle.

Name _____

Date _____

Section _____

The ⚠ corresponds to the indicated outcome(s) found at the beginning of the laboratory exercise.

Muscles of the Chest, Shoulder, and Upper Limb

Part A Assessments

Identify the muscles indicated in the chest, shoulder, arm, and forearm in figures 23.6, 23.7, and 23.8.

FIGURE 23.6 Label the anterior muscles of the chest, shoulder, and upper limb. ⚠

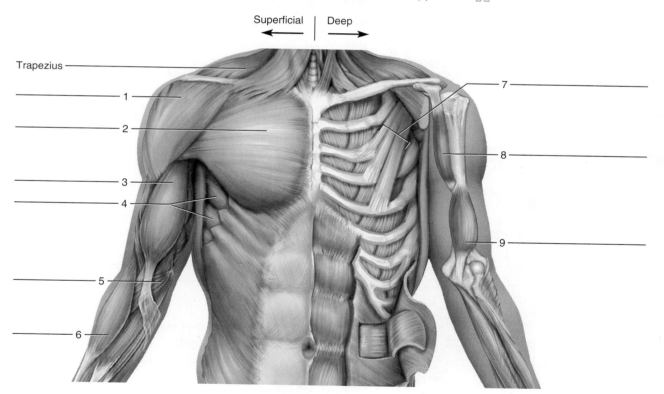

FIGURE 23.7 Identify the posterior muscles of the left shoulder and arm of a cadaver, using the terms provided. The deltoid and trapezius muscles have been removed. ⚠️

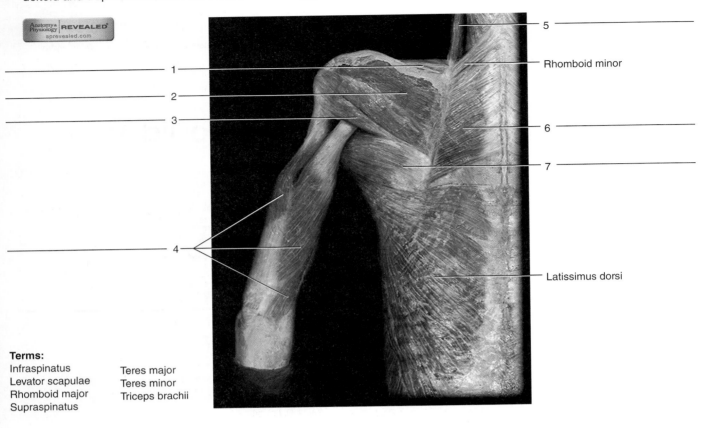

5
Rhomboid minor
6
7
Latissimus dorsi

Terms:

Infraspinatus   Teres major
Levator scapulae   Teres minor
Rhomboid major   Triceps brachii
Supraspinatus

FIGURE 23.8 Identify the anterior muscles of the right forearm of a cadaver, using the terms provided. ⚠️

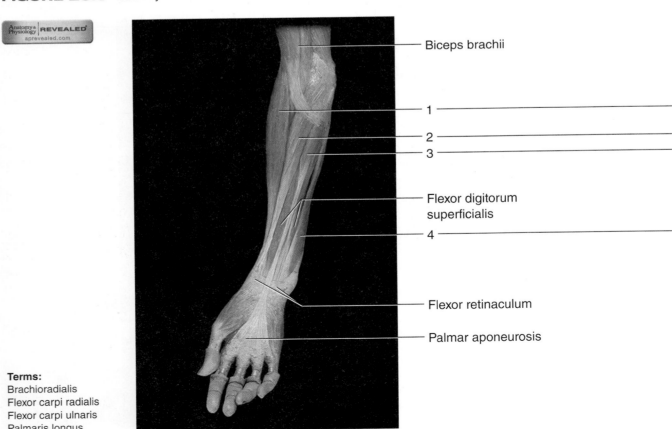

Biceps brachii
1
2
3
Flexor digitorum
superficialis
4
Flexor retinaculum
Palmar aponeurosis

Terms:

Brachioradialis
Flexor carpi radialis
Flexor carpi ulnaris
Palmaris longus

Part B Assessments

Match the muscles in column A with the actions in column B. Place the letter of your choice in the space provided. **2**

Column A

a. Brachialis
b. Coracobrachialis
c. Deltoid
d. Extensor carpi ulnaris
e. Flexor carpi ulnaris
f. Infraspinatus
g. Pectoralis major
h. Pectoralis minor
i. Rhomboid major
j. Serratus anterior
k. Teres major
l. Triceps brachii

Column B

_____ 1. Abducts, extends, and flexes arm

_____ 2. Pulls arm forward (flexion) and across chest (adduction) and rotates arm medially

_____ 3. Flexes wrist and adducts hand

_____ 4. Raises and adducts (retracts) scapula

_____ 5. Raises ribs in forceful inhalation or pulls scapula forward and downward

_____ 6. Used to thrust shoulder anteriorly (protraction), as when pushing something

_____ 7. Flexes the forearm at the elbow

_____ 8. Flexes and adducts arm at the shoulder along with pectoralis major

_____ 9. Extends the forearm at the elbow

_____ 10. Extends, adducts, and rotates arm medially

_____ 11. Extends and adducts hand at the wrist

_____ 12. Rotates arm laterally

Part C Assessments

Name the muscle indicated by the following combinations of origin and insertion. **3**

Origin	Insertion	Muscle
1. Spinous processes of thoracic vertebrae	Medial border of scapula	_____
2. Lateral surfaces of most ribs	Medial border of scapula	_____
3. Sternal ends of ribs 3–5	Coracoid process of scapula	_____
4. Coracoid process of scapula	Shaft of humerus	_____
5. Lateral border of scapula	Intertubercular sulcus of humerus	_____
6. Subscapular fossa of scapula	Lesser tubercle of humerus	_____
7. Lateral border of scapula	Greater tubercle of humerus	_____
8. Distal anterior shaft of humerus	Coronoid process of ulna	_____
9. Medial epicondyle of humerus	Palmar aponeurosis	_____
10. Anterior radius and interosseous membrane	Distal phalanx of thumb	_____
11. Posterior ulna and interosseous membrane	Middle and distal phalanges of finger 2	_____
12. Clavicle and spine and acromion of scapula	Deltoid tuberosity of humerus	_____

Part D Assessments

Critical Thinking Assessment

Identify the muscles indicated in figure 23.9.

FIGURE 23.9 Label these muscles that appear as body surface features in these photographs (a, b, and c) by placing the correct numbers in the spaces provided. 🄰

_____ Biceps brachii
_____ Deltoid
_____ External oblique
_____ Pectoralis major
_____ Rectus abdominis
_____ Serratus anterior
_____ Sternocleidomastoid
_____ Trapezius

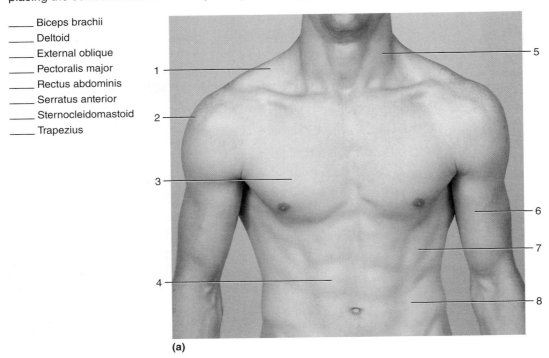

(a)

_____ Biceps brachii
_____ Deltoid
_____ Infraspinatus
_____ Latissimus dorsi
_____ Trapezius
_____ Triceps brachii

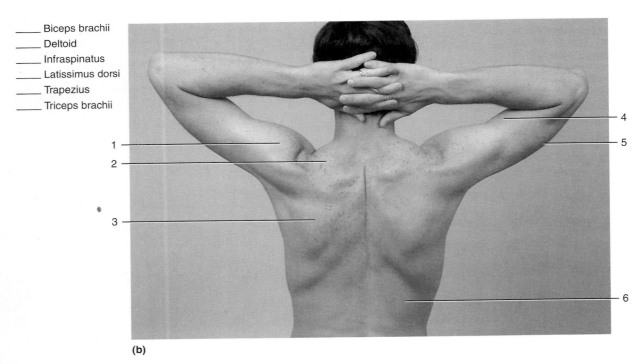

(b)

FIGURE 23.9 *Continued.*

_____ Biceps brachii
_____ Brachioradialis
_____ Deltoid
_____ Pectoralis major
_____ Serratus anterior
_____ Trapezius
_____ Triceps brachii

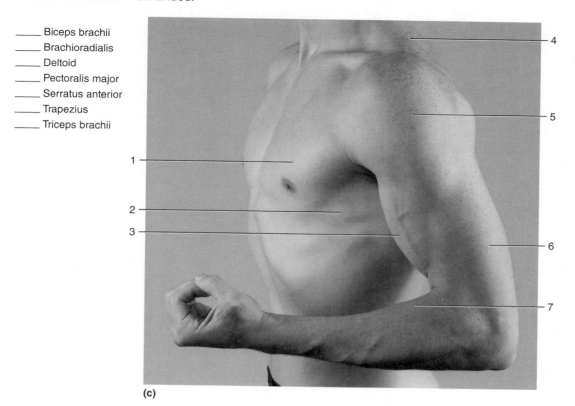

(c)

NOTES

Muscles of the Vertebral Column, Abdominal Wall, and Pelvic Floor

Purpose of the Exercise

To review the actions, origins, and insertions of the muscles of the vertebral column, abdominal wall, and pelvic floor.

Materials Needed

Human torso model with musculature
Human skelton, articulated
Muscular models of male and female pelves

Learning Outcomes

After completing this exercise, you should be able to

1. Locate and identify the muscles of the vertebral column, abdominal wall, and pelvic floor.
2. Describe the action of each of these muscles.
3. Locate the origin and insertion of each of these muscles in a human skeleton or on muscular models.

Pre-Lab

Carefully read the introductory material and examine the entire lab. Be familiar with the locations, actions, origins, and insertions of the muscles of the vertebral column, abdominal wall, and the pelvic floor from lecture or the textbook. Answer the pre-lab questions.

Pre-Lab Questions: Select the correct answer for each of the following questions:

1. Which of the following muscles is *not* part of the erector spinae group?
   - **a.** iliocostalis
   - **b.** longissimus
   - **c.** spinalis
   - **d.** quadratus lumborum

2. The abdominal wall muscles possess the following functions *except*
   - **a.** extend the spine.
   - **b.** compress the abdominal visceral organs.
   - **c.** assist in forceful exhalation.
   - **d.** help maintain posture.

3. The abdominal muscle nearest the midline is the
   - **a.** external oblique.
   - **b.** internal oblique.
   - **c.** rectus abdominis.
   - **d.** transversus abdominis.

4. Which of the following muscles would be most important in relation to a vaginal delivery?
   - **a.** bulbospongiosus
   - **b.** quadratus lumborum
   - **c.** spinalis
   - **d.** erector spinae

5. The four muscles of the abdominal wall
   - **a.** contain muscle fibers in the same direction.
   - **b.** contain muscle fibers in four different directions.
   - **c.** are responsible for trunk extension.
   - **d.** fully overlap each other.

6. The iliocostalis represents the lateral muscle group of the erector spinae.
   True _____   False _____

7. The pelvic floor muscles are all arranged in a single superficial layer.
   True _____   False _____

The deep muscles of the back extend the vertebral column. Because the muscles have many origins, insertions, and subgroups, the muscles overlap each other. The deep back muscles can extend the spine when contracting as a group but also help to maintain posture and normal spine curvatures.

The anterior and lateral walls of the abdomen contain broad, flattened muscles arranged in layers. These muscles connect the rib cage and vertebral column to the pelvic girdle. The abdominal wall muscles compress the abdominal visceral organs, help maintain posture, assist in forceful exhalation, and contribute to trunk flexion and waist rotation.

The muscles in the region of the perineum form the pelvic floor (outlet) and can be divided into two triangles: a urogenital triangle and an anal triangle. The anterior urogenital triangle contains superficial and deeper muscle layers associated with the external genitalia. The posterior anal triangle contains the pelvic diaphragm and the anal sphincter. The pelvic floor is penetrated by the urethra, vagina, and anus in a female, and thus pelvic floor muscles are important in obstetrics.

Procedure A—Muscles of the Vertebral Column and Abdominal Wall

The muscles that move the vertebral column are located deep in the back from the pelvic girdle to the skull. The erector spinae group includes three major columns of muscles. Each group is composed of several muscles. For the purpose of this exercise, the entire erector spinae group is considered to have the common action of extending the vertebral column when they contract bilaterally; if they contract unilaterally, lateral flexion of the trunk occurs. A familiar lifting injury, known as a muscle strain (pulled muscle), can occur especially in the lumbar region of the erector spinae group. The quadratus lumborum is another major muscle of the inferior vertebral column and functions for extension and lateral flexion of the vertebral column.

The splenius capitis and the semispinalis capitis muscles, located in the posterior neck extending from the vertebral column to the skull, are large muscles associated with the vertebral column and act in extensions and rotations of the head and neck. These muscles were previously included in Laboratory Exercise 22. Several smaller muscles of the vertebral column, including the serratus posterior, semi-

spinalis thoracis, and multifidus, assist in either breathing, extension, or rotation functions.

The abdominal wall is composed of four muscles situated as broad, flat muscle layers. Their muscle fibers are oriented in different directions much like plywood. This provides additional strength to support and compress the abdominal viscera. The muscle layers attach to structures such as the linea alba, inguinal ligament, or aponeuroses, since the abdominal region lacks skeletal attachments for these muscles.

1. Study figure 24.1 and table 24.1.
2. Locate the following muscles in the human torso model:
   erector spinae group
   - iliocostalis (lateral group)
   - longissimus (intermediate group)
   - spinalis (medial group)
   quadratus lumborum
3. Demonstrate the actions of these muscles in your body.
4. Locate the origins and insertions of these muscles in the human skeleton.
5. Study figures 24.2, 24.3, and table 24.2.
6. Locate the following muscles in the human torso model:
   external oblique
   internal oblique
   transversus abdominis (transverse abdominal)
   rectus abdominis
7. Demonstrate the actions of these muscles in your body.
8. Locate the origins and insertions of these muscles in the human skeleton.
9. Complete Parts A and B of Laboratory Assessment 24.

Procedure B—Muscles of the Pelvic Floor

The diamond-shaped region between the lower limbs is the *perineum*. It is located between the ischial tuberosities on the lateral sides, the anterior pubic symphysis, and the posterior coccyx. The perineum possesses an anterior urogenital triangle and a posterior anal triangle. The urogenital triangle includes the external genitalia, the urogenital openings, and associated muscles. The anal triangle includes the anus as the digestive opening and associated muscles. The pelvic floor is composed of three layers of muscles. The term *diaphragm* is used in conjunction with some of these muscle

TABLE 24.1 Muscles of the Vertebral Column

Muscle	Origin	Insertion	Action
Erector spinae (3 columns of muscles)	Numerous sites including the sacrum, lumbar vertebrae, thoracic vertebrae, and most ribs	Numerous sites on thoracic vertebrae, cervical vertebrae, all ribs, and mastoid process	Extends vertebral column; lateral flexion
Quadratus lumborum	Iliac crest	Transverse processes of L1–L4 and rib 12	Extends vertebral column; lateral flexion

FIGURE 24.1 Muscles associated with the vertebral column. Muscles shown on the right side are deeper than those on the left side. (*Note:* Superficial muscles are not included on this figure.)

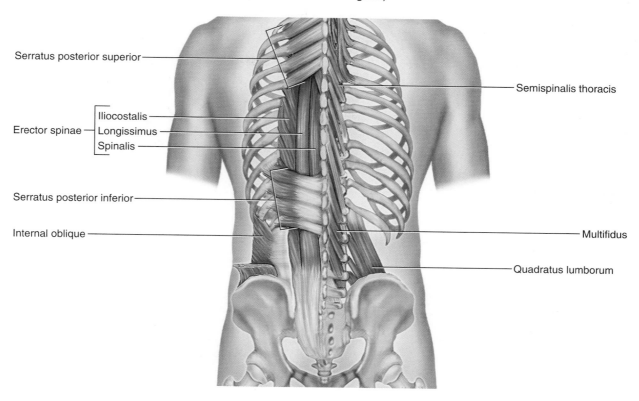

Serratus posterior superior

Semispinalis thoracis

Erector spinae
- Iliocostalis
- Longissimus
- Spinalis

Serratus posterior inferior

Internal oblique

Multifidus

Quadratus lumborum

FIGURE 24.2   Muscles of the abdominal wall.

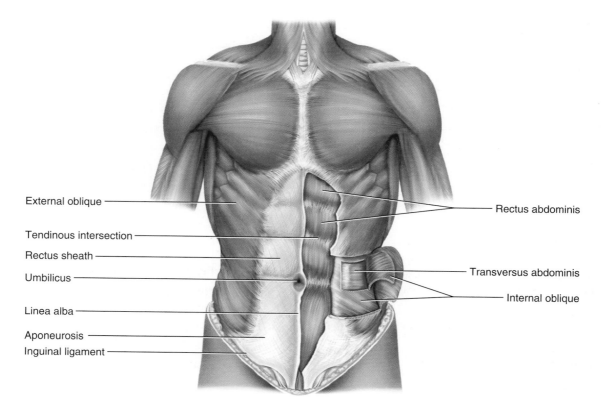

External oblique

Rectus abdominis

Tendinous intersection

Rectus sheath

Umbilicus

Transversus abdominis

Internal oblique

Linea alba

Aponeurosis

Inguinal ligament

FIGURE 24.3 Transverse view of the midsection of the abdominal wall.

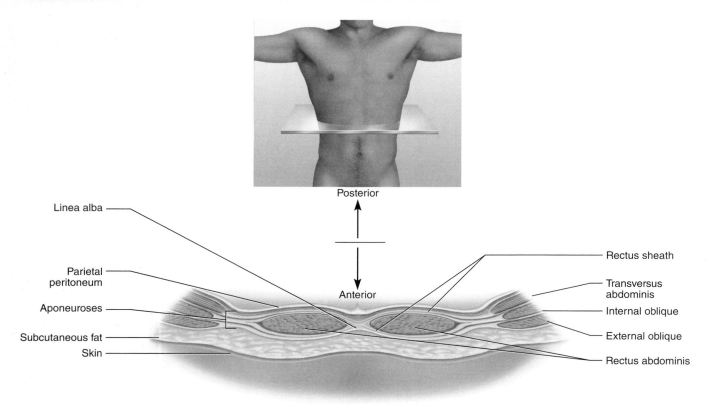

TABLE 24.2 Muscles of the Abdominal Wall

Muscle	Origin	Insertion	Action
External oblique	Anterior surfaces of ribs 5–12	Anterior iliac crest and linea alba by aponeurosis	Tenses abdominal wall; compresses abdominal contents; aids in trunk rotation
Internal oblique	Iliac crest and inguinal ligament	Cartilages of lower ribs, linea alba, and crest of pubis	Same as external oblique
Transversus abdominis	Costal cartilages of lower ribs, iliac crest, and inguinal ligament	Linea alba and pubic crest	Compresses abdominal contents
Rectus abdominis	Pubic crest and pubic symphysis	Xiphoid process of sternum and costal cartilages of ribs 5–7	Flexes vertebral column; compresses abdominal contents

layers, forming a type of partition or wall, and referred to as a urogenital diaphragm and a pelvic diaphragm.

1. Study figure 24.4 and table 24.3.
2. Locate the following muscles in the models of the male and female pelves:

    levator ani

    coccygeus

    superficial transverse perineal muscle

    bulbospongiosus

    ischiocavernosus

3. Locate the origins and insertions of these muscles in the human skeleton.

4. Complete Part C of the laboratory assessment.

FIGURE 24.4 Muscles of the pelvic floor: (a) superficial perineal muscles of male and female; (b) urogenital diaphragm (next deeper perineal muscles) of male and female; (c) pelvic diaphragm (deepest layer) of female.

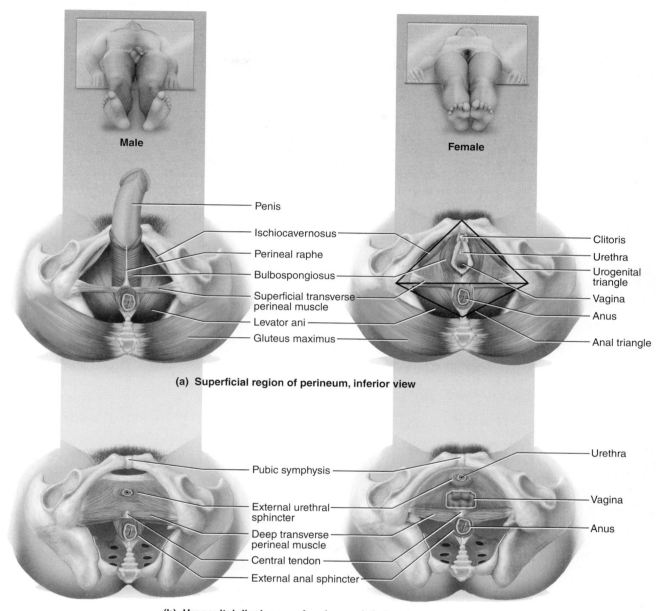

Male

Female

Penis

Ischiocavernosus

Perineal raphe

Bulbospongiosus

Superficial transverse perineal muscle

Levator ani

Gluteus maximus

Clitoris

Urethra

Urogenital triangle

Vagina

Anus

Anal triangle

(a) Superficial region of perineum, inferior view

Pubic symphysis

External urethral sphincter

Deep transverse perineal muscle

Central tendon

External anal sphincter

Urethra

Vagina

Anus

(b) Urogenital diaphragm of perineum, inferior view

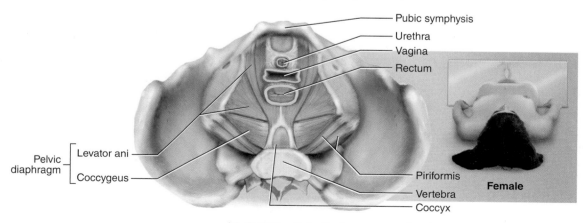

Pubic symphysis

Urethra

Vagina

Rectum

Pelvic diaphragm
  Levator ani
  Coccygeus

Piriformis

Vertebra

Coccyx

Female

(c) Female pelvic diaphragm, superior view

247

TABLE 24.3 Muscles of the Pelvic Floor

Muscle	Origin	Insertion	Action
Levator ani	Pubis and ischial spine	Coccyx	Supports pelvic viscera and provides sphincterlike action in anal canal and vagina
Coccygeus	Ischial spine	Sacrum and coccyx	Supports pelvic viscera
Superficial transverse perineal muscle	Ischial tuberosity	Central tendon of perineum	Supports pelvic viscera
Bulbospongiosus	Central tendon of perineum	Males: Ensheaths root of penis Females: Ensheaths root of clitoris	Males: Assists emptying of urethra and assists in erection of penis Females: Constricts vagina and assists in erection of clitoris
Ischiocavernosus	Ischial tuberosity	Pubic symphysis near base of penis or clitoris	Males: Erects penis Females: Erects clitoris

Name _____

Date _____

Section _____

The Ⓐ corresponds to the indicated outcome(s) found at the beginning of the laboratory exercise.

Muscles of the Vertebral Column, Abdominal Wall, and Pelvic Floor

Part A Assessments

Identify the muscles indicated in figures 24.5 and 24.6.

FIGURE 24.5 Label the deep back muscles of a cadaver, using the terms provided. The trapezius, latissimus dorsi, rhomboids, and serratus posterior have been removed. Ⓐ

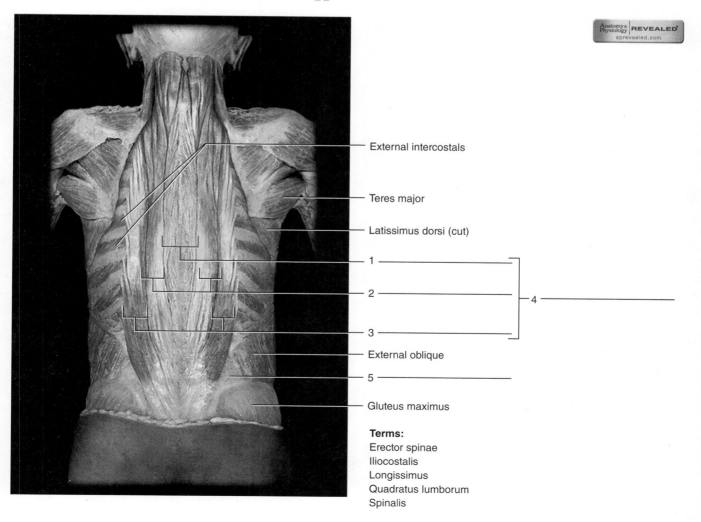

External intercostals

Teres major

Latissimus dorsi (cut)

1

2

4

3

External oblique

5

Gluteus maximus

Terms:
Erector spinae
Iliocostalis
Longissimus
Quadratus lumborum
Spinalis

FIGURE 24.6   Label the muscles of the abdominal wall.

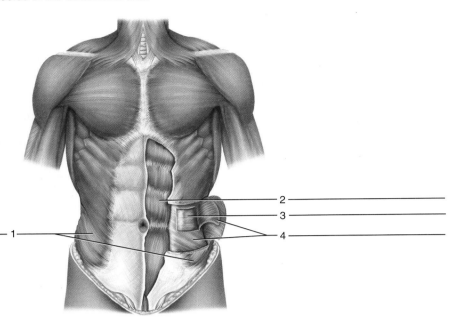

Critical Thinking Assessment

List the muscles from superficial to deep for an appendectomy incision. ⚠**1**

_____

_____

_____

_____

Part B Assessments

Complete the following statements:

1. A band of tough connective tissue in the midline of the anterior abdominal wall called the _____ serves as a muscle attachment. **1**

2. The _____ muscle spans from the costal cartilages and xiphoid process to the pubic bones. **3**

3. The _____ forms the third layer (deepest layer) of the abdominal wall muscles. **1**

4. The action of the external oblique muscle is to _____. **2**

5. The action of the rectus abdominis is to _____. **2**

6. The iliocostalis, longissimus, and spinalis muscles together form the _____. **1**

7. The _____ has the origin on the iliac crest and the insertion on rib 12 and L1–L4 vertebrae. **3**

8. The origin of the _____ is on the iliac crest and inguinal ligament. **3**

9. The insertion of the _____ is on the xiphoid process and costal cartilages. **3**

10. Name four muscles that compress the abdominal contents. **2**

    _____

    _____

    _____

    _____

Part C Assessments

Complete the following statements:

1. The levator ani and coccygeus together form the _____ diaphragm. **1**

2. The posterior portion of the perineum is called the _____ triangle. **1**

3. The _____ ensheaths the root of the penis. **1**

4. In females, the bulbospongiosus acts to constrict the _____. **2**

5. In the female, the pelvic floor (outlet) is penetrated by the urethra, vagina, and _____. **1**

6. In the female, the _____ muscles are separated by the vagina, urethra, and anal canal. **1**

7. The insertion of the levator ani is on the _____. **3**

8. The coccygeus muscle has an origin on the _____. **3**

9. The superficial transverse perineal muscle acts to support the _____. **2**

10. Name five muscles of the pelvic floor that males and females both have in common. **1** (There are more than five correct answers.)

    _____

    _____

    _____

    _____

    _____

Muscles of the Hip and Lower Limb

Purpose of the Exercise

To review the actions, origins, and insertions of the muscles that move the thigh, leg, and foot.

Materials Needed

Human torso model with musculature
Human skeleton, articulated
Muscular models of the lower limb

For Learning Extension Activity:
Long rubber bands

Learning Outcomes

After completing this exercise, you should be able to

1. Locate and identify the muscles that move the thigh, leg, and foot.

2. Describe and demonstrate the actions of each of these muscles.

3. Locate the origin and insertion of each of these muscles in a human skeleton and on muscular models.

Pre-Lab

Carefully read the introductory material and examine the entire lab. Be familiar with the locations, actions, origins, and insertions of the muscles of the hip and lower limb from lecture or the textbook. Visit www.mhhe.com/martinseries2 for LabCam videos. Answer the pre-lab questions.

Pre-Lab Questions: Select the correct answer for each of the following questions:

1. Muscles that move the thigh at the hip joint have origins on the
   a. pelvis.   b. femur.
   c. tibia.   d. patella.

2. Anterior thigh muscles serve as prime movers of
   a. flexion of the leg at the knee joint.
   b. abduction of the thigh at the hip joint.
   c. adduction of the thigh at the hip joint.
   d. extension of the leg at the knee joint.

3. The muscles of the lower limb _____ than those of the upper limb.
   a. are more numerous   b. are larger
   c. conduct actions with d. are less powerful
      more precision

4. Which of the following muscles is *not* part of the quadriceps group?
   a. rectus femoris   b. vastus lateralis
   c. semitendinosus   d. vastus medialis

5. Which of the following muscles is *not* part of the hamstring group?
   a. rectus femoris   b. biceps femoris
   c. semitendinosus   d. semimembranosus

6. Muscles located in the lateral leg include actions of
   a. extension of the   b. eversion of the foot.
      knee joint.
   c. flexion of the toes. d. extension of the toes.

7. Contractions of the sartorius muscle involve movements across two different joints.
   True _____   False _____

8. The insertions of the four quadriceps femoris muscles are on the femur.
   True _____   False _____

The muscles that move the thigh at the hip joint have their origins on the pelvis and insertions usually on the femur. Those muscles attached on the anterior pelvis act to flex the thigh at the hip joint; those attached on the posterior pelvis act to extend the thigh at the hip joint; those attached on the lateral side of the pelvis act as abductors and rotators of the thigh at the hip joint; and those attached on the medial pelvis act as adductors of the thigh at the hip joint.

The muscles that move the leg at the knee joint have their origins on the femur or the hip bone. The anterior thigh muscles have their insertions on the proximal tibia and serve as prime movers of extension of the leg at the knee joint. The posterior thigh muscles have their insertions on the tibia or fibula and act as prime movers of flexion of the leg at the knee joint.

The muscles that move the foot have their origins on the tibia, fibula, or femur. The anterior leg muscles have actions of dorsiflexion and extension of the toes; the posterior leg muscles have actions of plantar flexion and flexion of the toes; the lateral leg muscles have actions of eversion of the foot.

In contrast to the upper limb muscles, those in the lower limb tend to be much larger and are involved in powerful contractions for walking or running as well as isometric contractions in standing. As compared to the forearm, the leg has fewer muscles because the foot is not as involved in actions that require precision movements as the hand.

As you demonstrate the actions of muscles in the lower limb, refer to standard anatomical regions of the limb. The *thigh* refers to hip to knee; the *leg* refers to knee to ankle; and the *foot* refers to ankle, metatarsal area, and toes. Several muscles in the hip and lower limb cross more than one joint and involve movements of both joints. Examples of muscles crossing more than one joint include the sartorius and gastrocnemius.

Procedure—Muscles of the Hip and Lower Limb

1. Study figures 25.1, 25.2, 25.3, and 25.4, and table 25.1.
2. Locate the following muscles in the human torso model and in the lower limb models. Also locate as many of them as possible in your body.

   muscles that move the thigh/hip joint
   - anterior hip muscles
     - iliopsoas group
       psoas major
       iliacus
   - posterior and lateral hip muscles
     - gluteus maximus
     - gluteus medius
     - gluteus minimus
     - tensor fasciae latae
     - piriformis

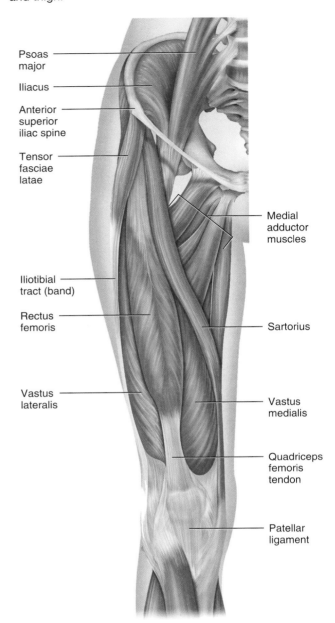

FIGURE 25.1 Muscles of the anterior right hip and thigh.

Psoas major
Iliacus
Anterior superior iliac spine
Tensor fasciae latae
Medial adductor muscles
Iliotibial tract (band)
Rectus femoris
Sartorius
Vastus lateralis
Vastus medialis
Quadriceps femoris tendon
Patellar ligament

   - medial adductor muscles
     - pectineus
     - adductor longus
     - adductor magnus
     - adductor brevis
     - gracilis
3. Demonstrate the actions of these muscles in your body.

FIGURE 25.2 Selected individual muscles of the anterior right hip and medial thigh.

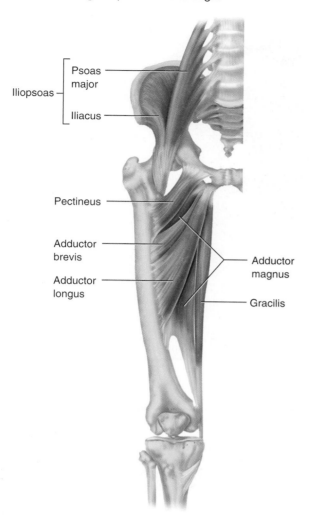

Iliopsoas
- Psoas major
- Iliacus

Pectineus

Adductor brevis

Adductor longus

Adductor magnus

Gracilis

FIGURE 25.3 Muscles of the posterior right hip and thigh.

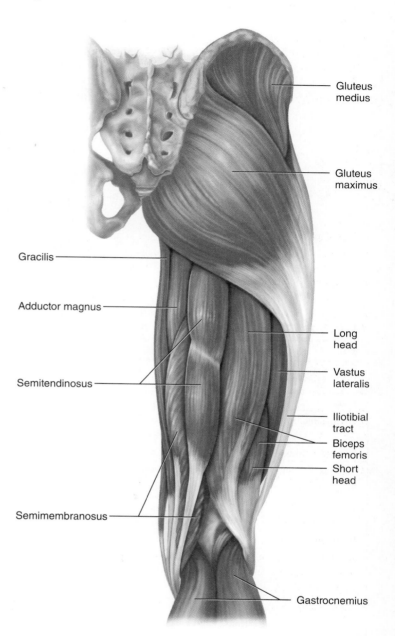

Gluteus medius

Gluteus maximus

Gracilis

Adductor magnus

Semitendinosus

Long head

Vastus lateralis

Iliotibial tract

Biceps femoris

Short head

Semimembranosus

Gastrocnemius

255

FIGURE 25.4 Superficial and deep posterior hip muscles.

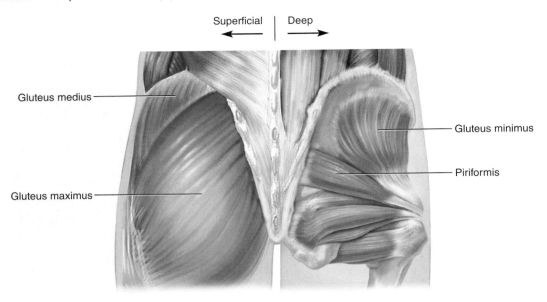

TABLE 25.1 Muscles That Move the Thigh

Muscle	Origin	Insertion	Action
Psoas major	Intervertebral discs, bodies, and transverse processes of T12–L5	Lesser trochanter of femur	Flexes thigh at hip; flexes trunk at hip (when thigh is fixed)
Iliacus	Iliac fossa of ilium	Lesser trochanter of femur	Same as psoas major
Gluteus maximus	Sacrum, coccyx, and posterior iliac crest	Gluteal tuberosity of femur and iliotibial tract	Extends thigh at hip
Gluteus medius	Lateral surface of ilium	Greater trochanter of femur	Abducts and medially rotates thigh
Gluteus minimus	Lateral surface of ilium	Greater trochanter of femur	Same as gluteus medius
Tensor fasciae latae	Iliac crest and anterior superior iliac spine	Iliotibial tract	Abducts and medially rotates thigh
Piriformis	Anterior surface of sacrum	Greater trochanter of femur	Abducts and laterally rotates thigh
Pectineus	Pubis	Femur distal to lesser trochanter	Adducts and flexes thigh
Adductor longus	Pubis near pubic symphysis	Linea aspera of femur	Adducts and flexes thigh
Adductor magnus	Pubis and ischial tuberosity	Linea aspera of femur	Adducts thigh; posterior portion extends thigh, and anterior portion flexes thigh
Adductor brevis	Pubis	Linea aspera of femur	Adducts thigh
Gracilis	Pubis	Medial surface of tibia	Adducts thigh and flexes leg at knee

4. Locate the origins and insertions of these muscles in the human skeleton.
5. Study figures 25.1, 25.3, and table 25.2.
6. Locate the following muscles in the human torso model and in the lower limb models. Also locate as many of them as possible in your body.

muscles that move the leg/knee joint
- anterior thigh muscles
  - sartorius
  - quadriceps femoris group
    rectus femoris
    vastus lateralis
    vastus medialis
    vastus intermedius
- posterior thigh muscles
  - hamstring group
    biceps femoris
    semitendinosus
    semimembranosus

7. Demonstrate the actions of these muscles in your body.
8. Locate the origins and insertions of these muscles in the human skeleton.
9. Complete Part A of Laboratory Assessment 25.
10. Study figures 25.5, 25.6, and 25.7, and table 25.3.

TABLE 25.2   Muscles That Move the Leg

Muscle	Origin	Insertion	Action
Sartorius	Anterior superior iliac spine	Proximal medial surface of tibia	Flexes, abducts, and laterally rotates thigh at hip; flexes leg at knee
Quadriceps Femoris Group			
Rectus femoris	Anterior inferior iliac spine and superior margin of acetabulum	Patella by common quadriceps tendon, which continues as patellar ligament to tibial tuberosity	Extends leg at knee; flexes thigh at hip
Vastus lateralis	Greater trochanter and linea aspera of femur		Extends leg at knee
Vastus medialis	Linea aspera of femur		Extends leg at knee
Vastus intermedius	Anterior and lateral surfaces of femur		Extends leg at knee
Hamstring Group			
Biceps femoris	Ischial tuberosity (long head) and linea aspera of femur (short head)	Head of fibula and lateral condyle of tibia	Flexes leg at knee; rotates leg laterally; extends thigh
Semitendinosus	Ischial tuberosity	Proximal medial surface of tibia	Flexes leg at knee; rotates leg medially; extends thigh
Semimembranosus	Ischial tuberosity	Posterior medial condyle of tibia	Flexes leg at knee; rotates leg medially; extends thigh

FIGURE 25.5   Muscles of the anterior right leg (a and b).

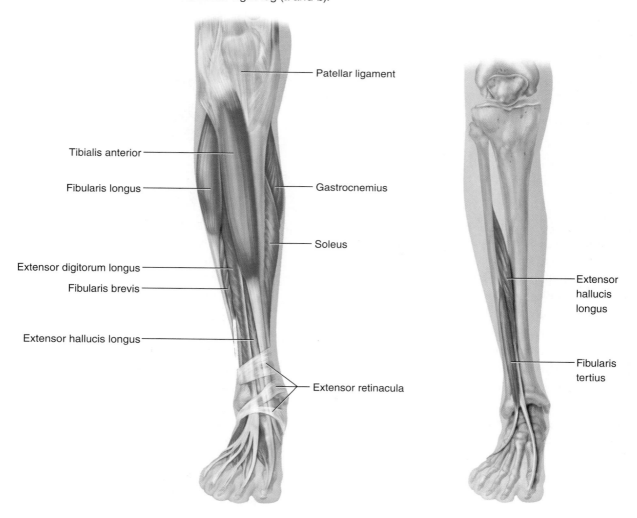

(a) Superficial view

(b) Selected individual muscles

FIGURE 25.6 Muscles of the posterior right leg (a and b).

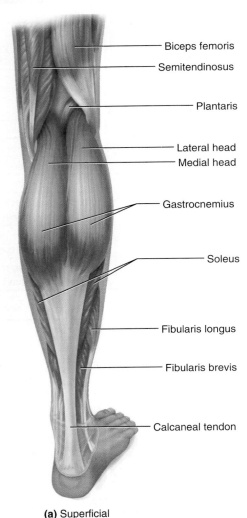

- Biceps femoris
- Semitendinosus
- Plantaris
- Lateral head
- Medial head
- Gastrocnemius
- Soleus
- Fibularis longus
- Fibularis brevis
- Calcaneal tendon

(a) Superficial

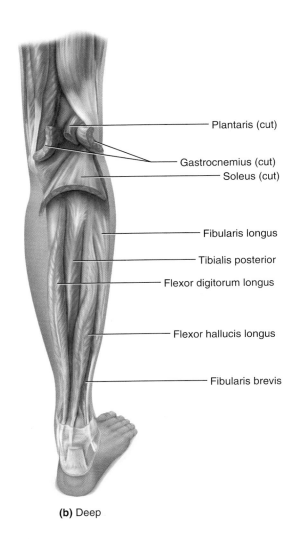

- Plantaris (cut)
- Gastrocnemius (cut)
- Soleus (cut)
- Fibularis longus
- Tibialis posterior
- Flexor digitorum longus
- Flexor hallucis longus
- Fibularis brevis

(b) Deep

11. Locate the following muscles in the human torso model and in the lower limb models. Also locate as many of them as possible in your body.

   muscles that move the foot
   - anterior leg muscles
     - tibialis anterior
     - extensor digitorum longus
     - extensor hallucis longus
     - fibularis (peroneus) tertius
   - posterior leg muscles
     - superficial group
       gastrocnemius
       soleus
       plantaris
     - deep group
       tibialis posterior
       flexor digitorum longus
       flexor hallucis longus

   - lateral leg muscles
     - fibularis (peroneus) longus
     - fibularis (peroneus) brevis

12. Demonstrate the actions of these muscles in your body.
13. Locate the origins and insertions of these muscles in the human skeleton.
14. Complete Parts B, C, and D of the laboratory assessment.

Learning Extension Activity

A long rubber band can be used to simulate muscle locations, origins, insertions, and actions on muscular models, the skeleton, or on a laboratory partner. Hold one end of the rubber band firmly on the origin location of a muscle, then slightly stretch the rubber band and hold the other end on the insertion site. Allow the insertion end to slowly move toward the origin end to simulate the contraction and action of the muscle.

FIGURE 25.7 Muscles of the lateral right leg.

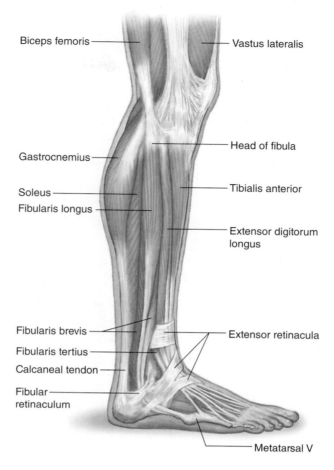

Biceps femoris
Vastus lateralis
Head of fibula
Gastrocnemius
Tibialis anterior
Soleus
Fibularis longus
Extensor digitorum longus
Fibularis brevis
Extensor retinacula
Fibularis tertius
Calcaneal tendon
Fibular retinaculum
Metatarsal V

TABLE 25.3 Muscles That Move the Foot

Muscle	Origin	Insertion	Action
Tibialis anterior	Lateral condyle and proximal tibia	Medial cuneiform and metatarsal I	Dorsiflexion and inversion of foot
Extensor digitorum longus	Lateral condyle of tibia and anterior surface of fibula	Dorsal surfaces of second and third phalanges of toes 2–5	Extends toes 2–5 and dorsiflexion
Extensor hallucis longus	Anterior surface of fibula	Distal phalanx of the great toe	Extends great toe and dorsiflexion
Fibularis tertius	Anterior distal surface of fibula	Dorsal surface of metatarsal V	Dorsiflexion and eversion of foot
Gastrocnemius	Lateral and medial condyles of femur	Calcaneus via calcaneal tendon	Plantar flexion of foot and flexes knee
Soleus	Head and shaft of fibula and posterior surface of tibia	Calcaneus via calcaneal tendon	Plantar flexion of foot
Plantaris	Superior to lateral condyle of femur	Calcaneus	Plantar flexion of foot and flexes knee
Tibialis posterior	Posterior tibia and fibula	Tarsals (several) and metatarsals II–IV	Plantar flexion and inversion of foot
Flexor digitorum longus	Posterior surface of tibia	Distal phalanges of toes 2–5	Flexes toes 2–5 and plantar flexion and inversion of foot
Flexor hallucis longus	Posterior distal fibula	Distal phalanx of great toe	Flexes great toe and plantar flexes foot
Fibularis longus	Head and shaft of fibula and lateral condyle of tibia	Medial cuneiform and metatarsal I	Plantar flexion and eversion of foot; also supports arch
Fibularis brevis	Distal fibula	Metatarsal V	Plantar flexion and eversion of foot

NOTES

Name _____

Date _____

Section _____

The Ⓐ corresponds to the indicated outcome(s) found at the beginning of the laboratory exercise.

Muscles of the Hip and Lower Limb

Part A Assessments

Identify the muscles indicated in the hip and thigh in figures 25.8 and 25.9.

FIGURE 25.8 Label the right anterior muscles of the hip and thigh. Ⓐ

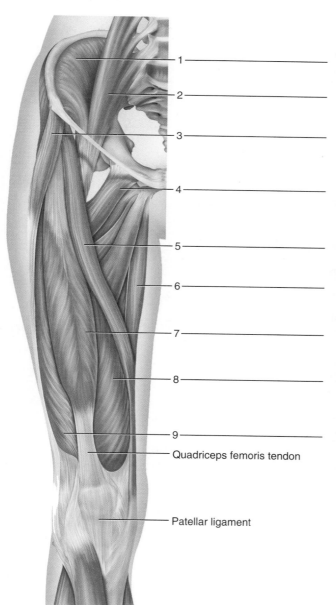

1 _____

2 _____

3 _____

4 _____

5 _____

6 _____

7 _____

8 _____

9 _____

Quadriceps femoris tendon

Patellar ligament

FIGURE 25.9 Label the right posterior hip and thigh muscles of a cadaver, using the terms provided. The gluteus maximus has been removed. Ⓐ

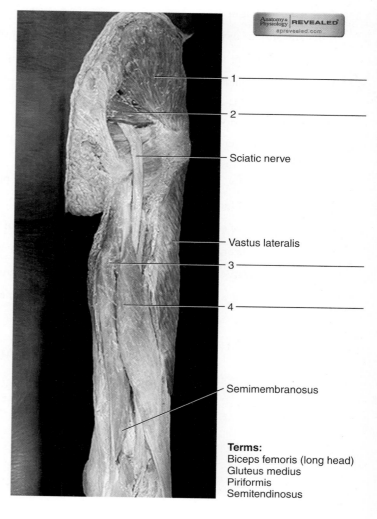

Anatomy & Physiology REVEALED
apsrevealed.com

1 _____

2 _____

Sciatic nerve

Vastus lateralis

3 _____

4 _____

Semimembranosus

Terms:
Biceps femoris (long head)
Gluteus medius
Piriformis
Semitendinosus

Part B Assessments

Match the muscles in column A with the actions in column B. Place the letter of your choice in the space provided. 🔼

Column A

a. Biceps femoris
b. Fibularis (peroneus) longus
c. Gluteus medius
d. Gracilis
e. Psoas major and iliacus
f. Quadriceps femoris group
g. Sartorius
h. Tibialis anterior
i. Tibialis posterior

Column B

_____ 1. Adducts thigh and flexes knee

_____ 2. Plantar flexion and eversion of foot

_____ 3. Flexes thigh at the hip

_____ 4. Abducts and flexes thigh and rotates it laterally; flexes leg at knee

_____ 5. Abducts thigh and rotates it medially

_____ 6. Plantar flexion and inversion of foot

_____ 7. Flexes leg at the knee and laterally rotates leg

_____ 8. Extends leg at the knee

_____ 9. Dorsiflexion and inversion of foot

Part C Assessments

Name the muscle indicated by the following combinations of origin and insertion. 🔼

Origin	Insertion	Muscle
1. Lateral surface of ilium	Greater trochanter of femur	_____
2. Anterior superior iliac spine	Medial surface of proximal tibia	_____
3. Lateral and medial condyles of femur	Calcaneus	_____
4. Iliac crest and anterior superior iliac spine	Iliotibial tract	_____
5. Greater trochanter and linea aspera of femur	Patella to tibial tuberosity	_____
6. Ischial tuberosity	Medial surface of proximal tibia	_____
7. Linea aspera of femur	Patella to tibial tuberosity	_____
8. Posterior surface of tibia	Distal phalanges of four lateral toes	_____
9. Lateral condyle and proximal tibia	Medial cuneiform and first metatarsal	_____
10. Ischial tuberosity and pubis	Linea aspera of femur	_____
11. Anterior sacrum	Greater trochanter of femur	_____

Part D Assessments

Identify the muscles indicated in figure 25.10.

FIGURE 25.10 Label these muscles that appear as lower limb surface features in these photographs (a and b), by placing the correct numbers in the spaces provided. ⚠

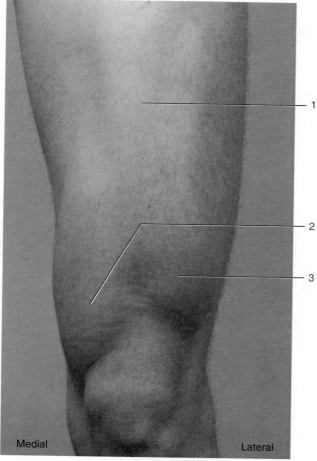

(a) Left thigh, anterior view

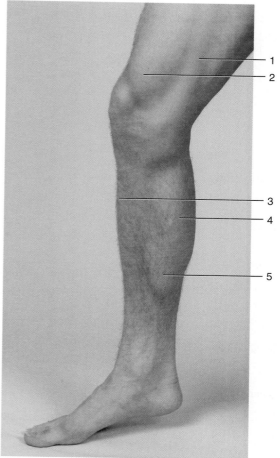

(b) Right lower limb, medial view

_____ Rectus femoris

_____ Vastus lateralis

_____ Vastus medialis

_____ Gastrocnemius

_____ Sartorius

_____ Soleus

_____ Tibialis anterior

_____ Vastus medialis

NOTES

Surface Anatomy

Purpose of the Exercise

To examine the surface features of the human body and the terms used to describe them.

Materials Needed

Small round stickers
Colored pencils (black and red)

Learning Outcomes

After completing this exercise you should be able to

1. Locate and identify major surface features of the human body.

2. Arrange surface features by body region.

3. Distinguish between surface features that are bony land-marks and soft tissue.

Pre-Lab

Carefully read the introductory material and examine the entire lab. Be familiar with body regions, the skeletal system, and the muscular system from lecture or the textbook. Answer the pre-lab questions.

Pre-Lab Questions: Select the correct answer for each of the following questions:

1. The technique of examining surface features with hands and fingers is called
   a. exploration.   **b.** touching.
   c. palpation.   **d.** feeling.

2. Soft tissues include the following *except*
   a. bone.   **b.** muscle.
   c. connective tissue.   **d.** skin.

3. Which of the following is a bony feature of the upper torso?
   a. pectoralis major   **b.** anterior axillary fold
   c. jugular notch   **d.** deltoid

4. Which of the following is a soft tissue feature of the lower limb?
   a. tibial tuberosity   **b.** head of the fibula
   c. medial malleolus   **d.** vastus lateralis

5. Palpation enables us to determine all of the following characteristics of a structure *except*
   a. location.   **b.** function.
   c. size.   **d.** texture.

6. The laryngeal prominence can be palpated in the anterior neck.
   True _____   False _____

7. The linea alba is a feature of the posterior torso (trunk).
   True _____   False _____

External landmarks, called surface anatomy, located on the human body provide an opportunity to examine surface features and to help locate other internal structures. This exercise will focus on bony landmarks and superficial soft tissue structures, primarily related to the skeletal and muscular systems. The technique used for the examination of surface features is called *palpation,* accomplished by touching with the hands or fingers. This enables us to determine the location, size, and texture of the structure. Some of the surface features can be visible without the need of palpation, especially on body builders and very thin individuals. If the individual involved has considerable subcutaneous adipose tissue, additional pressure might be necessary to palpate some of the structures.

These surface features will help us to locate additional surface features and deeper structures during the study of the additional systems. Some of the respiratory structures in the anterior neck can be palpated. Surface features will help us to determine the pulse locations of superficial arteries, to assess injection sites, to describe pain and injury sites, to insert examination tubes, and to listen to heart, lung, and bowel sounds with a stethoscope. For those who choose careers in emergency medical services, nurses, physicians, physical educators, physical therapists, chiropractors, and massage therapists, surface anatomy is especially valuable.

Many clues to a person's overall health can be revealed from the body surface. The general nature of the surface texture and coloration can be observed and examined quickly. Recall how quickly you can sometimes determine if a person appears healthy or ill! Palpation of various surface features can provide additional clues to possible ailments. With the use of different pressures during palpations, clues to the depth of organs and possible abnormalities become more evident. An external examination is an important aspect of a physical exam. The usefulness of this laboratory exercise will become even more apparent as you continue the study of anatomy and physiology and related careers.

Procedure—Surface Anatomy

1. Review the earlier laboratory manual exercises and figures on body regions, skeletal system, and muscular system.
2. Examine the surface features labeled in figures 26.1 through 26.8. Features in **boldface** are bony features. Other labeled features (not boldface) are soft tissue features. A bony feature is a part of a bone and the skeletal system. Soft tissue features are comprised of tissue other than bone such as muscle, connective tissue, or skin. Palpate the labeled surface features to distinguish the texture and degree of firmness of the bony features and the soft tissue features. Most of the surface features included in this laboratory exercise can be palpated on your own body.

FIGURE 26.1 Surface features of the posterior shoulder and torso. (**Boldface** indicates bony features; not boldface indicates soft tissue features.)

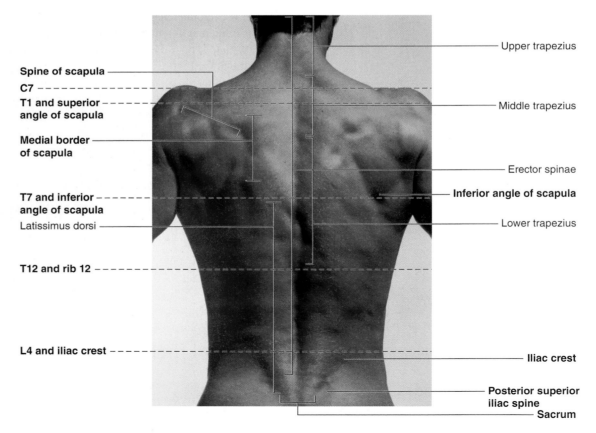

Spine of scapula
C7
T1 and superior angle of scapula
Medial border of scapula
T7 and inferior angle of scapula
Latissimus dorsi
T12 and rib 12
L4 and iliac crest

Upper trapezius
Middle trapezius
Erector spinae
Inferior angle of scapula
Lower trapezius
Iliac crest
Posterior superior iliac spine
Sacrum

FIGURE 26.2 Surface features of (a) the anterior head and neck, (b) the lateral head and neck, and (c) the posterior head and neck. (**Boldface** indicates bony features; not boldface indicates soft tissue features.)

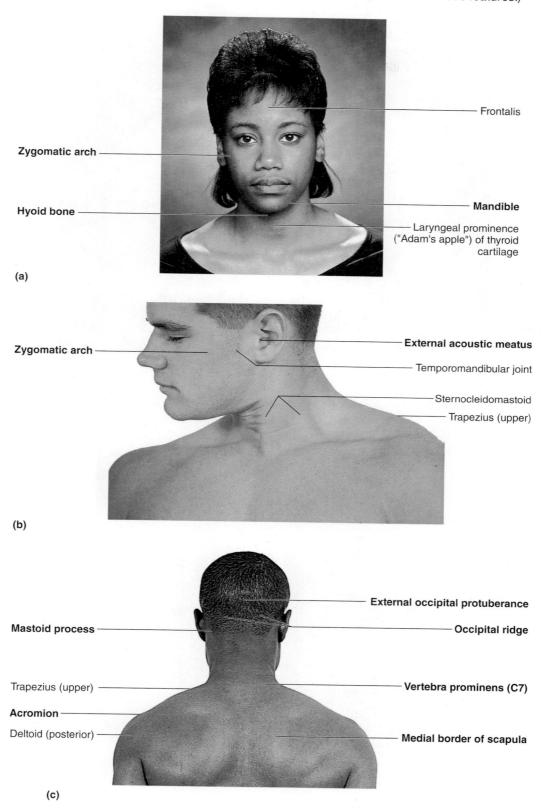

(a)

Frontalis

Zygomatic arch

Hyoid bone

Mandible

Laryngeal prominence ("Adam's apple") of thyroid cartilage

(b)

Zygomatic arch

External acoustic meatus

Temporomandibular joint

Sternocleidomastoid

Trapezius (upper)

(c)

External occipital protuberance

Mastoid process

Occipital ridge

Trapezius (upper)

Vertebra prominens (C7)

Acromion

Deltoid (posterior)

Medial border of scapula

FIGURE 26.3 Anterior surface features of (a) upper torso and (b) lower torso. (**Boldface** indicates bony features; not boldface indicates soft tissue features.)

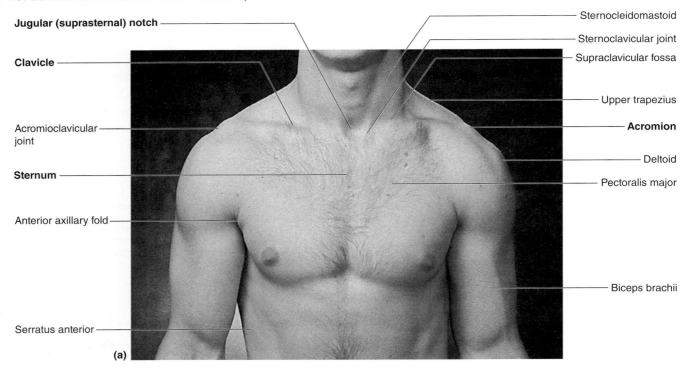

Jugular (suprasternal) notch

Clavicle

Acromioclavicular joint

Sternum

Anterior axillary fold

Serratus anterior

(a)

Sternocleidomastoid

Sternoclavicular joint

Supraclavicular fossa

Upper trapezius

Acromion

Deltoid

Pectoralis major

Biceps brachii

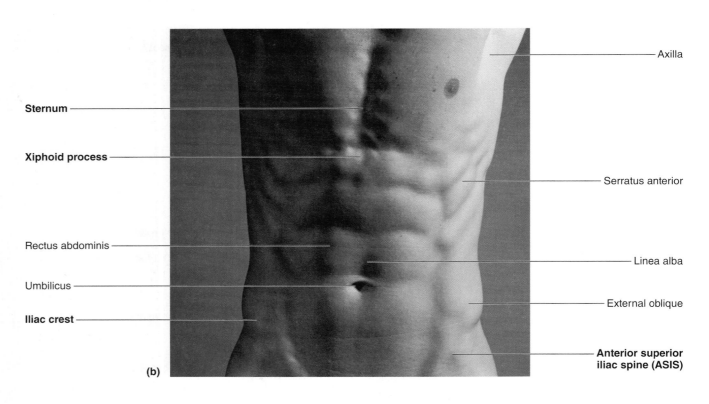

Sternum

Xiphoid process

Rectus abdominis

Umbilicus

Iliac crest

(b)

Axilla

Serratus anterior

Linea alba

External oblique

Anterior superior iliac spine (ASIS)

FIGURE 26.4 Surface features of the lateral shoulder and upper limb. (**Boldface** indicates bony features; not boldface indicates soft tissue features.)

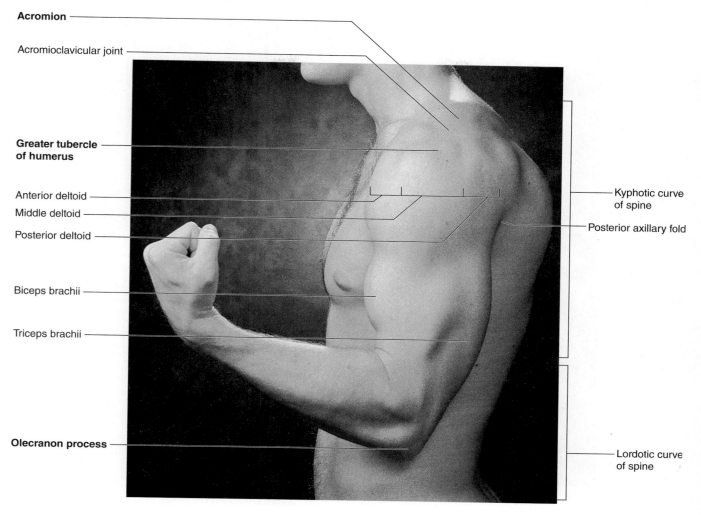

Acromion

Acromioclavicular joint

Greater tubercle of humerus

Anterior deltoid

Middle deltoid

Posterior deltoid

Biceps brachii

Triceps brachii

Olecranon process

Kyphotic curve of spine

Posterior axillary fold

Lordotic curve of spine

FIGURE 26.5 Surface features of the upper limb (a) anterior view and (b) posterior view. (**Boldface** indicates bony features; not boldface indicates soft tissue features.)

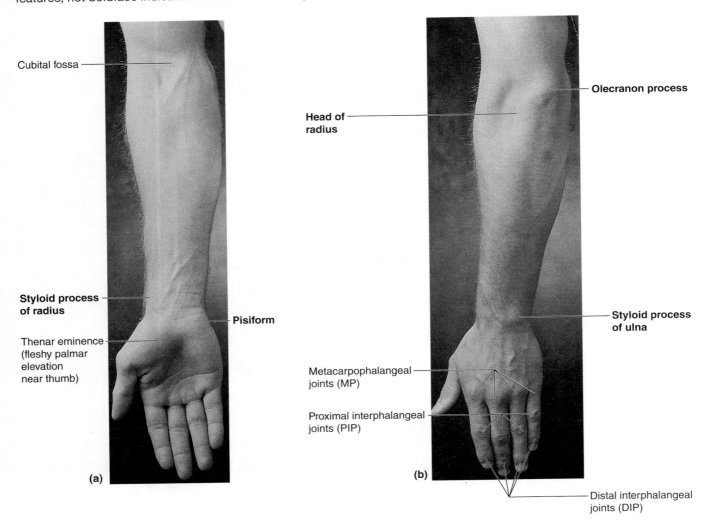

Cubital fossa

Olecranon process

Head of radius

Styloid process of radius

Pisiform

Thenar eminence (fleshy palmar elevation near thumb)

Styloid process of ulna

Metacarpophalangeal joints (MP)

Proximal interphalangeal joints (PIP)

(a)

(b)

Distal interphalangeal joints (DIP)

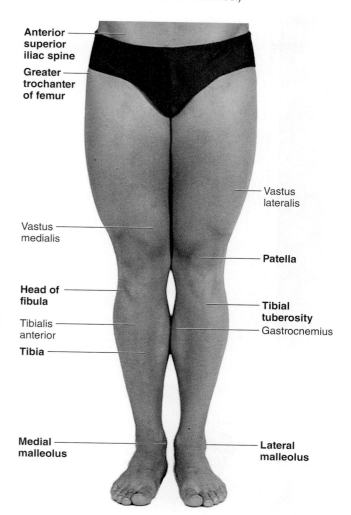

Anterior superior iliac spine

Greater trochanter of femur

Vastus lateralis

Vastus medialis

Patella

Head of fibula

Tibialis anterior

Tibial tuberosity

Gastrocnemius

Tibia

Medial malleolus

Lateral malleolus

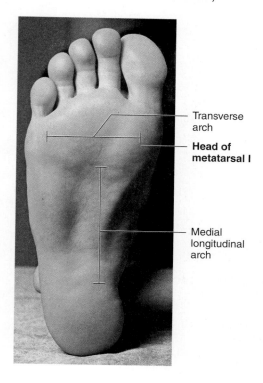

Transverse arch

Head of metatarsal I

Medial longitudinal arch

271

FIGURE 26.8 Surface features of the (a) posterior lower limb, (b) lateral lower limb, and (c) medial lower limb. (**Boldface** indicates bony features; not boldface indicates soft tissue features.)

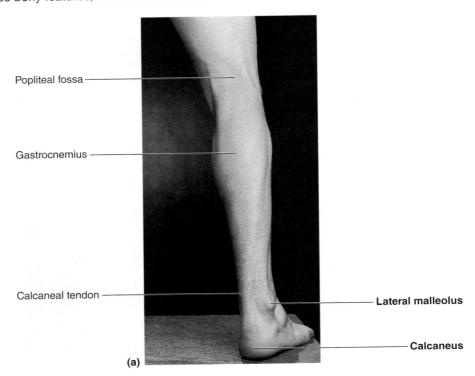

Popliteal fossa

Gastrocnemius

Calcaneal tendon

Lateral malleolus

Calcaneus

(a)

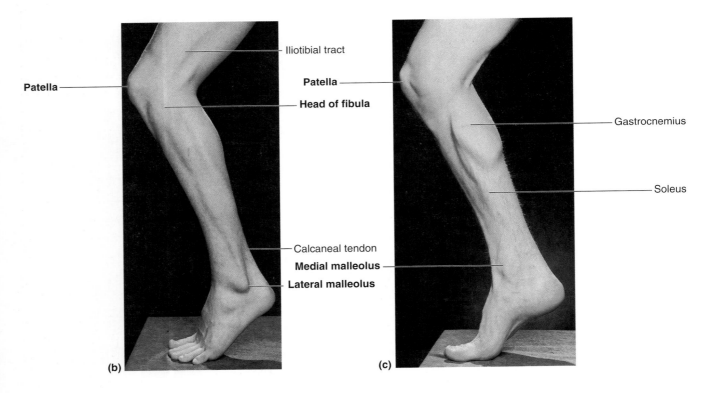

Iliotibial tract

Patella

Head of fibula

Calcaneal tendon

Lateral malleolus

(b)

Patella

Gastrocnemius

Soleus

Medial malleolus

(c)

3. Complete Parts A and B of Laboratory Assessment 26.
4. Use figures 26.1 through 26.8 to determine which items listed in table 26.1 are bony features and which items are soft tissue features. Examine the list of bony and soft tissue surface features in the first column. If the feature listed is a bony surface feature, place an "X" in the second column; if the feature listed is a soft tissue feature, place an "X" in the third column.
5. Complete Part C of the laboratory assessment.

Learning Extension Activity

Select fifteen surface features presented in the laboratory exercise figures. Using small, round stickers, write the name of a surface feature on each sticker. Working with a lab partner, locate the selected surface features on yourself and on your lab partner. Use the stickers to accurately "label" your partner.

- Were you able to find all the features on both of you? _____

- Was it easier to find the surface features on yourself or your lab partner? _____

TABLE 26.1 Representative Bony and Soft Tissue Surface Features

(*Note:* After indicating your "X" in the appropriate column, use the extra space in columns 2 and 3 to add personal descriptions of the nature of the bony feature or the soft tissue feature palpation.)

Bony and Soft Tissues	Bony Features	Soft Tissue Features
Biceps brachii		
Calcaneal tendon		
Deltoid		
Gastrocnemius		
Greater trochanter		
Head of fibula		
Iliac crest		
Iliotibial tract		
Lateral malleolus		
Mandible		
Mastoid process		
Medial border of scapula		
Rectus abdominis		
Sacrum		
Serratus anterior		
Sternocleidomastoid		
Sternum		
Thenar eminence		
Tibialis anterior		
Zygomatic arch		

NOTES

Name _____

Date _____

Section _____

The Ⓐ corresponds to the indicated outcome(s) found at the beginning of the laboratory exercise.

Surface Anatomy

Part A Assessments

Match the body region in column A with the surface feature in column B. Place the letter of your choice in the space provided. ⚊1 ⚊2

Column A

a. Head

b. Trunk (torso)

c. Upper limb

d. Lower limb

Column B

_____ 1. Umbilicus

_____ 2. Medial malleolus

_____ 3. Iliac crest

_____ 4. Transverse arch

_____ 5. Spine of scapula

_____ 6. External occipital protuberance

_____ 7. Cubital fossa

_____ 8. Sternum

_____ 9. Olecranon process

_____ 10. Zygomatic arch

_____ 11. Mastoid process

_____ 12. Thenar eminence

_____ 13. Popliteal fossa

_____ 14. Metacarpophalangeal joints

Label figure 26.9 with the surface features provided.

FIGURE 26.9   Using the terms provided, label the surface features of the (a) anterior view of the body and (b) posterior view of the body. (**Boldface** indicates bony features; not boldface indicates soft tissue features.) ⚠

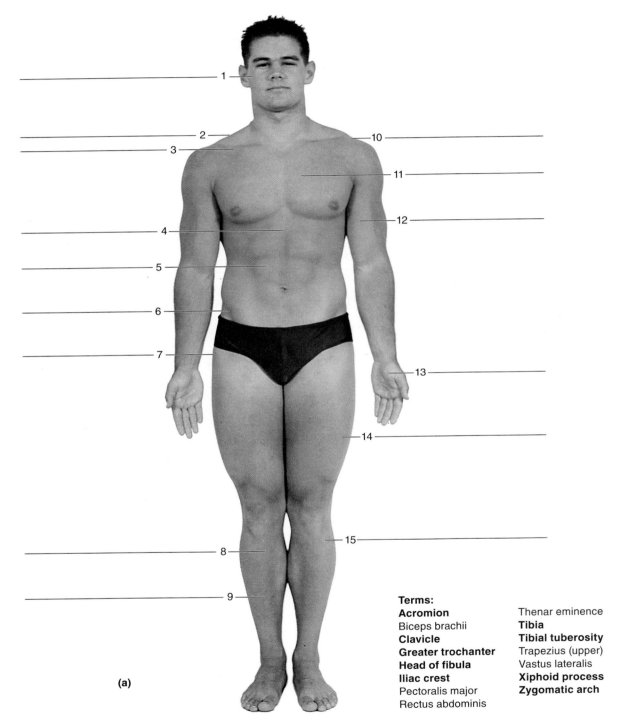

(a)

Terms:

Acromion
Biceps brachii
Clavicle
Greater trochanter
Head of fibula
Iliac crest
Pectoralis major
Rectus abdominis

Thenar eminence
Tibia
Tibial tuberosity
Trapezius (upper)
Vastus lateralis
Xiphoid process
Zygomatic arch

FIGURE 26.9 *Continued.*

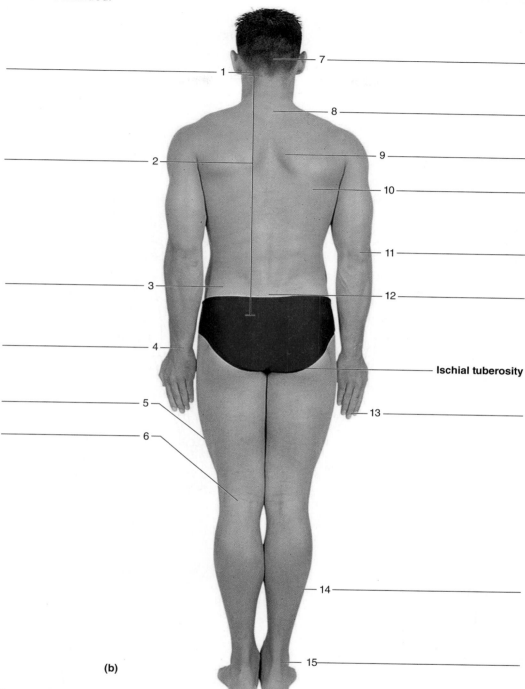

Ischial tuberosity

(b)

Terms:

C7
Calcaneal tendon
Distal interphalageal joint (DIP)
Erector spinae
External occipital protuberance

Iliac crest
Iliotibial tract
Inferior angle of scapula
Mastoid process
Medial border of scapula

Olecranon process
Popliteal fossa
Sacrum
Soleus
Styloid process of ulna

Part C Assessments

Using table 26.1 from Laboratory Exercise 26, indicate the locations of the surface features with an "X" on figure 26.10. Use a black "X" for the bony features and a red "X" for the soft tissue features.

FIGURE 26.10 Indicate bony surface features with a black "X" and soft tissue surface features with a red "X" using the results from table 26.1, on the (a) anterior and (b) posterior diagrams of the body. 3

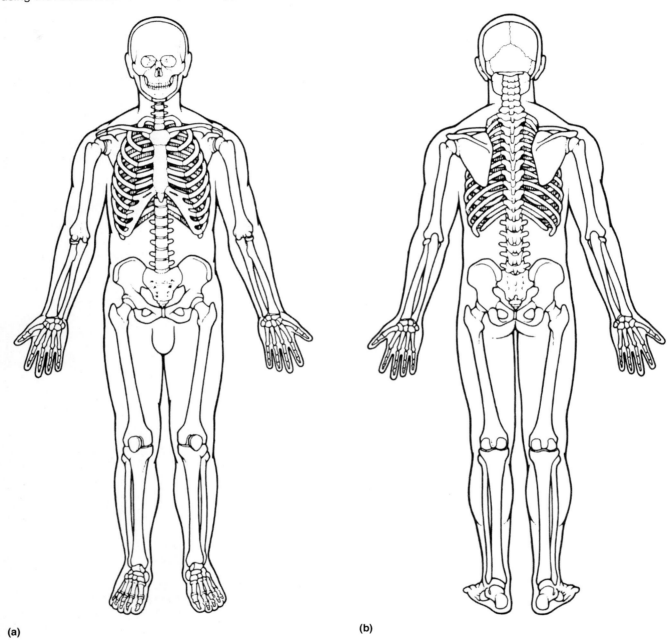

(a)   (b)

Terms from table 26.1:

Biceps brachii	Iliac crest	Sacrum
Calcaneal tendon	Iliotibial tract	Serratus anterior
Deltoid	Lateral malleolus	Sternocleidomastoid
Gastrocnemius	Mandible	Sternum
Greater trochanter	Mastoid process	Thenar eminence
Head of fibula	Medial border of scapula	Tibialis anterior
	Rectus abdominis	Zygomatic arch

Nervous Tissue and Nerves

Purpose of the Exercise

To review the characteristics of nervous tissue and to observe neurons, neuroglia, and various features of the nerves.

Materials Needed

Compound light microscope
Prepared microscope slides of the following:
    Spinal cord (smear)
    Dorsal root ganglion (section)
    Neuroglia (astrocytes)
    Peripheral nerve (cross section and longitudinal section)
Neuron model

For Learning Extension Activity:
Prepared microscope slide of Purkinje cells from cerebellum

Learning Outcomes

After completing this exercise, you should be able to

1. Describe and locate the characteristics of nervous tissue.
2. Distinguish structural and functional characteristics between neurons and neuroglia.
3. Identify and sketch the major structures of a neuron and a nerve.

Pre-Lab

Carefully read the introductory material and examine the entire lab. Be familiar with the structures and functions of nervous tissue and nerves from lecture or the textbook. Answer the pre-lab questions.

Pre-Lab Questions: Select the correct answer for each of the following questions:

1. The cell body of a neuron contains the
    a. nucleus.   **b.** dendrites.
    c. axon.   **d.** neuroglia.

2. A multipolar neuron contains
    a. one dendrite and many axons.
    b. many dendrites and one axon.
    c. one dendrite and one axon.
    d. a single process with the dendrite and axon.

3. Neuroglia that produce myelin insulation in the CNS are
    a. microglia.   **b.** astrocytes.
    c. ependymal cells.   **d.** oligodendrocytes.

4. The PNS contains
    a. 12 pairs of cranial nerves only.
    b. 31 pairs of spinal nerves only.
    c. 12 pairs of cranial nerves and 31 pairs of spinal nerves.
    d. 43 pairs of spinal nerves.

5. Schwann cells
    a. are only in the brain.
    b. are only in the spinal cord.
    c. are throughout the CNS.
    d. have a myelin sheath and neurilemma.

6. A _____ neuron is the most common structural neuron in the brain and spinal cord.
    a. multipolar   **b.** tripolar
    c. bipolar   **d.** unipolar

7. Sensory neurons conduct impulses from the spinal cord to a muscle or a gland.
    True _____   False _____

8. Astrocytes have contacts between blood vessels and neurons in the CNS.
    True _____   False _____

Nervous tissue, which occurs in the brain, spinal cord, and peripheral nerves, contains *neurons* (nerve cells) and *neuroglia* (neuroglial cells; glial cells). Neurons are irritable (excitable) and easily respond to stimuli, resulting in the conduction of an action potential (nerve impulse). A neuron contains a cell body with a nucleus and most of the cytoplasm, and elongated cell processes (dendrites and axons) along which impulse conductions occur. Neurons can be classified according to structural variations of their cell processes: multipolar with many dendrites and one axon (nerve fiber), bipolar with one dendrite and one axon, and unipolar with a single process where the dendrite leads directly into the axon. Functional classifications are used according to the direction of the impulse conduction: sensory (afferent) neurons conduct impulses from receptors to the central nervous system (CNS), interneurons (association neurons) conduct impulses within the CNS, and motor (efferent) neurons conduct impulses away from the CNS to effectors (muscles or glands).

Neuroglia (supportive cells) are located in close association with neurons. Four types of neuroglia are in the CNS: astrocytes, microglia, oligodendrocytes, and ependymal cells. Astrocytes are numerous, and their branches support neurons and blood vessels. Microglia can phagocytize microorganisms and nerve tissue debris. Oligodendrocytes produce the myelin insulation in the CNS. Ependymal cells line the brain ventricles and secrete CSF (cerebrospinal fluid), and the cilia on their apical surfaces aid the circulation of the CSF.

Two types of neuroglia are located in the peripheral nervous system (PNS): Schwann cells and satellite cells. Schwann cells surround nerve fibers numerous times. The wrappings nearest the nerve fiber represent the myelin sheath and serve to insulate and increase the impulse speed, while the last wrapping, called the neurilemma, contains most of the cytoplasm and the nucleus and functions in nerve fiber regeneration in PNS neurons. Satellite cells surround and support the cell body regions, called ganglia, of peripheral neurons.

The PNS contains 12 pairs of cranial nerves and 31 pairs of spinal nerves, all containing parallel axons of neurons and neuroglia representing the nervous tissue components. Most nerves are mixed in that they contain both sensory and motor neurons, but some contain only sensory or motor components. The nerves represent an organ structure as they also contain small blood vessels and fibrous connective tissue. The fibrous connective tissue around each nerve fiber and the Schwann cell is the endoneurium; a bundle of nerve fibers, representing a fascicle, is surrounded by a perineurium; and the entire nerve is surrounded by an epineurium. The fibrous connective components of a nerve provide protection and stretching during body movements.

Procedure A—Nervous Tissue

In this procedure you will concentrate on more detail of nervous tissue than was observed when nervous tissue was first examined in the earlier tissue lab. Two tables of nervous tissue include the structural and functional characteristics of neurons, and neuroglia within the central nervous system (CNS) and the peripheral nervous system (PNS). Diagrams of neurons and neuroglia are provided as sources of comparison for your microscopic observations using the compound microscope. When you make sketches in the laboratory assessment, include all the observed structures with labels.

1. Examine tables 27.1 and 27.2 and figures 27.1, 27.2, and 27.3.
2. Complete Parts A and B of Laboratory Assessment 27.
3. Obtain a prepared microscope slide of a spinal cord smear. Using low-power magnification, search the slide

TABLE 27.1 Structural and Functional Types of Neurons

Classified by Structure		
Type	**Structural Characteristics**	**Location**
Multipolar neuron	Cell body with one axon and multiple dendrites	Most common type of neuron in the brain and spinal cord
Bipolar neuron	Cell body with one axon and one dendrite	In receptor parts of the eyes, nose, and ears; rare type
Unipolar neuron	Cell body with a single process that divides into two branches and functions as an axon; only the receptor ends of the peripheral (distal) process function as dendrites	Most sensory neurons
Classified by Function		
Type	**Functional Characteristics**	**Structural Characteristics**
Sensory (afferent) neuron	Conducts impulses from receptors in peripheral body parts into the brain or spinal cord	Most unipolar; some bipolar
Interneuron	Transmits impulses between neurons in the brain and spinal cord	Multipolar
Motor (efferent) neuron	Conducts impulses from the brain or spinal cord out to muscles or glands (effectors)	Multipolar

TABLE 27.2   Types of Neuroglia in the CNS and PNS

Type	Structural Characteristics	Functions
CNS (brain and spinal cord)		
Astrocytes	Star-shaped cells contacting neurons and blood vessels; most abundant type	Structural support; formation of scar tissue to replace damaged neurons; transport of substances between blood vessels and neurons; communicate with one another and with neurons; remove excess ions and neurotransmitters
Microglia	Small cells with slender cellular processes	Structural support; phagocytosis of microorganisms and damaged tissue
Oligodendrocytes	Shaped like astrocytes, but with fewer cellular processes; processes wrap around axons	Form myelin sheaths in the brain and spinal cord
Ependyma	Cuboidal cells lining cavities of the brain and spinal cord	Secrete and assist in circulation of cerebrospinal fluid (CSF)
PNS (peripheral nerves)		
Schwann cells	Cells that wrap tightly around the axons of peripheral neurons	Speed neurotransmission
Satellite cells	Flattened cells that surround cell bodies of neurons in ganglia	Support ganglia

FIGURE 27.1   Structural types of neurons (a) multipolar neuron, (b) bipolar neuron, and (c) unipolar neuron.

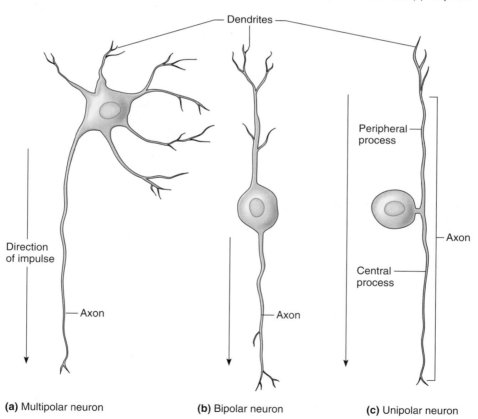

(a) Multipolar neuron   (b) Bipolar neuron   (c) Unipolar neuron

FIGURE 27.2 Diagram of a multipolar motor neuron.

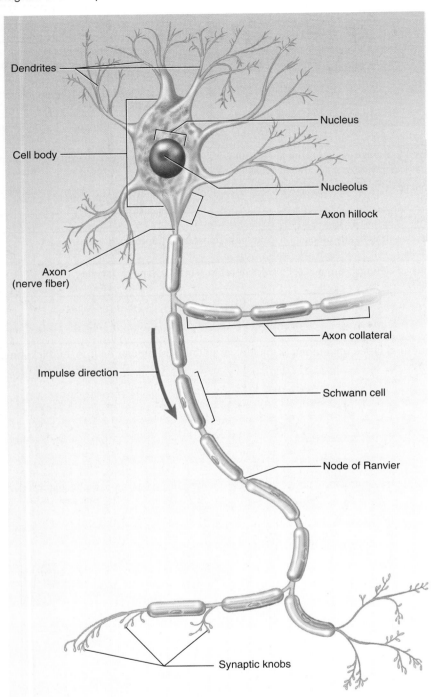

Dendrites

Nucleus

Cell body

Nucleolus

Axon hillock

Axon
(nerve fiber)

Axon collateral

Impulse direction

Schwann cell

Node of Ranvier

Synaptic knobs

FIGURE 27.3 Diagram of a cross section of a myelinated axon (nerve fiber) of a spinal nerve.

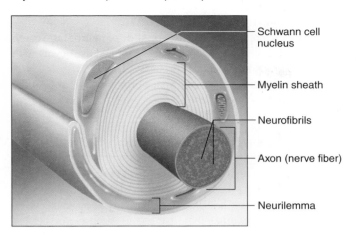

— Schwann cell nucleus

— Myelin sheath

— Neurofibrils

— Axon (nerve fiber)

— Neurilemma

and locate the relatively large, deeply stained cell bodies of multipolar motor neurons (fig. 27.4).

4. Observe a single multipolar motor neuron, using high-power magnification, and note the following features:

    cell body (soma)

- nucleus
- nucleolus
- Nissl bodies (chromatophilic substance)—a type of rough ER in neurons
- neurofibrils—threadlike structures extending into axon
- axon hillock—origin region of axon

    dendrites—conduct impulses toward cell body

    axon (nerve fiber)—conducts impulse away from cell body

Compare the slide to the neuron model and to figure 27.4. Also note small, darkly stained nuclei of neuroglia around the motor neuron.

FIGURE 27.4 Micrograph of a multipolar neuron and neuroglia from a spinal cord smear (600×).

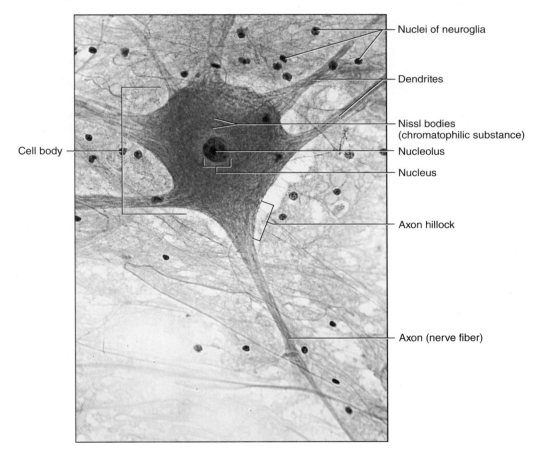

Nuclei of neuroglia

Dendrites

Nissl bodies (chromatophilic substance)

Nucleolus

Nucleus

Axon hillock

Axon (nerve fiber)

Cell body

5. Sketch and label a motor (efferent) neuron in the space provided in Part C of the laboratory assessment.

6. Obtain a prepared microscope slide of a dorsal root ganglion. Search the slide and locate a cluster of sensory neuron cell bodies. You also may note bundles of nerve fibers passing among groups of neuron cell bodies (fig. 27.5).

7. Sketch and label a sensory (afferent) neuron cell body in the space provided in Part C of the laboratory assessment.

8. Examine figure 27.6 and table 27.2.

9. Obtain a prepared microscope slide of neuroglia. Search the slide and locate some darkly stained astrocytes with numerous long, slender processes (fig. 27.7).

10. Sketch a neuroglia in the space provided in Part C of the laboratory assessment.

FIGURE 27.5 Micrograph of a dorsal root ganglion (100×).

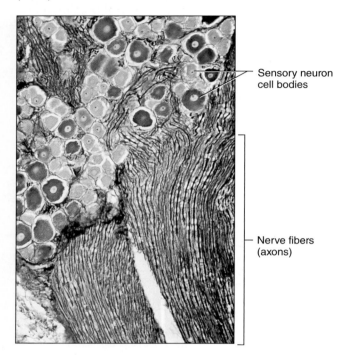

Sensory neuron cell bodies

Nerve fibers (axons)

Learning Extension Activity

Obtain a prepared microscope slide of Purkinje cells (fig. 27.8). To locate these neurons, search the slide for large, flask-shaped cell bodies. Each cell body has one or two large, thick dendrites that give rise to extensive branching networks of dendrites. These large cells are located in a particular region of the brain (cerebellar cortex).

Procedure B—Nerves

In this procedure you will compare an illustration of a peripheral spinal nerve with actual microscopic views of nerves. A nerve is an organ of the nervous system and contains various tissues in combination with the nervous tissue. The nerve contains the nerve fibers (axons) of the neurons and the Schwann cells representing some neuroglia in the PNS.

1. Study figure 27.9 of a cross section of a spinal nerve. Examine the fibrous connective tissue pattern including the epineurium around the entire nerve, the perineurium around fascicles, and the endoneurium around an

FIGURE 27.6 Types of neuroglia in the CNS.

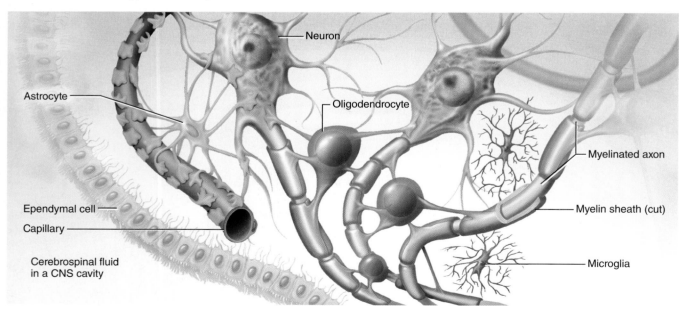

Neuron

Astrocyte

Oligodendrocyte

Myelinated axon

Myelin sheath (cut)

Microglia

Ependymal cell

Capillary

Cerebrospinal fluid in a CNS cavity

FIGURE 27.7 Micrograph of astrocytes (1,000×).

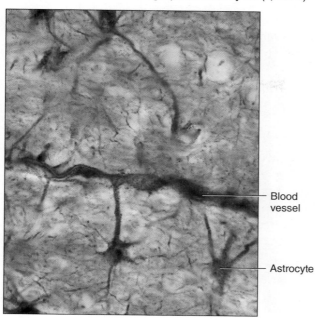

Blood vessel

Astrocyte

FIGURE 27.8 Large multipolar Purkinje cell from cerebellum of the brain (400×).

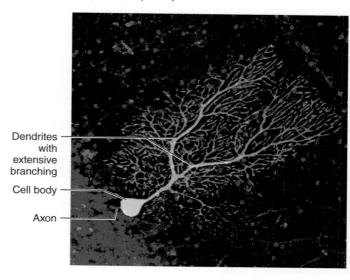

Dendrites with extensive branching

Cell body

Axon

FIGURE 27.9 Diagram of a cross section of a peripheral spinal nerve.

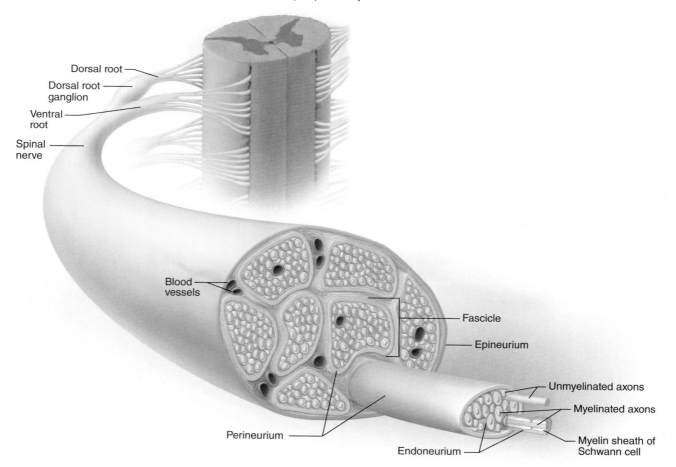

Dorsal root

Dorsal root ganglion

Ventral root

Spinal nerve

Blood vessels

Fascicle

Epineurium

Unmyelinated axons

Myelinated axons

Myelin sheath of Schwann cell

Perineurium

Endoneurium

FIGURE 27.10 Cross section of a bundle of neurons within a nerve (400×).

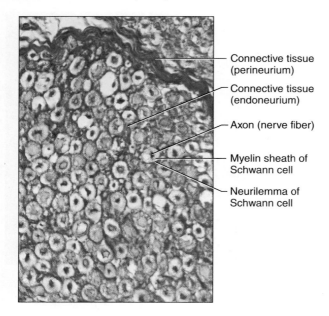

- Connective tissue (perineurium)
- Connective tissue (endoneurium)
- Axon (nerve fiber)
- Myelin sheath of Schwann cell
- Neurilemma of Schwann cell

FIGURE 27.11 Longitudinal section of a nerve (2,000×).

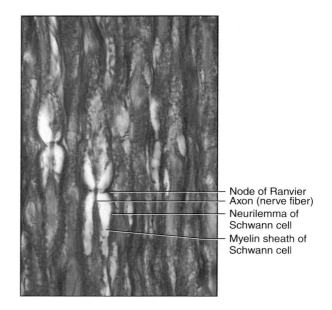

- Node of Ranvier
- Axon (nerve fiber)
- Neurilemma of Schwann cell
- Myelin sheath of Schwann cell

individual axon and Schwann cell. Reexamine figure 27.3 to note additional cross-sectional detail of the Schwann cell layers.

2. Obtain a prepared microscope slide of a nerve. Locate the cross section of the nerve and note the many round nerve fibers inside. Also note the dense layer of connective tissue (perineurium) that encircles a fascicle of nerve fibers and holds them together in a bundle. The individual nerve fibers are surrounded by a layer of more delicate connective tissue (endoneurium) (fig. 27.10).

3. Using high-power magnification, observe a single nerve fiber and note the following features:

   axon

   myelin sheath of Schwann cell around the axon (most of the myelin may have been dissolved and lost during the slide preparation)

   neurilemma of Schwann cell

4. Sketch and label a nerve fiber with Schwann cell (cross section) in the space provided in Part D of the laboratory assessment.

5. Locate the longitudinal section of the nerve on the slide (fig. 27.11). Note the following:

   axons

   myelin sheath of Schwann cells

   neurilemma of Schwann cells

   nodes of Ranvier—narrow gaps between the Schwann cells (figs. 27.2 and 27.11)

6. Sketch and label a nerve fiber with Schwann cell (longitudinal section) in the space provided in Part D of the laboratory assessment.

Name _____

Date _____

Section _____

The A corresponds to the indicated outcome(s) found at the beginning of the laboratory exercise.

Nervous Tissue and Nerves

Part A Assessments

Match the terms in column A with the descriptions in column B. Place the letter of your choice in the space provided. 1 2

Column A		Column B
a. Astrocyte	_____	**1.** Sheath of Schwann cell containing cytoplasm and nucleus that encloses myelin
b. Axon		
c. Collateral	_____	**2.** Corresponds to rough endoplasmic reticulum in other cells
d. Dendrite	_____	**3.** Network of threadlike structures within cell body and extending into axon
e. Myelin		
f. Neurilemma	_____	**4.** Substance of Schwann cell composed of lipoprotein that insulates axons and increases impulse speed
g. Neurofibrils		
h. Nissl bodies (chromatophilic substance)	_____	**5.** Neuron process with many branches that conducts an action potential (impulse) toward the cell body
	_____	**6.** Branch of an axon
i. Unipolar neuron	_____	**7.** Star-shaped neuroglia between neurons and blood vessels
	_____	**8.** Nerve fiber arising from a slight elevation of the cell body that conducts an action potential (impulse) away from the cell body
	_____	**9.** Possesses a single process from the cell body

Part B Assessments

Match the terms in column A with the descriptions in column B. Place the letter of your choice in the space provided. 1 2

Column A		Column B
a. Effector	_____	**1.** Transmits impulse from sensory to motor neuron within central nervous system
b. Ependymal cell		
c. Ganglion	_____	**2.** Transmits impulse out of the brain or spinal cord to effectors (muscles and glands)
d. Interneuron (association neuron)		
e. Microglia	_____	**3.** Transmits impulse into brain or spinal cord from receptors
f. Motor (efferent) neuron	_____	**4.** Myelin-forming neuroglia in brain and spinal cord
g. Oligodendrocyte	_____	**5.** Phagocytic neuroglia
h. Sensory (afferent) neuron	_____	**6.** Structure capable of responding to motor impulse
	_____	**7.** Specialized mass of neuron cell bodies outside the brain or spinal cord
	_____	**8.** Cells that line cavities of the brain and secrete cerebrospinal fluid

Part C Assessments

In the space that follows, sketch the indicated cells. Label any of the cellular structures observed, and indicate the magnification of each sketch. 🄰 🄱

Motor neuron (_____×)

Sensory neuron cell body (_____×)

Neuroglia (_____×)

Part D Assessments

In the space that follows, sketch the indicated view of a nerve fiber (axon). Label any structures observed, and indicate the magnification of each sketch. 🄰 🄱

Nerve fiber cross section with Schwann cell (_____×)

Nerve fiber longitudinal section with Schwann cell (_____×)

Spinal Cord, Spinal Nerves, and Meninges

Purpose of the Exercise

To review the characteristics of the spinal cord, spinal nerves, and meninges and to observe the major features of these structures.

Materials Needed

Compound light microscope
Prepared microscope slide of a spinal cord cross section with spinal nerve roots
Spinal cord model with meninges
Vertebral column model with spinal nerves

For Demonstration Activity:
Preserved spinal cord with meninges intact

Learning Outcomes

After completing this exercise, you should be able to

1. Identify the major features and functions of the spinal cord.

2. Locate the distribution and features of the spinal nerves.

3. Arrange the layers of the meninges and describe the structure of each.

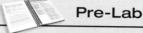

Pre-Lab

Carefully read the introductory material and examine the entire lab. Be familiar with the structures and functions of the spinal cord, spinal nerves, and meninges from lecture or the textbook. Answer the pre-lab questions.

Pre-Lab Questions: Select the correct answer for each of the following questions:

1. The inferior end of the adult spinal cord ends
   a. inferior to L1.   **b.** inferior to L5.
   c. in the sacrum.   **d.** in the coccyx.

2. The dorsal root of spinal nerves contains
   a. interneurons.   **b.** sensory neurons.
   c. motor neurons.   **d.** sensory and motor neurons.

3. The _____ is the most superficial membrane of the meninges.
   a. subarachnoid space   **b.** pia mater
   c. arachnoid mater   **d.** dura mater

4. Which of the following is *not* part of the gray matter of the spinal cord?
   a. gray commissure   **b.** posterior horn
   c. lateral funiculus   **d.** lateral horn

5. The central canal of the spinal cord is located within the
   a. white matter.   **b.** epidural space.
   c. gray commissure.   **d.** subarachnoid space.

6. The brachial plexus is formed from components of spinal nerves C5–T1.
   True _____ False _____

7. There are eight pairs of cervical spinal nerves.
   True _____ False _____

8. The major ascending (sensory) and descending (motor) tracts compose the gray matter of the spinal cord.
   True _____ False _____

The spinal cord is a column of nerve fibers that extends down through the vertebral canal. Together with the brain, it makes up the central nervous system. In the cervical and lumbar regions of the spinal cord, enlargements give rise to spinal nerves to the upper limbs and lower limbs, respectively. The inferior end portion of the spinal cord, the conus medullaris, is located just inferior to lumbar vertebra L1 in the adult.

There are 31 pairs of spinal nerves attached to the spinal cord. At close proximity to the cord, the spinal nerve has two branches: a dorsal (posterior) root containing sensory neurons and dorsal root ganglion, and a ventral (anterior) root containing motor neurons. The spinal nerves emerge through the nearby intervertebral foramina; however, most lumbar and sacral nerves extend inferiorly through the vertebral canal as the cauda equina to emerge in their respective regions of the vertebral column.

The spinal cord has a central region, the *gray matter,* with paired posterior, lateral, and anterior horns connected by a gray commissure containing the central canal. The gray matter is a processing center for spinal reflexes and synaptic integration. The *white matter* of the spinal cord, represented by paired posterior, lateral, and anterior funiculi (columns), contains ascending (sensory) and descending (motor) tracts. The specific names of the spinal tracts often reflect their respective origins and the destinations of the fibers. At various levels of the spinal cord, many of the tracts cross over (decussate) to the opposite side of the spinal cord or the brainstem.

The meninges consist of three layers of fibrous connective tissue membranes located between the bones of the skull and vertebral column and the soft tissues of the central nervous system. The most superficial layer, the *dura mater,* is a tough membrane with an epidural space containing blood vessels, loose connective tissue, and adipose tissue between the membrane and the vertebrae. A weblike *arachnoid mater* adheres to the inside of the dura mater. The subarachnoid space, located beneath the arachnoid mater, contains cerebrospinal fluid (CSF) and serves as a protective cushion for the spinal cord and brain. The delicate innermost membrane, the *pia mater,* adheres to the surface of the spinal cord and brain. The denticulate ligaments, extending from the pia mater to the dura mater, anchor the spinal cord. An inferior extension of the pia mater, the filum terminale, anchors the spinal cord to the coccyx.

Procedure A—Structure of the Spinal Cord

1. Study figures 28.1, 28.2, and 28.3
2. Obtain a prepared microscope slide of a spinal cord cross section. Use the low power of the microscope to locate the following features:

   posterior median sulcus
   anterior median fissure
   central canal
   gray matter
   - gray commissure
   - posterior (dorsal) horn
   - lateral horn
   - anterior (ventral) horn
   white matter
   - posterior (dorsal) funiculus (column)
   - lateral funiculus (column)
   - anterior (ventral) funiculus (column)

FIGURE 28.1 Features of the spinal cord cross section and surrounding structures.

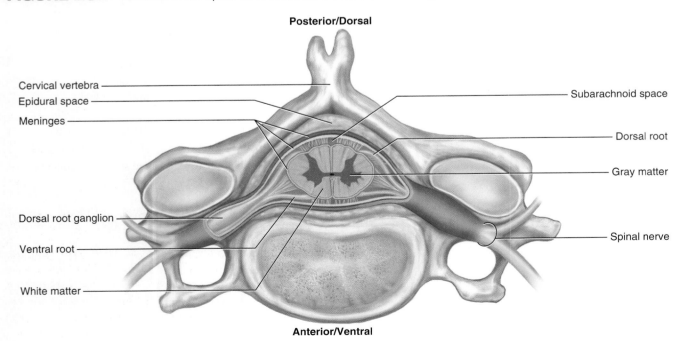

FIGURE 28.2 Cross section of the spinal cord, including the features of the white and gray matter.

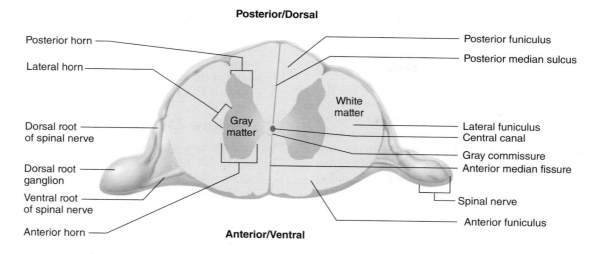

Posterior/Dorsal

Posterior horn
Lateral horn
Dorsal root of spinal nerve
Dorsal root ganglion
Ventral root of spinal nerve
Anterior horn

Gray matter
White matter

Posterior funiculus
Posterior median sulcus
Lateral funiculus
Central canal
Gray commissure
Anterior median fissure
Spinal nerve
Anterior funiculus

Anterior/Ventral

FIGURE 28.3 Cross section of a spinal cord with the three funiculi of the white matter shown on one side. Each funiculus contains specific tracts. Major ascending (sensory) and descending (motor) tracts (pathways) are shown only on one side of the spinal cord, but are located on both sides. Ascending tracts are in pink; descending tracts are in rust. This pattern varies with the level of the spinal cord. This pattern is representative of the midcervical region. (*Note:* These tracts are not visible as individually stained structures on microscope slides.)

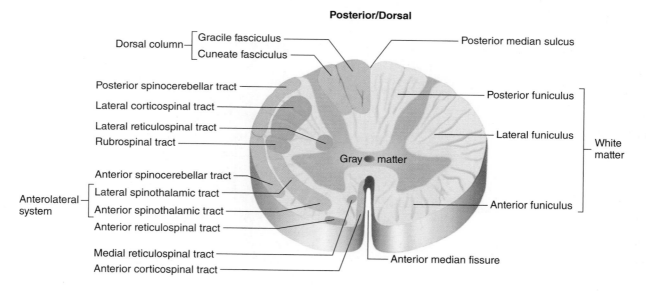

Posterior/Dorsal

Dorsal column
— Gracile fasciculus
— Cuneate fasciculus
Posterior spinocerebellar tract
Lateral corticospinal tract
Lateral reticulospinal tract
Rubrospinal tract
Anterior spinocerebellar tract
Anterolateral system
— Lateral spinothalamic tract
— Anterior spinothalamic tract
Anterior reticulospinal tract
Medial reticulospinal tract
Anterior corticospinal tract

Gray matter

Posterior median sulcus
Posterior funiculus
Lateral funiculus
White matter
Anterior funiculus
Anterior median fissure

Anterior/Ventral

roots of spinal nerve
- dorsal roots
- dorsal root ganglia
- ventral roots

3. Observe the model of the spinal cord, and locate the features listed in step 2.

4. Complete Part A of Laboratory Assessment 28.

Procedure B—Spinal Nerves

There are 31 pairs of spinal nerves that emerge between bones of the vertebral column. Cervical spinal nerve pair C1 emerges between the occipital bone of the skull and the atlas. The other cervical nerves emerge inferior to the 7 cervical vertebrae, resulting in 8 pairs of cervical nerves. The rest of the spinal nerves emerge inferiorly to the corresponding vertebrae, resulting in 12 pairs of thoracic nerves, 5 pairs of lumbar nerves, 5 pairs of sacral nerves, and 1 coccygeal pair. Recall that the single sacrum of the adult is a result of fusions of 5 sacral vertebrae. Because the inferior end of the spinal cord of the adult terminates just inferior to L1, many of the spinal nerves give rise to a bundle of nerve roots, the cauda equina, extending inferiorly through the remainder of the lumbar vertebrae and the sacrum.

Each spinal nerve possesses a dorsal root and a ventral root adjacent to the spinal cord. The dorsal root contains the sensory neurons; the ventral root contains the motor neurons. The resulting main spinal nerve, formed from a fusion of the dorsal and ventral roots, is referred to as a mixed nerve because sensory and motor signals (action potentials) exist within the same nerve. A short distance from the spinal cord, some of spinal nerves merge (anastomose) and form a weblike plexus. Major anastomoses include the cervical, brachial, lumbar, and sacral nerve plexuses.

1. Examine figures 28.4 and 28.5.
2. Observe a vertebral column model with spinal nerves extending from the intervertebral foramina. Compare the emerging locations of the spinal nerves with the names of the vertebrae.
3. Complete Part B of the laboratory assessment.

Procedure C—Meninges

Three membranes, or meninges, surround the entire CNS. The superficial, tough dura mater is a single meningeal layer around the spinal cord, but it is a double layer in the cranial cavity due to the fusion of the periosteal and meningeal layers. The dura mater of the cranial cavity adheres directly to the skull bones; however, the dura mater of the

FIGURE 28.4 Posterior view of the origins and categories of the 31 pairs of spinal nerves on the right. The plexuses formed by various spinal nerves are illustrated on the left.

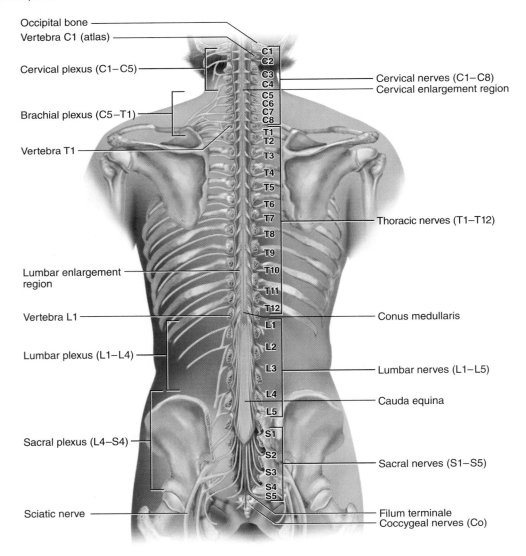

Occipital bone
Vertebra C1 (atlas)
Cervical plexus (C1–C5)
Brachial plexus (C5–T1)
Vertebra T1
Lumbar enlargement region
Vertebra L1
Lumbar plexus (L1–L4)
Sacral plexus (L4–S4)
Sciatic nerve

Cervical nerves (C1–C8)
Cervical enlargement region
Thoracic nerves (T1–T12)
Conus medullaris
Lumbar nerves (L1–L5)
Cauda equina
Sacral nerves (S1–S5)
Filum terminale
Coccygeal nerves (Co)

FIGURE 28.5 Posterior view of cervical region of spinal cord and associated nerves and meninges of a cadaver.

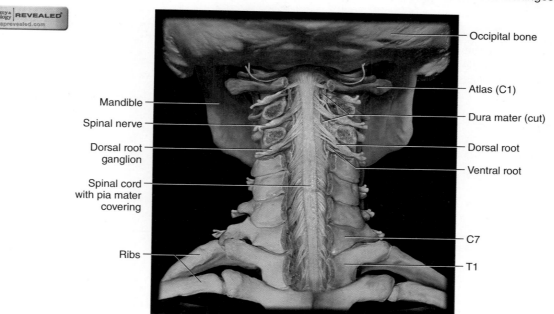

- Occipital bone
- Atlas (C1)
- Dura mater (cut)
- Dorsal root
- Ventral root
- C7
- T1
- Mandible
- Spinal nerve
- Dorsal root ganglion
- Spinal cord with pia mater covering
- Ribs

spinal cord has an epidural space between the membrane and the vertebrae. A subarachnoid space exists between the middle arachnoid mater and the pia mater that contains cerebrospinal fluid (CSF). The delicate pia mater closely adheres to the surface of the brain and spinal cord. Collectively, the meninges enclose and provide physical protective layers around the delicate brain and spinal cord. The CSF provides an additional protective cushion from sudden jolts to the head or the back.

1. Study figures 28.6 and 28.7. Review figures 28.1 and 28.5.
2. Observe the spinal cord model with meninges and locate the following features:

   dura mater (dural sheath)
   arachnoid mater (membrane)
   subarachnoid space
   denticulate ligament
   pia mater

3. Complete Part C of the laboratory assessment.

FIGURE 28.6 The meninges, associated near the spinal cord and spinal nerves.

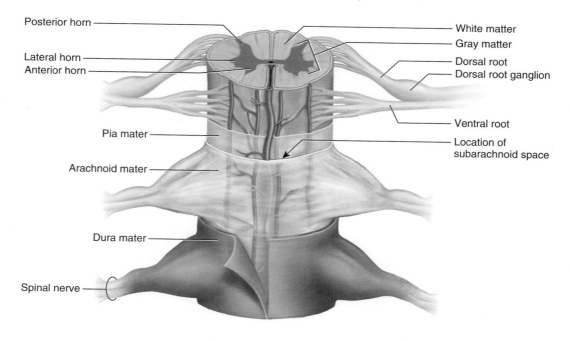

- Posterior horn
- Lateral horn
- Anterior horn
- Pia mater
- Arachnoid mater
- Dura mater
- Spinal nerve
- White matter
- Gray matter
- Dorsal root
- Dorsal root ganglion
- Ventral root
- Location of subarachnoid space

Observe the preserved section of spinal cord. Note the heavy covering of dura mater, firmly attached to the cord on each side by a set of ligaments (denticulate ligaments) originating in the pia mater. The intermediate layer of meninges, the arachnoid mater, is devoid of blood vessels, but in a live human being, the space beneath this layer contains cerebrospinal fluid. The pia mater, closely attached to the surface of the spinal cord, contains many blood vessels. What are the functions of these layers?

_____

_____

_____

FIGURE 28.7 Meninges associated with the brain.

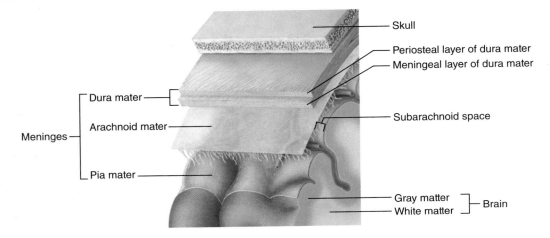

Name _____

Date _____

Section _____

The A corresponds to the indicated outcome(s) found at the beginning of the laboratory exercise.

Spinal Cord, Spinal Nerves, and Meninges

Part A Assessments

Identify the features indicated in the spinal cord cross section of figure 28.8.

FIGURE 28.8 Micrograph of a spinal cord cross section with spinal nerve roots (7.5×). Label the features by placing the correct numbers in the spaces provided. A A

Posterior/Dorsal

Dura mater

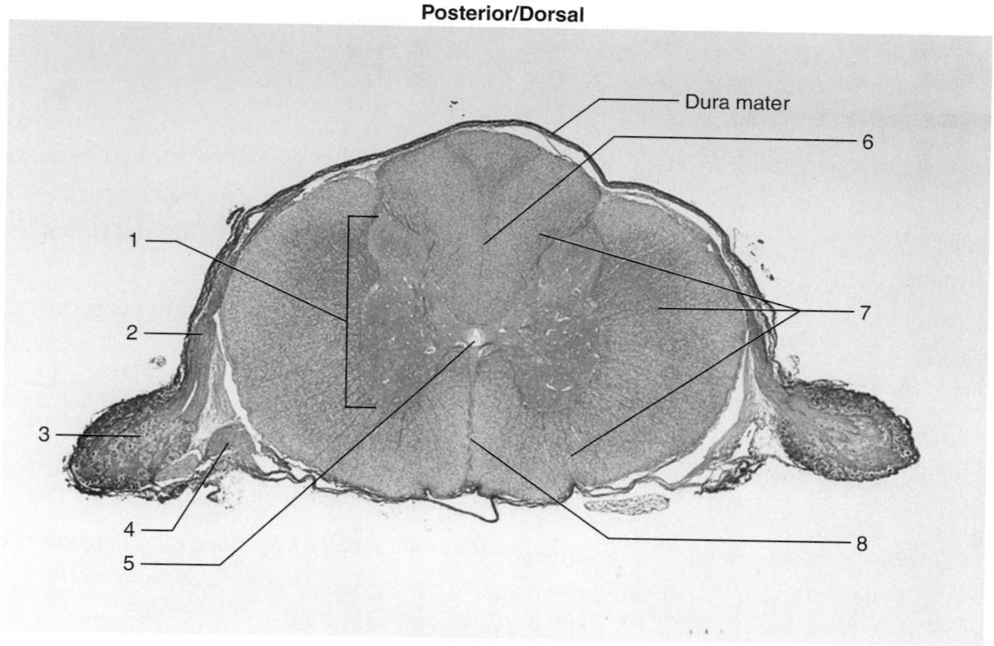

Anterior/Ventral

_____ Anterior median fissure   _____ Dorsal root of spinal nerve   _____ Ventral root of spinal nerve

_____ Central canal   _____ Gray matter   _____ White matter

_____ Dorsal root ganglion   _____ Posterior median sulcus

Part B Assessments

Complete the following statements:

1. The spinal cord gives rise to 31 pairs of _____. 1️⃣

2. The bulge in the spinal cord that gives off nerves to the upper limbs is called the _____ enlargement. 1️⃣

3. The bulge in the spinal cord that gives off nerves to the lower limbs is called the _____ enlargement. 1️⃣

4. The _____ is a groove that extends the length of the spinal cord posteriorly. 1️⃣

5. In a spinal cord cross section, the posterior _____ of the gray matter resemble the upper wings of a butterfly. 1️⃣

6. The motor neurons are found in the _____ roots of spinal nerves. 2️⃣

7. The _____ connects the gray matter on the left and right sides of the spinal cord. 1️⃣

8. The _____ in the gray commissure of the spinal cord contains cerebrospinal fluid and is continuous with the cavities of the brain. 1️⃣

9. The white matter of the spinal cord is divided into anterior, lateral, and posterior _____. 1️⃣

10. There are _____ pairs of cervical spinal nerves. 2️⃣

11. There are _____ pairs of sacral spinal nerves. 2️⃣

12. Cervical spinal nerve pair C1 originates between the occipital bone and the _____. 2️⃣

13. Spinal nerves L4 through S4 form a _____ plexus. 2️⃣

14. The gray matter of the spinal cord is divided into anterior, lateral, and posterior _____. 1️⃣

Part C Assessments

Match the terms in column A with the descriptions in column B. Place the letter of your choice in the space provided. 3️⃣

Column A	Column B
a. Arachnoid mater	_____ 1. Connections from pia mater to dura mater that anchor the spinal cord
b. Denticulate ligaments	
c. Dura mater	_____ 2. Inferior continuation of pia mater to the coccyx
d. Epidural space	_____ 3. Outermost layer of meninges
e. Filum terminale	_____ 4. Follows irregular contours of spinal cord surface
f. Pia mater	_____ 5. Contains cerebrospinal fluid
g. Subarachnoid space	_____ 6. Thin, weblike middle membrane
	_____ 7. Separates dura mater from bone of vertebra or skull

Reflex Arc and Reflexes

Purpose of the Exercise

To review the characteristics of reflex arcs and reflex behavior and to demonstrate some of the reflexes that occur in the human body.

Materials Needed

Rubber percussion hammer

Learning Outcomes

After completing this exercise, you should be able to

1. Demonstrate and record stretch reflexes that occur in humans.

2. Describe the components of a reflex arc.

3. Analyze the components and patterns of stretch reflexes.

Pre-Lab

Carefully read the introductory material and examine the entire lab. Be familiar with reflexes from lecture or the textbook. Answer the pre-lab questions.

Pre-Lab Questions: Select the correct answer for each of the following questions:

1. The impulse over a motor neuron will lead to
   a. an interneuron.   **b.** the spinal cord.
   c. a receptor.   **d.** an effector.

2. Stretch reflex receptors are called
   a. effectors.   **b.** muscle spindles.
   c. interneurons.   **d.** motor neurons.

3. Stretch reflexes include all of the following *except* the _____ reflex.
   a. withdrawal   **b.** patellar
   c. calcaneal   **d.** biceps

4. A withdrawal reflex could occur from
   a. striking the patellar ligament.
   b. striking the calcaneal tendon.
   c. striking the triceps tendon.
   d. touching a hot object.

5. The calcaneal reflex response is
   a. pain interpretation.
   b. flexion of the leg at the knee joint.
   c. plantar flexion of the foot.
   d. a separation of toes.

6. The quadriceps femoris is the effector muscle of the patellar reflex.
   True _____   False _____

7. The dorsal roots of spinal nerves contain the axons of the motor neurons.
   True _____   False _____

8. The normal patellar reflex response involves extension of the leg at the knee joint.
   True _____   False _____

A reflex arc represents the simplest type of nerve pathway found in the nervous system. This pathway begins with a receptor at the dendrite end of a sensory (afferent) neuron. The sensory neuron leads into the central nervous system and may communicate with one or more interneurons. Some of these interneurons, in turn, communicate with motor (efferent) neurons, whose axons (nerve fibers) lead outward to effectors. Thus, when a sensory receptor is stimulated by a change occurring inside or outside the body, impulses may pass through a reflex arc, and, as a result, effectors may respond. Such an automatic, subconscious response is called a *reflex*.

A *stretch reflex* involves a single synapse (monosynaptic) between a sensory and a motor neuron within the gray matter of the spinal cord. Examples of stretch reflexes include the patellar, calcaneal, biceps, triceps, and plantar reflexes. Other more complex *withdrawal reflexes* involve interneurons (association neurons) in combination with sensory and motor neurons; thus they are polysynaptic. Examples of withdrawal reflexes include responses to touching hot objects or stepping on sharp objects.

Reflexes demonstrated in this lab are stretch reflexes. When a muscle is stretched by a tap over its tendon, stretch receptors (proprioceptors) called *muscle spindles* are stretched within the muscle, which initiates an impulse over a reflex arc. A sensory neuron conducts an impulse from the muscle spindle into the gray matter of the spinal cord, where it synapses with a motor neuron, which conducts the impulse to the effector muscle. The stretched muscle responds by contracting to resist or reverse further stretching. These stretch reflexes are important to maintaining proper posture, balance, and movements. Observations of many of these reflexes in clinical tests on patients may indicate damage to a level of the spinal cord or peripheral nerves of the particular reflex arc.

Procedure—Reflex Arc and Reflexes

1. Study figure 29.1 as an example of a withdrawal reflex. Compare the withdrawal reflex with the stretch reflex shown in figure 29.2.
2. Work with a laboratory partner to demonstrate each of the reflexes listed. (See fig. 29.3a–e also.) *It is important that muscles involved in the reflexes be totally relaxed to observe proper responses.* If a person is trying too hard to experience the reflex or is trying to suppress the reflex, assign a multitasking activity while the stimulus with the rubber percussion hammer occurs. For example, assign a physical task with upper limbs along with a

FIGURE 29.1 Diagram of a withdrawal (polysynaptic) reflex arc.

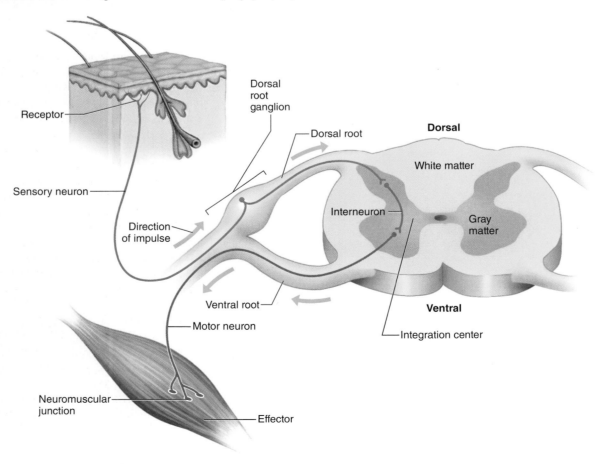

FIGURE 29.2 Diagram of a stretch (monosynaptic) reflex arc. The patellar reflex represents a specific example of a stretch reflex.

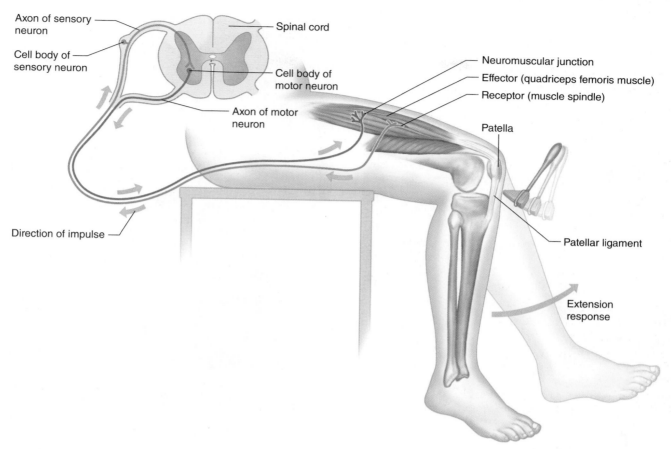

complex mental activity during the patellar reflex. After each demonstration, record your observations in the table provided in Part A of Laboratory Assessment 29.

a. *Patellar (knee-jerk) reflex.* Have your laboratory partner sit on a table (or sturdy chair) with legs relaxed and hanging freely over the edge without touching the floor. Gently strike your partner's patellar ligament (just below the patella) with the blunt side of a rubber percussion hammer (fig. 29.3a). The reflex arc involves the femoral nerve and the spinal cord. The normal response is a moderate extension of the leg at the knee joint.

b. *Calcaneal (ankle-jerk) reflex.* Have your partner kneel on a chair with back toward you and with feet slightly dorsiflexed over the edge and relaxed. Gently strike the calcaneal tendon (just above its insertion on the calcaneus) with the blunt side of the rubber hammer (fig. 29.3b). The reflex arc involves the tibial nerve and the spinal cord. The normal response is plantar flexion of the foot.

c. *Biceps (biceps-jerk) reflex.* Have your partner place a bare arm bent about 90° at the elbow on the table. Press your thumb on the inside of the elbow over the tendon of the biceps brachii, and gently strike your thumb with the rubber hammer (fig. 29.3c). The reflex arc involves the musculocutaneous nerve and the spi-

nal cord. Watch the biceps brachii for a response. The response might be a slight twitch of the muscle or flexion of the forearm at the elbow joint.

d. *Triceps (triceps-jerk) reflex.* Have your partner lie supine with an upper limb bent about 90° across the abdomen. Gently strike the tendon of the triceps brachii near its insertion just proximal to the olecranon process at the tip of the elbow (fig. 29.3d). The reflex arc involves the radial nerve and the spinal cord. Watch the triceps brachii for a response. The response might be a slight twitch of the muscle or extension of the forearm at the elbow joint.

e. *Plantar reflex.* Have your partner remove a shoe and sock and lie supine with the lateral surface of the foot resting on the table. Draw the metal tip of the rubber hammer, applying firm pressure, over the sole from the heel to the base of the large toe (fig. 29.3e). The normal response is flexion (curling) of the toes and plantar flexion of the foot. If the toes spread apart and dorsiflexion of the great toe occurs, the reflex is the abnormal *Babinski reflex* response (normal in infants until the nerve fibers have complete myelinization). If the Babinski reflex occurs later in life, it may indicate damage to the corticospinal tract of the CNS.

3. Complete Part B of the laboratory assessment.

FIGURE 29.3 Demonstrate each of the following reflexes: (a) patellar reflex; (b) calcaneal reflex; (c) biceps reflex; (d) triceps reflex; and (e) plantar reflex.

(a) Patellar reflex

(b) Calcaneal reflex

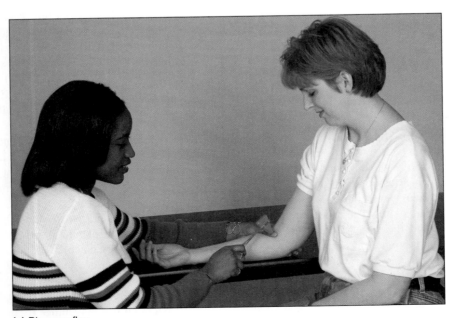

(c) Biceps reflex

FIGURE 29.3 *Continued.*

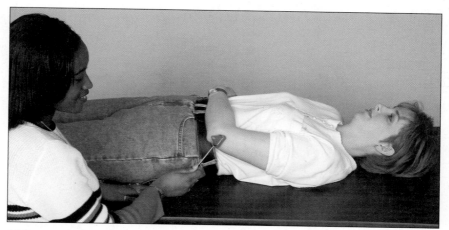

(d) Triceps reflex

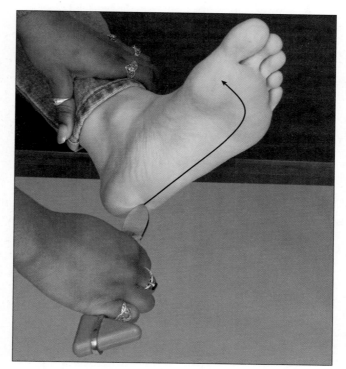

(e) Plantar reflex

NOTES

Laboratory Assessment

29

Name _____

Date _____

Section _____

The ⚠ corresponds to the indicated outcome(s) found at the beginning of the laboratory exercise.

Reflex Arc and Reflexes

Part A Assessments

Complete the following table: ⚠1

Reflex Tested	Response Observed	Degree of Response (hypoactive, normal, or hyperactive)	Effector Muscle Involved
Patellar			
Calcaneal			
Biceps			
Triceps			
Plantar			

Part B Assessments

Complete the following statements:

1. A withdrawal reflex employs _____ neurons in conjunction with sensory and motor neurons. ⚠2

2. Interneurons in a withdrawal reflex are located in the _____. ⚠2

3. A reflex arc begins with the stimulation of a _____ at the dendrite end of a sensory neuron. ⚠2

4. Effectors of a reflex arc are glands and _____. ⚠2

5. A patellar reflex employs only _____ and motor neurons. ⚠2

6. The effector muscle of the patellar reflex is the _____. ⚠2

7. The sensory stretch receptors (muscle spindles) of the patellar reflex are located in the _____ muscle. ⚠2

8. The dorsal root of a spinal nerve contains the _____ neurons. ⚠2

9. The normal plantar reflex results in _____ of toes. ⚠2

10. Stroking the sole of the foot in infants results in dorsiflexion and toes that spread apart, called the _____ reflex. ⚠2

11. List the major events that occur in the patellar reflex, from the striking of the patellar ligament to the resulting response. ⚠2 ⚠3

_____

_____

_____

Critical Thinking Assessment

What characteristics do the reflexes you demonstrated have in common?

_____

_____

_____

Brain and Cranial Nerves

Purpose of the Exercise

To review the structural and functional characteristics of the human brain and cranial nerves.

Materials Needed

Dissectible model of the human brain
Preserved human brain
Anatomical charts of the human brain

Learning Outcomes

After completing this exercise, you should be able to

1 Identify the major external and internal structures in the human brain.

2 Locate the major functional regions of the brain.

3 Identify each of the 12 pairs of cranial nerves.

4 Differentiate the functions of each cranial nerve.

Pre-Lab

Carefully read the introductory material and examine the entire lab. Be familiar with the brain and cranial nerves from lecture or the textbook. Answer the pre-lab questions.

Pre-Lab Questions: Select the correct answer for each of the following questions:

1. Each hemisphere of the cerebrum regulates
    a. motor functions on the opposite side of the body.
    b. motor functions on the same side of the body.
    c. only functions within the brain.
    d. only functions within the spinal cord.

2. There are _____ pairs of cranial nerves.
    a. 2　　　　　　　b. 12
    c. 31　　　　　　 d. 43

3. Which of the following is *not* part of the brainstem?
    a. midbrain　　　　b. pons
    c. medulla oblongata　d. thalamus

4. The _____ is the deep lobe of the cerebrum.
    a. temporal　　　　b. insula
    c. parietal　　　　 d. occipital

5. The _____ separates the precentral and postcentral gyrus.
    a. lateral sulcus
    b. parieto-occipital sulcus
    c. central sulcus
    d. longitudinal fissure

6. Primary vesicles of embryonic brain development include the forebrain, midbrain, and the hindbrain.
    True _____　　 False _____

7. Ventricles of the brain contain cerebrospinal fluid.
    True _____　　 False _____

8. Functions of the cerebellum include reasoning, memory, and regulation of body temperature.
    True _____　　 False _____

The brain, the largest and most complex part of the nervous system, contains nerve centers associated with sensory functions and is responsible for sensations and perceptions. It issues motor commands to skeletal muscles and carries on higher mental activities. It also functions to coordinate muscular movements, and it contains centers and nerve pathways necessary for the regulation of internal organs.

The cerebral cortex, comprised of gray matter and billions of interneurons, represents areas for conscious awareness and decision-making processes. Sensory areas receive information from various receptors, association areas interpret sensory input, and motor areas involve planning and controlling muscle movements. All of these functional regions are influenced and integrated together in making complex decisions. Each hemisphere primarily interprets sensory and regulates motor functions on the opposite (contralateral) side of the body.

Twelve pairs of cranial nerves arise from the ventral surface of the brain and are designated by number and name. The cranial nerves are part of the PNS; most arise from the brainstem region of the brain. Although most of these nerves conduct both sensory and motor impulses, some contain only sensory fibers associated with special sense organs. Others are primarily composed of motor fibers and are involved with the activities of muscles and glands.

Procedure A—Human Brain

The early development of the brain includes three primary vesicles formed from the neural tube: the forebrain (prosencephalon), midbrain (mesencephalon), and hindbrain (rhombencephalon). Five secondary vesicles form, including the telencephalon and diencephalon from the forebrain; the mesencephalon is retained; and the metencephalon and the myelencephalon from the hindbrain. The spaces that develop in the adult CNS from the vesicles include the central canal of the spinal cord and the ventricles of the brain. Various

TABLE 30.1 Structural Development of the Brain

Primary and Secondary Brain Vesicles	Adult Spaces Produced	Adult Brain Structures
Forebrain (prosencephalon)		
Anterior portion (telencephalon)	Lateral ventricles	Cerebrum (white matter, cortex, basal nuclei)
Posterior portion (diencephalon)	Third ventricle	Thalamus Hypothalamus Pineal gland (from epithalamus)
Midbrain (mesencephalon)	Cerebral aqueduct	Midbrain
Hindbrain (rhombencephalon)		
Anterior portion (metencephalon)	Fourth ventricle	Cerebellum Pons
Posterior portion (myelencephalon)	Fourth ventricle	Medulla oblongata

enlargements from the brain vesicles develop into the adult brain structures. See table 30.1 for a summary of the brain development.

The study of the brain will include several regional categories: the ventricles, external surface features, cerebral hemispheres, diencephalon, brainstem, and cerebellum. During each regional study, examine available anatomical charts, dissectible models, and a human brain.

1. Examine figures 30.1 and 30.2 illustrating the ventricles of the brain. The four ventricles contain a clear cerebrospinal fluid (CSF) that was secreted by blood capillaries named choroid plexuses. The CSF flows from the lateral

FIGURE 30.1 Four ventricles of the brain and their connections: (a) right lateral view and (b) anterior view.

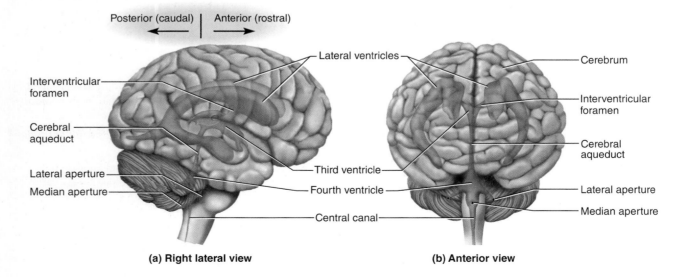

Posterior (caudal) | Anterior (rostral)

Interventricular foramen
Cerebral aqueduct
Lateral aperture
Median aperture
Lateral ventricles
Third ventricle
Fourth ventricle
Central canal

Cerebrum
Interventricular foramen
Cerebral aqueduct
Lateral aperture
Median aperture

(a) Right lateral view

(b) Anterior view

FIGURE 30.2 Transverse section of the human brain showing some ventricles, surface features, and internal structures of the cerebrum (superior view). This section exposes the anterior and posterior horns of both lateral ventricles.

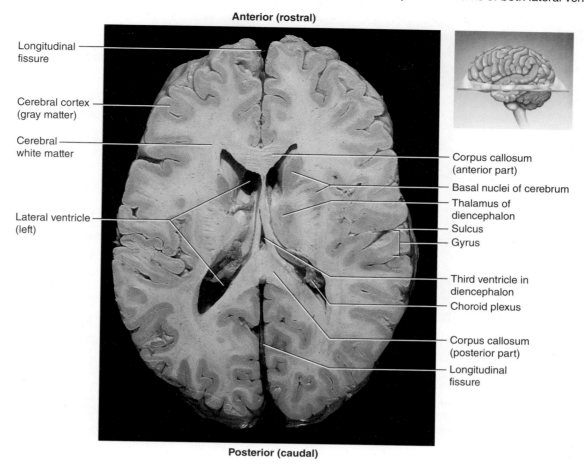

Anterior (rostral)

Longitudinal fissure

Cerebral cortex (gray matter)

Cerebral white matter

Lateral ventricle (left)

Corpus callosum (anterior part)

Basal nuclei of cerebrum

Thalamus of diencephalon

Sulcus

Gyrus

Third ventricle in diencephalon

Choroid plexus

Corpus callosum (posterior part)

Longitudinal fissure

Posterior (caudal)

ventricles into the third ventricle and the fourth ventricle and then into the central canal of the spinal cord and through pores into the subarachnoid space. Locate each of the following features using dissectible brain models:

ventricles

- lateral ventricles—the largest ventricles and one located in each cerebral hemisphere
- third ventricle—located within the diencephalon inferior to the corpus callosum
- fourth ventricle—located between the pons and the cerebellum

2. Examine figures 30.2, 30.3 and 30.4 for a study of the external surface features of the brain. Locate the following features using dissectible models and the human brain:

gyri—elevated surface ridges

- precentral gyrus
- postcentral gyrus

sulci—shallow grooves

- central sulcus—divides the frontal from the parietal lobe
- lateral sulcus (fissure)—divides the temporal from the parietal lobe

- parieto-occipital sulcus—divides the occipital from the parietal lobe

fissures—deep grooves

- longitudinal fissure—separates the cerebral hemispheres
- transverse fissure—separates the cerebrum from the cerebellum

lobes of cerebrum—names associated with bones of the cranium

- frontal lobe
- parietal lobe
- temporal lobe
- occipital lobe
- insula (insular lobe)—deep within cerebrum; not visible on surface

3. Examine figures 30.2, 30.3, 30.4, and 30.5 and table 30.2 during the study of the cerebral hemispheres. The cerebrum is the largest part of the brain and has two hemispheres connected by the corpus callosum. The lobes of the cerebrum were included in the surface features because they are also visible on the surface, except the deep insula. The functions associated with these

FIGURE 30.3 Superior view of the surface of the brain within the skull of a cadaver.

Anterior

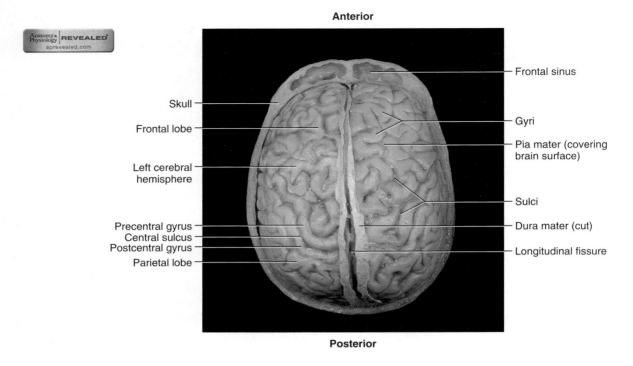

Skull

Frontal lobe

Left cerebral hemisphere

Precentral gyrus
Central sulcus
Postcentral gyrus
Parietal lobe

Frontal sinus

Gyri

Pia mater (covering brain surface)

Sulci

Dura mater (cut)

Longitudinal fissure

Posterior

FIGURE 30.4 Lobes of the cerebrum. Retractors are used to expose the deep insula.

Anterior (rostral) | Posterior (caudal)

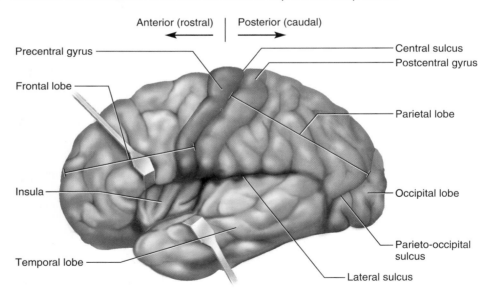

Precentral gyrus

Frontal lobe

Insula

Temporal lobe

Central sulcus
Postcentral gyrus

Parietal lobe

Occipital lobe

Parieto-occipital sulcus

Lateral sulcus

FIGURE 30.5 Diagram of a sagittal (median) section of the brain.

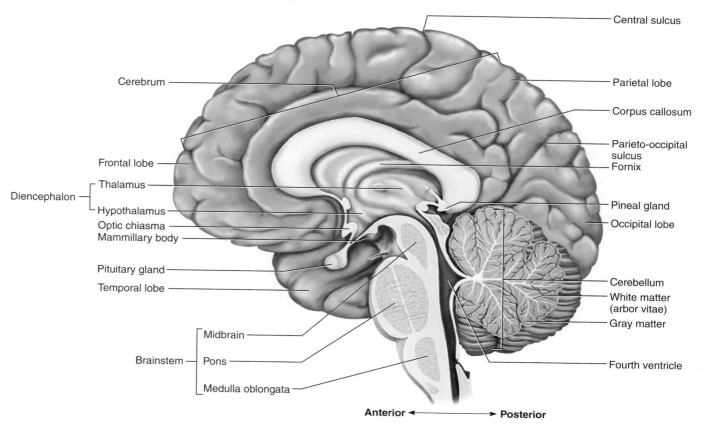

TABLE 30.2 Major Regions and Functions of the Brain

Region	Functions
1. Cerebrum	Controls higher brain functions, including sensory perception, storing memory, reasoning, and determining intelligence; initiates voluntary muscle movements
2. Diencephalon	
a. Thalamus	Relay station for sensory impulses ascending from other parts of the nervous system to the cerebral cortex
b. Hypothalamus	Helps maintain homeostasis by regulating visceral activities and by linking the nervous and endocrine systems; regulates body temperature, sleep cycles, emotions, and autonomic nervous system control
3. Brainstem	
a. Midbrain	Contains reflex centers that move the eyes and head
b. Pons	Relays impulses between higher and lower brain regions; helps regulate rate and depth of breathing
c. Medulla oblongata	Conducts ascending and descending impulses between the brain and spinal cord; contains cardiac, vasomotor, and respiratory control centers and various nonvital reflex control centers
4. Cerebellum	Processes information from other parts of the CNS by nerve tracts; integrates sensory information concerning the position of body parts; and coordinates muscle activities and maintains posture

lobes will be included in step 4. Locate the following features using dissectible models and a human brain:

cerebral cortex—thin surface layer of gray matter containing very little myelin; area of conscious awareness and processing information

cerebral white matter—largest portion of the cerebrum containing myelinated nerve fibers; transmits impulses between cerebral areas and lower brain centers

basal nuclei—masses of gray matter deep within the white matter; sometimes called basal ganglia as a clinical term; relays motor impulses from the cerebral cortex to the brainstem and spinal cord; the main structures include the caudate nucleus, putamen, and globus pallidus

4. Examine figures 30.4 and 30.6 to compare the structural lobes of the cerebrum with the functional regions of the lobes. Functional areas are not visible as distinct parts on an actual human brain. Examine the labeled areas in figure 30.6 that represent the following functional regions of the cerebrum:

sensory areas

- primary somatosensory cortex—receives information from skin receptors and proprioceptors
- somatosensory association cortex—integrates sensory information from primary cortex
- Wernicke's area—processes spoken and written language
- visual areas—consists of a primary and association (interpretation) area for vision
- auditory areas—consists of a primary and association area for hearing
- olfactory association area—interpretation of odors
- gustatory cortex—perceptions of taste

motor areas

- primary motor cortex—controls skeletal muscles
- motor association (premotor) area—planning body movements
- Broca's area—planning speech movements

5. Examine figures 30.5 and 30.7 and tables 30.1 and 30.2 during the study of the diencephalon, brainstem, and cerebellum. Locate the following features using anatomical charts, dissectible models, and a human brain:

diencephalon

- thalamus—largest portion
- hypothalamus—inferior portion
- optic chiasma (chiasm)—optic nerves meet
- mammillary bodies—pair of small humps
- pineal gland—formed from epithalamus

brainstem

- midbrain—superior region of brainstem
  - cerebral peduncles—connects pons to cerebrum
  - corpora quadrigemina—four bulges
- pons—bulge on underside of brainstem
- medulla oblongata—inferior region of brainstem

FIGURE 30.6 Some structural and functional areas of the left cerebral hemisphere. (*Note:* These areas are not visible as distinct parts of the brain.)

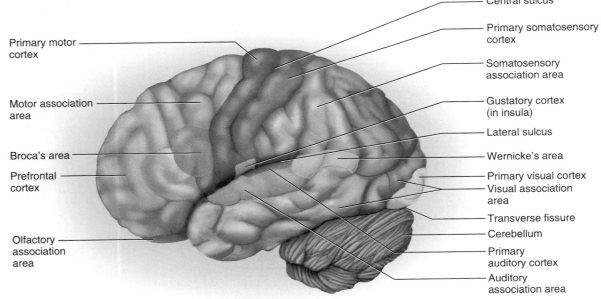

Primary motor cortex

Motor association area

Broca's area

Prefrontal cortex

Olfactory association area

Central sulcus

Primary somatosensory cortex

Somatosensory association area

Gustatory cortex (in insula)

Lateral sulcus

Wernicke's area

Primary visual cortex

Visual association area

Transverse fissure

Cerebellum

Primary auditory cortex

Auditory association area

FIGURE 30.7 Cerebellum and brainstem (a) median section and (b) superior view of cerebellum.

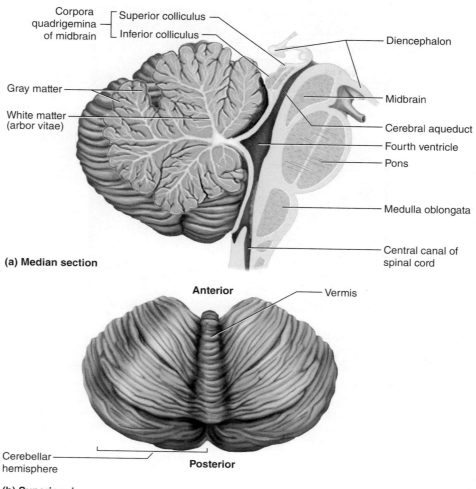

Corpora quadrigemina of midbrain {Superior colliculus — Inferior colliculus —

Diencephalon

Gray matter

Midbrain

White matter (arbor vitae)

Cerebral aqueduct

Fourth ventricle

Pons

Medulla oblongata

Central canal of spinal cord

(a) Median section

Anterior

Vermis

Cerebellar hemisphere

Posterior

(b) Superior view

cerebellum—cauliflower-like appearance
- right and left hemispheres
- vermis—connects the two hemispheres
- cerebellar cortex—gray matter portion
- arbor vitae—deeper branching pattern of white matter

6. Complete Parts A, B, and C of Laboratory Assessment 30.

Procedure B—Cranial Nerves

1. The cranial nerves are part of the PNS. Examine figure 30.8 and table 30.3.
2. Observe the model and preserved specimen of the human brain, and locate as many of the following cranial nerves as possible as you differentiate their associated functions:

   olfactory nerves (I)
   optic nerves (II)
   oculomotor nerves (III)
   trochlear nerves (IV)

   trigeminal nerves (V)
   abducens nerves (VI)
   facial nerves (VII)
   vestibulocochlear nerves (VIII)
   glossopharyngeal nerves (IX)
   vagus nerves (X)
   accessory nerves (XI)
   hypoglossal nerves (XII)

The following mnemonic device will help you learn the twelve pairs of cranial nerves in the proper order:

Old **Op**ie **oc**casionally **tr**ies **trig**onometry, **a**nd **f**eels **v**ery **glo**omy, **vagu**e, **a**nd **hypo**active.[1]

3. Complete Parts D and E of the laboratory assessment.

[1]From *HAPS-Educator,* Winter 2002. An official publication of the Human Anatomy & Physiology Society (HAPS).

FIGURE 30.8 Photograph of the cranial nerves attached to the base of the human brain.

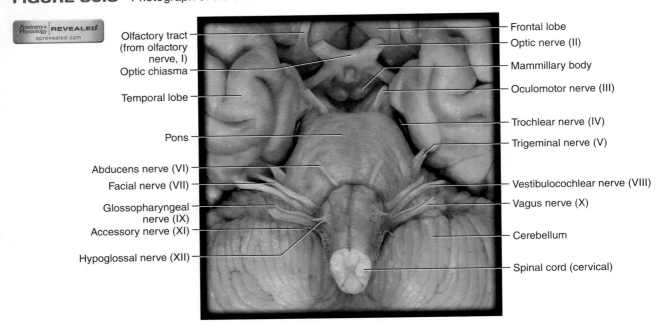

TABLE 30.3 The Cranial Nerves and Functions

Number and Name		Type	Function
I	Olfactory	Sensory	Sensory impulses associated with smell
II	Optic	Sensory	Sensory impulses associated with vision
III	Oculomotor	Primarily motor*	Motor impulses to superior, inferior, and medial rectus and inferior oblique muscles that move the eyes, adjust the amount of light entering the eyes, focus the lenses, and raise the eyelids
IV	Trochlear	Primarily motor*	Motor impulses to superior oblique muscles that move the eyes
V	Trigeminal	Mixed	
	Ophthalmic division		Sensory impulses from the surface of the eyes, tear glands, scalp, forehead, and upper eyelids
	Maxillary division		Sensory impulses from the upper teeth, upper gum, upper lip, lining of the palate, and skin of the face
	Mandibular division		Sensory impulses from the scalp, skin of the jaw, lower teeth, lower gum, and lower lip Motor impulses to muscles of mastication and to muscles in the floor of the mouth
VI	Abducens	Primarily motor*	Motor impulses to lateral rectus muscles that move the eyes laterally
VII	Facial	Mixed	Sensory impulses associated with taste receptors of the anterior tongue Motor impulses to muscles of facial expression, tear glands, and salivary glands
VIII	Vestibulocochlear	Sensory	
	Vestibular branch		Sensory impulses associated with equilibrium
	Cochlear branch		Sensory impulses associated with hearing
IX	Glossopharyngeal	Mixed	Sensory impulses from the pharynx, tonsils, posterior tongue, and carotid arteries Motor impulses to salivary glands and to muscles of the pharynx used in swallowing
X	Vagus	Mixed	Somatic motor impulses to muscles associated with speech and swallowing; autonomic motor impulses to the viscera of the thorax and abdomen Sensory impulses from the pharynx, larynx, esophagus, and viscera of the thorax and abdomen
XI	Accessory	Primarily motor*	
	Cranial branch		Motor impulses to muscles of the soft palate, pharynx, and larynx
	Spinal branch		Motor impulses to muscles of the neck and shoulder
XII	Hypoglossal	Primarily motor*	Motor impulses to muscles that move the tongue

*These nerves contain a small number of sensory impulses from proprioceptors.

312

Laboratory Assessment

30

Name _____

Date _____

Section _____

The ⚠ corresponds to the indicated outcome(s) found at the beginning of the laboratory exercise.

Brain and Cranial Nerves

Part A Assessments

Match the terms in column A with the descriptions in column B. Place the letter of your choice in the space provided. ⚠

Column A

a. Central sulcus
b. Cerebral cortex
c. Corpus callosum
d. Gyrus
e. Hypothalamus
f. Insula
g. Medulla oblongata
h. Midbrain
i. Optic chiasma
j. Pineal gland
k. Pons
l. Ventricle

Column B

_____ 1. Structure formed by the crossing-over of the optic nerves

_____ 2. Part of diencephalon that forms lower walls and floor of third ventricle

_____ 3. Cone-shaped gland in the upper posterior portion of diencephalon

_____ 4. Connects cerebral hemispheres

_____ 5. Ridge on surface of cerebrum

_____ 6. Separates frontal and parietal lobes

_____ 7. Part of brainstem between diencephalon and pons

_____ 8. Rounded bulge on underside of brainstem

_____ 9. Part of brainstem continuous with the spinal cord

_____ 10. Internal brain chamber filled with CSF

_____ 11. Cerebral lobe located deep within lateral sulcus

_____ 12. Thin layer of gray matter on surface of cerebrum

Part B Assessments

Complete the following statements:

1. The cerebral cortex contains the _____ matter. ⚠

2. Grooves on the surface of the brain are sulci; ridges on the surface are _____. ⚠

3. The auditory areas of the brain are part of the _____ lobe. ⚠

4. The vision areas of the brain are part of the _____ lobe. ⚠

5. The left cerebral hemisphere primarily controls the _____ side of the body. ⚠

6. The brainstem includes the pons, the midbrain, and the _____. ⚠

7. The delicate _____ membrane is located on the surface of the brain. ⚠

8. The _____ fissure separates the two cerebral hemispheres. ⚠

9. The primary motor cortex is located within the _____ gyrus. ⚠

10. Arbor vitae and vermis are components of the _____. ⚠

11. The _____ ventricle is located between the pons and the cerebellum. ⚠

12. The _____ connects the two hemispheres of the cerebellum. ⚠

Part C Assessments

Identify the features indicated in the median section of the right half of the human brain in figure 30.9.

FIGURE 30.9 Label the features on this median section of the right half of the human brain by placing the correct numbers in the spaces provided. 🔺

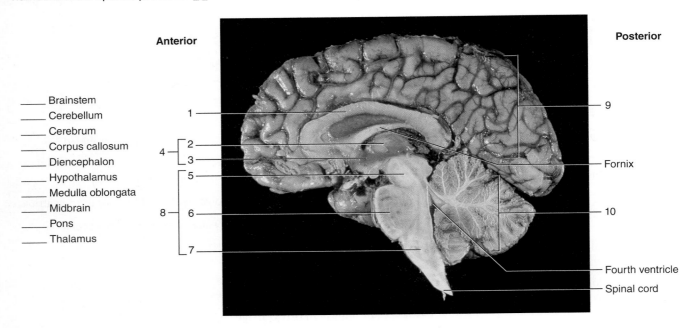

_____ Brainstem
_____ Cerebellum
_____ Cerebrum
_____ Corpus callosum
_____ Diencephalon
_____ Hypothalamus
_____ Medulla oblongata
_____ Midbrain
_____ Pons
_____ Thalamus

Part D Assessments

Identify the cranial nerves that arise from the base of the brain in figure 30.10.

FIGURE 30.10 Complete the labeling of the 12 pairs of cranial nerves as viewed from the base of the brain. The Roman numerals indicated are also often used to reference a cranial nerve. ⒊

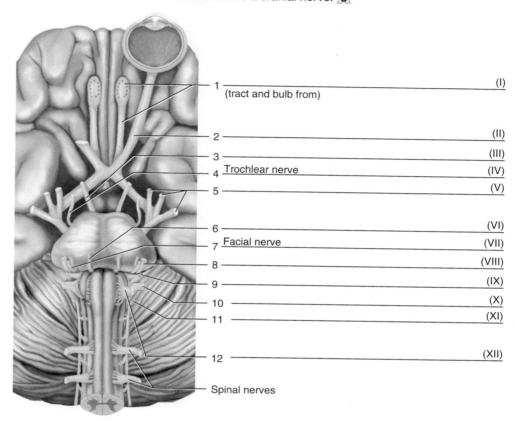

1 _____ (I)
(tract and bulb from)

2 _____ (II)

3 _____ (III)

4 Trochlear nerve _____ (IV)

5 _____ (V)

6 _____ (VI)

7 Facial nerve _____ (VII)

8 _____ (VIII)

9 _____ (IX)

10 _____ (X)

11 _____ (XI)

12 _____ (XII)

Spinal nerves

Part E Assessments

Match the cranial nerves in column A with the associated functions in column B. Place the letter of your choice in the space provided. ⒋

Column A

a. Abducens
b. Accessory
c. Facial
d. Glossopharyngeal
e. Hypoglossal
f. Oculomotor
g. Olfactory
h. Optic
i. Trigeminal
j. Trochlear
k. Vagus
l. Vestibulocochlear

Column B

_____ 1. Regulates thoracic and abdominal viscera

_____ 2. Equilibrium and hearing

_____ 3. Stimulates superior oblique muscle of eye

_____ 4. Sensory impulses from teeth and face

_____ 5. Adjusts light entering eyes and eyelid opening

_____ 6. Smell

_____ 7. Controls neck and shoulder movements

_____ 8. Controls tongue movements

_____ 9. Vision

_____ 10. Stimulates lateral rectus muscle of eye

_____ 11. Sensory from anterior tongue and controls salivation and secretion of tears

_____ 12. Sensory from posterior tongue and controls salivation and swallowing

315

NOTES

Electroencephalography I: BIOPAC© Exercise

Purpose of the Exercise

To record an EEG from an awake, resting subject with eyes open and closed, and to identify and examine alpha, beta, delta, and theta components of an EEG.

Learning Outcomes

After completing this exercise, you should be able to:

1. Identify alpha, beta, delta, and theta components of an EEG.

2. Distinguish variations in these components because of differences in the level of stimulation, that is, eyes closed or open.

Materials Needed

Computer system (MAC OS X 10.4–10.6, or PC running Windows XP, Vista, or 7)

BIOPAC Student Lab Software ver. 3.7.7 or above

BIOPAC Data Acquisition Unit MP36, MP35, MP30, or MP45 (Note: The MP30 is being phased out.) with (AC100A) transformer

BIOPAC serial cable (CBLSERA)

BIOPAC electrode lead set (SS2L)

BIOPAC disposable vinyl electrodes (EL503), 3 electrodes per subject

BIOPAC electrode gel (GEL1) and abrasive pad (ELPAD) or alcohol wipes

Exercise mat, cot, or clean lab bench and small pillow

Suggested: Lycra swim cap or supportive wrap (such as 3M Coban™ self-adhering support wrap) to press electrodes against head for improved contact

Pre-Lab

Carefully read introductory material and examine the entire lab. Be familiar with structures and functions of the brain, and the physiology of action potentials from lecture or the textbook.

Safety

▶ The BIOPAC electrode lead set is safe and easy to use and should be used only as described in the procedures section of the laboratory exercise.

▶ The electrode lead clips are color-coded. Make sure they are connected to the properly placed electrode as demonstrated in figure 31.2.

▶ The vinyl electrodes are disposable and meant to be used only once. Each subject should use a new set.

Nerve impulses generated in the neurons of the nervous system are responsible for controlling many vital functions in the body, such as thought processes, responses to stimuli, higher mental functions (like problem solving), and the control of muscle contraction. Nerve impulses are electrical events that transmit information from neurons to other neurons, muscle cells, or glandular cells. This information is then used to accomplish various functional tasks throughout the body.

Neurons respond to certain types of stimuli by generating *action potentials*, which are accomplished by transporting sodium (Na^+) and potassium (K^+) ions across their plasma membranes in a particular sequence. This neural response leads to changes in the charge both inside the neurons and in the extracellular fluid surrounding them. This movement of ions in the extracellular fluid of the cerebral cortex can be detected by placing electrodes on the surface of the scalp and measuring the voltage between them. The recording of electrical activity in the cerebral cortex, via electrodes placed on the scalp, is called an *electroencephalogram (EEG)*. An EEG is a safe, noninvasive, and painless method of recording the sum of electrical activity in many thousands of neurons in the cerebral cortex. It is important to realize that an EEG is not a record of action potentials occurring in an individual neuron.

An EEG can be used to assess states of consciousness (awake, concentrating, asleep) and various brain disorders, and to establish brain death. The major brain waves found in EEGs are of four types, called *alpha, beta, delta,* and *theta* waves. Each type of brain wave is characterized by a specific range of frequencies and amplitudes. Figure 31.1 shows the typical appearance of the four types of brain waves.

The *frequency* of a brain wave is the number of complete wave cycles that occurs each second; it is measured in Hertz (Hz, or cycles per second). Frequency is an indication of the state of alertness; therefore, the higher the wave frequency, the more alert and attentive the person is.

The *amplitude* of a brain wave is the height, or magnitude, of the wave; it is measured in microvolts (μV, or one millionth of a volt). Amplitude indicates the degree of synchrony, or simultaneous firing of many neurons underneath the recording electrodes. Therefore, the higher the amplitude, the more neurons are firing simultaneously (in synchrony). It is important to understand that synchrony is not a measure of alertness; in fact, some of the highest amplitudes are found in delta waves, which occur mainly during the deeper stages of sleep. The lowest amplitudes are usually found in beta waves, which are dominant during high levels of attentiveness. This can be explained by the fact that asynchronous firing of neurons leads to a "canceling out" of positive and negative voltages occurring in different neurons at the same time.

All four types of brain waves can usually be observed in the EEG of a normal adult, although different wave forms become predominant in the various states of consciousness,

FIGURE 31.1 The four types of brain waves of the EEG.

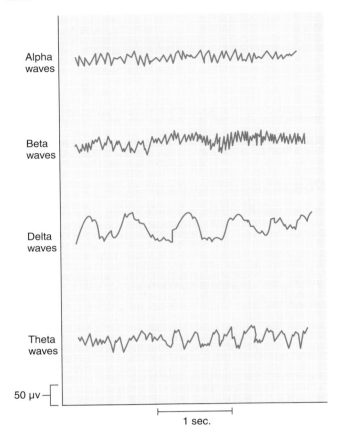

including the different stages of sleep. There is also quite a bit of individual variation in the characteristics of brain waves. This variation is due to a variety of factors, including gender, age, personality, the type of activity a person is engaged in, the area of the brain being used for the EEG, and various disease states of the brain. Table 31.1 shows the average frequencies and amplitudes of the four types of brain waves, along with a short description of the state(s) of consciousness in which they are predominant.

In this laboratory exercise, you will be using the bipolar recording method. In this method, two electrodes are placed over the same region of the cerebral cortex, and the difference in the electrical potential (voltage) between the electrodes is measured. A third electrode is placed on the earlobe as a "ground," or reference point for the body's baseline voltage.

The BIOPAC program will separate your EEG into the four types of brain waves. You will be able to observe all four types and measure the frequencies of the waves, as well as the standard deviation (variability) of wave amplitudes. You will also demonstrate *alpha block,* a replacement of alpha waves with lower-amplitude beta waves that occurs when closed eyes are suddenly opened.

TABLE 31.1 Frequencies, Amplitudes, and State(s) of Consciousness Associated with the Four Types of Brain Waves

Brain Wave	Average Frequency (Hz)	Average Amplitude (µV)	State(s) of Consciousness
Alpha	8–13	~50 (can vary from 20–200)	Awake, resting, with eyes closed; decrease when concentrating
Beta	14–30 (sometimes higher)	5–10 (can be as high as 30)	Awake, alert, with eyes open, paying attention or concentrating
Delta	1–4	~100 (can vary from 20–200)	Deep stages of sleep; some forms of severe brain disease
Theta	4–7	5–10 (can be as high as 60–70)	More commonly observed in children, but often seen in adults, especially while concentrating or under stress, or during sleep; some forms of brain disease

Procedure A—Setup

1. With your computer turned **ON** and the BIOPAC MP3X or MP45 unit turned **OFF,** plug the electrode lead set (SSL2) into Channel 1.
2. Turn on the MP3X or MP45 Data Acquisition Unit.
3. Selection of subject is important; choose someone with easy access to his/her scalp so that you can be sure you can make good contact between the skin and the electrode. With the subject lying down, face-up, on the cot, mat, or clean lab bench, position the electrodes (leads) on the same side of the head, as in figure 31.2. The ground electrode (black) is attached to the earlobe and can be folded under for better adhesion. (The ground electrode may also be attached to the neck just below the ear, if that is easier.) To help ensure good contact, make sure the area to be in contact with the electrodes is clean by wiping with the abrasive pad (ELPAD) or alcohol wipe. A small amount of BIOPAC electrode gel (GEL1) may also be used to make better contact between the sensor in the electrode and the skin. For optimal adhesion, hold the electrodes against the skin for about a minute after placement. Wrap the subject's head with a swim cap or supportive wrap to secure electrode placement.
4. Start the BIOPAC Student Lab Program for Electroencephalography I, Lesson 3 (LO3-EEG-1).
5. Designate a filename to be used to save the subject's data.

Procedure B—Calibration

1. Click on **Calibrate.** A warning will pop up asking you to check electrode placement.
2. The subject should be in a supine position with the head resting comfortably, and tilted to one side. After checking the electrode attachments, click **OK.** The calibration will stop automatically after 8 seconds.
3. The calibration recording should look similar to figure 31.3. If not, click **Redo Calibration.**

FIGURE 31.2 Subject setup.

Source: Courtesy of and © *BIOPAC* Systems Inc.

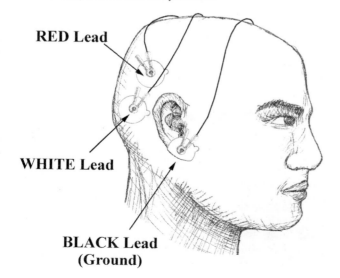

RED Lead

WHITE Lead

BLACK Lead
(Ground)

FIGURE 31.3 Calibration recording.

Source: Courtesy of and © *BIOPAC* Systems Inc.

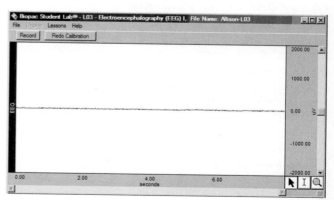

Procedure C—Recording

1. Before beginning the recording phase, you should designate a **Director** (who will instruct the subject to open or close eyes) and a **Recorder** (who will insert the markers by pressing the F4 key when eyes are opened, and the F5 key when eyes are closed.)

2. The data for this exercise will be recorded over a 1-minute period. During the first 20 seconds, the subject will have his/her eyes closed. During the next 20 seconds, the subject should have eyes open, and for the last 20 seconds, eyes closed again. Make sure all group members understand the procedure before proceeding to step 3.

3. Click on **Record.** The Director should then instruct the subject to remain relaxed and change eye condition as follows:
   a. 0–20 seconds, eyes closed
   b. 21–40 seconds, eyes open
   c. 41–60 seconds, eyes closed
   The Recorder will insert markers at these times: 21 seconds when eyes are opened (F4 key), and 41 seconds when eyes are closed again (F5 key).

4. Click on **Suspend.** If the recording is similar to figure 31.4, click **Done,** and go to step 5. If there are large fluctuations or baseline drift in the recording, or if the EEG is flat, click on **Redo.** This will erase the current data. Check electrode placement, and make sure that the subject does not make any excessive movements during the new recording. After completing the new recording, click **Suspend,** and then **Done.**

5. Click **Yes.** After you click **Yes,** the program will separate the EEG into the alpha, beta, delta and theta waves and display them in the data window. The data window should now look similar to figure 31.5. If similar, the subject may now remove the electrodes.

6. A pop-up window will appear with five options: **"record from another subject; analyze current data file; analyze another data file; copy to another location; quit."** Make your choice and continue as directed.

FIGURE 31.4 Raw EEG recording.

Source: Courtesy of and © *BIOPAC* Systems Inc.

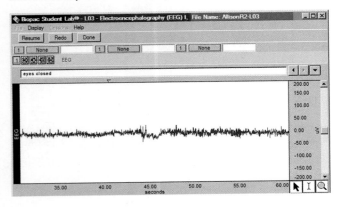

FIGURE 31.5 EEG recording showing four different types of brain waves.

Source: Courtesy of and © *BIOPAC* Systems Inc.

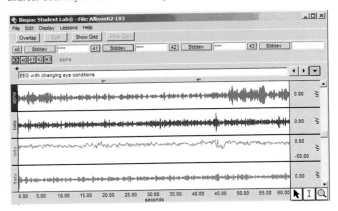

Procedure D—Data Analysis

1. You can either analyze the current file now or save it and do the analysis later, using the **Review Saved Data** mode.

2. Note the channel number designations: CH 1—EEG (hidden), CH 40—alpha, CH 41—beta, CH 42—delta, CH 43—theta.

3. Across the top of the recordings you will see three boxes for each channel: channel number, measurement type, and the result. Note the settings for the measurement boxes: CH 40–Ch 43 should all be set on **Stddev** and SC on **Freq.** The program will automatically calculate the standard deviation or frequency for the selected area.

4. Using the I-beam tool, in the right bottom corner, select the area from time 0–20 seconds (to the first marker). See figure 31.6 for an example of how to select the area for the first 20 seconds. Record the result for each channel in table 31.2 of Part A of Laboratory Assessment 31. The standard deviation of amplitude will not provide

FIGURE 31.6 Proper selection of the area from 0 to 20 seconds for Standard Deviation measurement.

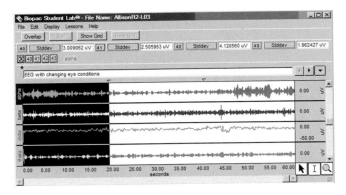

actual wave amplitudes, but rather the variability in those amplitudes.

5. Repeat step 4 for the time period from 20–40 seconds (between the first and second markers) and record the results in table 31.2 of the laboratory assessment.

6. Repeat step 4 for the time period from 40–60 seconds (from the second marker to the end) and record the results in table 31.2 of the laboratory assessment.

7. Now set up the data window, so you can just see a 4- to 5-second interval in the first 20 seconds of the recording. Do this by clicking anywhere in the horizontal scale (the numbers just above the word "seconds" at the bottom of the data window). A box will pop up in the middle of the screen, called **Horizontal Scale.** Under **Scale Range,** change the lower scale to 2, and the upper scale to 6. Then click **OK.** This will spread out the data so you can see individual waves and measure their frequency.

8. Now begin to measure wave frequencies. In order for the program to measure **FREQ** of each type of wave form, you have to select each type and measure it individually. First select alpha waves for **FREQ** measurement by clicking on the word "alpha" along the left side of the data window. Now SC (Selected Channel) will display only alpha wave frequencies.

9. Use the I-beam cursor to select an area that represents one cycle in the alpha wave. One cycle runs from one peak to the next peak (refer to fig. 31.7). Record the result in table 31.3 of the laboratory assessment.

10. Repeat for two other nearby alpha cycles and record the results. Calculate the mean and record it in table 31.3 of the laboratory assessment.

FIGURE 31.7 Recording with one alpha wave selected.

Source: Courtesy of and © *BIOPAC* Systems Inc.

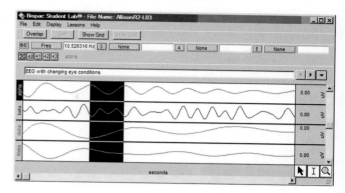

11. Repeat steps 9 and 10 for each of the other waveforms (beta, delta, theta) and record the results in table 31.3 of the laboratory assessment. (Remember to click on the word "beta," delta," or "theta" before beginning to measure **FREQ** for each of these wave types.

12. If you would like to view the entire recording after completing **FREQ** measurements, open the **Display** menu. Click on **Autoscale Horizontal** and then **Autoscale Waveforms.** Now the entire one-minute recording will be visible in the data window.

13. When finished, you can save the data on the hard drive, a disk, or a network drive, and then exit the system.

14. Complete Part B of the laboratory assessment.

NOTES

Name _____

Date _____

Section _____

The Ⓐ corresponds to the indicated outcome(s) found at the beginning of the laboratory exercise.

Electroencephalography I: BIOPAC© Exercise

Subject Profile

Name _____   Height _____

Age _____   Weight _____

Gender: Male / Female

Part A   Data and Calculations Assessments

1. EEG Variability in Amplitude Measurements Ⓐ1 Ⓐ2

TABLE 31.2   Standard Deviation [stddev]

Rhythm	CH Measurement	Eyes Closed	Eyes Open	Eyes Re-closed
Alpha	40 Stddev			
Beta	41 Stddev			
Delta	42 Stddev			
Theta	43 Stddev			

2. EEG Frequency Measurements Ⓐ1 Ⓐ2

TABLE 31.3   Frequency [Hz] Measurements

Rhythm	CH Measurement	Cycle 1	Cycle 2	Cycle 3	Mean
Alpha	SC Freq				
Beta	SC Freq				
Delta	SC Freq				
Theta	SC Freq				

Part B Assessments

Complete the following:

1. Examine the alpha and beta waveforms for change between the "eyes closed" state and the "eyes open" state. ⟋2⟍

    a. Does desynchronization of the alpha rhythm occur when the eyes are open? How do you know?

    _____

    _____

    _____

    _____

    _____

    _____

    b. Does the beta rhythm become more pronounced in the "eyes open" state? ⟋2⟍

    _____

2. Examine the delta and theta rhythms. Is there an increase in delta and theta activity when the eyes are open? Explain your observation. ⟋1⟍

    _____

    _____

    _____

3. Define the following terms: ⟋1⟍

    a. Alpha rhythm

    _____

    _____

    b. Beta rhythm

    _____

    _____

    c. Delta rhythm

    _____

    _____

    d. Theta rhythm

    _____

    _____

Dissection of the Sheep Brain

Purpose of the Exercise

To observe the major features of the sheep brain and to compare these features with those of the human brain.

Materials Needed

Dissectible model of human brain
Preserved sheep brain
Dissecting tray
Dissection instruments
Long knife

For Demonstration Activity:
Frontal sections of sheep brains

Safety

▶ Wear disposable gloves when handling the sheep brains.
▶ Save or dispose of the brains as instructed.
▶ Wash your hands before leaving the laboratory.

Learning Outcomes

After completing this exercise, you should be able to

① Locate the major structures of the sheep brain.
② Summarize differences and similarities between the sheep brain and the human brain.
③ Locate the larger cranial nerves of the sheep brain.

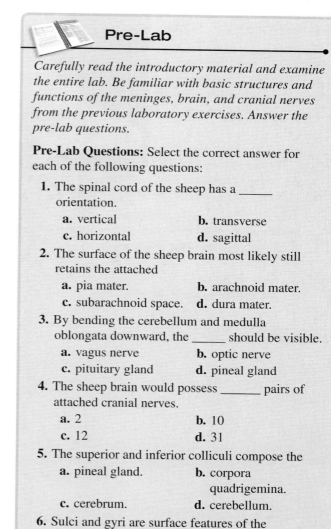
Pre-Lab

Carefully read the introductory material and examine the entire lab. Be familiar with basic structures and functions of the meninges, brain, and cranial nerves from the previous laboratory exercises. Answer the pre-lab questions.

Pre-Lab Questions: Select the correct answer for each of the following questions:

1. The spinal cord of the sheep has a _____ orientation.
   a. vertical   **b.** transverse
   c. horizontal   **d.** sagittal

2. The surface of the sheep brain most likely still retains the attached
   a. pia mater.   **b.** arachnoid mater.
   c. subarachnoid space.   **d.** dura mater.

3. By bending the cerebellum and medulla oblongata downward, the _____ should be visible.
   a. vagus nerve   **b.** optic nerve
   c. pituitary gland   **d.** pineal gland

4. The sheep brain would possess _____ pairs of attached cranial nerves.
   a. 2   **b.** 10
   c. 12   **d.** 31

5. The superior and inferior colliculi compose the
   a. pineal gland.   **b.** corpora quadrigemina.
   c. cerebrum.   **d.** cerebellum.

6. Sulci and gyri are surface features of the cerebrum of the sheep brain.
   True_____   False_____

7. The cerebellum of the sheep brain has deep gray matter and superficial white matter.
   True_____   False_____

Mammalian brains have many features in common. Human brains may not be available, so sheep brains often are dissected as an aid to understanding mammalian brain structure. However, the adaptations of the sheep differ from the adaptations of the human, so comparisons of their structural features may not be precise. The sheep is a quadruped, therefore the spinal cord is horizontal, unlike the vertical orientation in a bipedal human. Preserved sheep brains have a different appearance and are firmer than those that are removed directly from the cranial cavity because of the preservatives used.

Procedure—Dissection of the Sheep Brain

As you examine a sheep brain, contemplate any differences and similarities between the sheep brain and the human brain. Several questions in the laboratory assessment address comparisons of sheep and human brains.

1. Obtain a preserved sheep brain and rinse it thoroughly in water to remove as much of the preserving fluid as possible.
2. Examine the surface of the brain for the presence of meninges. (The outermost layers of these membranes may have been lost during removal of the brain from the cranial cavity.) If meninges are present, locate the following:

   dura mater—the thick, opaque outer layer

   arachnoid mater—the delicate, transparent middle layer attached to the undersurface of the dura mater

   pia mater—the thin, vascular layer that adheres to the surface of the brain (should be present)

3. Remove any remaining dura mater by cutting and pulling it gently from the surface of the brain.
4. Position the brain with its ventral surface down in the dissecting tray. Study figure 32.1, and locate the following structures on the specimen:

   longitudinal fissure
   cerebral hemispheres
   gyri
   sulci
   frontal lobe
   parietal lobe
   temporal lobe
   occipital lobe
   cerebellum
   medulla oblongata
   spinal cord

5. Gently separate the cerebral hemispheres along the longitudinal fissure and expose the transverse band of white fibers within the fissure that connects the hemispheres. This band is the *corpus callosum.*
6. Bend the cerebellum and medulla oblongata slightly downward and away from the cerebrum (fig. 32.2). This will expose the *pineal gland* in the upper midline and the *corpora quadrigemina,* which consists of four rounded structures, called *colliculi,* associated with the midbrain.

FIGURE 32.1 Dorsal surface of the sheep brain.

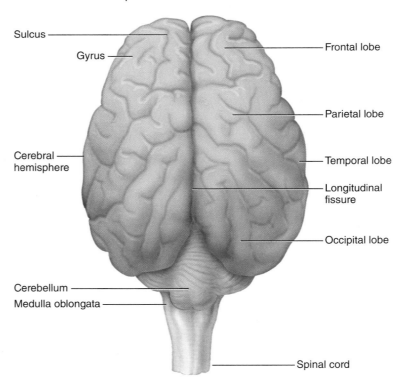

326

FIGURE 32.2 Gently bend the cerebellum and medulla oblongata away from the cerebrum to expose the pineal gland of the diencephlon and the corpora quadrigemina of the midbrain.

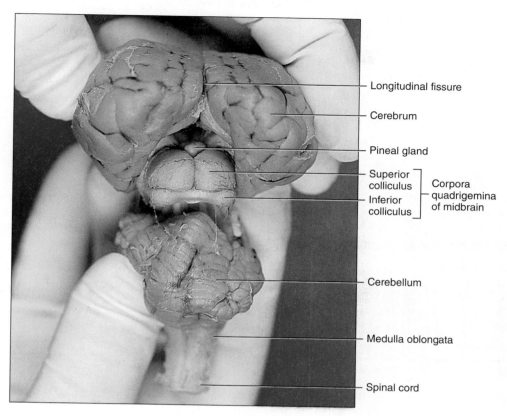

- Longitudinal fissure
- Cerebrum
- Pineal gland
- Superior colliculus ⎤
- Inferior colliculus ⎦ Corpora quadrigemina of midbrain
- Cerebellum
- Medulla oblongata
- Spinal cord

7. Position the brain with its ventral surface upward. Study figures 32.3 and 32.4, and locate the following structures on the specimen:

> **longitudinal fissure**
> **olfactory bulbs**
> **optic nerves**
> **optic chiasma**
> **optic tract**
> **mammillary bodies**
> **infundibulum (pituitary stalk)**
> **midbrain**
> **pons**

8. Although some of the cranial nerves may be missing or are quite small and difficult to find, locate as many of the following as possible, using figures 32.3 and 32.4 as references:

> **oculomotor nerves**
> **trochlear nerves**
> **trigeminal nerves**
> **abducens nerves**
> **facial nerves**
> **vestibulocochlear nerves**
> **glossopharyngeal nerves**
> **vagus nerves**
> **accessory nerves**
> **hypoglossal nerves**

FIGURE 32.3 Lateral surface of the sheep brain.

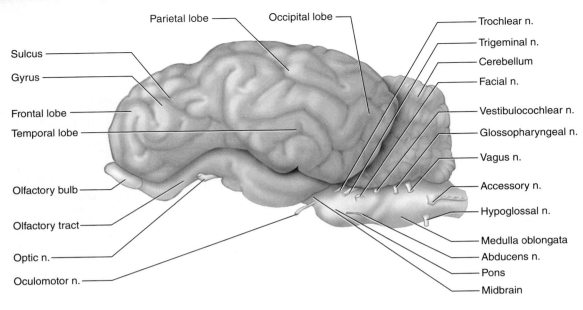

FIGURE 32.4 Ventral surface of the sheep brain.

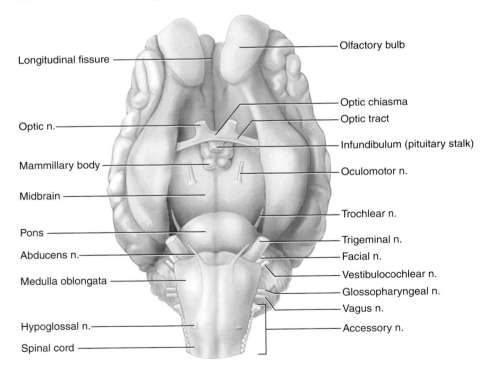

9. Using a long, sharp knife, cut the sheep brain along the midline to produce a median section. Study figures 32.2 and 32.5, and locate the following structures on the specimen:

cerebrum

olfactory bulb

corpus callosum

cerebellum

- white matter
- gray matter

lateral ventricle—one in each cerebral hemisphere

third ventricle—within diencephalon

fourth ventricle—between brainstem and cerebellum

diencephalon

- optic chiasma
- infundibulum
- pituitary gland—this structure may be missing
- mammillary bodies
- thalamus
- hypothalamus
- pineal gland of epithalmus

midbrain

- corpora quadrigemina
    - superior colliculus
    - inferior colliculus

pons

medulla oblongata

Demonstration Activity

Observe a frontal section from a sheep brain (fig. 32.6). Note the longitudinal fissure, gray matter, white matter, corpus callosum, lateral ventricles, third ventricle, and thalamus.

10. Dispose of the sheep brain as directed by the laboratory instructor.
11. Complete Parts A, B, and C of Laboratory Assessment 32.

FIGURE 32.5 Median section of the right half of the sheep brain dissection.

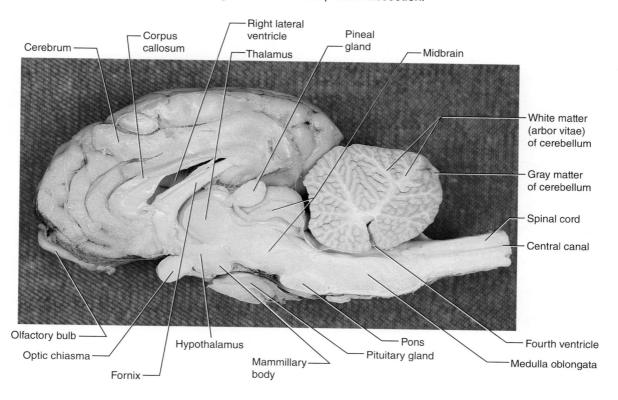

Cerebrum

Corpus callosum

Right lateral ventricle

Thalamus

Pineal gland

Midbrain

White matter (arbor vitae) of cerebellum

Gray matter of cerebellum

Spinal cord

Central canal

Olfactory bulb

Optic chiasma

Fornix

Hypothalamus

Mammillary body

Pituitary gland

Pons

Fourth ventricle

Medulla oblongata

FIGURE 32.6 Frontal section of a sheep brain.

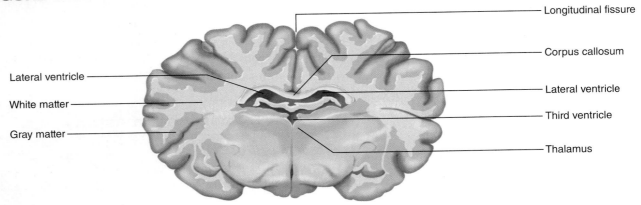

Longitudinal fissure

Corpus callosum

Lateral ventricle

Lateral ventricle

White matter

Third ventricle

Gray matter

Thalamus

Name _____

Date _____

Section _____

The A corresponds to the indicated outcome(s) found at the beginning of the laboratory exercise.

Dissection of the Sheep Brain

Part A Assessments

Answer the following questions:

1. Describe the location of any meninges observed to be associated with the sheep brain. 1 _____

_____

_____

2. How do the relative sizes of the sheep and human cerebral hemispheres differ? 2 _____

_____

_____

3. How do the gyri and sulci of the sheep cerebrum compare with the human cerebrum in numbers? 2 _____

_____

_____

_____

4. What is the significance of the differences you noted in your answers for questions 2 and 3? 2 _____

_____

_____

5. What difference did you note in the structures of the sheep cerebellum and the human cerebellum? 2 _____

_____

_____

6. How do the sizes of the olfactory bulbs of the sheep brain compare with those of the human brain? 2 _____

_____

_____

7. Based on their relative sizes, which of the cranial nerves seems to be most highly developed in the sheep brain? 3

_____

_____

_____

8. What is the significance of the observations you noted in your answers for questions 6 and 7? 2 _____

_____

_____

Part B Assessments

Critical Thinking Assessment

Prepare a list of at least six features to illustrate ways in which the brains of sheep and humans are similar. 2

1. _____

2. _____

3. _____

4. _____

5. _____

6. _____

Interpret the significance of these similarities. 2 _____

_____

_____

Part C Assessments

Identify the features indicated in the median section of the sheep brain in figure 32.7.

FIGURE 32.7 Label the features of this median section of the sheep brain by placing the correct numbers in the spaces provided.

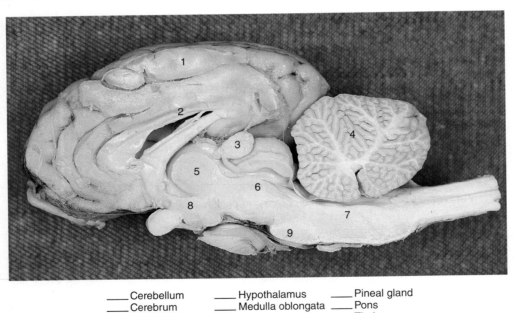

____ Cerebellum   ____ Hypothalamus   ____ Pineal gland
____ Cerebrum   ____ Medulla oblongata ____ Pons
____ Corpus callosum ____ Midbrain   ____ Thalamus

Laboratory Exercise 33

General Senses

Purpose of the Exercise

To review the characteristics of sensory receptors and general senses and to investigate some of the general senses associated with the skin.

Materials Needed

Marking pen (washable)
Millimeter ruler
Bristle or sharp pencil
Forceps with fine points or another two-point
   discriminator device
Blunt metal probes
Three beakers (250 mL)
Warm tap water or 45°C (113°F) water bath
Cold water (ice water)
Thermometer

For Demonstration Activity:

Prepared microscope slides of tactile (Meissner's) and
   lamellated (Pacinian) corpuscles
Compound light microscope

Learning Outcomes

After completing this exercise, you should be able to

1. Associate types of sensory receptors with general senses throughout the body.

2. Determine and record the distribution of touch, warm, and cold receptors in various regions of the skin.

3. Measure the two-point threshold of various regions of the skin.

Pre-Lab

Carefully read the introductory material and examine the entire lab. Be familiar with basic structures and functions of the receptors associated with general senses from the lecture or the textbook. Answer the pre-lab questions.

Pre-Lab Questions: Select the correct answer for each of the following questions:

1. General senses include all of the following *except*
   a. touch.   b. vision.
   c. temperature.   d. pain.

2. Thermoreceptors are associated with
   a. deep pressure. b. light touch.
   c. tissue trauma. d. temperature changes.

3. When receptors are continuously stimulated, the sensations may fade away; this phenomenon is known as
   a. tolerance.   b. perception.
   c. sensory adaptation. d. fatigue.

4. Encapsulated nerve endings include
   a. tactile corpuscles. b. pain receptors.
   c. cold receptors.   d. warm receptors.

5. A lamellated corpuscle is stimulated by
   a. light touch.   b. deep pressure.
   c. warm temperatures. d. cold temperatures.

6. Free nerve endings function as pain, warm, and cold receptors.
   True _____   False _____

7. Lamellated corpuscles are located in the epidermis of the skin.
   True _____   False _____

Sensory receptors are sensitive to changes that occur within the body and its surroundings. Each type of receptor is particularly sensitive to a distinct kind of environmental change and is much less sensitive to other forms of stimulation. When receptors are stimulated, they initiate nerve impulses that travel into the central nervous system. The raw form in which these receptors send information to the brain is called *sensation*. The way our brains interpret this information is called *perception*.

The sensory receptors found widely distributed throughout skin, muscles, joints, and visceral organs are associated with **general senses.** These senses include touch, pressure, temperature, pain, and the senses of muscle movement and body position. Receptors (muscle spindles) associated with muscle movements were included as part of Laboratory Exercise 29.

The general senses associated with the body surface can be classified according to the stimulus type. *Mechanoreceptors* include tactile (Meissner's) corpuscles, which are stimulated by light touch, stretch, or vibration, and lamellated (Pacinian) corpuscles, which are stimulated by deep pressure, stretch, or vibration. *Thermoreceptors* include those associated with temperature changes. Warm receptors are most sensitive to temperatures between 25°C (77°F) and 45°C (113°F). Cold receptors are most sensitive to temperatures between 10°C (50°F) and 20°C (68°F). *Nociceptors* are pain receptors that respond to tissue trauma, which may include a cut or pinch, extreme heat or cold, or excessive pressure. Tests for pain receptors are not included as part of this laboratory exercise.

Receptors possess structural differences at the dendrite ends of sensory neurons. Those with unencapsulated free nerve endings include pain, cold, and warm receptors; those with encapsulated nerve endings include tactile and lamellated corpuscles.

A sensation may fade away when receptors are continuously stimulated. This is known as *sensory adaptation,* like adjusting to a room temperature or an odor of our environment. However, sensory adaptation to pain is not as prevalent because impulses may continue into the CNS for longer periods of time. The importance of perceiving pain not only alerts us to a change in our environment and potential injury, but it also allows us to detect body abnormalities so that proper adjustments can be taken. Unfortunately, with some spinal cord injuries or diseases such as diabetes mellitus, where the sense of pain is lost or becomes less noticeable, unwarranted tissue damages can occur.

Sensory receptors that are more specialized and confined to the head are associated with **special senses.** Laboratory Exercises 34 through 38 describe the special senses.

Procedure A—Receptors and General Senses

1. Reexamine the introduction to this laboratory exercise and study table 33.1.
2. Complete Part A of Laboratory Assessment 33.

Demonstration Activity

Observe the tactile (Meissner's) corpuscle with the microscope set up by the laboratory instructor. This type of receptor is abundant in the superficial dermis in outer regions of the body, such as in the fingertips, soles, lips, and external genital organs. It is responsible for the sensation of light touch. (See fig. 33.1.)

Observe the lamellated (Pacinian) corpuscle in the second demonstration microscope. This corpuscle is composed of many layers of connective tissue cells and has a nerve fiber in its central core. Lamellated corpuscles are numerous in the hands, feet, joints, and external genital organs. They are responsible for the sense of deep pressure (fig. 33.2). How are tactile and lamellated corpuscles similar?

_____

_____

How are they different?

_____

_____

TABLE 33.1   Receptors Associated with General Senses of Skin

Receptor Category	Respond to	Receptor Examples
Mechanoreceptors	Touch, stretch, vibration, pressure	Encapsulated tactile (Meissner's) corpuscle; encapsulated lamellated (Pacinian) corpuscle
Thermoreceptors	Temperature changes	Free nerve endings for warm temperatures; free nerve endings for cold temperatures
Nociceptors	Intense mechanical, chemical, or temperature stimuli; tissue trauma	Free nerve endings for pain

FIGURE 33.1 Tactile (Meissner's) corpuscles, such as this one, are responsible for the sensation of light touch (250×).

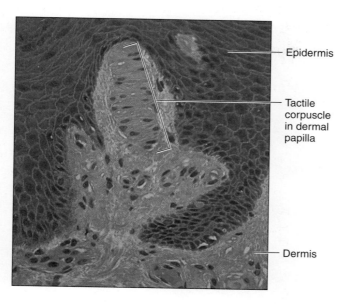

— Epidermis

— Tactile corpuscle in dermal papilla

— Dermis

FIGURE 33.2 Lamellated (Pacinian) corpuscles, such as this one, are responsible for the sensation of deep pressure (100×).

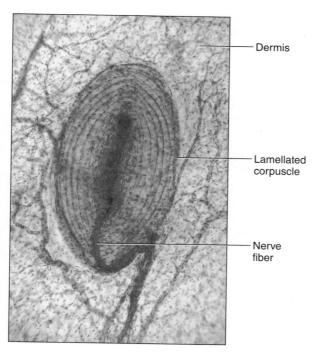

— Dermis

— Lamellated corpuscle

— Nerve fiber

Procedure B—Sense of Touch

1. Investigate the distribution of touch receptors in your laboratory partner's skin. To do this, follow these steps:
   a. Use a marking pen and a millimeter ruler to prepare a square with 2.5 cm on each side on the skin on your partner's inner wrist, near the palm.
   b. Divide the square into smaller squares with 0.5 cm on a side, producing a small grid.
   c. Ask your partner to rest the marked wrist on the tabletop and to keep his or her eyes closed throughout the remainder of the experiment.
   d. Press the end of a bristle on the skin in some part of the grid, using just enough pressure to cause the bristle to bend. A sharp pencil could be used as an alternate device.
   e. Ask your partner to report whenever the touch of the bristle is felt. Record the results in Part B of the laboratory assessment.
   f. Continue this procedure until you have tested twenty-five different locations on the grid. Move randomly through the grid to help prevent anticipation of the next stimulation site.
2. Test two other areas of exposed skin in the same manner, and record the results in Part B of the laboratory assessment.
3. Answer the questions in Part B of the laboratory assessment.

Procedure C—Two-Point Threshold

1. Test your partner's ability to recognize the difference between one and two points of skin being stimulated simultaneously. To do this, follow these steps:
   a. Have your partner place a hand with the palm up on the table and close his or her eyes.
   b. Hold the tips of a forceps tightly together and gently touch the skin on your partner's fingertip (fig. 33.3).
   c. Ask your partner to report if it feels like one or two points are touching the finger.
   d. Allow the tips of the forceps to spread so they are 1 mm apart, press both points against the skin simultaneously, and ask your partner to report as before.
   e. Repeat this procedure, allowing the tips of the forceps to spread more each time until your partner can feel both tips being pressed against the skin. The minimum distance between the tips of the forceps when both can be felt is called the *two-point threshold*. As soon as you are able to distinguish two points, two separate receptors are being stimulated instead of only one receptor (fig. 33.3).
   f. Record the two-point threshold for the skin of a fingertip in Part C of the laboratory assessment.

FIGURE 33.3 Two-point threshold determination test. Example (a) stimulates a single sensory neuron and (b) stimulates two different sensory neurons.

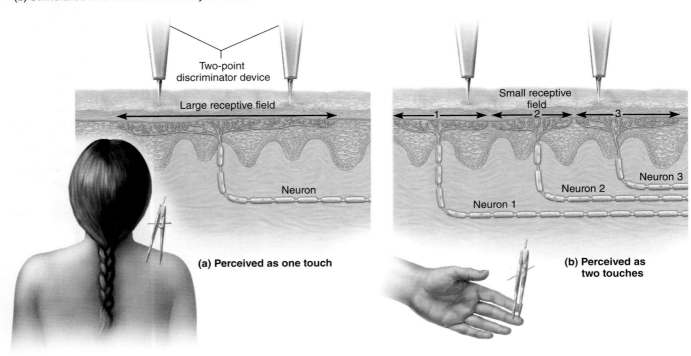

(a) Perceived as one touch

(b) Perceived as two touches

2. Repeat this procedure to determine the two-point threshold of the palm, the back of the hand, the back of the neck, the forearm, and the leg. Record the results in Part C of the laboratory assessment.

3. Answer the questions in Part C of the laboratory assessment.

Procedure D—Sense of Temperature

1. Investigate the distribution of *warm (heat) receptors* in your partner's skin. To do this, follow these steps:

   a. Mark a square with 2.5 cm sides on your partner's palm.

   b. Prepare a grid by dividing the square into smaller squares, 0.5 cm on a side.

   c. Have your partner rest the marked palm on the table and close his or her eyes.

   d. Heat a blunt metal probe by placing it in a beaker of warm (hot) water (about 40–45°C/ 104–113°F) for a minute or so. (*Be sure the probe does not get so hot that it burns the skin.*) Use a thermometer to monitor the appropriate warm water from the tap or the water bath.

   e. Wipe the probe dry and touch it to the skin on some part of the grid.

   f. Ask your partner to report if the probe feels warm. Then record the results in Part D of the laboratory assessment.

   g. Keep the probe warm, and repeat the procedure until you have randomly tested several different locations on the grid.

2. Investigate the distribution of *cold receptors* by repeating the procedure. Use a blunt metal probe that has been cooled by placing it in ice water for a minute or so. Record the results in Part D of the laboratory assessment.

3. Answer the questions in Part D of the laboratory assessment.

Learning Extension Activity

Prepare three beakers of water of different temperatures. One beaker should contain warm water (about 40°C/104°F), one should be room temperature (about 22°C/72°F), and one should contain cold water (about 10°C/50°F). Place the index finger of one hand in the warm water and, at the same time, place the index finger of the other hand in the cold water for about 2 minutes. Then, simultaneously move both index fingers into the water at room temperature. What temperature do you sense with each finger? How do you explain the resulting perceptions?

_____

_____

_____

_____

_____

_____

Name _____

Date _____

Section _____

The ⚠ corresponds to the indicated outcome(s) found at the beginning of the laboratory exercise.

General Senses

Part A—Receptors and General Senses Assessments

Complete the following statements:

1. Whenever tissues are damaged, _____ receptors are likely to be stimulated. ⚠

2. Receptors that are sensitive to temperature changes are called _____. ⚠

3. A sensation may seem to fade away when receptors are continuously stimulated as a result of _____ adaptation. ⚠

4. Tactile (Meissner's) corpuscles are responsible for the sense of light _____. ⚠

5. Lamellated (Pacinian) corpuscles are responsible for the sense of deep _____. ⚠

6. _____ receptors are most sensitive to temperatures between 25°C (77°F) and 45°C (113°F). ⚠

7. _____ receptors are most sensitive to temperatures between 10°C (50°F) and 20°C (68°F). ⚠

8. Widely distributed sensory receptors throughout the body are associated with _____ senses in contrast to special senses. ⚠

Part B—Sense of Touch Assessments

1. Record a + to indicate where the bristle was felt and a *0* to indicate where it was not felt. ⚠

|←——— 2.5 cm ———→|

Skin of wrist

2. Show the distribution of touch receptors in two other regions of skin. ⚠

Region tested _____   Region tested _____

3. Answer the following questions:

   a. How do you describe the pattern of distribution for touch receptors in the regions of the skin you tested? ⚠

   _____

   _____

   b. How does the concentration of touch receptors seem to vary from region to region? ⚠ _____

   _____

Part C—Two-Point Threshold Assessments

1. Record the two-point threshold in millimeters for skin in each of the following regions: ⚠️3

   Fingertip _____

   Palm _____

   Back of hand _____

   Back of neck _____

   Forearm _____

   Leg _____

2. Answer the following questions:

   a. What region of the skin tested has the greatest ability to discriminate two points? ⚠️3 _____

   b. What region of the skin tested has the least sensitivity to this test? ⚠️3 _____

   c. What is the significance of these observations in questions *a* and *b*? ⚠️3 _____

   _____

   _____

Part D—Sense of Temperature Assessments

1. Record a + to indicate where warm was felt and a *0* to indicate where it was not felt. ⚠️2

Skin of palm

2. Record a + to indicate where cold was felt and a *0* to indicate where it was not felt. ⚠️2

Skin of palm

3. Answer the following questions:

   a. How do temperature receptors appear to be distributed in the skin of the palm? ⚠️2 _____

   _____

   b. Compare the distribution and concentration of warm and cold receptors in the skin of the palm. ⚠️2 _____

   _____

   _____

34

Smell and Taste

Purpose of the Exercise

To review the structures of the organs of smell and taste and to investigate the abilities of smell and taste receptors to discriminate various chemical substances.

Materials Needed

For Procedure A—Sense of Smell (Olfaction)
Set of substances in stoppered bottles: cinnamon, sage, vanilla, garlic powder, oil of clove, oil of wintergreen, and perfume

For Procedure B—Sense of Taste (Gustation)
Paper cups (small)
Cotton swabs (sterile; disposable)
5% sucrose solution (sweet)
5% NaCl solution (salt)
1% acetic acid or unsweetened lemon juice (sour)
0.5% quinine sulfate solution or 0.1% Epsom salt solution (bitter)
1% monosodium glutamate (MSG) solution (umami)

For Demonstration Activities:
Compound light microscope
Prepared microscope slides of olfactory epithelium and of taste buds

For Learning Extension Activity:
Pieces of apple, potato, carrot, and onion or packages of mixed flavors of LifeSavers

Learning Outcomes

After completing this exercise, you should be able to

1. Compare the characteristics of the smell (olfactory) receptors and the taste (gustatory) receptors.
2. Explain how the senses of smell and taste function and are subsequently preceived by the brain.
3. Record recognized odors and the time needed for olfactory sensory adaptation to occur.
4. Locate the distribution of taste receptors on the surface of the tongue and mouth cavity.

Pre-Lab

Carefully read the introductory material and examine the entire lab. Be familiar with the basic structures and functions of the receptors associated with smell and taste from the lecture or the textbook. Answer the pre-lab questions.

Pre-Lab Questions: Select the correct answer for each of the following questions:

1. Receptor cells for taste are located
   a. only on the tongue.
   b. on the tongue, oral cavity, and pharynx.
   c. in the oral cavity except the tongue.
   d. on the tongue and nasal passages.

2. Olfactory interpretation centers are located in the
   a. oral cavity.   b. inferior nose.
   c. temporal and frontal   d. brainstem
      lobes of the cerebrum

3. Which of the following is *not* considered a recognized taste?
   a. mint   b. salt
   c. sweet   d. umami

4. Taste interpretation occurs in the _____ of the cerebrum.
   a. frontal   b. temporal
   c. occipital   d. insula

5. Sour sensations are produced from
   a. acids.   b. sugars.
   c. ionized inorganic salts.   d. alkaloids.

Safety

▶ Be aware of possible food allergies when selecting test solutions and foods to taste.
▶ Prepare fresh solutions for use in Procedure B.
▶ Wash your hands before starting the taste experiment.
▶ Wear disposable gloves when performing taste tests on your laboratory partner.
▶ Use a clean cotton swab for each test. Do not dip a used swab into a test solution.
▶ Dispose of used cotton swabs and paper towels as directed.
▶ Wash your hands before leaving the laboratory.

The senses of smell (olfaction) and taste (gustation) are dependent upon chemoreceptors that are stimulated by various chemicals dissolved in liquids. The receptors of smell are found in the olfactory organs, which are located in the superior parts of the nasal cavity and in a portion of the nasal septum. The receptors of taste occur in the taste buds, which are sensory organs primarily found on the surface of the tongue. Chemicals are considered odorless and tasteless if receptor sites for them are absent.

Olfactory receptor cells are actually neurons, surrounded by columnar epithelial cells, with their apical ends covered with cilia, also called hair cells, embedded in the mucus of the superior nasal cavity. These neurons form the olfactory cranial nerves. In order to detect an odor, the molecules must first dissolve in the mucus before they bind to the receptor sites of the cilia (olfactory hairs). When receptor cells temporarily bind to an odorant molecule, it results in an action potential over the olfactory neurons passing through the foramina of the cribriform plate and nerve fibers in the olfactory bulb. Eventually the impulses arrive at interpreting centers located deep within the temporal lobes and the inferior frontal lobes of the cerebrum. Although humans have the ability to distinguish nearly 10,000 different odors, coded by perhaps less than 1,000 genes, various odors can result from combinations of the receptor cells stimulated. We become sensory adapted to an odor very quickly, but exposure to a different substance can be quickly noticed.

Taste receptor cells are located in taste buds on the tongue, but receptor cells are also distributed in other areas of the oral cavity and pharynx. A taste bud contains taste cells with terminal microvilli, called taste hairs, projecting through a taste pore on the epithelium of the tongue. Taste sensations are grouped into five recognized categories: sweet, sour, salt, bitter, and umami. The *sweet* sensation is produced from sugars, the *sour* sensation from acids, the *salt* sensation from ionized inorganic salts, the *bitter* sensation from alkaloids and spoiled foods, and the *umami* sensation from aspartic and glutamic acids or a derivative such as monosodium glutamate. Three cranial nerves conduct impulses from the taste buds. The facial nerve (VII) conducts sensory impulses from the anterior two-thirds of the tongue; the glossopharyngeal nerve (IX) from the pos-

terior tongue; and the vagus nerve (X) from the pharyngeal region. When a molecule binds to a receptor cell, a neurotransmitter is secreted and an action potential occurs over a sensory neuron. The action potential continues through the medulla oblongata and the thalamus, and is interpreted in the insula of the cerebrum. Sensory adaptation also occurs rather quickly, which helps explain the reason the first few bites of a particular food have the most vivid flavor.

The senses of smell and taste function closely together, because substances that are tasted often are smelled at the same moment, and they play important roles in the selection of foods. The taste and aroma of foods are also influenced by appearance, texture, temperature, and the person's mood.

Procedure A—Sense of Smell (Olfaction)

1. Reexamine the introduction to this laboratory exercise and study figure 34.1.
2. Complete Part A of Laboratory Assessment 34.
3. Test your laboratory partner's ability to recognize the odors of the bottled substances available in the laboratory. To do this, follow these steps:
   a. Have your partner keep his or her eyes closed.
   b. Remove the stopper from one of the bottles, and hold it about 4 cm under your partner's nostrils for about 2 seconds.
   c. Ask your partner to identify the odor, and then replace the stopper.
   d. Record your partner's response in Part B of the laboratory assessment.
   e. Repeat steps *b–d* for each of the bottled substances.
4. Repeat the preceding procedure, using the same set of bottled substances, but present them to your partner in a different sequence. Record the results in Part B of the laboratory assessment.
5. Wait 10 minutes and then determine the time it takes for your partner to experience olfactory sensory adaptation. To do this, follow these steps:
   a. Ask your partner to breathe in through the nostrils and exhale through the mouth.
   b. Remove the stopper from one of the bottles, and hold it about 4 cm under your partner's nostrils.
   c. Keep track of the time that passes until your partner is no longer able to detect the odor of the substance.
   d. Record the result in Part B of the laboratory assessment.
   e. Wait 5 minutes and repeat this procedure, using a different bottled substance.
   f. Test a third substance in the same manner.
   g. Record the results as before.
6. Complete Part B of the laboratory assessment.

FIGURE 34.1 Structures associated with with smell.

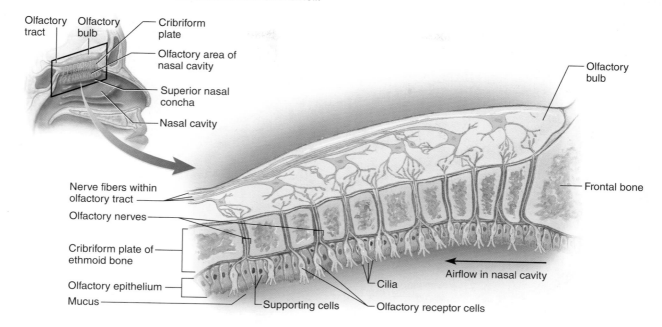

Demonstration Activity

Observe the olfactory epithelium in the demonstration microscope. The olfactory receptor cells are spindle-shaped, bipolar neurons with spherical nuclei. They also have six to eight cilia at their apical ends. The supporting cells are pseudostratified columnar epithelial cells. However, in this region the tissue lacks goblet cells. (See fig. 34.2.)

FIGURE 34.2 Olfactory receptors have cilia at their apical ends (250×).

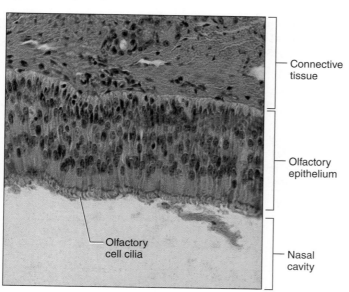

Procedure B—Sense of Taste (Gustation)

1. Reexamine the introduction to this laboratory exercise and study figure 34.3.
2. Complete Part C of the laboratory assessment.
3. Map the distribution of the receptors for the primary taste sensations on your partner's tongue. To do this, follow these steps:
   a. Ask your partner to rinse his or her mouth with water and then partially dry the surface of the tongue with a paper towel.
   b. Moisten a clean cotton swab with 5% sucrose solution, and touch several regions of your partner's tongue with the swab.
   c. Each time you touch the tongue, ask your partner to report what sensation is experienced.
   d. Test the tip, sides, and back of the tongue in this manner.
   e. Test some other representative areas inside the oral cavity such as the cheek, gums, and roof of the mouth (hard and soft palate) for a sweet sensation.
   f. Record your partner's responses in Part D of the laboratory assessment.
   g. Have your partner rinse his or her mouth and dry the tongue again, and repeat the preceding procedure, using each of the other four test solutions—NaCl, acetic acid, quinine or Epsom salt, and MSG solution. Be sure to use a fresh swab for each test substance and dispose of used swabs and paper towels as directed.
4. Complete Part D of the laboratory assessment.

FIGURE 34.3 Structures associated with taste: (a) tongue, (b) papillae, and (c) taste bud.

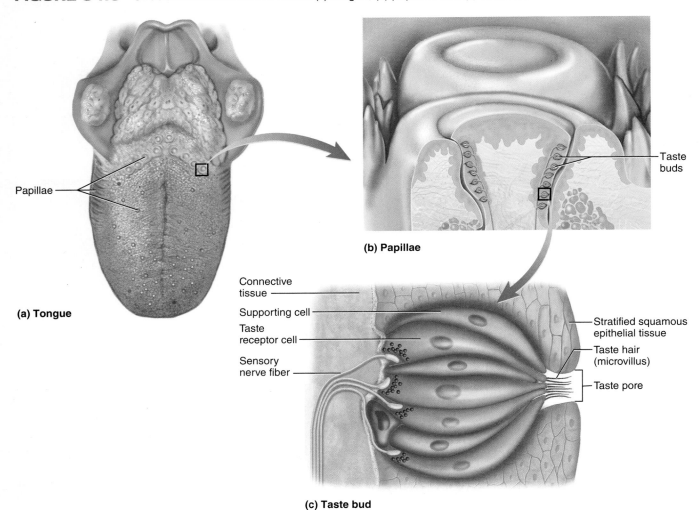

(a) Tongue

Papillae

(b) Papillae

Taste buds

Connective tissue

Supporting cell

Taste receptor cell

Sensory nerve fiber

Stratified squamous epithelial tissue

Taste hair (microvillus)

Taste pore

(c) Taste bud

Demonstration Activity

Observe the oval-shaped taste bud in the demonstration microscope. Note the surrounding epithelial cells. The taste pore, an opening into the taste bud, may be filled with taste hairs (microvilli). Within the taste bud there are supporting cells and thinner taste-receptor cells, which often have lightly stained nuclei (fig. 34.4).

Learning Extension Activity

Test your laboratory partner's ability to recognize the tastes of apple, potato, carrot, and onion. (A package of mixed flavors of LifeSavers is a good alternative.) To do this, follow these steps:

1. Have your partner close his or her eyes and hold the nostrils shut.
2. Place a small piece of one of the test substances on your partner's tongue.
3. Ask your partner to identify the substance without chewing or swallowing it.
4. Repeat the procedure for each of the other substances.

How do you explain the results of this experiment?

_____

_____

FIGURE 34.4 Taste receptors are found in taste buds (400×).

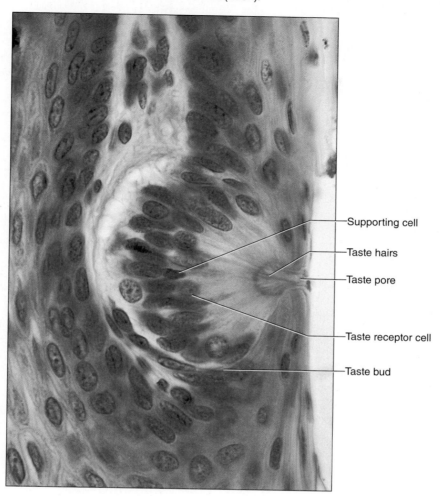

Supporting cell

Taste hairs

Taste pore

Taste receptor cell

Taste bud

NOTES

Name _____

Date _____

Section _____

The ⚠ corresponds to the indicated outcome(s) found at the beginning of the laboratory exercise.

Laboratory Assessment

34

Smell and Taste

Part A Assessments

Complete the following statements:

1. The distal ends of the olfactory neurons are covered with hairlike _____. ⚠

2. Before gaseous substances can stimulate the olfactory receptors, they must be dissolved in _____ that surrounds the cilia. ⚠

3. The axons of olfactory receptors pass through small openings in the _____ of the ethmoid bone. ⚠

4. The olfactory interpreting centers are located deep within the temporal lobes and at the base of the _____ lobes of the cerebrum. ⚠

5. Olfactory sensations usually fade rapidly as a result of _____. ⚠

6. A chemical would be considered _____ if a person lacks a particular receptor site on the cilia of the olfactory neurons. ⚠

Part B Sense of Smell Assessments

1. Record the results (as +, if recognized; as 0, if unrecognized) from the tests of odor recognition in the following table: ⚠

Substance Tested	Odor Reported	
	First Trial	Second Trial

2. Record the results of the olfactory sensory adaptation time in the following table: ⚠

Substance Tested	Adaptation Time in Seconds

345

3. Complete the following:

   a. How do you describe your partner's ability to recognize the odors of the substances you tested? **2**

   _____

   _____

   b. Compare your experimental results with those of others in the class. Did you find any evidence to indicate that individuals may vary in their abilities to recognize odors? Explain your answer. **2**

   _____

   _____

   _____

Critical Thinking Assessment

Does the time it takes for sensory adaptation to occur seem to vary with the substances tested? Explain your answer. **3**

_____

_____

_____

Part C Assessments

Complete the following statements:

1. Taste is interpreted in the _____ of the cerebrum. **2**

2. The opening to a taste bud is called a _____. **1**

3. The _____ of a taste cell are its sensitive part. **1**

4. The facial, _____, and vagus cranial nerves conduct impulses related to the sense of taste. **2**

5. Substances that stimulate taste cells bind with _____ sites on the surfaces of taste hairs. **1**

6. Sour receptors are mainly stimulated by _____. **1**

7. Salt receptors are mainly stimulated by ionized inorganic _____. **1**

8. Alkaloids usually have a _____ taste. **1**

346

Part D Sense of Taste Assessments

1. *Taste receptor distribution.* Record a + to indicate where a taste sensation seemed to originate and a *0* if no sensation occurred when the spot was stimulated.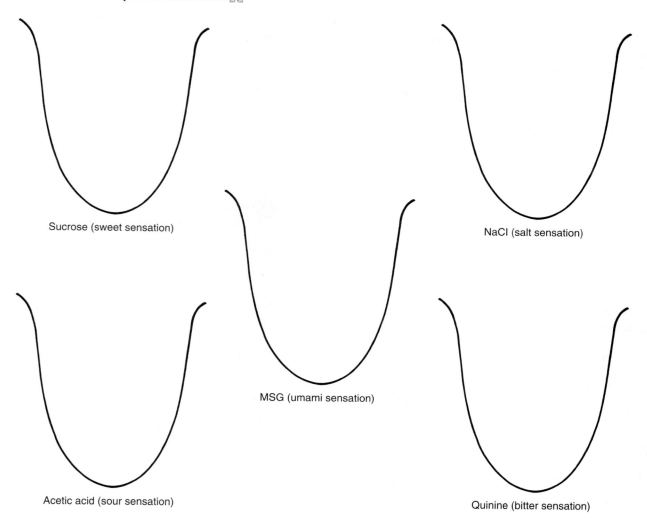

Sucrose (sweet sensation)

NaCl (salt sensation)

MSG (umami sensation)

Acetic acid (sour sensation)

Quinine (bitter sensation)

2. Complete the following:

    a. Describe how each type of taste receptor is distributed on the surface of your partner's tongue. A

    b. Describe other locations inside the mouth where any sensations of sweet, salt, sour, bitter, or umami were located. A

c. How does your taste distribution map on the tongue compare to those of other students in the class? 🔺

_____

_____

_____

_____

3. Identify the structures associated with a taste bud in figure 34.5. 🔺 🔺

FIGURE 34.5 Label this diagram of structures associated with a taste bud by placing the correct numbers in the spaces provided.

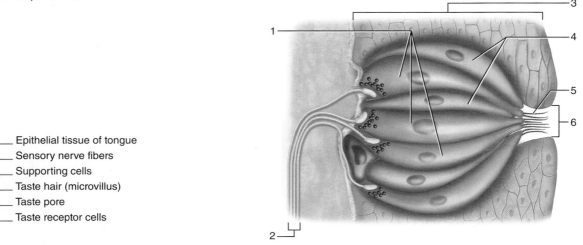

_____ Epithelial tissue of tongue
_____ Sensory nerve fibers
_____ Supporting cells
_____ Taste hair (microvillus)
_____ Taste pore
_____ Taste receptor cells

Eye Structure

Purpose of the Exercise

To review the structure and function of the eye and to dissect a mammalian eye.

Materials Needed

Dissectible eye model
Compound light microscope
Prepared microscope slide of a mammalian eye
  (sagittal section)
Sheep or beef eye (fresh or preserved)
Dissecting tray
Dissecting instruments—forceps, sharp scissors, and
  dissecting needle

For Learning Extension Activity:
Ophthalmoscope

⚠ Safety

▶ Do not allow light to reach the macula lutea during
  the eye exam for longer than one-second intervals.
▶ Wear disposable gloves when working on the eye
  dissection.
▶ Dispose of tissue remnants and gloves as instructed.
▶ Wash the dissecting tray and instruments as
  instructed.
▶ Wash your laboratory table.
▶ Wash your hands before leaving the laboratory.

Learning Outcomes

After completing this exercise, you should be able to

① Identify the major structures of an eye.
② Describe the functions of the structures of an eye.
③ Trace the structures through which light passes as it
  travels from the cornea to the retina.
④ Dissect a mammalian eye and locate its major features.

Pre-Lab

*Carefully read the introductory material and examine
the entire lab. Be familiar with the basic structures
and functions of parts of the eye from the lecture or
the textbook. Answer the pre-lab questions.*

Pre-Lab Questions: Select the correct answer for
each of the following questions:

1. The cornea and the sclera compose the _____
   layer.
   a. outer   b. middle
   c. inner   d. gel

2. We are able to see color because the eye contains
   a. melanin.   b. an optic disc.
   c. rods.   d. cones

3. The perception of vision occurs in the
   a. optic nerve.
   b. occipital lobe of the cerebrum.
   c. retina.
   d. frontal lobe of the cerebrum.

4. Which of the following is *not* part of the middle
   eye layer?
   a. choroid   b. conjunctiva
   c. ciliary body   d. iris

5. The area of our eye where visual acuity is best
   is the
   a. macula lutea.   b. optic disc.
   c. fovea centralis.   d. pupil.

6. Which of the following extrinsic skeletal muscles
   rotates the eyeball superiorly and medially?
   a. superior rectus   b. superior oblique
   c. inferior rectus   d. inferior oblique

7. The conjunctiva covers the superficial surface of
   the cornea.
   True _____   False _____

8. Tears from the lacrimal gland eventually flow
   through a nasolacrimal duct into the nasal cavity.
   True _____   False _____

The special sense of vision involves the complex structure of the eye, accessory structures associated with the orbit, the optic nerve, and the brain. The eyebrow, eyelids, conjunctiva, and lacrimal apparatus protect the eyes. The vascular conjunctiva, a mucous membrane, covers all of the anterior surface of the eyeball except the cornea, and it lines the eyelids. Its mucus prevents the eye from drying out, and the lacrimal fluid from the lacrimal (tear) gland cleanses the eye and prevents infections due to the antibacterial agent lysozyme in its secretions. Two voluntary muscles involve control of the eyelids, while six extrinsic muscles allow voluntary eye movements in multiple directions.

Three layers (tunics) compose the main structures of the eye. The outer (fibrous) layer includes the protective white sclera and a transparent anterior portion, the cornea, which allows light to enter the eye cavities. The middle (vascular) layer includes the heavily pigmented choroid, the ciliary body that controls the shape of the lens, and the iris with its smooth muscle fibers that control the amount of light passing through its central opening, the pupil. The inner layer includes pigmented and neural components of the retina and the optic nerve, which contains sensory neurons that carry nerve impulses from the rods and cones. A large posterior cavity is filled with a transparent gel, the vitreous humor. A smaller anterior cavity has anterior and posterior chambers separated at the iris, which is filled with a clear aqueous humor.

The photoreceptor cells, the rods and cones of the retina, can absorb light to produce visual images. The more numerous rods can be stimulated in dim light and are most abundant in the peripheral retina. The cones are stimulated only in brighter light, but they allow us to perceive sharper images, and they allow us to see them in color because there are three types of cones, with each type most sensitive to wavelengths peaking in the blue, green, or red portion of the visible light spectrum. Located at the posterior center of the retina is a region containing mostly cones, called the macula lutea ("yellow spot"). A tiny pit in the center of the macula lutea, the fovea centralis, contains only cones and represents our best image location. Slightly to the medial side of the macula lutea is the optic disc, where the nerve fibers of the retina leave the eye to become parts of the optic nerve. It is also referred to as the "blind spot" because rods and cones are absent at this location. Sensory impulses from the optic nerves pass through the thalamus and eventually arrive at the primary visual cortex in the occipital lobe of the cerebrum, where the perception of visual images occurs.

Procedure A—Structure and Function of the Eye

1. Reexamine the introduction to this laboratory exercise.
2. Use figures 35.1, 35.2, 35.3, and table 35.1 as references as you progress through this procedure.
3. Examine the dissectible model of the eye and locate the following features:

> **eyelid (palpebra)**—anterior protection of eyes
> **conjunctiva**—a mucous membrane
> **tarsal glands**—secrete oily lubricant
> **orbicularis oculi**—closes eyelids

FIGURE 35.1 Accessory structures of the eye (a) sagittal section of orbit (b) lacrimal apparatus.

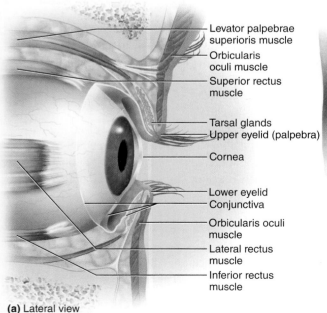

Levator palpebrae superioris muscle
Orbicularis oculi muscle
Superior rectus muscle
Tarsal glands
Upper eyelid (palpebra)
Cornea
Lower eyelid
Conjunctiva
Orbicularis oculi muscle
Lateral rectus muscle
Inferior rectus muscle

(a) Lateral view

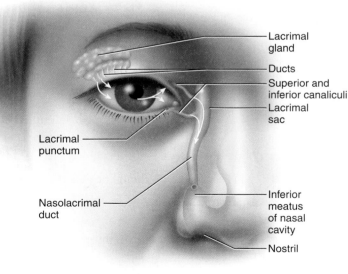

Lacrimal gland
Ducts
Superior and inferior canaliculi
Lacrimal sac
Lacrimal punctum
Nasolacrimal duct
Inferior meatus of nasal cavity
Nostril

(b) Anterior view

FIGURE 35.2 Extrinsic muscles of the right eyeball (anterior view). The arrows indicate the direction of the pull by each of the six muscles.

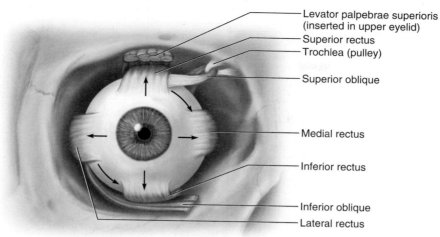

- Levator palpebrae superioris (inserted in upper eyelid)
- Superior rectus
- Trochlea (pulley)
- Superior oblique
- Medial rectus
- Inferior rectus
- Inferior oblique
- Lateral rectus

FIGURE 35.3 Structures of the eye (sagittal section).

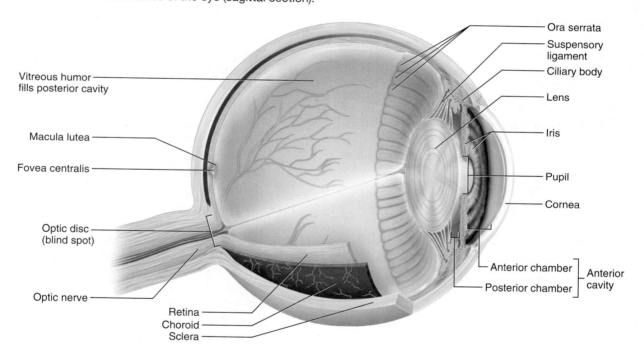

- Ora serrata
- Suspensory ligament
- Ciliary body
- Lens
- Iris
- Pupil
- Cornea
- Anterior chamber
- Posterior chamber
- Anterior cavity
- Vitreous humor fills posterior cavity
- Macula lutea
- Fovea centralis
- Optic disc (blind spot)
- Optic nerve
- Retina
- Choroid
- Sclera

levator palpebrae superioris—elevates upper eyelid

lacrimal apparatus (fig. 35.1b)
- lacrimal gland—secretes tears
- lacrimal puncta—two small pores into canaliculi
- canaliculi—passageways to lacrimal sac
- lacrimal sac—collects tears from cannaliculi
- nasolacrimal duct—drainage to nasal cavity

extrinsic muscles (fig. 35.2 and table 35.1)
- superior rectus
- inferior rectus
- medial rectus
- lateral rectus
- superior oblique
- inferior oblique

351

TABLE 35.1 Muscles Associated with the Eyelids and Eyes

Skeletal Muscles	Action	Innervation	Smooth Muscles	Action	Innervation
Muscles of the eyelids			Ciliary Muscles	Relax suspensory ligaments	Oculomotor nerve (III) parasympathetic fibers
Orbicularis oculi	Closes eye	Facial nerve (VII)	Iris, circular muscles	Constrict pupil	Oculomotor nerve (III) parasympathetic fibers
Levator palpebrae superioris	Opens eye	Oculomotor nerve (III)	Iris, radial muscles	Dilate pupil	Sympathetic fibers
Extrinsic muscles of the eyeballs					
Superior rectus	Rotates eyeball superiorly and medially	Oculomotor nerve (III)			
Inferior rectus	Rotates eyeball inferiorly and medially	Oculomotor nerve (III)			
Medial rectus	Rotates eyeball medially	Oculomotor nerve (III)			
Lateral rectus	Rotates eyeball laterally	Abducens nerve (VI)			
Superior oblique	Rotates eyeball inferiorly and laterally	Trochlear nerve (IV)			
Inferior oblique	Rotates eyeball superiorly and laterally	Oculomotor nerve (III)			

trochlea (pulley)—fibrocartilage ring for superior oblique muscle

outer (fibrous) layer (tunic)
- sclera—white protective layer of dense fibrous connective tissue
- cornea—transparent anterior portion continuous with sclera

middle (vascular) layer (tunic)
- choroid—rich blood vessel area; contains melanocytes
- ciliary body—thickest part of middle layer
  - ciliary processes—secrete aqueous humor
  - ciliary muscles—change lens shape
  - suspensory ligaments—hold lens in position
- iris—smooth muscles change pupil diameter
- pupil—opening in center of iris allowing light passage

inner layer (tunic)
- retina—sensory layer with rods and cones
  - macula lutea—high cone density area
  - fovea centralis—area of best visual acuity
  - optic disc—location where optic nerve leaves eye
  - orra serrata—serrated anterior margin of sensory retina
- optic nerve—contains sensory neurons

lens—transparent; focuses light onto retina
anterior cavity—contains aqueous humor
- anterior chamber—between cornea and iris
- posterior chamber—between iris and lens
- aqueous humor—clear, watery fluid; fills anterior and posterior chambers and pupil

posterior cavity
- vitreous humor—transparent, gel-like filler

4. Obtain a microscope slide of a mammalian eye section, and locate as many of the features in step 3 as possible.
5. Using high-power magnification, observe the posterior portion of the eye wall, and locate the sclera, choroid, and retina.
6. Using high-power magnification, examine the retina, and note its layered structure (fig. 35.4). Locate the following:

nerve fibers leading to the optic nerve— innermost layer of the retina

layer of ganglion cells

layer of bipolar neurons

nuclei of rods and cones

receptor ends of rods and cones

pigmented epithelium—outermost layer of the retina

7. Complete Part A of Laboratory Assessment 35.

FIGURE 35.4 The cells of the retina are arranged in distinct layers (75×).

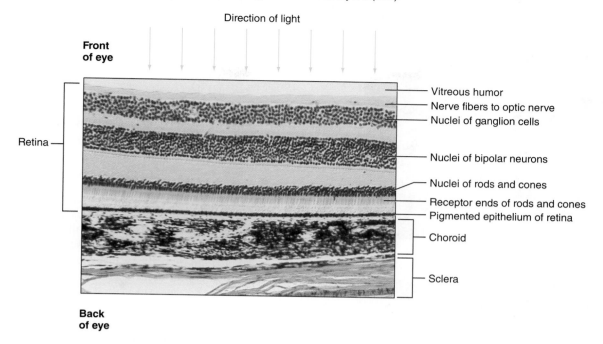

Direction of light

Front of eye

Retina

- Vitreous humor
- Nerve fibers to optic nerve
- Nuclei of ganglion cells
- Nuclei of bipolar neurons
- Nuclei of rods and cones
- Receptor ends of rods and cones
- Pigmented epithelium of retina
- Choroid
- Sclera

Back of eye

Learning Extension Activity

Use an ophthalmoscope to examine the interior of your laboratory partner's eye. This instrument consists of a set of lenses held in a rotating disc, a light source, and some mirrors that reflect the light into the test subject's eye.

The examination should be conducted in a dimly lit room. Have your partner seated and staring straight ahead at eye level. Move the rotating disc of the ophthalmoscope so that the *O* appears in the lens selection window. Hold the instrument in your right hand with the end of your index finger on the rotating disc (fig. 35.5). Direct the light at a slight angle from a distance of about 15 cm into the subject's right eye. The light beam should pass along the inner edge of the pupil. Look through the instrument, and you should see a reddish, circular area—the interior of the eye. Rotate the disc of lenses to higher values until sharp focus is achieved.

Move the ophthalmoscope to within about 5 cm of the eye being examined *being very careful that the instrument does not touch the eye,* and again rotate the lenses to sharpen the focus (fig. 35.6). Locate the optic disc and the blood vessels that pass through it. Also locate the yellowish macula lutea by having your partner stare directly into the light of the instrument (fig. 35.7). *Do not allow light to reach the macula lutea region for longer than one-second intervals.*

Examine the subject's iris by viewing it from the side and by using a lens with a +15 or +20 value.

FIGURE 35.5 The ophthalmoscope is used to examine the interior of the eye.

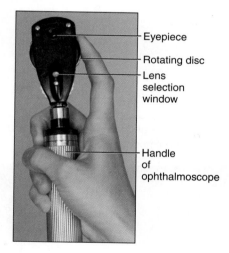

- Eyepiece
- Rotating disc
- Lens selection window
- Handle of ophthalmoscope

Procedure B—Eye Dissection

1. Obtain a mammalian eyeball, place it in a dissecting tray, and dissect it as follows:
   a. Trim away the fat and other connective tissues but leave the stubs of the *extrinsic muscles* and the *optic nerve.* This nerve projects outward from the posterior region of the eyeball.
   b. The *conjunctiva* lines the inside of the eyelid and is reflected over the anterior surface of the eye, except for the cornea. Lift some of this thin membrane away from the eye with forceps and examine it.

FIGURE 35.6 (a) Rotate the disc of lenses until sharp focus is achieved. (b) Move the ophthalmoscope to within 5 cm of the eye to examine the optic disc.

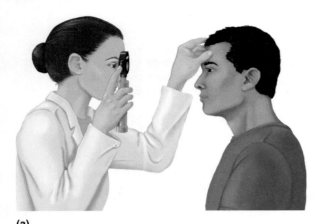

(a)

(b)

c. Locate and observe the *cornea, sclera,* and *iris.* Also note the *pupil* and its shape. The cornea from a fresh eye will be transparent; when preserved, it becomes opaque.

d. Use sharp scissors to make a frontal section of the eye. To do this, cut through the wall about 1 cm from the margin of the cornea and continue all the way around the eyeball. Try not to damage the internal structures of the eye (fig. 35.8).

e. Gently separate the eyeball into anterior and posterior portions. Usually the jellylike vitreous humor will remain in the posterior portion, and the lens may adhere to it. Place the parts in the dissecting tray with their contents facing upward (fig. 35.9).

f. Examine the anterior portion of the eye, and locate the *ciliary body,* which appears as a dark, circular structure. Also note the *iris* and the *lens* if it remained in the anterior portion. The lens is normally attached to the ciliary body by many *suspensory ligaments,* which appear as delicate, transparent threads (fig. 35.9).

g. Use a dissecting needle to gently remove the lens, and examine it. If the lens is still transparent, hold it up and look through it at something in the distance and note that the lens inverts the image. The lens of a preserved eye is usually too opaque for this experience. If the lens of the human eye becomes opaque, the defect is called a cataract (fig. 35.10).

h. Examine the posterior portion of the eye. Note the *vitreous humor.* This jellylike mass helps to hold the lens in place anteriorly and helps to hold the *retina* against the choroid.

FIGURE 35.7 Right retina as viewed through the pupil using an ophthalmoscope.

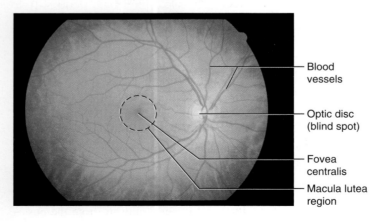

Blood vessels

Optic disc (blind spot)

Fovea centralis

Macula lutea region

FIGURE 35.8 Prepare a frontal section of the eye.

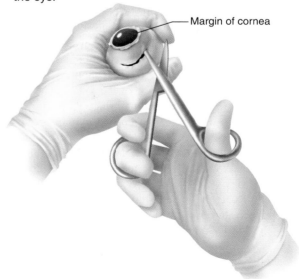

Margin of cornea

FIGURE 35.9 Internal structures of the preserved beef eye dissection.

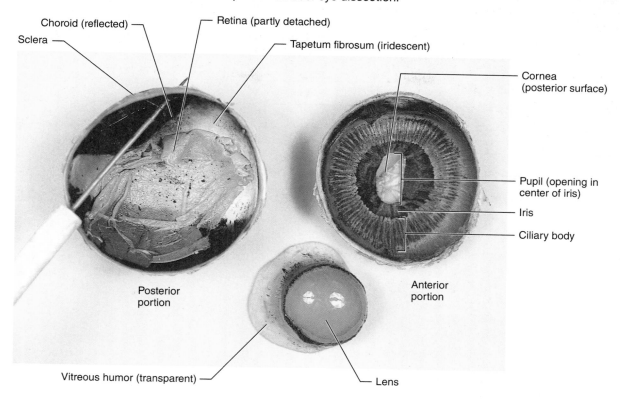

Choroid (reflected)

Sclera

Retina (partly detached)

Tapetum fibrosum (iridescent)

Cornea (posterior surface)

Pupil (opening in center of iris)

Iris

Ciliary body

Posterior portion

Anterior portion

Vitreous humor (transparent)

Lens

FIGURE 35.10 Human lens showing a cataract and one type of lens implant (intraocular lens replacement) on tips of fingers to show their relative size. A normal lens is transparent.

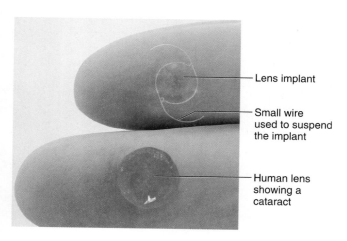

Lens implant

Small wire used to suspend the implant

Human lens showing a cataract

i. Carefully remove the vitreous humor and examine the retina. This layer will appear as a thin, nearly colorless to cream-colored membrane that detaches easily from the choroid coat. Compare the structures identified to figure 35.9.

j. Locate the *optic disc*—the point where the retina is attached to the posterior wall of the eyeball and where the optic nerve originates. There are no receptor cells in the optic disc, so this region is also called the "blind spot."

k. Note the iridescent area of the choroid beneath the retina. This colored surface in ungulates (mammals having hoofs) is called the *tapetum fibrosum*. It serves to reflect light back through the retina, an action thought to aid the night vision of some animals. The tapetum fibrosum is lacking in the human eye.

2. Discard the tissues of the eye as directed by the laboratory instructor.

3. Complete Parts B and C of the laboratory assessment.

Critical Thinking Activity

A strong blow to the head might cause the retina to detach. From observations made during the eye dissection, explain why this could happen.

_____

_____

_____

_____

Name _____

Date _____

Section _____

The 𝐀 corresponds to the indicated outcome(s) found at the beginning of the
laboratory exercise.

Eye Structure

Part A   Assessments

Identify the features of the eye indicated in figures 35.11 and 35.12. 𝟙

FIGURE 35.11   Label the structures in the sagittal section of the eye.

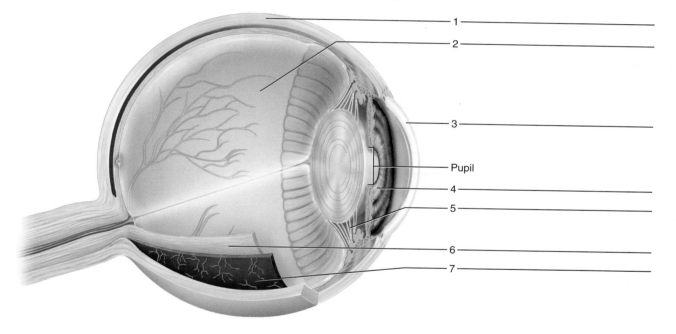

1 _____

2 _____

3 _____

Pupil

4 _____

5 _____

6 _____

7 _____

FIGURE 35.12 Sagittal section of the eye (5×). Identify the numbered features by placing the correct numbers in the spaces provided.

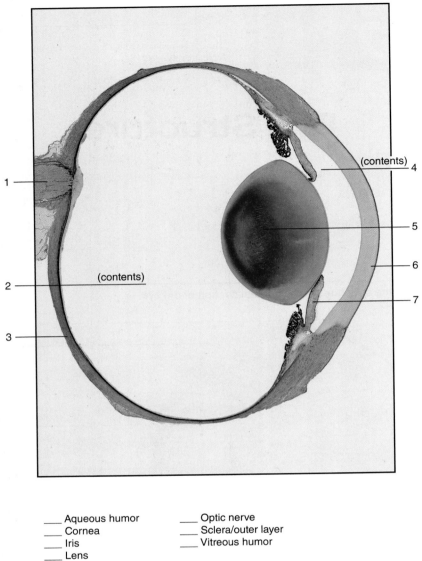

_____ Aqueous humor
_____ Cornea
_____ Iris
_____ Lens

_____ Optic nerve
_____ Sclera/outer layer
_____ Vitreous humor

Part B Assessments

Match the terms in column A with the descriptions in column B. Place the letter of your choice in the space provided. ⚠ ⚠

Column A

a. Aqueous humor
b. Choroid
c. Ciliary muscles
d. Conjunctiva
e. Cornea
f. Iris
g. Lacrimal gland
h. Optic disc
i. Retina
j. Sclera
k. Suspensory ligament
l. Vitreous humor

Column B

_____ 1. Posterior five-sixths of middle (vascular) layer

_____ 2. White part of outer (fibrous) layer

_____ 3. Transparent anterior portion of outer layer

_____ 4. Inner lining of eyelid

_____ 5. Secretes tears

_____ 6. Fills posterior cavity of eye

_____ 7. Area where optic nerve exits the eye

_____ 8. Smooth muscle that controls light entering the eye

_____ 9. Fills anterior and posterior chambers of the anterior cavity of the eye

_____ 10. Contains visual receptors called rods and cones

_____ 11. Connects lens to ciliary body

_____ 12. Cause lens to change shape

Complete the following:

13. List the structures and fluids through which light passes as it travels from the cornea to the retina. ⓷ _____

_____

_____

_____

14. List three ways in which rods and cones differ in structure or function. ⚠ _____

_____

_____

_____

_____

Part C Assessments

Complete the following:

1. Which layer/tunic of the eye was the most difficult to cut? **4** _____

2. What kind of tissue do you think is responsible for this quality of toughness? **4** _____

3. How do you compare the shape of the pupil in the dissected eye with your own pupil? **4** _____

   _____

   _____

4. Where was the aqueous humor in the dissected eye? **4** _____

   _____

5. What is the function of the dark pigment in the choroid? **2** _____

   _____

   _____

6. Describe the lens of the dissected eye. **4** _____

   _____

   _____

   _____

7. Describe the vitreous humor of the dissected eye. **4** _____

   _____

   _____

Visual Tests and Demonstrations

Purpose of the Exercise

To conduct and interpret the tests for visual acuity, astigmatism, accommodation, color vision, the blind spot, and certain reflexes of the eye.

Materials Needed

Snellen eye chart
3" × 5" card (plain)
3" × 5" card with any word typed in center
Astigmatism chart
Meterstick
Metric ruler
Pen flashlight
Ichikawa's or Ishihara's color plates for color
   blindness test

Learning Outcomes

After completing this exercise, you should be able to

1. Conduct and interpret the tests used to evaluate visual acuity, astigmatism, the ability to accommodate for close vision, and color vision.

2. Describe four conditions that can lead to defective vision.

3. Demonstrate and describe the blind spot, photopupillary reflex, accommodation pupillary reflex, and convergence reflex.

4. Summarize the results of the performed visual tests and demonstrations.

Pre-Lab

Carefully read the introductory material and examine the entire lab. Be familiar with common eye tests and defects from lecture or the textbook. Answer the pre-lab questions.

Pre-Lab Questions: Select the correct answer for each of the following questions:

1. As a person ages, the elasticity of the lens
   a. decreases.   **b.** increases.
   c. remains unchanged.   **d.** improves.

2. Visual acuity is measured using a
   a. metric ruler.   **b.** pen flashlight.
   c. Snellen eye chart.   **d.** astigmatism chart.

3. Color blindness is a characteristic in _____ males.
   a. 0%   **b.** 0.4%
   c. 7%   **d.** 10%

4. The blind spot is located at the _____ of the eye.
   a. pupil   **b.** vitreous humor
   c. aqueous humor   **d.** optic disc

5. Astigmatism results from a defect in the
   a. control of the pupil size.   **b.** curvature of the cornea or lens.
   c. rods of the retina.   **d.** cones of the retina.

6. Only males can inherit the color blindness condition.
   True _____   False _____

7. Nearsightedness is also called myopia.
   True _____   False _____

8. A convex lens can be used to correct hyperopia.
   True _____   False _____

Normal vision (emmetropia) results when light rays from objects in the external environment are refracted by the cornea and lens of the eye and focused onto the photoreceptors of the retina. If a person has a condition of *nearsightedness (myopia),* the image focuses in front of the retina because the eyeball is too long; nearby objects are clear, but distant objects are blurred. If a person has a condition of *farsightedness (hyperopia),* the image focuses behind the retina because the eyeball is too short; distant objects are clear, but nearby objects are blurred. A concave (diverging) lens is required to correct vision for nearsightedness; a convex (converging) lens is required to correct vision for farsightedness (fig. 36.1).

If a defect occurs from unequal curvatures of the cornea or lens, a condition called astigmatism results. Focusing in different planes cannot occur simultaneously. Corrective cylindrical lenses will allow proper refraction of light onto the retina to compensate for the unequal curvatures.

As part of a natural aging process, the lens's elasticity decreases, which degrades our ability to focus on nearby objects. The near point of accommodation test is used to determine the ability to accommodate. Often a person will use "reading glasses" to compensate for this condition. Hereditary defects in color vision result from a lack of certain cones needed to absorb certain wavelengths of light. Special color test plates are used to diagnose a condition of color blindness.

As you conduct this laboratory exercise, it does not matter in what order the visual tests or visual demonstra-tions are performed. The tests and demonstrations do not depend upon the results of any of the others. If you wear glasses, perform the tests with and without the corrective lenses. If you wear contact lenses, it is not necessary to run the tests under both conditions, but indicate in the laboratory assessment that all tests were performed with contact lenses in place.

Procedure A—Visual Tests

Perform the following visual tests, using your laboratory partner as a test subject. If your partner usually wears glasses, test each eye with and without the glasses.

1. *Visual acuity test.* Visual acuity (sharpness of vision) can be measured by using a Snellen eye chart (fig. 36.2). This chart consists of several sets of letters in different sizes printed on a white card. The letters near the top of the chart are relatively large, and those in each lower set become smaller. At one end of each set of letters is an acuity value in the form of a fraction. One of the sets near the bottom of the chart, for example, is marked 20/20. The normal eye can clearly see these letters from the standard distance of 20 feet and thus is said to have 20/20 vision. The letter at the top of the chart is marked 20/200. The normal eye can read letters of this size from a distance of 200 feet. Thus, an eye able to read only the top letter of the chart from a distance of 20 feet is said to have 20/200 vision. This person has less than normal

FIGURE 36.1 Normal eye compared to two common refractive defects. (a) The focal point is on the retina in the normal eye (emmetropia); (b) the focal point is behind the retina in hyperopia; (c) the focal point is before the retina in myopia. A convex lens can correct hyperopia; a concave lens can correct myopia.

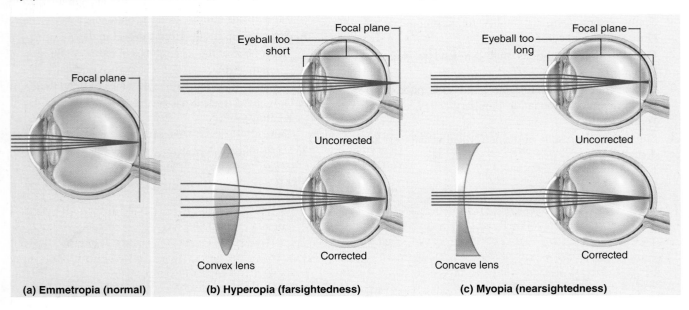

(a) Emmetropia (normal)   (b) Hyperopia (farsightedness)   (c) Myopia (nearsightedness)

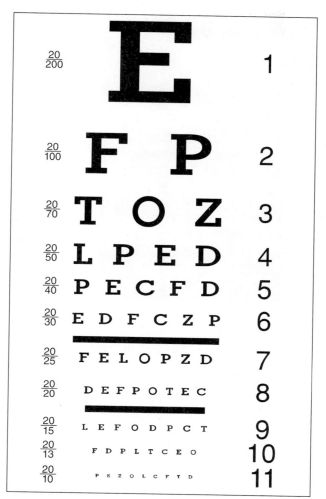

FIGURE 36.3 Astigmatism is evaluated using a chart such as this one.

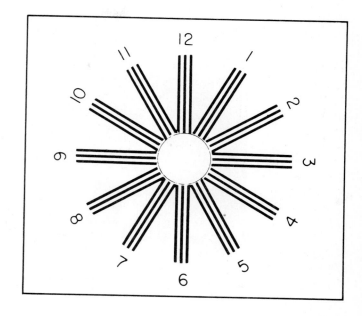

on the retina are sharply focused, and other portions are blurred. Astigmatism can be evaluated by using an astigmatism chart (fig. 36.3). This chart consists of sets of black lines radiating from a central spot like the spokes of a wheel. To a normal eye, these lines appear sharply focused and equally dark; however, if the eye has an astigmatism some sets of lines appear sharply focused and dark, while others are blurred and less dark.

To conduct the astigmatism test, follow these steps:

a. Hang the astigmatism chart on a well-illuminated wall at eye level.

b. Have your partner stand 20 feet in front of the chart, and gently cover your partner's left eye with a 3" × 5" card. Ask your partner to focus on the spot in the center of the radiating lines using their right eye and report which lines, if any, appear more sharply focused and darker.

c. Repeat the procedure, using the left eye for the test.

d. Record the results in Part A of the laboratory assessment.

3. *Accommodation test.* Accommodation is the changing of the shape of the lens that occurs when the normal eye is focused for close vision. It involves a reflex in which muscles of the ciliary body are stimulated to contract, releasing tension on the suspensory ligaments that are fastened to the lens capsule. This allows the capsule to rebound elastically, causing the surface of the lens to become more convex. The ability to accommodate is likely to decrease with age because the tissues involved tend to lose their elasticity. This common aging condition,

vision. A line of letters near the bottom of the chart is marked 20/15. The normal eye can read letters of this size from a distance of 15 feet, but a person might be able to read it from 20 feet. This person has better than normal vision.

To conduct the visual acuity test, follow these steps:

a. Hang the Snellen eye chart on a well-illuminated wall at eye level.

b. Have your partner stand 20 feet in front of the chart, gently cover your partner's left eye with a 3" × 5" card, and ask your partner to read the smallest set of letters possible using their right eye.

c. Record the visual acuity value for that set of letters in Part A of Laboratory Assessment 36.

d. Repeat the procedure, using the left eye for the test.

2. *Astigmatism test.* Astigmatism is a condition that results from a defect in the curvature of the cornea or lens. As a consequence, some portions of the image projected

FIGURE 36.4 To determine the near point of accommodation, slide the 3" × 5" card along the meterstick toward your partner's open eye until it reaches the closest location where your partner can still see the word sharply focused.

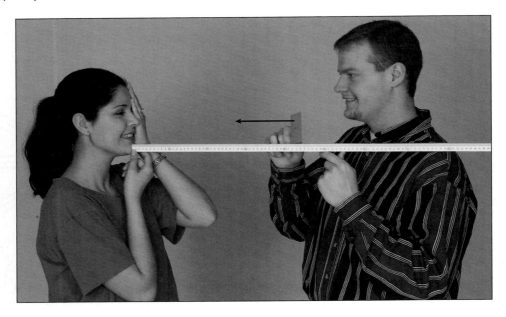

called presbyopia, can be corrected with reading glasses or bifocal lenses.

To evaluate the ability to accommodate, follow these steps:

a. Hold the end of a meterstick against your partner's chin so that the stick extends outward at a right angle to the plane of the face (fig. 36.4).

b. Have your partner close their left eye.

c. Hold a 3" × 5" card with a word typed in the center at the distal end of the meterstick.

d. Slide the card along the stick toward your partner's open right eye, and locate the *point closest to the eye* where your partner can still see the letters of the word sharply focused. This distance is called the *near point of accommodation,* and it tends to increase with age (table 36.1).

e. Repeat the procedure with the right eye closed.

f. Record the results in Part A of the laboratory assessment.

4. *Color vision test.* Some people exhibit defective color vision because they lack certain cones, usually those sensitive to the reds or greens. This trait is an X-linked (sex-linked) inheritance, so the condition is more prevalent in males (7%) than in females (0.4%). People who lack or possess decreased sensitivity to the red-sensitive cones possess protanopia color blindness; those who lack or possess decreased sensitivity to green-sensitive cones possess deuteranopia color blindness. The color blindness condition is often more of a deficiency or a weakness than one of blindness. Laboratory Exercise 61 describes the genetics for color blindness.

To conduct the color vision test, follow these steps:

a. Examine the color test plates in Ichikawa's or Ishihara's book to test for any red-green color vision deficiency. Also examine figure 36.5.

b. Hold the test plates approximately 30 inches from the subject in bright light. All responses should occur within 3 seconds.

c. Compare your responses with the correct answers in Ichikawa's or Ishihara's book. Determine the percentage of males and females in your class who exhibit any color-deficient vision. If an individual exhibits color-deficient vision, determine if the condition is protanopia or deuteranopia.

d. Record the class results in Part A of the laboratory assessment.

5. Complete Part A of the laboratory assessment.

TABLE 36.1 Near Point of Accommodation

Age (years)	Average Near Point (cm)
10	7
20	10
30	13
40	20
50	45
60	90

FIGURE 36.5 Samples of Ishihara's color plates. These plates are reproduced from *Ishihara's Tests for Colour Blindness* published by KANEHARA & CO., LTD., Tokyo, Japan, but tests for color blindness cannot be conducted with this material. For accurate testing, the original plates should be used.

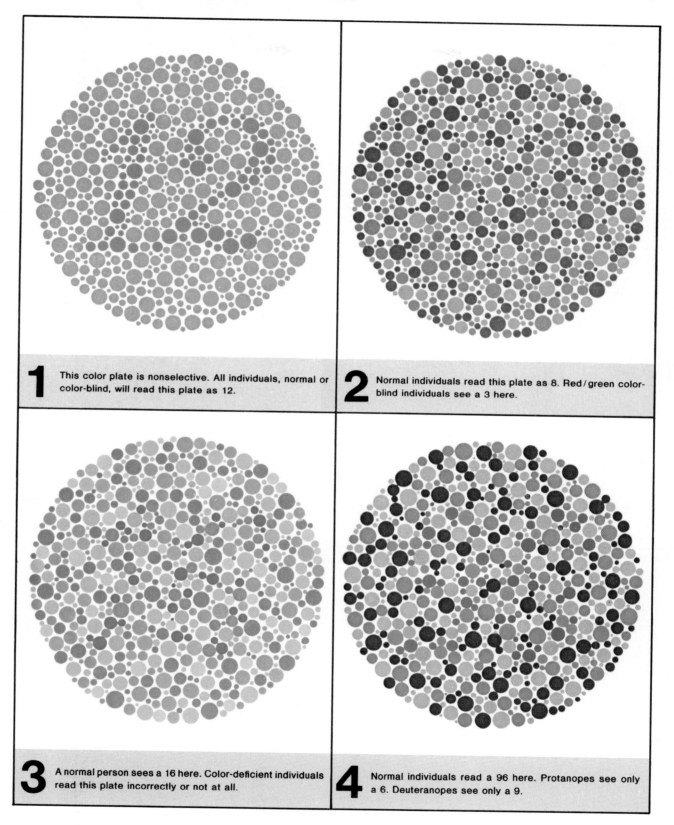

1 This color plate is nonselective. All individuals, normal or color-blind, will read this plate as 12.

2 Normal individuals read this plate as 8. Red/green color-blind individuals see a 3 here.

3 A normal person sees a 16 here. Color-deficient individuals read this plate incorrectly or not at all.

4 Normal individuals read a 96 here. Protanopes see only a 6. Deuteranopes see only a 9.

Procedure B—Visual Demonstrations

Perform the following demonstrations with the help of your laboratory partner.

1. *Blind-spot demonstration.* There are no photoreceptors in the optic disc, located where the nerve fibers of the retina leave the eye and enter the optic nerve. Consequently, this region of the retina is commonly called the *blind spot.*

   To demonstrate the blind spot, follow these steps:

   a. Close your left eye, hold figure 36.6 about 34 cm away from your face, and stare at the + sign in the figure with your right eye.

   b. Move the figure closer to your face as you continue to stare at the + until the dot on the figure suddenly disappears. This happens when the image of the dot is focused on the optic disc. Measure the right eye distance using a metric ruler or a meterstick.

   c. Repeat the procedures with your right eye closed. This time stare at the dot with your left eye, and the + will disappear when the image falls on the optic disc. Measure the distance.

   d. Record the results in Part B of the laboratory assessment.

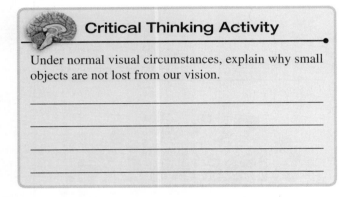

Critical Thinking Activity

Under normal visual circumstances, explain why small objects are not lost from our vision.

_____

_____

_____

_____

FIGURE 36.6 Blind-spot demonstration.

2. *Photopupillary reflex.* The smooth muscles of the iris control the size of the pupil. For example, when the intensity of light entering the eye increases, a photopupillary reflex is triggered, and the circular muscles of the iris are stimulated to contract. As a result, the size of the pupil decreases, and less light enters the eye. This reflex is an autonomic response and protects the photoreceptor cells of the retina from damage.

   To demonstrate this photopupillary reflex, follow these steps:

   a. Ask your partner to sit with his or her hands thoroughly covering his or her eyes for 2 minutes.

   b. Position a pen flashlight close to one eye with the light shining on the hand that covers the eye.

   c. Ask your partner to remove the hand quickly.

   d. Observe the pupil and note any change in its size.

   e. Have your partner remove the other hand, but keep that uncovered eye shielded from extra light.

   f. Observe both pupils and note any difference in their sizes.

3. *Accommodation pupillary reflex.* The pupil constricts as a normal accommodation reflex response to focusing on close objects. This reduces peripheral light rays and allows the refraction of light to occur closer to the center portion of the lens, resulting in a clearer image.

   To demonstrate the accommodation pupillary reflex, follow these steps:

   a. Have your partner stare for several seconds at some dimly illuminated object in the room that is more than 20 feet away.

   b. Observe the size of the pupil of one eye. Then hold a pencil about 25 cm in front of your partner's face, and have your partner stare at it.

   c. Note any change in the size of the pupil.

4. *Convergence reflex.* The eyes converge as a normal convergence response to focusing on close objects. The convergence reflex orients the eyeballs toward the object so the image falls on the fovea centralis of both eyes.

   To demonstrate the convergence reflex, follow these steps:

   a. Repeat the procedure outlined for the accommodation pupillary reflex.

   b. Note any change in the position of the eyeballs as your partner changes focus from the distant object to the pencil.

5. Complete Part B of the laboratory assessment.

Name _____

Date _____

Section _____

The Ⓐ corresponds to the indicated outcome(s) found at the beginning of the laboratory exercise.

Visual Tests and Demonstrations

Part A   Assessments

1. Visual acuity test results: Ⓐ

Eye Tested	Acuity Values
Right eye	
Right eye with glasses (if applicable)	
Left eye	
Left eye with glasses (if applicable)	

2. Astigmatism test results: Ⓐ

Eye Tested	Darker Lines
Right eye	
Right eye with glasses (if applicable)	
Left eye	
Left eye with glasses (if applicable)	

3. Accommodation test results: Ⓐ

Eye Tested	Near Point (cm)
Right eye	
Right eye with glasses (if applicable)	
Left eye	
Left eye with glasses (if applicable)	

4. Color vision test results: ⚊1⚊

Condition	Males			Females		
	Class Number	**Class Percentage**	**Expected Percentage**	**Class Number**	**Class Percentage**	**Expected Percentage**
Normal color vision			93			99.6
Deficient red-green color vision			7			0.4
Protanopia (lack red-sensitive cones)			less-frequent type			less-frequent type
Deuteranopia (lack green-sensitive cones)			more-frequent type			more-frequent type

5. Complete the following:

   a. What is meant by 20/70 vision? ⚋2⚋ _____

_____

   b. What is meant by 20/10 vision? ⚋2⚋ _____

_____

   c. What visual problem is created by astigmatism? ⚋2⚋ _____

_____

   d. Why does the near point of accommodation often increase with age? ⚋2⚋ _____

_____

   e. Describe the eye defect that causes color-deficient vision. ⚋2⚋ _____

_____

Part B Assessments

1. Blind-spot results: ⚋3⚋

   a. Right eye distance _____

   b. Left eye distance _____

2. Complete the following:

   a. Explain why an eye has a blind spot. ⚋3⚋ _____

_____

   b. Describe the photopupillary reflex. ⚋3⚋ _____

_____

c. What difference did you note in the size of the pupils when one eye was exposed to bright light and the other eye was shielded from the light? **3** _____

_____

_____

d. Describe the accommodation pupillary reflex. **3** _____

_____

e. Describe the convergence reflex. **3** _____

_____

3. Summarize the vision of the person tested based upon the visual tests and demonstrations conducted in this laboratory exercise. Include information on suspected structural defects, genetic disorders, and aging conditions. **A** _____

_____

_____

_____

_____

_____

NOTES

Ear and Hearing

Purpose of the Exercise

To review the structural and functional characteristics of the ear and to conduct some ordinary hearing tests.

Materials Needed

Dissectible ear model
Watch that ticks
Tuning fork (128 or 256 cps)
Rubber hammer
Cotton
Meterstick

For Demonstration Activities:
Compound light microscope
Prepared microscope slide of cochlea (section)
Audiometer

Learning Outcomes

After completing this exercise, you should be able to

1. Identify the major structures of the ear.
2. Trace the pathway of sound vibrations from the tympanic membrane to the hearing receptors.
3. Describe the functions of the structures of the ear.
4. Conduct four ordinary hearing tests and summarize the results.

Pre-Lab

Carefully read the introductory material and examine the entire lab. Be familiar with the basic structures and functions of the ear associated with hearing from lecture or the textbook. Answer the pre-lab questions.

Pre-Lab Questions: Select the correct answer for each of the following questions:

1. Hearing is interpreted in the _____ lobe of the cerebrum.
   - **a.** frontal
   - **b.** occipital
   - **c.** parietal
   - **d.** temporal

2. Sound loudness is measured in
   - **a.** pitch.
   - **b.** frequencies.
   - **c.** decibels.
   - **d.** hertz.

3. The middle ear bones articulate from tympanic membrane to oval window in which of the following sequences?
   - **a.** malleus, incus, stapes
   - **b.** incus, stapes, malleus
   - **c.** stapes, malleus, incus
   - **d.** malleus, stapes, incus

4. The _____ test is done to assess possible conduction deafness by comparing bone and air conduction.
   - **a.** Weber
   - **b.** Rinne
   - **c.** auditory acuity
   - **d.** sound localization

5. Which of the following structures is part of the inner ear?
   - **a.** tympanic membrane
   - **b.** cochlea
   - **c.** stapes
   - **d.** pharyngotympanic tube

6. The pharyngotympanic tube connects the outer ear to the inner ear.
   True _____ False _____

7. The cochlear nerve serves as the hearing branch of the vestibulocochlear nerve.
   True _____ False _____

8. Endolymph is located within the cochlear duct of the cochlea.
   True _____ False _____

The ear is composed of outer (external), middle, and inner (internal) parts. The large external auricle gathers sound waves (vibrations) of air and directs them into the external acoustic meatus to the tympanic membrane. The middle ear auditory ossicles conduct and amplify the waves from the tympanic membrane to the oval window. To allow the same pressure in the outer and middle ear, the pharyngotympanic (auditory) tube connects the middle ear to the pharynx. The sound waves are further conducted through the fluids (perilymph and endolymph) of the cochlea in the inner ear. Perilymph is located within the scala vestibuli and the scala tympani; endolymph occupies the cochlear duct, which contains the spiral organ. Finally, some hair cells of the spiral organ are stimulated. As certain hair cells (receptor cells) are stimulated, these receptors initiate nerve impulses to pass over the cochlear branch of the vestibulocochlear nerve into the brain. When nerve impulses arrive at the auditory cortex of the temporal lobe of the cerebrum, the impulses are interpreted and the sensations of hearing are created.

The human hearing range is from approximately 20 to 20,000 waves per second, known as hertz (Hz). We perceive these different frequencies as pitch. Different frequencies stimulate particular hair cells in specific regions of the spiral organ, allowing us to interpret different pitches. An additional quality of sound is loudness. The intensity of the sound is measured in units called decibels (dB). Permanent damage can happen to the hair cells of the spiral organ if frequent and prolonged exposures occur over 90 decibels.

Although the ear is considered here as a hearing organ, it also has important functions for equilibrium (balance). The inner ear semicircular ducts and vestibule have equilibrium functions. The ear and equilibrium will be studied in Laboratory Exercise 38.

Procedure A—Structure and Function of the Ear

1. Use figures 37.1, 37.2, and 37.3 as references as you progress through this procedure.
2. Examine the dissectible model of the ear and locate the following features:

   outer (external) ear
   - auricle (pinna)—funnel for sound waves
   - external acoustic meatus (auditory canal)—2.5 cm S-shaped tube
   - tympanic membrane (eardrum)—semitransparent; vibrates upon sound wave arrival

   middle ear
   - tympanic cavity—air-filled chamber in temporal bone
   - auditory ossicles (fig. 37.4)—transmits vibrations to oval window
     - malleus (hammer)—first ossicle to vibrate
     - incus (anvil)—middle ossicle to vibrate
     - stapes (stirrup)—final ossicle to vibrate at oval window

   - oval window—opening at lateral wall of inner ear
   - round window—membrane-covered opening of inner ear wall
   - pharyngotympanic tube (auditory tube; eustachian tube)—connects middle ear to pharynx

   inner (internal) ear
   - bony labyrinth—bony chambers filled with perilymph; includes the cochlea, semicircular canals, and vestibule
   - membranous labyrinth—membranous chambers filled with endolymph; follows pattern of bony labyrinth
   - cochlea—converts sound waves to nerve impulses at spiral organ
   - semicircular ducts—equilibrium organ
   - vestibule—equilibrium organ with two chambers
     - utricle
     - saccule

   vestibulocochlear nerve VIII—cranial nerve
   - vestibular nerve—balance branch
   - cochlear nerve—hearing branch

3. Complete Parts A, B, and C of Laboratory Assessment 37.

Demonstration Activity

Observe the section of the cochlea in the demonstration microscope. Locate one of the turns of the cochlea, and using figures 37.3 and 37.5 as a guide, identify the *scala vestibuli, cochlear duct (scala media), scala tympani, vestibular membrane, basilar membrane,* and the *spiral organ (organ of Corti).*

Procedure B—Hearing Tests

Perform the following tests in a quiet room, using your laboratory partner as the test subject.

1. *Auditory acuity test.* To conduct this test, follow these steps:
   a. Have the test subject sit with eyes closed.
   b. Pack one of the subject's ears with cotton.
   c. Hold a ticking watch close to the open ear, and slowly move it straight out and away from the ear.
   d. Have the subject indicate when the sound of the ticking can no longer be heard.
   e. Use a meterstick to measure the distance in centimeters from the ear to the position of the watch.
   f. Repeat this procedure to test the acuity of the other ear.
   g. Record the test results in Part D of the laboratory assessment.

FIGURE 37.1 Major structures of the ear. The three divisions of the ear are roughly indicated by the dashed lines.

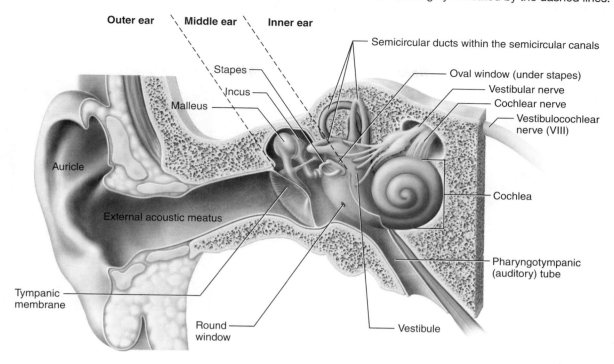

FIGURE 37.2 Structures of the middle and inner ear.

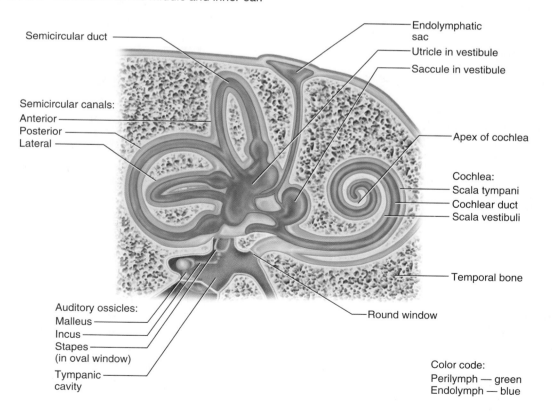

FIGURE 37.3 Anatomy of the cochlea: (a) section through cochlea; (b) view of one turn of cochlea; (c) view of spiral organ (organ of Corti).

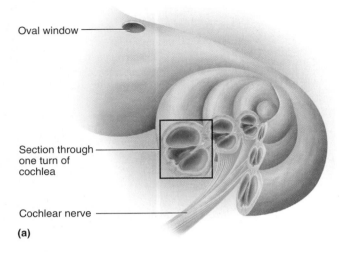

Oval window

Section through one turn of cochlea

Cochlear nerve

(a)

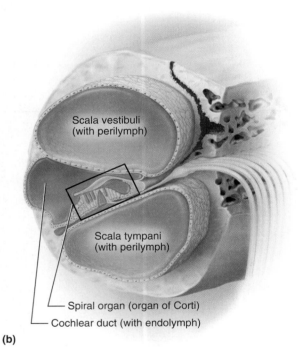

Scala vestibuli (with perilymph)

Scala tympani (with perilymph)

Spiral organ (organ of Corti)

Cochlear duct (with endolymph)

(b)

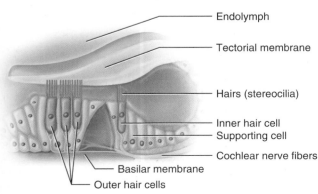

Endolymph

Tectorial membrane

Hairs (stereocilia)

Inner hair cell

Supporting cell

Cochlear nerve fibers

Basilar membrane

Outer hair cells

(c)

FIGURE 37.4 Middle ear bones (auditory ossicles) superimposed on a penny to show their relative size. The arrangement of the malleus, incus, and stapes in the enlarged photograph resembles the articulations in the middle ear.

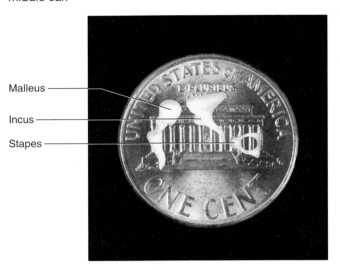

Malleus

Incus

Stapes

FIGURE 37.5 A section through the cochlea (22×).

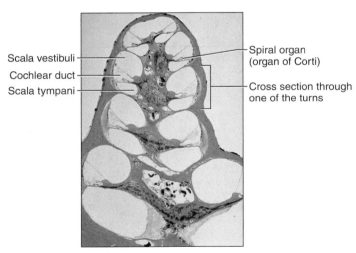

Scala vestibuli

Cochlear duct

Scala tympani

Spiral organ (organ of Corti)

Cross section through one of the turns

2. *Sound localization test.* To conduct this test, follow these steps:
   a. Have the subject sit with eyes closed.
   b. Hold the ticking watch somewhere within the audible range of the subject's ears, and ask the subject to point to the watch.
   c. Move the watch to another position and repeat the request. In this manner, determine how accurately the subject can locate the watch when it is in each of the following positions: in front of the head, behind the head, above the head, on the right side of the head, and on the left side of the head.
   d. Record the test results in Part D of the laboratory assessment.

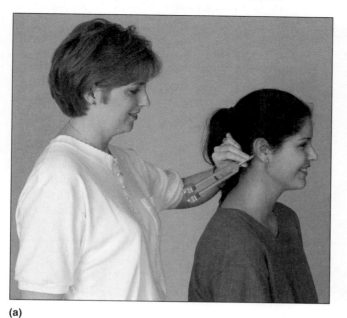

(a)

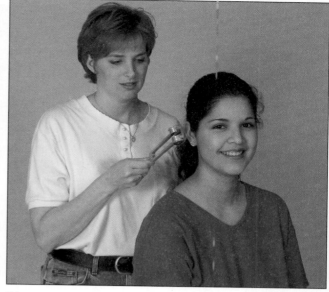

(b)

3. *Rinne test.* This test is done to assess possible conduction deafness by comparing bone and air conduction. To conduct this test, follow these steps:
   a. Obtain a tuning fork and strike it with a rubber hammer, or on the heel of your hand, causing it to vibrate.
   b. Place the end of the fork's handle against the subject's mastoid process behind one ear. Have the prongs of the fork pointed downward and away from the ear, and be sure nothing is touching them (fig. 37.6*a*). The sound sensation is that of bone conduction. If no sound is experienced, nerve deafness exists.
   c. Ask the subject to indicate when the sound is no longer heard.
   d. Then quickly remove the fork from the mastoid process and position it in the air close to the opening of the nearby external acoustic meatus (fig. 37.6*b*).

   If hearing is normal, the sound (from air conduction) will be heard again; if there is conductive impairment, the sound will not be heard. Conductive impairment involves outer or middle ear defects. Hearing aids can improve hearing for conductive deafness because bone conduction transmits the sound into the inner ear. Surgery could possibly correct this type of defect.
   e. Record the test results in Part D of the laboratory assessment.
4. *Weber test.* This test is used to distinguish possible conduction or sensory deafness. To conduct this test, follow these steps:
   a. Strike the tuning fork with the rubber hammer.
   b. Place the handle of the fork against the subject's forehead in the midline (fig. 37.7).

FIGURE 37.7 Weber test.

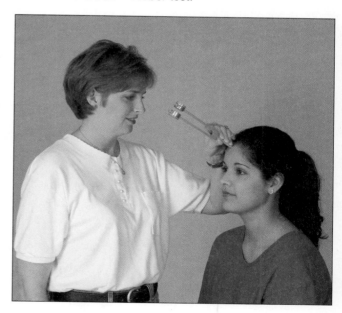

   c. Ask the subject to indicate if the sound is louder in one ear than in the other or if it is equally loud in both ears.

   If hearing is normal, the sound will be equally loud in both ears. If there is conductive impairment, the sound will appear louder in the affected ear. If some degree of sensory (nerve) deafness exists, the sound will be louder in the normal ear. The impairment involves the spiral organ or the cochlear nerve. Hearing aids will not improve sensory deafness.

d. Have the subject experience the effects of conductive impairment by packing one ear with cotton and repeating the Weber test. Usually the sound appears louder in the plugged (or impaired) ear because extraneous sounds from the room are blocked out.

e. Record the test results in Part D of the laboratory assessment.

5. Complete Part D of the laboratory assessment.

 Critical Thinking Activity

Ear structures from the outer ear into the inner ear are progressively smaller. Using results obtained from the hearing tests, explain this advantage.

_____

_____

_____

Demonstration Activity

Ask the laboratory instructor to demonstrate the use of the audiometer. This instrument produces sound vibrations of known frequencies transmitted to one or both ears of a test subject through earphones. The audiometer can be used to determine the threshold of hearing for different sound frequencies, and, in the case of hearing impairment, it can be used to determine the percentage of hearing loss for each frequency.

Name _____

Date _____

Section _____

The <u>A</u> corresponds to the indicated outcome(s) found at the beginning of the laboratory exercise.

Ear and Hearing

Part A   Assessments

1. Identify the features of the ear indicated in figures 37.8 and 37.9.

FIGURE 37.8   Label the structures associated with the ear. <u>A</u>

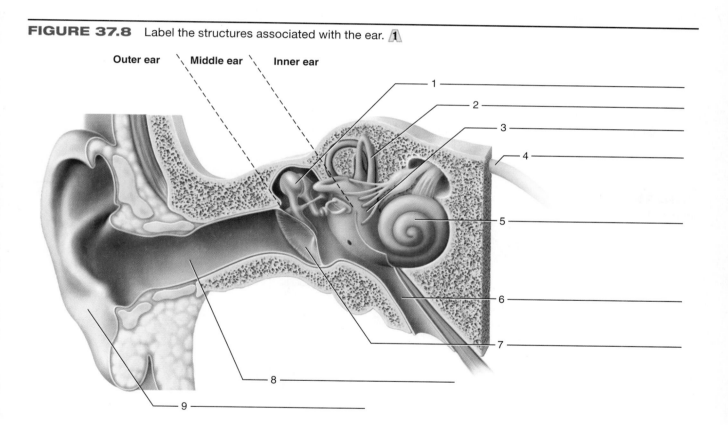

Outer ear   Middle ear   Inner ear

1 _____

2 _____

3 _____

4 _____

5 _____

6 _____

7 _____

8 _____

9 _____

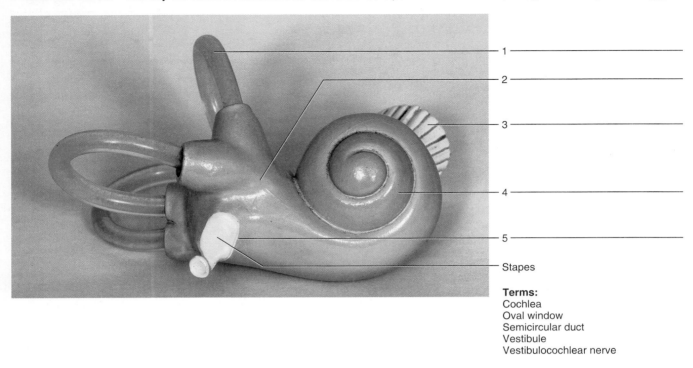

1
2
3
4
5
Stapes

Terms:
Cochlea
Oval window
Semicircular duct
Vestibule
Vestibulocochlear nerve

2. Label the structures indicated in the micrograph of the spiral organ in figure 37.10.

FIGURE 37.10   Label the structures associated with this spiral organ region of a cochlea, using the terms provided. (300×). ⚠

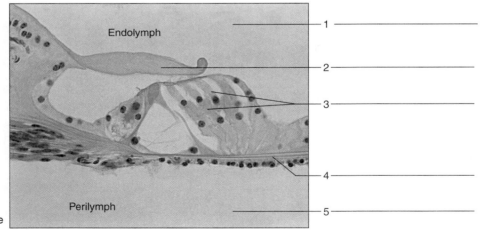

Endolymph

1
2
3
4
Perilymph
5

Terms:
Basilar membrane
Cochlear duct
Hair cells
Scala tympani
Tectorial membrane

Part B Assessments

Number the following structures (1–9) to indicate their respective positions in relation to the pathway of the sound vibrations. Assign number 1 to the most superficial portion of the outer ear. 🔼2

_____ Auricle (air vibrations within)

_____ External acoustic meatus (air vibrations within)

_____ Basilar membrane of spiral organ within cochlea

_____ Hair cells of spiral organ

_____ Incus

_____ Malleus

_____ Oval window

_____ Stapes

_____ Tympanic membrane

Part C Assessments

Match the terms in column A with the descriptions in column B. Place the letter of your choice in the space provided. 🔼1 🔼3

Column A	Column B
a. Bony labyrinth	_____ 1. Auditory ossicle attached to tympanic membrane
b. Cochlear duct	_____ 2. Air-filled space containing auditory ossicles within middle ear
c. External acoustic meatus	_____ 3. Contacts hairs of hearing receptors
d. Malleus	_____ 4. Leads from oval window to apex of cochlea and contains perilymph
e. Membranous labyrinth	_____ 5. S-shaped tube leading to tympanic membrane
f. Pharyngotympanic (auditory) tube	_____ 6. Tube within cochlea containing spiral organ and endolymph
g. Scala tympani	_____ 7. Cone-shaped, semitransparent membrane attached to malleus
h. Scala vestibuli	_____ 8. Auditory ossicle attached to oval window
i. Stapes	_____ 9. Chambers containing endolymph within bony labyrinth
j. Tectorial membrane	_____ 10. Bony chambers of inner ear in temporal bone
k. Tympanic cavity	_____ 11. Connects middle ear and pharynx
l. Tympanic membrane (eardrum)	_____ 12. Extends from apex of cochlea to round window and contains perilymph

Part D Assessments

1. Results of auditory acuity test: 🄰

Ear Tested	Audible Distance (cm)
Right	
Left	

2. Results of sound localization test: 🄰

Actual Location	Reported Location
Front of the head	
Behind the head	
Above the head	
Right side of the head	
Left side of the head	

3. Results of experiments using tuning forks: 🄰

Test	Left Ear (Normal or Impaired)	Right Ear (Normal or Impaired)
Rinne		
Weber		

4. Summarize the results of the hearing tests you conducted on your laboratory partner. 🄰

_____

_____

_____

_____

_____

_____

_____

Laboratory Exercise 38

Ear and Equilibrium

Purpose of the Exercise

To review the structure and function of the organs of equilibrium and to conduct some tests of equilibrium.

Materials Needed

Swivel chair
Bright light

For Demonstration Activity:
Compound light microscope
Prepared microscope slide of semicircular duct (cross section through ampulla)

Safety

▶ Do not pick subjects who have frequent motion sickness.
▶ Have four people surround the subject in the swivel chair in case the person falls from vertigo or loss of balance.
▶ Stop your experiment if the subject becomes nauseated.

Learning Outcomes

After completing this exercise, you should be able to

1. Locate the organs of static and dynamic equilibrium and describe their functions.
2. Explain the role of vision in the maintenance of equilibrium.
3. Conduct and record the results of the Romberg and Bárány tests of equilibrium.

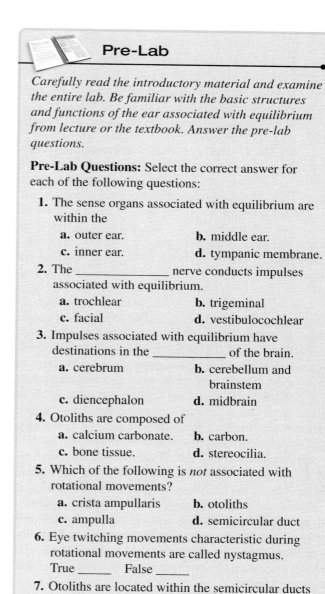

Pre-Lab

Carefully read the introductory material and examine the entire lab. Be familiar with the basic structures and functions of the ear associated with equilibrium from lecture or the textbook. Answer the pre-lab questions.

Pre-Lab Questions: Select the correct answer for each of the following questions:

1. The sense organs associated with equilibrium are within the
   a. outer ear.   b. middle ear.
   c. inner ear.   d. tympanic membrane.

2. The _____ nerve conducts impulses associated with equilibrium.
   a. trochlear   b. trigeminal
   c. facial   d. vestibulocochlear

3. Impulses associated with equilibrium have destinations in the _____ of the brain.
   a. cerebrum   b. cerebellum and brainstem
   c. diencephalon   d. midbrain

4. Otoliths are composed of
   a. calcium carbonate.   b. carbon.
   c. bone tissue.   d. stereocilia.

5. Which of the following is *not* associated with rotational movements?
   a. crista ampullaris   b. otoliths
   c. ampulla   d. semicircular duct

6. Eye twitching movements characteristic during rotational movements are called nystagmus.
   True _____   False _____

7. Otoliths are located within the semicircular ducts of the inner ear.
   True _____   False _____

The sense of equilibrium involves two sets of sensory organs. One set helps to maintain the stability of the head and body when they are motionless or during linear acceleration and produces a sense of static (gravitational) equilibrium. The other set is concerned with balancing the head and body when angular acceleration produces a sense of dynamic (rotational) equilibrium.

The sense organs associated with the sense of static equilibrium are located within the vestibules of the inner ears. Two chambers, the utricle and saccule, contain receptors called maculae. Each macula is composed of hair cells embedded within a gelatinous otolithic membrane. Embedded within the otolithic membrane are numerous tiny calcium carbonate "ear stones" called otoliths. As we change our head position, gravitational forces allow the shift of the otoliths to bend the stereocilia of the hair cells of the macula, resulting in stimulation of sensory vestibular neurons. Stimulation of the maculae also occurs during linear acceleration. Horizontal acceleration (as occurs when riding in a car) involves the utricle; vertical acceleration (as occurs when riding in an elevator) involves the saccule. As a result of static equilibrium, recognition of movements such as falling are detected and postural adjustments are accomplished.

The sense organs associated with the sense of dynamic equilibrium are located within the ampullae of the three semicircular ducts of the inner ear. Each membranous semicircular duct is located within a semicircular canal of the temporal bone. A small elevation within each ampulla possesses the crista ampullaris. Each crista ampullaris contains hair cells with stereocilia embedded within a gelatinous cap called the cupula. During rotational movements, the cupula is bent, which stimulates hair cells and sensory neurons of the vestibular nerve. Because the three semicircular ducts are in different planes, rotational movements in any direction result in stimulation of the associated hair cells. Impulses from vestibular neurons of the semicircular ducts result in reflex movements of the eye. During rotational movements, characteristic twitching movements of the eyes called nystagmus occur, often accompanied by dizziness (vertigo).

Impulses from inner ear receptors travel over the vestibular neurons of the vestibulocochlear nerve and include destinations in the brainstem and cerebellum. The sense of equilibrium works subconsciously to initiate appropriate corrections to body position and movements. Additional senses work in conjunction with the inner ear and equilibrium. Stretch receptors in muscles and tendons, touch, and vision complement the inner ear for equilibrium and proper body adjustments.

Procedure A—Organs of Equilibrium

1. Re-examine the introduction to this laboratory exercise.
2. Study figures 38.1 and 38.2, which present the structures and functions of static equilibrium.
3. Study figures 38.3 and 38.4, which present the structures and functions of dynamic equilibrium.
4. Complete Part A of Laboratory Assessment 38.

Demonstration Activity

Observe the cross section of the semicircular duct through the ampulla in the demonstration microscope. Note the crista ampullaris (fig. 38.5) projecting into the lumen of the membranous labyrinth, which in a living person is filled with endolymph. The space between the membranous and bony labyrinths is filled with perilymph.

Procedure B—Tests of Equilibrium

Use figures 38.1 and 38.3 as references as you progress through the various tests of equilibrium. Perform the following tests, using a person as a test subject who is not easily disturbed by dizziness or rotational movement. Also have some other students standing close by to help prevent the test subject from falling during the tests. *The tests should be stopped immediately if the test subject begins to feel uncomfortable or nauseated.*

1. *Vision and equilibrium test.* To demonstrate the importance of vision in the maintenance of equilibrium, follow these steps:
   a. Have the test subject stand erect on one foot for 1 minute with his or her eyes open.
   b. Observe the subject's degree of unsteadiness.
   c. Repeat the procedure with the subject's eyes closed. *Be prepared to prevent the subject from falling.*
   d. Answer the questions related to the vision and equilibrium test in Part B of the laboratory assessment.
2. *Romberg test.* The purpose of this test is to evaluate how the organs of static equilibrium in the vestibule enable one to maintain balance (fig. 38.1). To conduct this test, follow these steps:
   a. Position the test subject close to a chalkboard with the back toward the board.
   b. Place a bright light in front of the subject so that a shadow of the body is cast on the board.
   c. Have the subject stand erect with feet close together and eyes staring straight ahead for 3 minutes.
   d. During the test, make marks on the chalkboard along the edge of the shadow of the subject's shoulders to indicate the range of side-to-side swaying.
   e. Measure the maximum sway in centimeters and record the results in Part B of the laboratory assessment.
   f. Repeat the procedure with the subject's eyes closed.
   g. Position the subject so one side is toward the chalkboard.
   h. Repeat the procedure with the eyes open.
   i. Repeat the procedure with the eyes closed.
      The Romberg test is used to evaluate a person's ability to integrate sensory information from proprioceptors and receptors within the organs of static equilibrium and to relay appropriate motor impulses

FIGURE 38.1 Structures of static equilibrium; (a) utricle and saccule each contain a macula; (b) macula and otolithic membrane orientation when body is upright; (c) macula and otolithic membrane changes when the body is tilted.

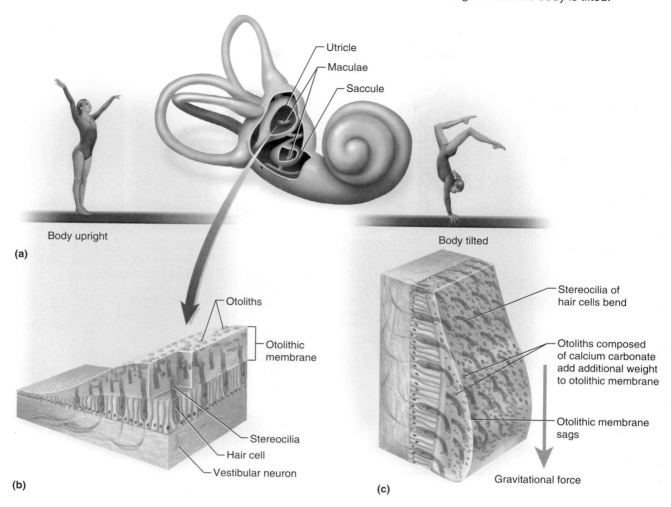

(a) Body upright

- Utricle
- Maculae
- Saccule

(b) Body tilted

- Otoliths
- Otolithic membrane
- Stereocilia
- Hair cell
- Vestibular neuron

(c)

- Stereocilia of hair cells bend
- Otoliths composed of calcium carbonate add additional weight to otolithic membrane
- Otolithic membrane sags
- Gravitational force

FIGURE 38.2 Particularly large otoliths are found in a freshwater drum (*Aplodinotus grunniens*). They have been used for lucky charms and jewelry. Human otoliths are microscopic size.

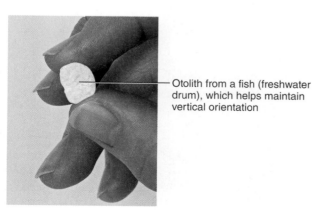

Otolith from a fish (freshwater drum), which helps maintain vertical orientation

to postural muscles. A person who shows little unsteadiness when standing with feet together and eyes open, but who becomes unsteady when the eyes are closed, has a positive Romberg test.

3. *Bárány test.* The purpose of this test is to evaluate the effects of rotational acceleration on the semicircular ducts and dynamic equilibrium (fig. 38.3). To conduct this test, follow these steps:

   a. Have the test subject sit on a swivel chair with his or her eyes open and focused on a distant object, the head tilted forward about 30°, and the hands gripped firmly to the seat. Position four people around the chair for safety. *Be prepared to prevent the subject and the chair from tipping over.*

   b. Rotate the chair ten rotations within 20 seconds.

   c. Abruptly stop the movement of the chair. The subject will still have the sensation of continuous movement and might experience some dizziness (vertigo).

FIGURE 38.3 Structures of dynamic equilibrium; (a) each semicircular duct has an ampulla containing a crista ampullaris; (b) crista ampullaris when the body is stationary; (c) changes in crista ampullaris during body rotation.

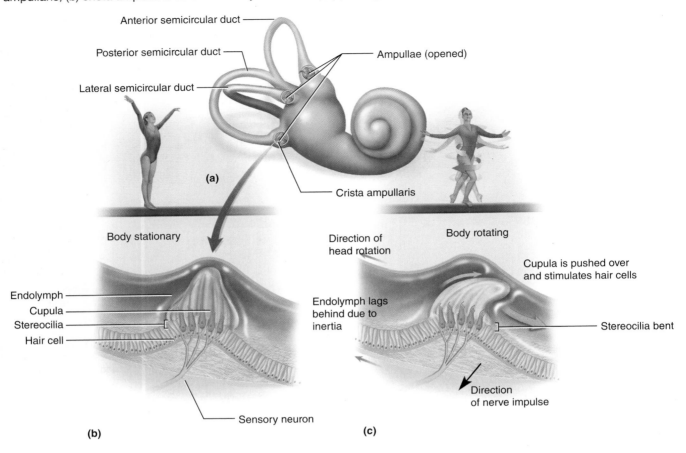

Anterior semicircular duct

Posterior semicircular duct

Lateral semicircular duct

Ampullae (opened)

(a)

Crista ampullaris

Body stationary

Endolymph
Cupula
Stereocilia
Hair cell

Sensory neuron

(b)

Body rotating

Direction of head rotation

Cupula is pushed over and stimulates hair cells

Endolymph lags behind due to inertia

Stereocilia bent

Direction of nerve impulse

(c)

FIGURE 38.4 Semicircular canal superimposed on a penny to show its relative size. A semicircular duct of the same shape would occupy the semicircular canal.

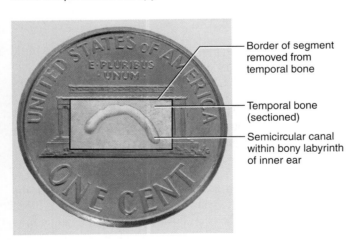

Border of segment removed from temporal bone

Temporal bone (sectioned)

Semicircular canal within bony labyrinth of inner ear

FIGURE 38.5 A micrograph of a crista ampullaris (1,400×). The crista ampullaris is located within the ampulla of each semicircular duct.

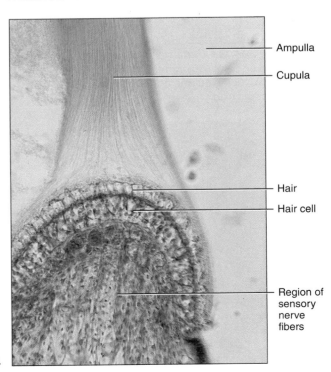

Ampulla

Cupula

Hair

Hair cell

Region of sensory nerve fibers

d. Have the subject look forward and immediately note the nature of the eye movements and their direction. (Such reflex eye twitching movements are called *nystagmus*.) Also note the time it takes for the nystagmus to cease. Nystagmus will continue until the cupula is returned to an original position.

e. Record your observations in Part B of the laboratory assessment.

f. Allow the subject several minutes of rest, then repeat the procedure with the subject's head tilted nearly 90° onto one shoulder.

g. After another rest period, repeat the procedure with the subject's head bent forward so that the chin is resting on the chest.

In this test, when the head is tilted about 30°, the lateral semicircular ducts receive maximal stimulation, and the nystagmus is normally from side to side. When the head is tilted at 90°, the superior ducts are stimulated, and the nystagmus is up and down. When the head is bent forward with the chin on the chest, the posterior ducts are stimulated, and the nystagmus is rotary.

4. Complete Part B of the laboratory assessment.

NOTES

Laboratory Assessment

38

Name _____

Date _____

Section _____

The A corresponds to the indicated outcome(s) found at the beginning of the laboratory exercise.

Ear and Equilibrium

Part A Assessments

Complete the following statements:

1. The organs of static equilibrium are located within two expanded chambers within the vestibule called the _____ and the saccule. **A**

2. All of the balance organs are found within the _____ bone of the skull. **A**

3. Otoliths are small grains composed of _____. **A**

4. Sensory impulses travel from the organs of equilibrium to the brain on vestibular neurons of the _____ nerve. **A**

5. The sensory organ of a semicircular duct lies within a swelling called the _____. **A**

6. The sensory organ within the ampulla of a semicircular duct is called a _____. **A**

7. The _____ of this sensory organ consists of a dome-shaped gelatinous cap. **A**

8. Parts of the brainstem and the _____ of the brain process impulses from the equilibrium receptors. **A**

Part B Tests of Equilibrium Assessments

1. Vision and equilibrium test results:

   a. When the eyes are open, what sensory organs provide information needed to maintain equilibrium? **2**

   _____

   _____

   b. When the eyes are closed, what sensory organs provide such information? **2**

   _____

   _____

2. Romberg test results:

   a. Record the test results in the following table: ⚠️3

Conditions	Maximal Movement (cm)
Back toward board, eyes open	
Back toward board, eyes closed	
Side toward board, eyes open	
Side toward board, eyes closed	

   b. Did the test subject's unsteadiness increase when the eyes were closed? _____ What is the significance of

    this observation? ⚠️2 _____

    _____

    _____

   c. Why would you expect a person with impairment of the organs of equilibrium to become more unsteady when the

    eyes are closed? ⚠️2 _____

    _____

    _____

3. Bárány test results:

   a. Record the test results in the following table: ⚠️3

Position of Head	Description of Eye Movements	Time for Movement to Cease
Tilted 30° forward		
Tilted 90° onto shoulder		
Tilted forward, chin on chest		

   b. Summarize the results of this test. ⚠️3

    _____

    _____

    _____

Critical Thinking Assessment

What additional sensory information would you expect persons with impairment of organs of equilibrium to use to supplement their relative lack of some sensory information?

_____

_____

_____

Endocrine Structure and Function

Purpose of the Exercise

To review the structure and function of major endocrine glands and to examine microscopically the tissues of these glands. To measure and compare the effect of cooling on metabolic rate in animals that differ in their ability to release thyroid hormone.

Materials Needed

Human torso model
Compound light microscope
Prepared microscope slides of the following:
    Pituitary gland
    Thyroid gland
    Parathyroid gland
    Adrenal gland
    Pancreas
Ph.I.L.S. 4.0

For Learning Extension Activity:

Water bath equipped with temperature control
    mechanism set at 37°C (98.6°F)
Laboratory thermometer

Learning Outcomes

After completing this exercise, you should be able to

1. Sketch and label tissue sections from the pituitary gland, thyroid gland, parathyroid glands, adrenal glands, and pancreas.

2. Name and locate the major endocrine glands.

3. Associate the principal hormones secreted by each of the major glands.

4. Explain the role of the hypothalamus in controlling the release of thyroid hormone.

5. Explain the relationship of thyroid hormone levels, metabolic rate, body temperature, and oxygen consumption.

6. Interpret changes in oxygen consumption that occur with decreases in environmental temperature.

7. Apply the concepts of thyroid hormone levels and their effects on metabolic rate, oxygen consumption, and temperature regulation to the human body.

Pre-Lab

Carefully read the introductory material and examine the entire lab. Be familiar with basic structures and functions of the pituitary, thyroid, parathyroid, adrenal, and pancreas glands from lecture or the textbook. Answer the pre-lab questions.

Pre-Lab Questions: Select the correct answer for each of the following questions:

1. Endocrine glands secrete
    a. neurotransmitters.
    b. true hormones.
    c. paracrine substances.
    d. autocrine substances.

2. True hormones influence
    a. blood cells.
    b. the secreting cells.
    c. neighboring cells.
    d. target cells some distance from the gland.

3. Which of the following endocrine glands has ducts as well as ductless functional sections?
    a. pancreas   **b.** thyroid
    c. thymus   **d.** adrenal

4. Which two endocrine glands have the closest proximity to each other?
    a. pancreas; thymus   **b.** thymus; thyroid
    c. thyroid; parathyroid   **d.** pituitary; thyroid

5. Which endocrine gland stores hormones synthesized by the hypothalamus?
    a. posterior pituitary   **b.** anterior pituitary
    c. adrenal cortex   **d.** adrenal medulla

6. The hypothalamus, pineal gland, and pituitary gland are located within the cranial cavity.
    True _____   False _____

7. Insulin is secreted by beta cells of the pancreatic islets.
    True _____   False _____

8. The adrenal medulla secretes the hormones aldosterone and cortisol.
    True _____   False _____

The endocrine system consists of ductless glands that act together with parts of the nervous system to help control body activities. The endocrine glands secrete regulatory molecules, called hormones, into the internal environment of interstitial fluid. Hormones are then transported in the blood and influence target cells, usually some distance from the original endocrine gland. At target cells, hormones bind to receptors, resulting in the activation of processes leading to functional changes within the target cells. In this way, hormones influence the rate of metabolic reactions, the transport of substances through cell membranes, the regulation of water and electrolyte balances, and many other functions.

Although some of the glands included in this laboratory exercise have ducts, the endocrine portion of the organ is ductless, and blood still transports the hormones. The pancreas, ovaries, and testes have ducts and function within systems other than the endocrine system. Additional secreted substances that are regulatory, although not considered true hormones, include paracrine and autocrine secretions. Paracrine secretions influence only neighboring cells; autocrine secretions influence only the secreting cell itself.

The nervous system and the endocrine system are both regulatory and involve chemical secretions and receptor sites. The neurons of the nervous system secrete neurotransmitters. Receptor sites for neurotransmitters are on the postsynaptic cell, resulting in rapid responses of a brief duration unless additional stimulation occurs. In contrast, the endocrine cells secrete hormones transported by the blood to receptor sites on target cells. The responses from the hormones last longer and might even continue for hours to days even without additional hormonal secretions. By controlling cellular processes, endocrine glands play important roles in the maintenance of homeostasis.

You walk outside on a cold winter day. As your body temperature begins to drop, specific physiological processes will be activated to help maintain your body temperature. Initially, increased amounts of thyroid hormone will be released from the thyroid gland. (Temperature receptors in your skin send neural signals to the hypothalamus to trigger the release of thyrotropin-releasing hormone [TRH]; TRH stimulates the anterior pituitary to release thyroid-stimulating hormone [TSH]; and TSH stimulates the thyroid gland to release the thyroid hormones, T_3 and T_4.) Thyroid hormone is able to help maintain body temperature by increasing the metabolic rate of almost all cells, especially neurons. Neurons are specifically stimulated to increase their number of Na^+/K^+ pumps. Since there are more Na^+/K^+ pumps, more energy is converted (chemical energy of ATP is converted to mechanical energy of pumping Na^+ and K^+), more heat is produced (second law of thermodynamics), and body temperature is maintained. If your body temperature continues to decrease, the hypothalamus will initiate shivering, involuntary contractions of skeletal muscle. Shivering increases body temperature as the muscles contract and generate heat. However, what occurs when you walk outside on a cold winter day if your thyroid gland is impaired and insufficient thyroid hormone is released? Will your body temperature drop or will you begin to shiver sooner to maintain body temperature?

Procedure A—Endocrine Gland Histology

1. Study the location of the major endocrine glands indicated in figure 39.1.
2. Examine the human torso model and locate the following:

   hypothalamus
   pituitary stalk (infundibulum)
   pituitary gland (hypophysis)
   - anterior lobe (adenohypophysis)
   - posterior lobe (neurohypophysis)

   thyroid gland
   parathyroid glands
   adrenal glands (suprarenal glands)
   - adrenal medulla
   - adrenal cortex

   pancreas
   pineal gland
   thymus
   ovaries
   testes

Learning Extension Activity

The secretions of endocrine glands are usually controlled by negative feedback systems. As a result, the concentrations of hormones in body fluids remain relatively stable, although they will fluctuate slightly within a normal range.

Similarly, the mechanism used to maintain the temperature of a laboratory water bath involves negative feedback. In this case, a temperature-sensitive thermostat in the water allows a water heater to operate whenever the water temperature drops below the thermostat's set point. Then, when the water temperature reaches the set point, the thermostat causes the water heater to turn off (a negative effect), and the water bath begins to cool again.

Use a laboratory thermometer to monitor the temperature of the water bath in the laboratory. Measure the temperature at regular intervals until you have recorded ten readings. What was the lowest temperature you recorded? _____ The highest temperature? _____ What was the average temperature of the water bath? _____ How is the water bath temperature control mechanism similar to a hormonal control mechanism in the body? _____

_____

_____

_____

_____

FIGURE 39.1 Major endocrine glands.

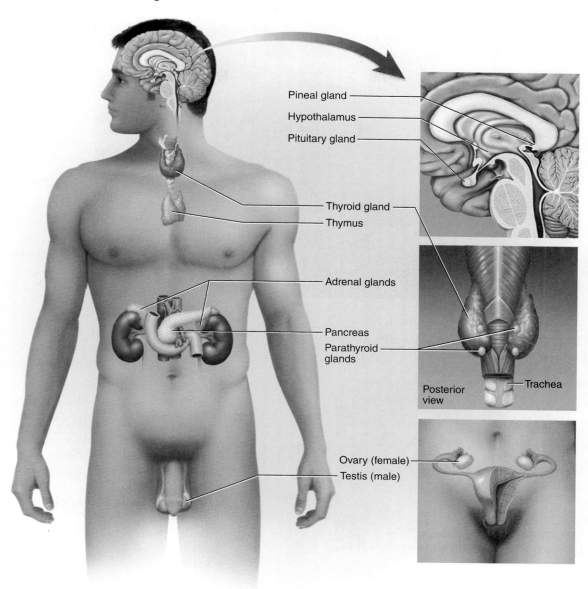

3. Examine the microscopic tissue sections of the following glands, and identify the features described:

Pituitary gland. The pituitary gland is located in the sella turcica of the sphenoid bone. To examine the pituitary tissue, follow these steps:

a. Observe the tissues using low-power magnification (fig. 39.2).

b. Locate the *pituitary stalk,* the *anterior lobe* (the largest part of the gland), and the *posterior lobe.* The anterior lobe makes hormones and appears glandular; the posterior lobe stores and releases hormones made by the hypothalamus and is nervous in appearance.

c. Observe an area of the anterior lobe with high-power magnification (fig. 39.3). Locate a cluster of

relatively large cells and identify some *acidophil cells,* which contain pink-stained granules, and some *basophil cells,* which contain blue-stained granules. These acidophil and basophil cells are the hormone-secreting cells. The hormones secreted include thyroid-stimulating hormone (TSH), adrenocorticotropic hormone (ACTH), prolactin (PRL), growth hormone (GH), follicle-stimulating hormone (FSH), and luteinizing hormone (LH).

d. Observe an area of the posterior lobe with high-power magnification (fig. 39.4). Note the numerous unmyelinated nerve fibers present in this lobe. Also locate some *pituicytes,* a type of glial cell, scattered among the nerve fibers. The posterior lobe stores and releases antidiuretic hormone (ADH) and oxytocin (OT), which are actually synthesized in the hypothalamus.

FIGURE 39.2 Micrograph of the pituitary gland (6×).

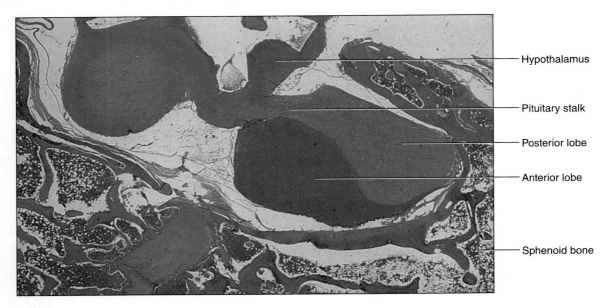

- Hypothalamus
- Pituitary stalk
- Posterior lobe
- Anterior lobe
- Sphenoid bone

FIGURE 39.3 Micrograph of the anterior lobe of the pituitary gland (240×).

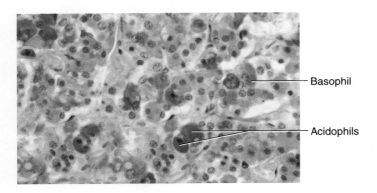

- Basophil
- Acidophils

FIGURE 39.4 Micrograph of the posterior lobe of the pituitary gland (240×).

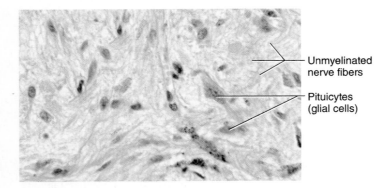

- Unmyelinated nerve fibers
- Pituicytes (glial cells)

e. Prepare labeled sketches of representative portions of the anterior and posterior lobes of the pituitary gland in Part A of Laboratory Assessment 39.

Thyroid gland. Locate the thyroid gland and associated structures (fig. 39.5). To examine the thyroid tissue, follow these steps:

a. Use low-power magnification to observe the tissue (fig. 39.6). Note the numerous *follicles,* each of which consists of a layer of cells surrounding a colloid-filled cavity.

b. Observe the tissue using high-power magnification. The cells forming the wall of a follicle are simple cuboidal epithelial cells. These cells secrete thyroid hormone (T_3 and T_4). The parafollicular cells secrete calcitonin.

FIGURE 39.5 Anterior view of thyroid gland and associated structures.

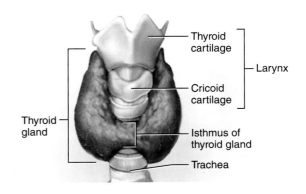

- Thyroid cartilage
- Larynx
- Cricoid cartilage
- Thyroid gland
- Isthmus of thyroid gland
- Trachea

FIGURE 39.6 Micrograph of thyroid gland (300×).

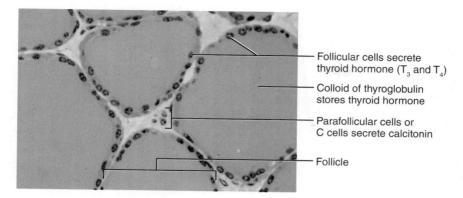

Follicular cells secrete thyroid hormone (T_3 and T_4)

Colloid of thyroglobulin stores thyroid hormone

Parafollicular cells or C cells secrete calcitonin

Follicle

c. Prepare a labeled sketch of a representative portion of the thyroid gland in Part A of the laboratory assessment.

Parathyroid gland. Typically, four parathyroid glands are attached on the posterior surface of the thyroid gland. Locate the parathyroid glands and associated structures (fig. 39.7). To examine the parathyroid tissue, follow these steps:

a. Use low-power magnification to observe the tissue (fig. 39.8). The gland consists of numerous tightly packed secretory cells.

b. Switch to high-power magnification and locate two types of cells—a smaller form (chief cells) arranged in cordlike patterns and a larger form (oxyphil cells) that have distinct cell boundaries and are present in clusters. The *chief cells* secrete parathyroid hormone

(PTH). The function of the *oxyphil cells* is not well understood.

c. Prepare a labeled sketch of a representative portion of the parathyroid gland in Part A of the laboratory assessment.

Adrenal gland. Locate an adrenal gland and associated structures (fig. 39.9). To examine the adrenal tissue, follow these steps:

a. Use low-power magnification to observe the tissue (fig. 39.10). Note the thin capsule that covers the gland. Just beneath the capsule, there is a relatively thick *adrenal cortex*. The central portion of the gland is the *adrenal medulla*. The cells of the cortex are in three poorly defined layers. Those of the outer layer (zona glomerulosa) are arranged irregularly; those of the middle layer (zona fasciculata) are in long

FIGURE 39.7 Posterior view of four parathyroid glands and associated structures.

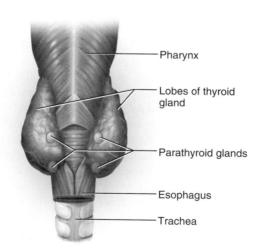

Pharynx

Lobes of thyroid gland

Parathyroid glands

Esophagus

Trachea

FIGURE 39.8 Micrograph of the parathyroid gland (65×).

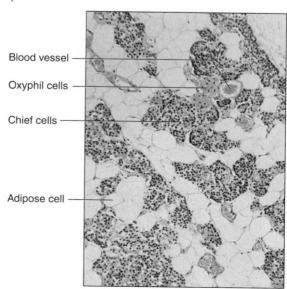

Blood vessel

Oxyphil cells

Chief cells

Adipose cell

FIGURE 39.9 Adrenal gland: (a) frontal section with associated structures; (b) diagram of the layers of the adrenal cortex and adrenal medulla.

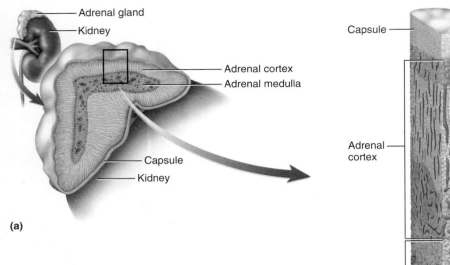

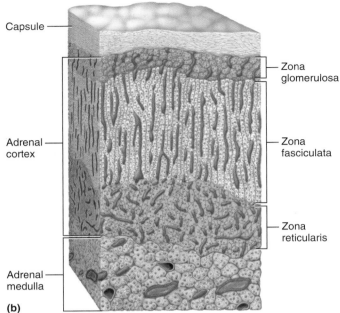

(a)

(b)

FIGURE 39.10 Micrograph of the adrenal cortex and the adrenal medulla (75×).

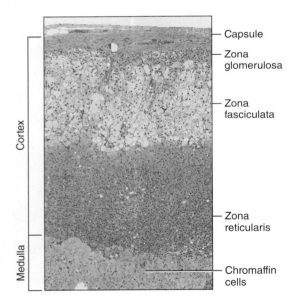

cords; and those of the inner layer (zona reticularis) are arranged in an interconnected network of cords. The most noted corticosteroids secreted from the cortex include aldosterone from the zona glomerulosa, cortisol from the zona fasciculata, and estrogens and androgens from the zona reticularis. The

cells of the medulla are relatively large, irregularly shaped, and often occur in clusters. The medulla cells secrete epinephrine and norepinephrine.

b. Using high-power magnification, observe each of the layers of the cortex and the cells of the medulla.

c. Prepare labeled sketches of representative portions of the adrenal cortex and medulla in Part A of the laboratory assessment.

Pancreas. Locate the pancreas and associated structures (fig. 39.11). To examine the tissues of the pancreas, follow these steps:

a. Use low-power magnification to observe the tissue (fig. 39.12). The gland largely consists of deeply stained exocrine cells arranged in clusters around secretory ducts. These exocrine cells (acinar cells) secrete pancreatic juice rich in digestive enzymes. There are circular masses of lightly stained cells scattered throughout the gland. These clumps of cells constitute the *pancreatic islets (islets of Langerhans),* and they represent the endocrine portion of the pancreas. The beta cells secrete insulin, the alpha cells secrete glucagon, and the delta cells secrete somatostatin.

b. Examine an islet, using high-power magnification. Unless special stains are used, it is not possible to distinguish alpha, beta, and delta cells.

c. Prepare a labeled sketch of a representative portion of the pancreas in Part A of the laboratory assessment.

4. Complete Parts B and C of the laboratory assessment.

FIGURE 39.11   (a) Pancreas and associated structures. (b) Diagram of a pancreatic islet.

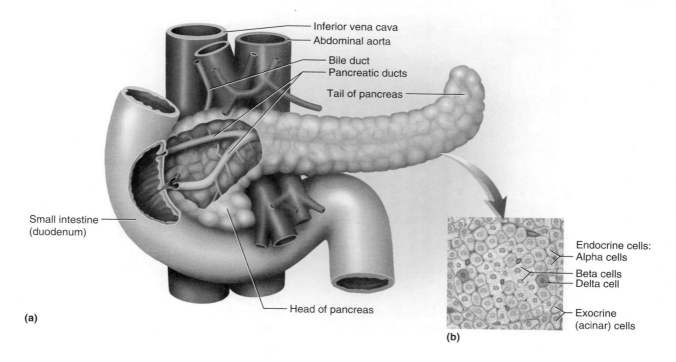

- Inferior vena cava
- Abdominal aorta
- Bile duct
- Pancreatic ducts
- Tail of pancreas
- Small intestine (duodenum)
- Head of pancreas

Endocrine cells:
- Alpha cells
- Beta cells
- Delta cell

Exocrine (acinar) cells

(a)

(b)

FIGURE 39.12   Micrograph of the pancreas (200×).

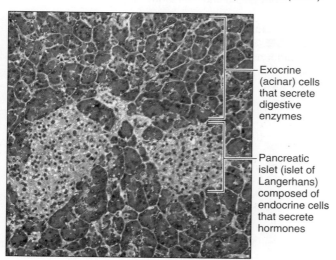

- Exocrine (acinar) cells that secrete digestive enzymes
- Pancreatic islet (islet of Langerhans) composed of endocrine cells that secrete hormones

Procedure B—Ph.I.L.S. Lesson 19 Endocrine Function: Thyroid Gland and Metabolic Rate

1. Open Exercise 19, Endocrine Function: Thyroid Gland and Metabolic Rate.
2. Read the objectives and introduction and take the pre-lab quiz.

3. After completing the pre-lab quiz, read through the wet lab. *Be sure to click open and view the videos that are indicated in red.*
4. The lab exercise will open when you click Continue after completing the wet lab (fig. 39.13).

Weighing the Mouse

5. Click the power switch to turn on the scale and then click the power switch to turn on the thermostat.
6. Click Tare to set scale to zero.
7. To weigh a mouse, click on one of the mice and drag it to the scale.

Setting Up the Chamber

8. After weighing the mouse, place it in the chamber by clicking and dragging it to the chamber.
9. Place bubbles on the end of the calibrated tube. Be sure to touch the pipette to the calibration tube.

Measuring the Bubbles

10. Measure the initial position of the bubbles (if at the end of the tube it will be 10 mm) and record at 0:00 time in the data table. (Notice that the data table includes a change in time every 15 seconds between 0:00 and 2:00.)
11. After clicking the Start button, measure the position of the soap bubble within the calibration tube at 15-second

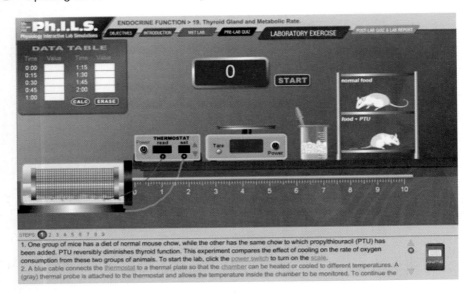

intervals (you will hear a beep) and record it in the data table. (The timer will begin when you hit Start.) You may find it more manageable to click Pause after each 15-second interval, measure, record, and then click Start to begin again.

12. When finished click Pause.

13. Click Calc to see a graph (linear regression). The Journal will open. See figure 39.14 as an example of a final graph.

Graph on left: Examine the linear regression graph that relates time and oxygen consumption. Does the amount of oxygen consumed for this mouse increase or decrease over time (i.e., the longer the mouse is in the chamber)?

_____ Is that because the mouse is becoming more relaxed the longer it is in the chamber?

_____

Final graph on right: Includes a graph and a table. Notice the table (with temperatures ranging from 8°C to 24°C) will have the first data point entered for the average oxygen consumed per minute (total amount of oxygen divided by 2 minutes). But no points will appear on the graph until all data points are entered.

Repeating the Experiment

14. To complete with the same animal, repeat numbers 9–13 (Ph.I.L.S. steps 3–9) but decrease the temperature 2 degrees (by clicking on the down arrow on the thermostat) prior to beginning the steps until data is recorded for *all temperatures.*

15. After returning the first animal to the cage, repeat numbers 5–13 (Ph.I.L.S. steps 3–9) for the second animal.

Interpreting Results

16. *With the graph still on the screen,* answer the questions in Part D of the laboratory assessment. If you acciden-

tally close the graph, click on the Journal panel (red rectangle at bottom right of screen).

17. Complete the Post-Lab Quiz by answering the ten questions on the computer screen.

18. Read the conclusion on the computer screen.

19. You may print the Lab Report for Endocrine Function: Thyroid Gland and Metabolic Rate.

FIGURE 39.14 Example: Graph to show the effect of thyroid function and temperature on oxygen consumption in mice.

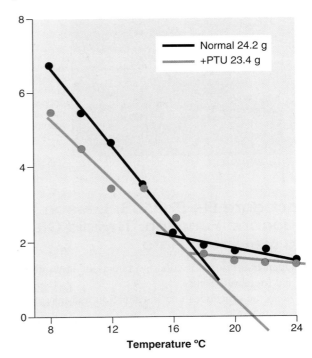

Name _____

Date _____

Section _____

The Ⓐ corresponds to the indicated outcome(s) found at the beginning of the laboratory exercise.

Endocrine Structure and Function

Part A   Assessments

Sketch and label representative portions of the following endocrine glands: Ⓐ

Pituitary gland (_____×) (anterior lobe)	Pituitary gland (_____×) (posterior lobe)
Thyroid gland (_____×)	Parathyroid gland (_____×)
Adrenal gland (_____×) (cortex and medulla)	Pancreas (_____×)

Part B Assessments

Match the endocrine gland in column A with a characteristic of the gland in column B. Place the letter of your choice in the space provided. **2** **3**

Column A

a. Adrenal cortex
b. Adrenal medulla
c. Hypothalamus
d. Pancreatic islets
e. Parathyroid gland
f. Pituitary gland
g. Thymus
h. Thyroid gland

Column B

_____ **1.** Located in sella turcica of sphenoid bone

_____ **2.** Contains alpha, beta, and delta cells

_____ **3.** Contains colloid-filled cavities

_____ **4.** Attached to posterior surface of thyroid gland

_____ **5.** Secretes corticosteroids

_____ **6.** Attached to pituitary gland by a stalk

_____ **7.** Gland inside another gland near kidneys

_____ **8.** Located in mediastinum

Part C Assessments

Complete the following:

1. Name six hormones secreted by the anterior lobe of the pituitary gland. **3** _____

_____

2. Name two hormones secreted by the posterior lobe of the pituitary gland. **3** _____

3. Name two thyroid hormones secreted from the follicular cells. **3** _____

4. Name the hormone secreted from the parafollicular cells of the thyroid gland. **3** _____

5. Name two hormones secreted by the adrenal medulla. **3** _____

6. Name the most important corticosteroid secreted by the zona fasciculata cells of the adrenal cortex. **3** _____

Part D Ph.I.L.S. Lesson 19, Endocrine Function: Thyroid Gland and Metabolic Rate Assessments

Interpreting Results

1. Click the Norm button to display the graph generated for the animal on the normal diet (normal functioning thyroid gland) to answer these questions. **4** **5** **6**

    a. How many lines are on the graph? _____

    b. Examine the change in oxygen consumption as the environmental temperature decreases from 24°C to 18°C. Is oxygen consumption increasing, decreasing, or staying the same? _____

    c. Given the relationship of oxygen consumption and metabolic rate, metabolic rate is _____ (increasing, decreasing, or staying the same).

    d. This part of the graph can be explained by _____ (the release of thyroid hormone or shivering).

    e. Examine the change in oxygen consumption as the environmental temperature decreases from 18°C to 8°C. Is oxygen consumption increasing, decreasing, or staying the same? _____

f. Given the relationship of oxygen consumption and metabolic rate, metabolic rate is _____ (increasing, decreasing, or staying the same).

g. This part of the graph can be explained by _____ (the release of thyroid hormone or shivering).

2. Click the PTU button to display the graph generated for the animal on the PTU diet (malfunctioning thyroid gland) to answer these questions. ◢4◣ ◢5◣ ◢6◣

   a. How many lines are on the graph? _____

   b. Examine the change in oxygen consumption as the environmental temperature decreases from 24°C to 18°C. Is oxygen consumption increasing, decreasing, or staying the same? _____

   c. Given the relationship of oxygen consumption and metabolic rate, metabolic rate is _____ (increasing, decreasing, or staying the same).

   d. This part of the graph can be explained by _____ (the impaired release of thyroid hormone or shivering).

   e. Examine the change in oxygen consumption as the environmental temperature decreases from 18°C to 8°C. Is oxygen consumption increasing, decreasing, or staying the same? _____

   f. Given the relationship of oxygen consumption and metabolic rate, metabolic rate is _____ (increasing, decreasing, or staying the same).

   g. This part of the graph can be explained by _____ (the impaired release of thyroid hormone or shivering).

3. Click the Both button to display the graphs generated for both groups of animals. ◢4◣ ◢5◣ ◢6◣

   a. For both animals, explain the relationship of temperature and oxygen consumption.

   _____
   _____
   _____
   _____
   _____

   When is the greatest amount of oxygen consumed—at cooler or warmer temperatures? _____

   b. Compare the red and black lines between 18°C and 24°C. Which is steeper? _____

   Which animal has a greater oxygen consumption? _____

   Which animal has a higher metabolic rate? _____

   c. The point of intersection of the two black lines (results from the mice with normal functioning thyroid gland) represents the point at which the animal begins to shiver. At what temperature does this occur? _____

   d. The point of intersection of the two red lines (results from the mice with malfunctioning thyroid gland) represents the point at which the animal begins to shiver. At what temperature does this occur? _____

   e. Which animals begin to shiver first? _____

   Explain _____
   _____
   _____
   _____

Critical Thinking Assessment 7

Would you predict your thyroid hormone level to increase, decrease, or stay the same in the winter? _____
Predict the results (increased, decreased, stay the same) for the following for a person with a hyperthyroid condition (over-active thyroid) and one with a hypothyroid condition (underactive thyroid):

Variable	Change Predicted with Hyperthyroidism	Change Predicted with Hypothyroidism
Thyroid hormone level		
Metabolic rate		
Respiratory rate		
Body temperature		
Body weight		

Which individual, one with a hyperthyroid or a hypothyroid condition, would be more likely to shiver when exposed to

decreases in environmental temperature? _____

Diabetic Physiology

Purpose of the Exercise

To observe behavior changes that occur during insulin shock and recovery from insulin shock, and to examine microscopically the pancreatic islets.

Materials Needed

500 mL beakers
Live small fish (1" to 1.5" goldfish, guppy, rosy red feeder, or other)
Small fish net
Insulin (regular U-100) (HumulinR in 10 mL vials has 100 units/mL; store in refrigerator—do not freeze)
Syringes for U-100 insulin
10% glucose solution
Clock with second hand or timer
Compound light microscope
Prepared microscope slides of the following:
   Normal human pancreas (stained for alpha and beta cells)
   Human pancreas of a diabetic

Safety

▶ Review all the safety guidelines inside the front cover.
▶ Wear disposable gloves when handling the fish and syringes.
▶ Dispose of the used syringe and needle in the puncture-resistant container.
▶ Wash your hands before leaving the laboratory.

Learning Outcomes

After completing this exercise, you should be able to

① Compare the causes, symptoms, and treatments for type 1 and type 2 diabetes mellitus.

② Examine and record the behavioral changes caused by insulin shock.

③ Examine and record the recovery from insulin shock when sugar is provided to cells.

④ Distinguish tissue sections of normal pancreatic islets from those indicating diabetes mellitus, and sketch both types.

Pre-Lab

Carefully read the introductory material and examine the entire lab. Be familiar with diabetes mellitus from lecture or the textbook. Answer the pre-lab questions.

Pre-Lab Questions: Select the correct answer for each of the following questions:

1. The _____ cells of pancreatic islets produce insulin.
     a. alpha   **b.** beta
     c. delta   **d.** F

2. Type 1 diabetes mellitus usually
     a. occurs in those over 40.
     b. afflicts 90% of diabetics.
     c. has a gradual onset.
     d. has a rapid onset.

3. Most of the insulin-secreting cells of a pancreas are
     a. scattered evenly throughout the pancreatic islet.
     b. near the periphery of the pancreatic islet.
     c. near the central portion of the pancreatic islet.
     d. among the exocrine cells of the pancreas.

4. Beta cells of a pancreatic islet occupy about _____ of the islet.
     a. 1%   **b.** 5%
     c. 15%   **d.** 80%

5. Which of the following has the *least* influence in the onset of type 2 diabetes mellitus?
     a. sleep patterns   **b.** obesity
     c. heredity   **d.** lack of exercise

6. Type 1 diabetes mellitus occurs in more people than type 2 diabetes mellitus.
   True _____   False _____

7. Obesity and lack of exercise increase the risk for the onset of type 2 diabetes mellitus.
   True _____   False _____

The primary "fuel" for the mitochondria of cells is sugar (glucose). Insulin produced by the beta cells of the pancreatic islets has a primary regulation role in the transport of blood sugar across the cell membranes. Some individuals do not produce enough insulin; others might not have adequate transport of glucose into the body cells when cells lose insulin receptors. This rather common disease is called *diabetes mellitus*. Diabetes mellitus is a functional disease that might have its onset either during childhood or later in life.

The protein hormone insulin has several functions: it stimulates the liver and skeletal muscles to form glycogen from glucose; it inhibits the conversion of noncarbohydrates into glucose; it promotes the facilitated diffusion of glucose across plasma membranes of cells possessing insulin receptors (adipose tissue, cardiac muscle, and resting skeletal muscle); it decreases blood sugar (glucose); it increases protein synthesis; and it promotes fat storage in adipose cells. In a normal person, as blood sugar increases during nutrient absorption after a meal, the rising glucose levels directly stimulate beta cells of the pancreas to secrete insulin. This increase in insulin prevents blood sugar from sudden surges (hyperglycemia) by promoting glycogen production in the liver and an increase in the entry of glucose into muscle and adipose cells.

Between meals and during sleep, blood glucose levels decrease and insulin also decreases. When insulin concentration decreases, more glucose is available to enter cells that lack insulin receptors. Brain cells, liver cells, kidney cells, and red blood cells either lack insulin receptors or express limited insulin receptors, and they absorb and utilize glucose without the need of insulin. These cells depend upon blood glucose concentrations to enable cellular respiration and ATP production. When insulin is given as a medication for diabetes mellitus, blood glucose levels may drop drastically (hypoglycemia) if the individual does not eat. This can have significant negative effects on the nervous system, resulting in insulin shock.

Type 1 diabetes mellitus usually has a rapid onset relatively early in life, but it can occur in adults. It occurs when insulin is no longer produced; hence sometimes it is referred to as *insulin-dependent diabetes mellitus (IDDM)*. This autoimmune disorder occurs when antibodies from the immune system destroy beta cells, resulting in a decrease in insulin production. This disorder affects about 10 to 15% of diabetics. The treatment for type 1 diabetes mellitus is to administer insulin, usually by injection.

Type 2 diabetes mellitus has a gradual onset, usually in those over age 40. It is sometimes called *non-insulin-dependent diabetes mellitus (NIDDM)*. This disease results as body cells become less sensitive to insulin or lose insulin receptors and thus cannot respond to insulin even though insulin levels may remain normal. This situation is called *insulin resistance*. Risk factors for the onset of type 2 diabetes mellitus, which afflicts about 85 to 90% of diabetics, include heredity, obesity, and lack of exercise. Obesity is a major indicator for predicting type 2 diabetes; a weight loss of a mere 10 pounds will increase the body's sensitivity to insulin as a means of prevention. The treatments for type 2 diabetes mellitus include a diet that avoids foods that stimulate insulin production, weight control, exercise, and medications. (The causes and treatments for type 2 diabetes closely resemble those for coronary artery disease!)

Procedure A—Insulin Shock

If too much insulin is present relative to the amount of glucose available to the cells, insulin shock can occur. In this procedure, insulin shock will be created in a fish by introducing insulin into the water with a fish. Insulin will be absorbed into the blood of the gills of the fish. When insulin in the blood reaches levels above normal, rapid hypoglycemia occurs. As hypoglycemia happens from the increased insulin absorption, the brain cells obtain less of the much-needed glucose, and epinephrine is secreted. Consequently, a rapid heart rate, anxiety, sweating, mental disorientation, impaired vision, dizziness, convulsions, and possible unconsciousness are complications during insulin shock. Although these complications take place in humans, some of them are visually noticeable if a fish is in insulin shock. The most observable components of insulin shock in the fish are related to behavioral changes, including rapid and irregular movements of the entire fish, along with increased gill cover and mouth movements.

1. Reexamine the introduction to this laboratory exercise.
2. Complete the table in Part A of Laboratory Assessment 40.
3. A fish can be used to observe normal behavior, induced insulin shock, and recovery from insulin shock. The pancreatic islets are located in the pyloric ceca or other scattered regions in fish. The behavioral changes of fish when they experience insulin shock and recovery mimic those of humans. Observe a fish in 200 mL of aquarium water in a 500 mL beaker. Make your observations of swimming, gill cover (operculum), and mouth movements for 5 minutes. Record your observations in Part B of the laboratory assessment.
4. Add 200 units of room-temperature insulin slowly, using the syringe, into the water with the fish to induce insulin shock. Record the time or start the timer when the insulin is added. Watch for changes of the swimming, gill cover, and mouth movements as insulin diffuses into the blood at the gills. Signs of insulin shock might include rapid and irregular movements. Record your observations of any behavioral changes and the time of the onset in Part B of the laboratory assessment.
5. After definite behavioral changes are observed, use the small fish net to move the fish into another 500 mL beaker containing 200 mL of a 10% room-temperature glucose solution. Record the time when the fish is transferred into this container. Make observations of any recovery of normal behaviors, including the amount of time involved. Record your observations in Part B of the laboratory assessment.
6. When the fish appears to be fully recovered, return the fish to its normal container. Dispose of the syringe according to directions from your laboratory instructor.
7. Complete Part B of the laboratory assessment.

Procedure B—Pancreatic Islets

1. Examine the normal specially stained pancreas under low-power and high-power magnification (fig. 40.1). Locate the clumps of cells that constitute the *pancreatic islets (islets of Langerhans)* that represent the endocrine portions of the organ. A pancreatic islet contains four distinct cells that secrete different hormones: *alpha cells,* which secrete glucagon, *beta cells,* which secrete insulin, *delta cells,* which secrete somatostatin, and *F cells,* which secrete pancreatic polypeptide. Some of the specially stained pancreatic islets allow for distinguishing the different types of cells (figs. 40.1 and 40.2). The normal pancreatic islet has about 80% beta cells concentrated in the central region of the islet, about 15% alpha cells near the periphery of an islet, about 5% delta cells, and a small number of F cells. There are complex interactions among the four hormones affecting blood sugar, but in general, glucagon will increase blood sugar while insulin will lower blood sugar. Somatostatin acts as a paracrine secretion to inhibit alpha and beta cell secretions, and

FIGURE 40.1   Micrograph of a specially stained pancreatic islet surrounded by exocrine (acinar) cells (500×). The insulin-containing beta cells are stained dark purple.

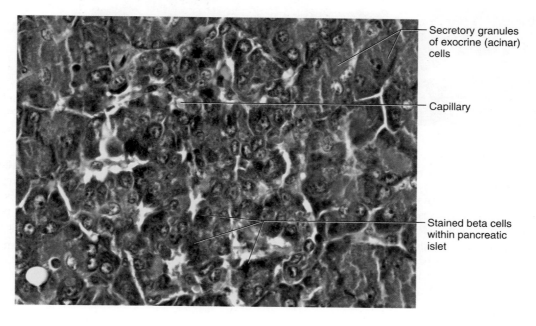

Secretory granules of exocrine (acinar) cells

Capillary

Stained beta cells within pancreatic islet

FIGURE 40.2   Micrograph of a pancreatic islet using immunofluorescence stains to visualize alpha and beta cells. Only the areas of interest are visible in the fluorescence microscope.

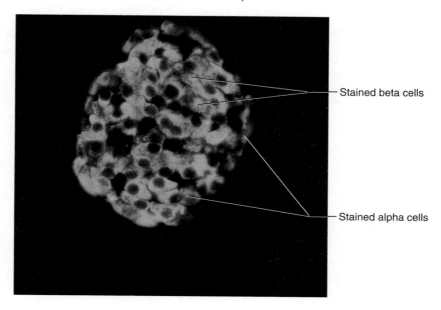

Stained beta cells

Stained alpha cells

FIGURE 40.3 Micrograph of pancreas showing indications of changes from diabetes mellitus (100×).

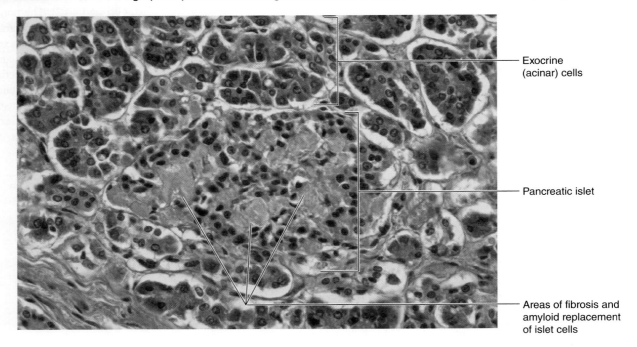

Exocrine (acinar) cells

Pancreatic islet

Areas of fibrosis and amyloid replacement of islet cells

pancreatic polypeptide inhibits secretions from delta cells and digestive enzyme secretions from the pancreatic acini.

2. Prepare a labeled sketch of a normal pancreatic islet in Part C of the laboratory assessment. Include labels for alpha cells and beta cells.

3. Examine the pancreas of a diabetic under high-power magnification (fig. 40.3). Locate a pancreatic islet and note the number of beta cells as compared to a normal number of beta cells in the normal pancreatic islet.

4. Prepare a labeled sketch of a pancreatic islet that shows some structural changes due to diabetes mellitus. Place the sketch in Part C of the laboratory assessment.

5. Complete Part C of the laboratory assessment.

Laboratory Assessment

40

Name _____

Date _____

Section _____

The ⒶThe corresponds to the indicated outcome(s) found at the beginning of the laboratory exercise.

Diabetic Physiology

Part A   Assessments

Complete the missing parts of the table of diabetes mellitus: Ⓐ1

Characteristic	Type 1 Diabetes	Type 2 Diabetes
Onset age	Early age or adult	
Onset of symptoms		Slow
Percentage of diabetics		85–90%
Natural insulin levels	Below normal	
Beta cells of pancreatic islets		Not destroyed
Pancreatic islet cell antibodies	Present	
Risk factors of having the disease	Heredity	
Typical treatments	Insulin administration	
Untreated blood sugar levels		Hyperglycemia

Part B   Assessments

1. Record your observations of normal behavior of the fish.

   a. Swimming movements: _____

   _____

   b. Gill cover movements: _____

   _____

   c. Mouth movements: _____

   _____

2. Record your observations of the fish after insulin was added to the water. Ⓐ2

   a. Recorded time that insulin was added: _____

   b. Swimming movements: _____

   _____

   c. Gill cover movements: _____

   _____

d. Mouth movements: _____

   _____

   e. Elapsed time until the insulin shock symptoms occurred: _____

3. Record your observations after the fish was transferred into a glucose solution. ⃟3

   a. Recorded time that fish was transferred into the glucose solution: _____

   b. Describe any changes in the behavior of the fish that indicates a recovery from insulin shock. _____

   _____

   _____

   _____

   c. Elapsed time until indications of a recovery from insulin shock occurred: _____

Part C Assessments

1. Sketch and label representative portions of the following pancreatic islets: ⃟4

Pancreatic islet (_____×) (normal)	Pancreatic islet (_____×) (showing changes from diabetes mellitus)

2. Describe the differences that you observed between a normal pancreatic islet and a pancreatic islet of a person with diabetes mellitus. ⃟4 _____

   _____

   _____

   _____

Critical Thinking Assessment

Justify the importance for a type 1 diabetic of regulating insulin administration, meals, and exercise. ⃟1

_____

_____

_____

_____

_____

Blood Cells

Purpose of the Exercise

To review the characteristics of blood cells, to examine them microscopically, and to perform a differential white blood cell count.

Materials Needed

Compound light microscope
Prepared microscope slides of human blood (Wright's stain)
Colored pencils

For Demonstration Activity:

Mammal blood other than human or contaminant-free human blood is suggested as a substitute for collected blood
Microscope slides (precleaned)
Sterile disposable blood lancets
Alcohol swabs (wipes)
Slide staining rack and tray
Wright's stain
Distilled water

For Learning Extension Activity:

Prepared slides of pathological blood, such as eosinophilia, leukocytosis, leukopenia, and lymphocytosis

Safety

▶ It is important that students learn and practice correct procedures for handling body fluids. Consider using either mammal blood other than human or contaminant-free blood that has been tested and is available from various laboratory supply houses. Some of the procedures might be accomplished as demonstrations only. If student blood is used, it is important that students handle only their own blood.
▶ Use an appropriate disinfectant to wash the laboratory tables before and after the procedures.
▶ Wear disposable gloves and safety glasses when handling blood samples.
▶ Clean the end of a finger with an alcohol swab before the puncture is performed.
▶ The sterile blood lancet should be used only once.

▶ Dispose of used lancets and blood-contaminated items in an appropriate container (never use the wastebasket).
▶ Wash your hands before leaving the laboratory.

Learning Outcomes

After completing this exercise, you should be able to

1 Identify and sketch red blood cells, five types of white blood cells, and platelets.

2 Describe the structure and function of red blood cells, white blood cells, and platelets.

3 Perform and interpret the results of a differential white blood cell count.

Pre-Lab

Carefully read the introductory material and examine the entire lab. Be familiar with RBCs, WBCs, and platelets from lecture or the textbook. Answer the pre-lab questions.

Pre-Lab Questions: Select the correct answer for each of the following questions:

1. Which of the following have significant functions mainly during bleeding?
   - **a.** red blood cells
   - **b.** white blood cells
   - **c.** platelets
   - **d.** plasma

2. Which of the following is among the agranulocytes?
   - **a.** monocyte
   - **b.** neutrophil
   - **c.** eosinophil
   - **d.** basophil

3. Which white blood cell has the greatest nuclear variations?
   - **a.** monocyte
   - **b.** neutrophil
   - **c.** eosinophil
   - **d.** basophil

4. A _____ lacks a nucleus.
   - **a.** red blood cell
   - **b.** lymphocyte
   - **c.** monocyte
   - **d.** basophil

5. Which cell has a large nucleus that fills most of the cell?
   - **a.** red blood cell
   - **b.** platelet
   - **c.** eosinophil
   - **d.** lymphocyte

6. Which leukocyte is the most abundant in a normal differential count?
   a. basophil   **b.** monocyte
   c. neutrophil   **d.** lymphocyte

7. Eosinophil numbers typically increase during allergic reactions.
   True _____   False _____

8. Erythrocytes are also called granulocytes because granules are visible in their cytoplasm when using Wright's stain.
   True _____   False _____

Warning

Because of the possibility of blood infections being transmitted from one student to another if blood slides are prepared in the classroom, it is suggested that commercially prepared blood slides be used in this exercise. The instructor, however, may wish to demonstrate the procedure for preparing such a slide. Observe all safety procedures for this lab.

Demonstration Activity

To prepare a stained blood slide, follow these steps:

1. Obtain two precleaned microscope slides. Avoid touching their flat surfaces.
2. Thoroughly wash hands with soap and water and dry them with paper towels.
3. Cleanse the end of the middle finger with an alcohol swab and let the finger dry in the air.
4. Remove a sterile disposable blood lancet from its package without touching the sharp end.
5. Puncture the skin on the side near the tip of the middle finger with the lancet and properly discard the lancet.
6. Wipe away the first drop of blood with the alcohol swab. Place a drop of blood about 2 cm from the end of a clean microscope slide. Cover the lanced finger location with a bandage.
7. Use a second slide to spread the blood across the first slide, as illustrated in figure 41.1. Discard the slide used for spreading the blood in the appropriate container.
8. Place the blood slide on a slide staining rack and let it dry in the air.
9. Put enough Wright's stain on the slide to cover the smear but not overflow the slide. Count the number of drops of stain that are used.
10. After 2–3 minutes, add an equal volume of distilled water to the stain and let the slide stand for

4 minutes. From time to time, gently blow on the liquid to mix the water and stain.
11. Flood the slide with distilled water until the blood smear appears light blue.
12. Tilt the slide to pour off the water, and let the slide dry in the air.
13. Examine the blood smear with low-power magnification, and locate an area where the blood cells are well distributed. Observe these cells, using high-power magnification and then an oil immersion objective if one is available.

FIGURE 41.1 To prepare a blood smear: (a) place a drop of blood about 2 cm from the end of a clean slide; (b) hold a second slide at about a 45° angle to the first one, allowing the blood to spread along its edge; (c) push the second slide over the surface of the first so that it pulls the blood with it; (d) observe the completed blood smear. The ideal smear should be 1.5 inches in length, be evenly distributed, and contain a smooth, feathered edge.

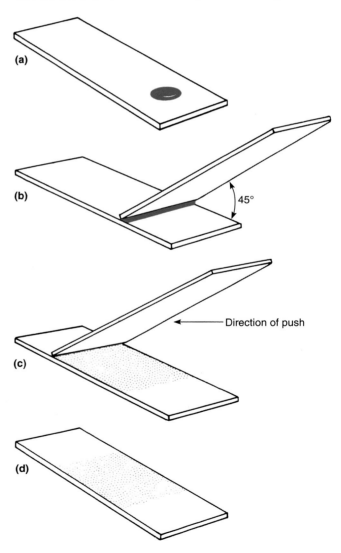

Blood is a type of connective tissue whose cells are suspended in a liquid extracellular matrix called *plasma*. Plasma is composed of water, proteins, nutrients, electrolytes, hormones, wastes, and gases. The cells, or formed elements, are mainly produced in red bone marrow, and they include *erythrocytes* (*red blood cells; RBCs*), *leukocytes* (*white blood cells; WBCs*), and some cellular fragments called *platelets* (*thrombocytes*). The formed elements compose about 45% of the total blood volume; the plasma composes approximately 55% of the blood volume.

Red blood cells contain hemoglobin and transport gases between the body cells and the lungs, white blood cells defend the body against infections, and platelets play an important role in stoppage of bleeding (hemostasis).

Clinics and hospitals test blood using a modern hematology blood analyzer. The more traditional procedures performed in these laboratory exercises will help you better understand each separate blood characteristic. On occasion, a doctor might question a blood test result from the hematology blood analyzer and request additional verification using traditional procedures.

Procedure A—Types of Blood Cells

1. Refer to figures 41.2 and 41.3 as an aid in identifying the various types of blood cells. Study the functions of the blood cells listed in table 41.1. Use the prepared slide of blood and locate each of the following:

red blood cell (erythrocyte)
white blood cell (leukocyte)
- granulocytes
  - neutrophil
  - eosinophil
  - basophil

TABLE 41.1   Cellular Components of Blood

Component	Function
Red blood cell (erythrocyte)	Transports oxygen and carbon dioxide
White blood cell (leukocyte)	Destroys pathogenic microorganisms and parasites, removes worn cells, and provides immunity
Granulocytes—have granular cytoplasm	
1. Neutrophil	Phagocytizes bacteria
2. Eosinophil	Destroys parasites and helps control inflammation and allergic reactions
3. Basophil	Releases heparin (an anticoagulant) and histamine (a blood vessel dilator)
Agranulocytes—lack granular cytoplasm	
1. Monocyte	Phagocytizes dead or dying cells and microorganisms
2. Lymphocyte	Provides immunity
Platelet (thrombocyte)	Helps control blood loss from injured blood vessels; needed for blood clotting

- agranulocytes
  - lymphocyte
  - monocyte

platelet (thrombocyte)

2. In Part A of Laboratory Assessment 41, prepare sketches of single blood cells to illustrate each type. Pay particular attention to relative size, nuclear shape, and color of granules in the cytoplasm (if present). The sketches should be accomplished using either the high-power

FIGURE 41.2   Micrograph of a blood smear using Wright's stain (500×).

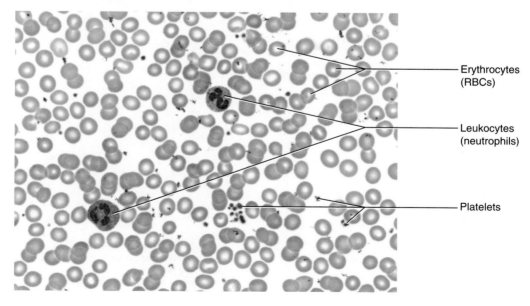

Erythrocytes (RBCs)

Leukocytes (neutrophils)

Platelets

FIGURE 41.3 Micrographs of blood cells illustrating some of the numerous variations of each type. Appearance characteristics for each cell pertain to a thin blood film using Wright's stain (1,000×).

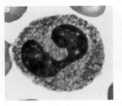 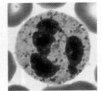 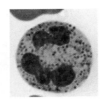

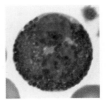 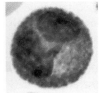

Neutrophils (3 of many variations)
- Fine light-purple granules
- Nucleus single to five lobes (highly variable)
- Immature neutrophils, called bands, have a single C-shaped nucleus
- Mature neutrophils, called segs, have a lobed nucleus
- Often called polymorphonuclear leukocytes when older

Eosinophils (3 of many variations)
- Coarse reddish granules
- Nucleus usually bilobed

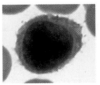 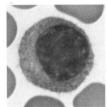

Basophils (3 of many variations)
- Coarse deep blue to almost black granules
- Nucleus often almost hidden by granules

Lymphocytes (3 of many variations)
- Slightly larger than RBCs
- Thin rim of nearly clear cytoplasm
- Nearly round nucleus appears to fill most of cell in smaller lymphocytes
- Larger lymphocytes hard to distinguish from monocytes

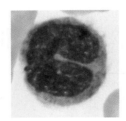 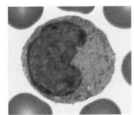 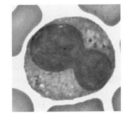

Monocytes (3 of many variations)
- Largest WBC; 2–3x larger than RBCs
- Cytoplasm nearly clear
- Nucleus round, kidney-shaped, oval, or lobed

Platelets (several variations)
- Cell fragments
- Single to small clusters

Erythrocytes (several variations)
- Lack nucleus (mature cell)
- Biconcave discs
- Thin centers appear almost hollow

objective or the oil immersion objective of the compound light microscope.

3. Complete Part B of the laboratory assessment.

Procedure B—Differential White Blood Cell Count

A differential white blood cell count is performed to determine the percentage of each of the various types of white blood cells present in a blood sample. The test is useful because the relative proportions of white blood cells may change in particular diseases as indicated in table 41.2. Neutrophils, for example, usually increase during bacterial infections, whereas eosinophils may increase during certain parasitic infections and allergic reactions.

1. To make a differential white blood cell count, follow these steps:

   a. Using high-power magnification or an oil immersion objective, focus on the cells at one end of a prepared blood slide where the cells are well distributed.

   b. Slowly move the blood slide back and forth, following a path that avoids passing over the same cells twice (fig. 41.4).

   c. Each time you encounter a white blood cell, identify its type and record it in Part C of the laboratory assessment.

   d. Continue searching for and identifying white blood cells until you have recorded 100 cells in the data table. *Percent* means "parts of 100" for each type of white blood cell, so the total number observed is equal to its percentage in the blood sample.

2. Complete Part C of the laboratory assessment.

FIGURE 41.4 Move the blood slide back and forth to avoid passing the same cells twice.

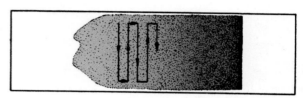

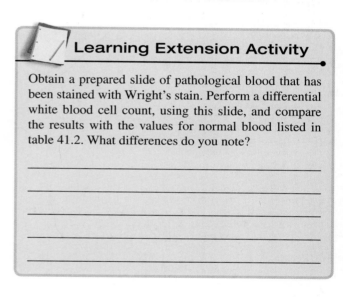

Learning Extension Activity

Obtain a prepared slide of pathological blood that has been stained with Wright's stain. Perform a differential white blood cell count, using this slide, and compare the results with the values for normal blood listed in table 41.2. What differences do you note?

_____

_____

_____

_____

_____

TABLE 41.2 Differential White Blood Cell Count

Cell Type	Normal Value (percent)	Elevated Levels May Indicate
Neutrophil	54–62	Bacterial infections, stress
Lymphocyte	25–33	Mononucleosis, whooping cough, viral infections
Monocyte	3–9	Malaria, tuberculosis, fungal infections
Eosinophil	1–3	Allergic reactions, autoimmune diseases, parasitic worms
Basophil	<1	Cancers, chicken pox, hypothyroidism

Laboratory Assessment

41

Name _____

Date _____

Section _____

The ⚠ corresponds to the indicated outcome(s) found at the beginning of the laboratory exercise.

Blood Cells

Part A   Assessments

Sketch a single blood cell of each type in the following spaces. Use colored pencils to represent the stained colors of the cells. Label any features that can be identified. ⚠

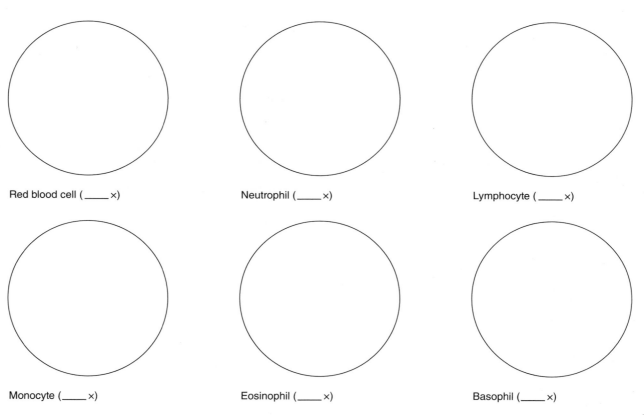

Red blood cell (_____ ×)   Neutrophil (_____ ×)   Lymphocyte (_____ ×)

Monocyte (_____ ×)   Eosinophil (_____ ×)   Basophil (_____ ×)

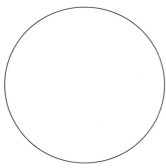

Platelet (_____ ×)

Part B Assessments

Complete the following statements:

1. Mature red blood cells are also called _____. **2**

2. The shape of a red blood cell can be described as a _____ disc. **2**

3. The functions of red blood cells are _____. **2**

4. _____ is the oxygen-carrying substance in a red blood cell. **2**

5. A mature red blood cell cannot reproduce because it lacks the _____ that was extruded during late development. **2**

6. White blood cells are also called _____. **2**

7. White blood cells with granular cytoplasm are called _____. **2**

8. White blood cells lacking granular cytoplasm are called _____. **2**

9. Polymorphonuclear leukocyte is another name for a _____ with a segmented nucleus. **2**

10. Normally, the most numerous white blood cells are _____. **2**

11. White blood cells with coarse reddish cytoplasmic granules are called _____. **2**

12. _____ are normally the least abundant of the white blood cells. **2**

13. _____ are the largest of the white blood cells. **2**

14. _____ are small agranulocytes that have relatively large, round nuclei with thin rims of cytoplasm. **2**

15. Small cell fragments that function to prevent blood loss from an injury site are called _____. **2**

Part C   Assessments

1. *Differential White Blood Cell Count Data Table.* As you identify white blood cells, record them on the table by using a tally system, such as ⊮⊪ ||. Place tally marks in the "Number Observed" column and total each of the five WBCs when the differential count is completed. Obtain a total of all five WBCs counted to determine the percent of each WBC type. ⓵

Type of WBC	Number Observed	Total	Percent
Neutrophil			
Lymphocyte			
Monocyte			
Eosinophil			
Basophil			
		Total of column	

2. How do the results of your differential white blood cell count compare with the normal values listed in table 41.2? ⓵

_____

_____

_____

_____

_____

Critical Thinking Assessment

What is the difference between a differential white blood cell count and a total white blood cell count?

_____

_____

_____

_____

_____

3. Identify the blood cells indicated in figure 41.5.

FIGURE 41.5 Label the specific blood cells on this micrograph of a stained blood smear (400×).

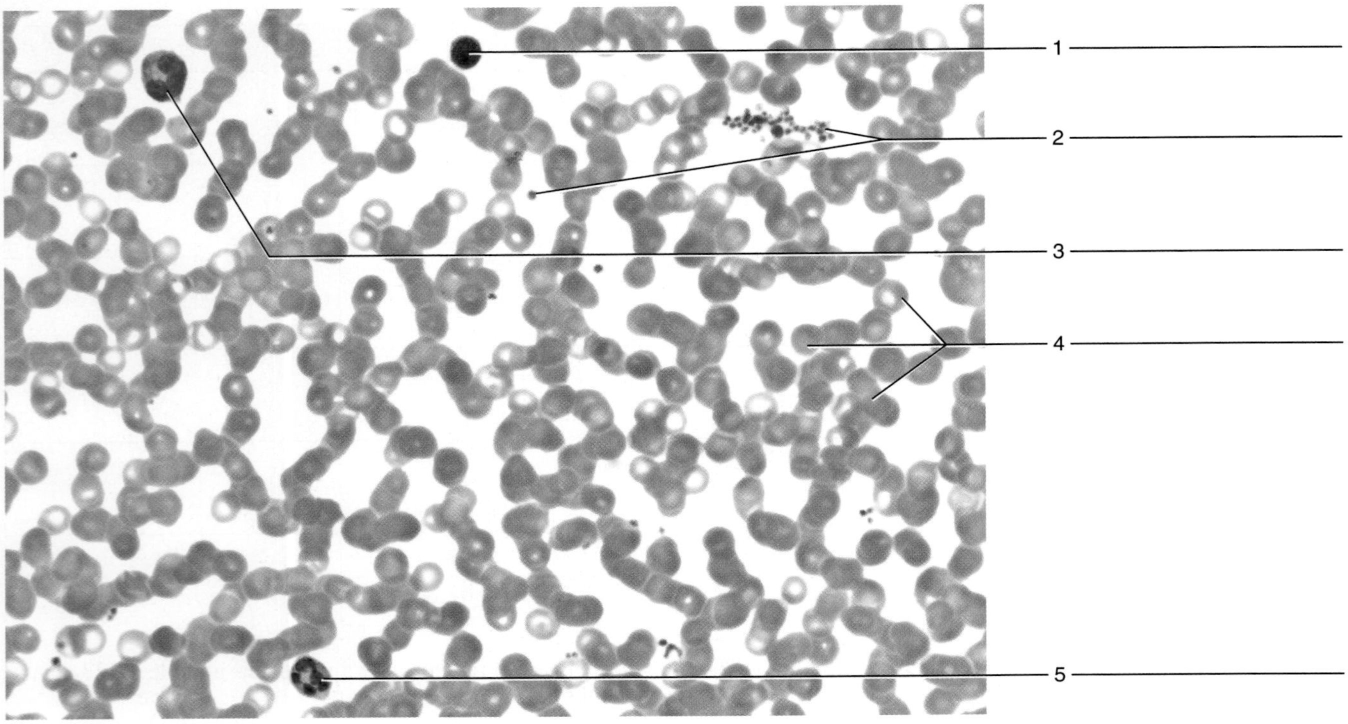

1

2

3

4

5

Blood Testing

Purpose of the Exercise

To observe and interpret the blood tests used to determine hematocrit, hemoglobin content, coagulation, and cholesterol level. To measure and compare the percent saturation of hemoglobin when the partial pressure of oxygen and the pH are changed.

Materials Needed

Mammal blood other than human or contaminant-free human blood is suggested as a substitute for collected blood

For Procedure A:
Heparinized microhematocrit capillary tube
Sealing clay (or Critocaps)
Microhematocrit centrifuge
Microhematocrit reader

For Procedure B:
Hemoglobinometer
Lens paper
Hemolysis applicator

For Alternative Hemoglobin Content Activity:
Tallquist test kit

Sterile disposable blood lancets
Alcohol swabs (wipes)
Ph.I.L.S. 4.0

For Procedure C:
Small triangular file
Capillary tube (nonheparinized)
Timer

For Procedure D:
Simulated blood kit for cholesterol determination*

* Kit contains all materials needed and is available from several biological supply companies.

⚠ Safety

▶ It is important that students learn and practice correct procedures for handling body fluids. Consider using either mammal blood other than human or contaminant-free blood that has been tested and is available from various laboratory supply houses. Some of the procedures might be accomplished as demonstrations only. If student blood is used, it is important that students handle only their own blood.
▶ Use an appropriate disinfectant to wash the laboratory tables before and after the procedures.
▶ Wear disposable gloves and safety goggles when handling blood samples.

▶ Clean the end of a finger with alcohol swabs before the puncture is performed.
▶ The sterile blood lancet should be used only once.
▶ Dispose of used lancets and blood-contaminated items in an appropriate container (never use the wastebasket).
▶ Wash your hands before leaving the laboratory.

Learning Outcomes

After completing this exercise, you should be able to

1. Determine the hematocrit, hemoglobin, coagulation, and cholesterol level in a blood sample.
2. Judge the results and significance of the blood tests compared to normal values.
3. Select the blood tests performed in this exercise that could indicate anemia.
4. Graph and explain the relationship of the percent transmittance of light to the partial pressure of oxygen.
5. Explain the relationship of percent transmittance of light to percent saturation of hemoglobin.
6. Explain the relationship of partial pressure of oxygen to percent saturation of hemoglobin.
7. Interpret the influence of pH on the oxygen dissociation curve.
8. Apply the concepts of how pH affects the percent saturation of hemoglobin to the human body.

Pre-Lab

Carefully read the introductory material and examine the entire lab. Be familiar with hematocrit, hemoglobin, coagulation, and cholesterol from lecture or the textbook. Visit www.mhhe.com/martinseries2 for LabCam videos. Answer the pre-lab questions.

Pre-Lab Questions: Select the correct answer for each of the following questions:

1. The _____ is the test to determine percentage of red blood cells.
   a. hemoglobin   **b.** hematocrit
   c. buffy coat   **d.** coagulation

As an aid in identifying various disease conditions, tests are often performed on blood to determine how its composition compares with normal values. These tests commonly include hematocrit (red blood cell percentage), hemoglobin content, coagulation, and cholesterol.

When a blood sample is collected in a heparinized capillary tube and left standing, the heavier cellular components settle to the bottom of the tube. Spinning the tube in a centrifuge can accelerate this process. The red layer at the bottom portion of the tube represents the compacted red blood cells (RBCs). This RBC portion of the entire volume of blood in the tube is called the *hematocrit,* normally about 45% of the entire volume. A thin, whitish layer on the surface of the RBCs, known as the *buffy coat,* contains the white blood cells and platelets, and it represents less than 1% of the total volume. The remaining straw-colored upper portion of the contents contains the plasma, nearly 55% of the volume.

The contents of a red blood cell are about one-third *hemoglobin,* a protein. Each hemoglobin molecule has four bound heme portions, each containing an atom of iron (Fe). An oxygen molecule can combine with each of the iron atoms, forming bright red *oxyhemoglobin.* When oxygen detaches from the hemoglobin in the capillaries of the tissues, it becomes a darker red and is called *deoxyhemoglobin,* which can appear bluish when viewed through the skin and blood vessel walls. Decreased values of hematocrits or hemoglobin levels could be attributed to hemorrhage, dietary deficiencies, infections, bone marrow cancer, or radiation.

Stoppage of bleeding, called *hemostasis,* involves three defense mechanisms: blood vessel spasm, platelet plug formation, and *coagulation.* The coagulation mechanism involves a series of chain reactions that result in a blood clot. The final stage of the series of reactions occurs when thrombin converts a soluble plasma protein called fibrinogen into insoluble *fibrin.* The fibrin mesh traps cellular components of the blood, stopping the blood loss. The fibrin formation of blood coagulation normally takes 2–10 minutes. Clotting deficiencies could be attributed to low platelet count, leukemia, hemophilia, liver disease, malnutrition, radiation, or anticoagulant drugs.

Blood cholesterol is an important component needed for the structure of cellular membranes and the formation of steroid hormones and bile components. However, blood cholesterol higher than recommended levels increases the chances of cardiovascular diseases and risks of heart attacks and strokes. High-cholesterol diets, lack of exercise, and heredity are among the factors that contribute to high cholesterol levels. Low-cholesterol diets, an increase in exercise, and various medications are some ways used to lower the levels of cholesterol.

The clinical significance or conclusive diagnosis of a disease is often difficult to achieve. Most often, several blood and urine characteristics along with other symptoms are assessed collectively in order to make a determination of the abnormality. A self-diagnosis should never be made as a result of a test result conducted in the biology laboratory. Always obtain proper medical exams and treatments from medical personnel.

You take in a deep breath of air into your lungs. The inspired oxygen diffuses from the alveoli of your lungs into your red blood cells, where it binds with a hemoglobin molecule. Oxygen is then transported to systemic capillaries and released from the hemoglobin to diffuse into your body tissues. The amount of oxygen transported and released depends in part on the molecular structure of hemoglobin. Each hemoglobin molecule is composed of four proteins (two alpha proteins and two beta proteins) each containing a heme pigment with an atom of iron. Each iron binds one oxygen molecule (O_2). As a result, each hemoglobin molecule can bind as many as four oxygen molecules. How much oxygen is bound to the hemoglobin molecules is expressed as the percent *(%) saturation of hemoglobin* (i.e., what percentage of the iron-binding sites of hemoglobin is bound with oxygen). The % saturation of hemoglobin is dependent on a number of variables. The most important is the partial pressure of oxygen. This relationship of % saturation of hemoglobin to partial pressure of oxygen can be graphed to produce an *oxygen dissociation curve.* Other variables that influence % saturation of hemoglobin include temperature, carbon dioxide levels, and pH, and the change of any of these variables will trigger a "shift" in the oxygen dissociation curve. A "shift right" indicates that for any given partial pressure of oxygen, there is a decrease in the % saturation of hemoglobin. A "shift left" indicates that for any given partial pressure of oxygen, there is an increase in the % saturation of hemoglobin.

Warning

Because of the possibility of blood infections being transmitted from one student to another during blood-testing procedures, it is suggested that the following procedures be performed as demonstrations by the instructor. Observe all safety procedures listed for this lab.

Procedure A—Hematocrit

To determine the hematocrit (percentage of red blood cells) in a whole blood sample, the cells must be separated from the liquid plasma. This separation can be rapidly accomplished by placing a tube of blood in a centrifuge. The force created by the spinning motion of the centrifuge causes the cells to be packed into the lower end of the tube. The quantities of cells and plasma can then be measured, and the percentage of cells (hematocrit or packed cell volume) can be calculated.

1. To determine the hematocrit in a blood sample, follow these steps:

   a. Thoroughly wash hands with soap and water and dry them with paper towels.

   b. Cleanse the end of the middle finger with an alcohol swab and let the finger dry in the air.

   c. Remove a sterile disposable blood lancet from its package without touching the sharp end.

   d. Puncture the skin on the side near the tip of the middle finger with the lancet and properly discard the lancet. Wipe away the first drop of blood with the alcohol swab.

   e. Touch a drop of blood with the colored end of a heparinized capillary tube. Hold the tube tilted slightly upward so that the blood will easily move into it by capillary action (fig. 42.1a). To prevent an air bubble, keep the tip in the blood until filled.

   f. Allow the blood to fill about two-thirds of the length of the tube. Cover the lanced finger location with a bandage.

   g. Hold a finger over the tip of the dry end so that blood will not drain out while you seal the blood end. Plug the blood end of the tube by pushing it with a rotating motion into sealing clay or by adding a plastic Critocap (fig. 42.1b).

   h. Place the sealed tube into one of the numbered grooves of a microhematocrit centrifuge. The tube's sealed end should point outward from the center and should touch the rubber lining on the rim of the centrifuge (fig. 42.1c).

   i. The centrifuge should be balanced by placing specimen tubes on opposite sides of the moving head, the inside cover should be tightened with the lock wrench, and the outside cover should be securely fastened.

   j. Run the centrifuge for 3–5 minutes.

FIGURE 42.1 Steps of the hematocrit (red blood cell percentage) procedure: (a) load a heparinized capillary tube with blood; (b) plug the blood end of the tube with sealing clay; (c) place the tube in a microhematocrit centrifuge.

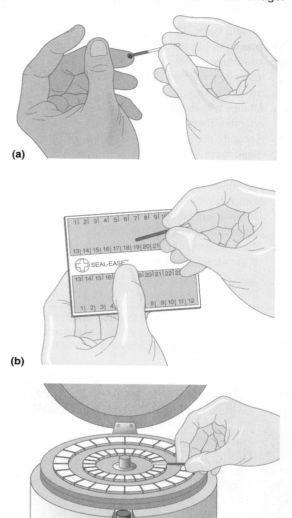

   k. After the centrifuge has stopped, remove the specimen tube. The red blood cells have been packed into the bottom of the tube. The clear liquid on top of the cells is plasma.

   l. Use a microhematocrit reader to determine the percentage of red blood cells in the tube. If a microhematocrit reader is not available, measure the total length of the blood column in millimeters (red cells plus plasma) and the length of the red blood cell column alone in millimeters. Divide the red blood cell length by the total blood column length and multiply the answer by 100 to calculate the percentage of red blood cells.

2. Record the test result in Part A of Laboratory Assessment 42.

Procedure B—Hemoglobin Content

Although the hemoglobin content of a blood sample can be measured in several ways, a common method uses a hemoglobinometer. This instrument is designed to compare the color of light passing through a hemolyzed blood sample with a standard color. The results of the test are expressed in grams of hemoglobin per 100 mL (g/dL) of blood or in percentage of normal values.

1. To measure the hemoglobin content of a blood sample, follow these steps:

   a. Obtain a hemoglobinometer and remove the blood chamber from the slot in its side.

   b. Separate the pieces of glass from the metal clip and clean them with alcohol swabs and lens paper. One of the pieces of glass has two broad, U-shaped areas surrounded by depressions. The other piece is flat on both sides.

   c. Obtain a large drop of blood from a finger, by following the directions in Procedure A.

   d. Place the drop of blood on one of the U-shaped areas of the blood chamber glass (fig. 42.2a).

   e. Stir the blood with the tip of a hemolysis applicator until the blood appears clear rather than cloudy. This usually takes about 45 seconds (fig. 42.2b).

   f. Place the flat piece of glass on top of the blood plate and slide both into the metal clip of the blood chamber.

   g. Push the blood chamber into the slot on the side of the hemoglobinometer, making sure that it is in all the way (fig. 42.2c).

   h. Hold the hemoglobinometer in the left hand with the thumb on the light switch on the underside (fig. 42.2d).

   i. Look into the eyepiece and note the green area split in half.

   j. Slowly move the slide on the side of the instrument back and forth with the right hand until the two halves of the green area look the same.

   k. Note the value in the upper scale (grams of hemoglobin per 100 mL of blood), indicated by the mark in the center of the movable slide.

2. Record the test result in Part A of the laboratory assessment.

FIGURE 42.2 Steps of the hemoglobin content procedure: (a) load the blood chamber with blood; (b) stir the blood with a hemolysis applicator; (c) place the blood chamber in the slot of the hemoglobinometer; (d) match the colors in the green area by moving the slide on the side of the instrument.

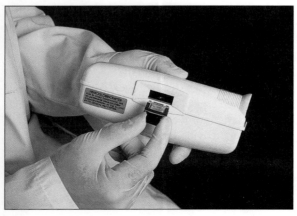

(a)

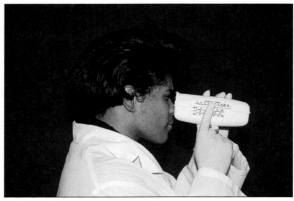

(b)

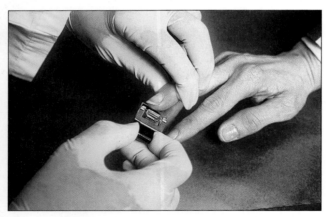

(c)

(d)

Procedure C—Coagulation

Coagulation time, often called clotting time, is the time from the onset of bleeding until the insoluble protein fibrin is formed. This happens when thrombin converts soluble fibrinogen into insoluble fibrin. This clotting time normally ranges from 2 to 10 minutes. This process is prolonged if the person has clotting deficiencies or is being treated with anticoagulants such as heparin (some is produced by basophils and mast cells), warfarin (Coumadin), or aspirin. In this laboratory exercise we will try to determine the time to the nearest minute.

1. To determine coagulation time, follow these steps:
    a. Prepare the finger to be lanced by following the directions in Procedure A. Lance the end of a finger to obtain a drop of blood. Wipe away the first drop of blood with the alcohol swab.
    b. Touch the next drop of blood with one end of a non-heparinized capillary tube. Hold the tube tilted slightly upward so that the blood will easily move into it by capillary action (fig. 42.1a). Keep the tip in the blood until the tube is nearly filled. If the tube is nearly filled, it will allow enough tube length for breaking it several times.
    c. Place the capillary tube on a paper towel. Cover the lanced finger location with a bandage. Record the time: _____
    d. At 1-minute intervals, use the small triangular file and make a scratch on the capillary tube starting near one end of the tube. Hold the tube with fingers on each side of the scratch, the weakened location of the tube, and break the tube away from you, being careful to keep the two pieces close together after the break. Gently pull the two ends of the tube apart while observing carefully to see if it breaks cleanly. If it breaks cleanly, fibrin has not formed yet (fig. 42.3a).
    e. Continue breaking the capillary tube each minute until fibrin is noted spanning the two parts of the capillary tube (fig. 42.3b). Note the time for coagulation.
2. Record the test results in Part A of the laboratory assessment.

Procedure D—Blood Cholesterol Level Determination

The recommended total cholesterol level is less than 200 mg/dL (dL = deciliter), which decreases the probability of cardiovascular diseases. Total blood cholesterol of less than 100 (hypocholesterolemia) is associated with possible conditions as malnutrition, hyperthyroidism, and depression.

FIGURE 42.3 Steps of the coagulation procedure: (a) clean break of capillary tube before any fibrin formed; (b) fibrin spans between the broken capillary tube segments when coagulation occurs.

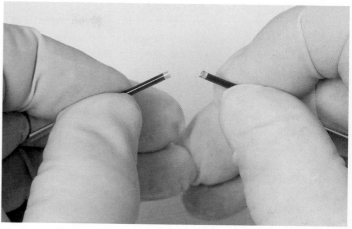

(a)

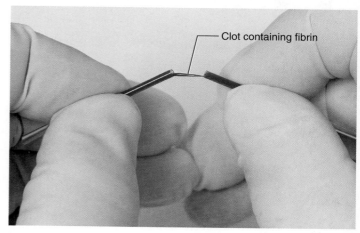

Clot containing fibrin

(b)

Since cholesterol is not water soluble, it is transported in the plasma incorporated with a protein as a low-density lipoprotein (LDL) and a high-density lipoprotein (HDL). The **HDLs** are transported to the liver, where they are degraded and removed from the body, and are therefore sometimes referred to as the "Healthy/good" form. The LDLs are transported to our tissue cells for cellular needs, but if the levels are excessive, there is an increased occurrence of atherosclerosis. **LDLs** are sometimes referred to as the "Lousy/bad" cholesterol. The recommended levels of HDLs are > 40 mg/dL; recommended levels of LDLs are < 130 mg/dL.

1. Each kit contains cholesterol samples of simulated blood, test strips, color charts, and directions for the particular kit being used. Some available kits contain samples that reflect the use of cholesterol-lowering medications, while others might contain samples that represent diets of various cholesterol levels. Dip a test strip into a sample; then compare the strip to a color chart to determine the cholesterol level of the sample.

2. Record the test results in Part A of the laboratory assessment.
3. Complete Part B of the laboratory assessment.

Alternative Activity

Laboratories in modern hospitals and clinics use updated hematology analyzers (fig. 42.4) for evaluations of the blood factors that are described in Laboratory Exercise 42. The more traditional procedures performed in these laboratory activities will help you to better understand each blood characteristic.

FIGURE 42.4 Modern hematology analyzer being used in the laboratory of a clinic. (*Note:* A tour of a laboratory in a hospital or clinic might be arranged.)

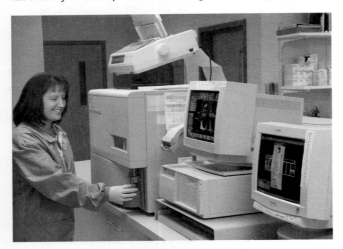

Procedure E—Ph.I.L.S. Lesson 34 Blood: pH & Hb-Oxygen Binding

1. Open Exercise 34, Blood: pH & Hb-Oxygen Binding.
2. Read the objectives and introduction and take the pre-lab quiz.
3. After completing the pre-lab quiz, read through the wet lab. *Be sure to click open and view the videos that are indicated in red in the wet lab.*
4. The lab exercise will open when you click Continue after completing the wet lab (fig. 42.5).

Setup

5. Click the power switch to turn on the spectrophotometer.
6. Set the wavelength at 620 nm (wavelength at which hemoglobin absorbs light if intact).

Calibrate Spectrophotometer

7. Set the transmittance value at 0 using the arrows at zero.
8. Open the lid of the "holder" of the spectrophotometer (if unsure, click on the word *spectrophotometer* in the instructions at the bottom of the screen).
9. Click on the globe and drag it to the top of one of the tubes (it will snap in place) to form the "tonometer." (To "click and drag," move the mouse to position the arrow on the globe; left-click on mouse and hold; drag the globe to the top of one of the tubes; and when positioned, release the left click.)
10. Place the tonometer into the spectrophotometer. (The Journal will open a table with the partial pressures of oxygen ranging from 160 mm Hg to 0 mm Hg for three different pH values—6.8, 7.4, and 8.0.) Set the transmittance at 100% using the arrows at Calibrate.

FIGURE 42.5 Opening screen for the laboratory exercise on Blood: pH & Hb-Oxygen Binding.

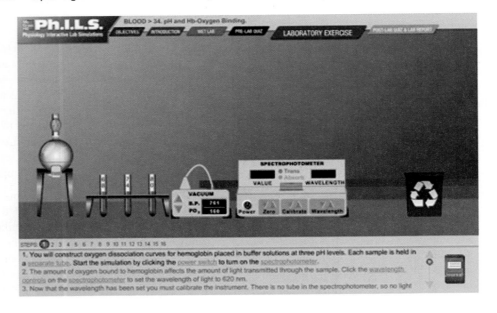

Measuring Transmittance

11. With the tonometer remaining in the spectrophotometer, record the percent transmittance by clicking open the Journal (red rectangle at bottom right of screen).
12. Click on the tonometer to remove it. (The tonometer will come up out of the spectrophotometer, and the Journal will close automatically.) Then drag it back to the rack.

Changing the Partial Pressure of O₂ (PO₂)

13. Position the end of the vacuum tube on the top of the "tonometer."
14. Click the down arrow once to reduce the barometric pressure (which simulates a decrease in partial pressure of oxygen).
15. Click on the vacuum tube to return it to its original position.
16. Place the tonometer in the spectrophotometer.
17. Click on the Journal to record.
18. *Repeat numbers 9–17 (Ph.I.L.S. steps 5–15) until all values for the different partial pressures of oxygen have been recorded in the table on the computer.*

Changing the pH

19. When all values have been recorded for a given sample, drag the tonometer to place the sample into the recycling can (three arrows in a circle in bottom right of screen).
20. Repeat numbers 9–19 (Ph.I.L.S. steps 5–15) but use one of the two other samples with a different pH.
21. Click on the Journal to see the complete table.
22. To view graph, click on "graph" in lower right of journal window. See figure 42.6 as an example of a final graph.

Interpreting Results

23. *With the graph still on the screen,* answer the questions in Part C of the laboratory assessment. If you acciden-

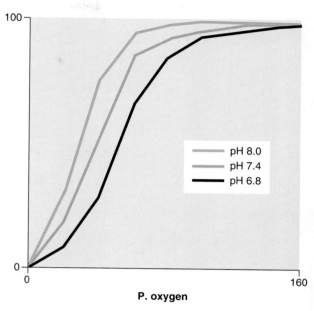

FIGURE 42.6 Example: Graph to show the effect of pH on the color of sheep hemoglobin set at different partial pressures of oxygen. Lower color values indicate that less oxygen is bound to hemoglobin.

tally closed the graph, click on the Journal panel (red rectangle at bottom right of screen).

24. Complete the Post-Lab Quiz (click open Post-Lab Quiz and Lab Report) by answering the ten questions on the computer screen.
25. Read the conclusion on the computer screen.
26. You may print the Lab Report for Blood: pH & Hb-Oxygen Binding.

Name _____

Date _____

Section _____

The A corresponds to the indicated outcome(s) found at the beginning of the laboratory exercise.

Blood Testing

Part A Assessments

Blood test data: A 1

Blood Test	Test Results	Normal Values
Hematocrit (mL per 100 mL blood)		Men: 40–54% Women: 37–47%
Hemoglobin content (g per 100 mL blood; g/dL)		Men: 14–18 g/100 mL (g/dL) Women: 12–16 g/100 mL (g/dL)
Coagulation		2–10 minutes
Cholesterol	Before reduction program: After reduction program:	Below 200 mg/dL

Part B Assessments

Complete the following:

1. How does the hematocrit from the blood test compare with the normal value? If the test results were abnormal, what condition would it possibly indicate? A 2

   _____

   _____

   _____

   _____

2. How does the hemoglobin content from the blood test compare with the normal value? If the test results were abnormal, what condition would it possibly indicate? A 2

   _____

   _____

   _____

   _____

3. How does the coagulation time from the blood test compare with the normal value? If the test results were abnormal, what condition would it possibly indicate? **2**

_____

_____

_____

_____

4. Assess the most beneficial method of lowering cholesterol from samples provided in the simulated blood kit. **2**

_____

_____

_____

_____

Critical Thinking Assessment

Which blood tests performed in this lab could be used to determine possible anemia? **3**

_____

_____

Part C Ph.I.L.S. Lesson 34, Blood: pH & Hb-Oxygen Binding Assessments

1. Use the data from the computer simulation collected at a pH of 7.4; graph the partial pressure of oxygen and the percent transmittance of light. (Include the title. Label each axis with units. Plot the points and connect). **4**

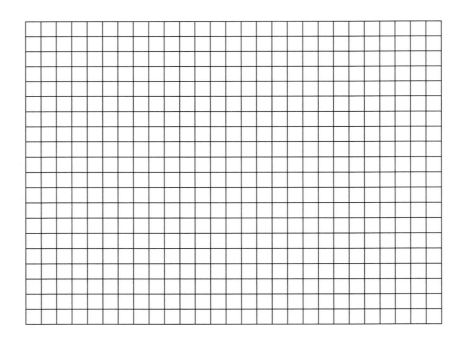

2. Examine the graph you produced in question 1. As the partial pressure of oxygen increases, the % transmittance of light _____. $\boxed{4}$

3. There is a direct relationship between percent transmittance of light and percent saturation of hemoglobin. So, as the percent transmittance of light increases, the percent saturation of hemoglobin _____ (increases or decreases). Thus, as the partial pressure of oxygen increases, the % saturation of hemoglobin _____ (increases, decreases, or stays the same). $\boxed{5}$ $\boxed{6}$

4. Compare the effect of pH on percent transmittance of light by examining the three dissociation curves produced at a pH of 6.8, 7.4, and 8.0 and answer the following questions: $\boxed{7}$

   a. When comparing the curve produced at 6.8 to the curve produced at 7.4, the curve produced when the pH is 6.8

   would be described as _____ (a "shift right" or a "shift left") and at any given partial pressure of

   oxygen, the % transmittance of light is _____ (higher or lower) and thus, the % saturation of hemoglobin

   would be _____ (higher or lower).

   b. At a given PO_2, hemoglobin has the highest affinity for oxygen (ability to bind oxygen) at this pH

   _____ and the lowest affinity for oxygen at this pH _____.

Critical Thinking Assessment $\boxed{8}$

Hydrogen ion (H^+) levels increase within the body from increased production of lactic acid (during anaerobic cellular respiration) and from increased production of carbon dioxide ($CO_2 + H_2O \longleftrightarrow H^+ + HCO_3^-$). The increased amounts of H^+ bind with the amino acids that compose the protein of hemoglobin, slightly altering hemoglobin's shape. This change in shape interferes with the ability of iron within the hemoglobin molecule to bind the oxygen (decreased affinity or attraction). (This is referred to as the Bohr effect.)

If the tissues of the body are metabolically active, _____ (more or less) H^+ are produced and the pH is

_____ (decreased or increased). As a result of the Bohr effect, when hemoglobin reaches the systemic

capillaries, the hemoglobin will bind the oxygen with _____ (greater or less) affinity; _____

(more or less) oxygen is released to the cells; the saturation of hemoglobin will be _____ (greater or less),

and hemoglobin returns to the lungs with _____ (more or less) oxygen than would occur at a normal pH.

Blood Typing

Purpose of the Exercise

To determine the ABO blood type of a blood sample and to observe an Rh blood-typing test.

Materials Needed

For Procedure A:
ABO blood-typing kit
Simulated blood-typing kits are suggested as a substitute for collected blood

For Procedure B:
Microscope slide
Alcohol swabs (wipes)
Sterile blood lancet
Toothpicks
Anti-D serum
Slide warming box (Rh blood-typing box or Rh view box)

Safety

► It is important that students learn and practice correct procedures for handling body fluids. Consider using simulated blood-typing kits or contaminant-free blood that has been tested and is available from various laboratory supply houses. Some of the procedures might be accomplished as demonstrations only. If student blood is used, it is important that students handle only their own blood.
► Use an appropriate disinfectant to wash the laboratory tables before and after the procedures.
► Wear disposable gloves and safety goggles when handling blood samples.
► Clean the end of a finger with an alcohol swab before the puncture is performed.
► The sterile blood lancet should be used only once.
► Dispose of used lancets and blood-contaminated items in an appropriate container (never use the wastebasket).
► Wash your hands before leaving the laboratory.

Learning Outcomes

After completing this exercise, you should be able to

1. Analyze the basis of ABO blood typing.
2. Interpret the ABO type of a blood sample.
3. Explain the basis of Rh blood typing.
4. Interpret the Rh type of a blood sample.

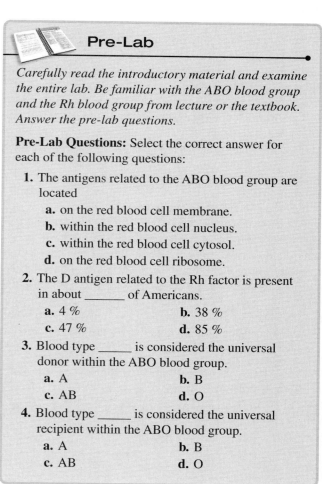

Pre-Lab

Carefully read the introductory material and examine the entire lab. Be familiar with the ABO blood group and the Rh blood group from lecture or the textbook. Answer the pre-lab questions.

Pre-Lab Questions: Select the correct answer for each of the following questions:

1. The antigens related to the ABO blood group are located
   a. on the red blood cell membrane.
   b. within the red blood cell nucleus.
   c. within the red blood cell cytosol.
   d. on the red blood cell ribosome.

2. The D antigen related to the Rh factor is present in about _____ of Americans.
   a. 4 % b. 38 %
   c. 47 % d. 85 %

3. Blood type _____ is considered the universal donor within the ABO blood group.
   a. A b. B
   c. AB d. O

4. Blood type _____ is considered the universal recipient within the ABO blood group.
   a. A b. B
   c. AB d. O

5. Hemolytic disease of the newborn could be of concern when an
   a. Rh-positive fetus and an Rh-positive mother condition occur.
   b. Rh-positive fetus and an Rh-negative mother condition occur.
   c. Rh-negative fetus and an Rh-positive mother condition occur.
   d. Rh-negative fetus and an Rh-negative mother condition occur.
6. Blood type B is the most common blood type found in the United States population.
   True _____ False _____
7. An individual with blood type O lacks both RBC antigens A and B.
   True _____ False _____

Blood typing involves identifying protein substances called *antigens* present in red blood cell membranes. Although there are many different antigens associated with human red blood cells, only a few of them are of clinical importance. These include the antigens of the ABO group and those of the Rh group.

To determine which antigens are present, a blood sample is mixed with blood-typing sera that contain known types of antibodies. If a particular antibody contacts a corresponding antigen, a reaction occurs, and the red blood cells clump together (agglutination). Thus, if blood cells are mixed with serum containing antibodies that react with antigen A and the cells clump together, antigen A must be present in those cells.

Although there are many antigens on the red blood cell membranes other than those of the ABO group and the Rh group, these antigens are of the greatest concern during transfusions. Most other antigens cause little if any transfusion reactions. As a final check on blood compatibility before transfusions, an important cross-matching test occurs. In this test, small samples of the blood from the donor and the recipient are mixed to be sure clumping (agglutination) does not create a transfusion reaction. If surgery is elective, autologous transfusions are promoted, allowing some of the person's own predonated blood from storage to be used during the surgery.

An antigen commonly found on the human RBC membrane was first identified in rhesus monkeys, and it became known as the Rh factor. Although many different types of antigens are related to the Rh factor, only the D antigen is checked using an anti-D reagent. About 85% of Americans possess the D antigen and are therefore considered to be Rh-positive (Rh⁺). Those lacking the D antigen are considered Rh-negative (Rh⁻). If an Rh-negative woman is pregnant carrying an Rh-positive fetus, the mother might obtain some RBCs from the fetus during the birth process or during a miscarriage. As a result, the mother would begin producing anti-D antibodies, creating complications for the second and future pregnancies. This condition is known as hemolytic disease of the fetus and newborn (erythroblastosis fetalis). This condition can be prevented with the proper administration of RhoGAM, which prevents the mother from producing anti-D antibodies.

Warning

Because of the possibility of blood infections being transmitted from one student to another if blood testing is performed in the classroom, it is suggested that commercially prepared blood-typing kits containing virus-free human blood be used for ABO blood typing. The instructor may wish to demonstrate Rh blood typing. Observe all of the safety procedures listed for this lab.

Procedure A—ABO Blood Typing

1. Reexamine the introductory material and study table 43.1.
2. Compete Part A of Laboratory Assessment 43.
3. Perform the ABO blood type test using the blood-typing kit. To do this, follow these steps:
   a. Obtain a clean microscope slide and mark across its center with a wax pencil to divide it into right and left halves. Also write "Anti-A" near the edge of the left half and "Anti-B" near the edge of the right half (fig. 43.1).

TABLE 43.1 Antigens and Antibodies of the ABO Blood Group and Preferred, Permissible, and Incompatible Donors

Blood Type	RBC Antigens (Agglutinogens)	Plasma Antibodies (Agglutinins)	Preferred Donor Type	Permissible Donor Type in Limited Amounts	Incompatible Donor
A	A	Anti-B	A	O	B, AB
B	B	Anti-A	B	O	A, AB
AB (universal recipient)	A and B	Neither anti-A nor anti-B	AB	A, B, O	None
O (universal donor)	Neither A nor B	Both anti-A and anti-B	O	No alternative types	A, B, AB

FIGURE 43.1 Slide prepared for ABO blood typing.

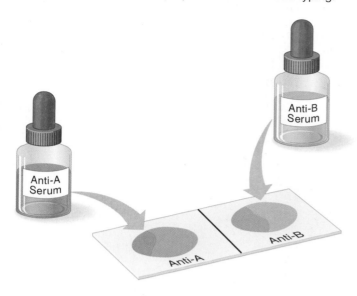

FIGURE 43.2 Four possible results of the ABO test.

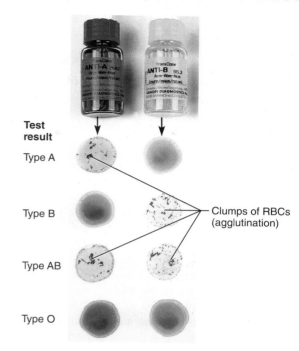

Clumps of RBCs (agglutination)

b. Place a small drop of blood on each half of the microscope slide. Work quickly so that the blood will not have time to clot.

c. Add a drop of anti-A serum to the blood on the left half and a drop of anti-B serum to the blood on the right half. Note the color coding of the anti-A and anti-B typing sera. To avoid contaminating the serum, avoid touching the blood with the serum while it is in the dropper; instead allow the serum to fall from the dropper onto the blood.

d. Use separate toothpicks to stir each sample of serum and blood together, and spread each over an area about as large as a quarter. Dispose of toothpicks in an appropriate container.

e. Examine the samples for clumping of blood cells (agglutination) after 2 minutes.

f. See table 43.2 and figure 43.2 for aid in interpreting the test results.

g. Discard contaminated materials as directed by the laboratory instructor.

4. Record your results and complete Part B of the laboratory assessment.

TABLE 43.2 ABO Blood-Typing Sera Reactions and U.S. Blood Type Frequency

Possible Reactions		Blood Type	U.S. Frequency (average)
Anti-A Serum	**Anti-B Serum**		
Clumping (agglutination)	No clumping	Type A	40%
No clumping	Clumping	Type B	10%
Clumping	Clumping	Type AB	4%
No clumping	No clumping	Type O	46%

Note: The inheritance of the ABO blood groups is described in Laboratory Exercise 61.

Critical Thinking Activity

Judging from your observations of the blood-typing results, suggest which components in the anti-A and anti-B sera caused clumping (agglutination).

_____

_____

_____

Procedure B—Rh Blood Typing

1. Reexamine the introductory material.
2. Complete Part C of the laboratory assessment.
3. To determine the Rh blood type of a blood sample, follow these steps:
   a. Lance the tip of a finger. (See the demonstration procedures in Laboratory Exercise 41 for directions.) Place a small drop of blood in the center of a clean microscope slide. Cover the lanced finger location with a bandage.
   b. Add a drop of anti-D serum to the blood and mix them together with a clean toothpick.

FIGURE 43.3 Slide warming box used for Rh blood typing.

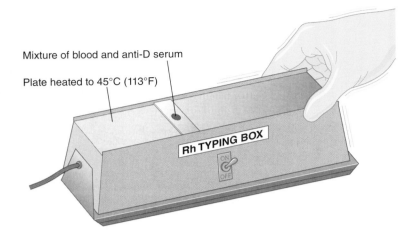

Mixture of blood and anti-D serum

Plate heated to 45°C (113°F)

Rh TYPING BOX

ON
OFF

c. Place the slide on the plate of a warming box (Rh blood-typing box or Rh view box) prewarmed to 45°C (113°F) (fig. 43.3).

d. Slowly rock the box back and forth to keep the mixture moving and watch for clumping (agglutination) of the blood cells. When clumping occurs in anti-D serum, the clumps usually are smaller than those that appear in anti-A or anti-B sera, so they may be less obvious. However, if clumping occurs, the blood is called Rh positive; if no clumping occurs *within 2 minutes,* the blood is called Rh negative.

e. Discard all contaminated materials in appropriate containers.

4. Record your results and complete Part D of the laboratory assessment.

Name _____

Date _____

Section _____

The ⚠ corresponds to the indicated outcome(s) found at the beginning of the laboratory exercise.

Blood Typing

Part A   Assessments

Complete the following statements:

1. The antigens of the ABO blood group are located in the red blood cell _____. ⚠

2. The blood of every person contains one of (how many possible?) _____ combinations of antigens. ⚠

3. Type A blood contains antigen _____. ⚠

4. Type B blood contains antigen _____. ⚠

5. Type A blood contains _____ antibody in the plasma.

6. Type B blood contains _____ antibody in the plasma. ⚠

7. Persons with ABO blood type _____ are often called universal recipients. ⚠

8. Persons with ABO blood type _____ are often called universal donors. ⚠

Part B   Assessments

Complete the following:

1. What was the ABO type of the blood tested? ⚠ _____

2. What ABO antigens are present in the red blood cells of this type of blood? ⚠ _____

3. What ABO antibodies are present in the plasma of this type of blood? ⚠ _____

4. If a person with this blood type needed a blood transfusion, what ABO type(s) of blood could be received safely? ⚠

_____

5. If a person with this blood type was serving as a blood donor, what ABO blood type(s) could receive the blood safely? ⚠

_____

Part C   Assessments

Complete the following statements:

1. The Rh blood group was named after the _____. ⚠

2. Of the antigens in the Rh group, the most important is _____. ⚠

3. If red blood cells lack Rh antigens, the blood is called _____. ⚠

4. If an Rh-negative person who is sensitive to Rh-positive blood receives a transfusion of Rh-positive blood, the donor's cells are likely to _____. ⚠

5. An Rh-negative woman who might be carrying an _____ fetus is given an injection of RhoGAM to prevent hemolytic disease of the fetus and newborn. ⚠

Part D Assessments

Complete the following:

1. What was the Rh type of the blood tested? **4** _____

2. What Rh antigen is present in the red blood cells of this type of blood? **4** _____

3. If a person with this blood type needed a blood transfusion, what type of blood could be received safely? **3**

   _____

4. If a person with this blood type was serving as a blood donor, a person with what type of blood could receive the blood safely? **3** _____

5. Observe the test results shown in figure 43.4 from an individual being tested for the ABO type and the Rh type.

   a. This individual has ABO type _____. **2**

   b. This individual has Rh type _____. **4**

FIGURE 43.4   Blood test results after adding anti-A serum, anti-B serum, and anti-D serum to separate drops of blood. An unaltered drop of blood is shown as a control; it does not indicate any clumping (agglutination) reaction.

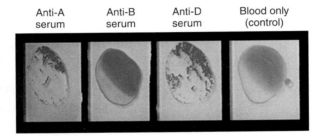

Anti-A serum   Anti-B serum   Anti-D serum   Blood only (control)

Heart Structure

Purpose of the Exercise

To review the structural characteristics of the human heart and to examine the major features of a mammalian heart.

Materials Needed

Dissectible human heart model
Dissectible human torso model
Preserved sheep or other mammalian heart
Dissecting tray
Dissecting instruments
For Learning Extension Assessment:
Colored pencils

⚠ Safety

▸ Wear disposable gloves when working on the heart dissection.
▸ Save or dispose of the dissected heart as instructed.
▸ Wash the dissecting tray and instruments as instructed.
▸ Wash your laboratory table.
▸ Wash your hands before leaving the laboratory.

Learning Outcomes

After completing this exercise, you should be able to

1 Identify and label the major structural features of the human heart and closely associated blood vessels.

2 Trace blood flow through the heart chambers and the systemic and pulmonary circuits.

3 Match heart structures with appropriate locations and functions.

4 Compare the features of the human heart with those of another mammal.

Pre-Lab

Carefully read the introductory material and examine the entire lab. Be familiar with the structure of the heart from lecture or the textbook. Answer the pre-lab questions.

Pre-Lab Questions: Select the correct answer for each of the following questions:

1. The _____ is the inferior end of the heart that is bluntly pointed.
   a. base   **b.** apex
   c. auricle   **d.** sulcus

2. Oxygen-rich blood is located in the
   a. left-side chambers.   **b.** right-side chambers.
   c. superior chambers.   **d.** inferior chambers.

3. The two superior heart chambers are the
   a. left atrium and left ventricle.
   b. right atrium and right ventricle.
   c. atria.
   d. ventricles.

4. Which of the following is an atrioventricular (AV) valve?
   a. aortic   **b.** pulmonary
   c. semilunar   **d.** mitral

5. The _____ lines the heart chambers.
   a. endocardium   **b.** fibrous pericardium
   c. serous pericardium   **d.** epicardium

6. Which heart valve has two cusps instead of three cusps?
   a. aortic   **b.** mitral
   c. tricuspid   **d.** pulmonary

7. Chordae tendineae connect the cusps of the AV valves to papillary muscles of the ventricles.
   True _____   False _____

8. The systemic circuit delivers blood to the lungs and back to the heart.
   True _____   False _____

9. The right and left coronary arteries containing oxygen-rich blood originate from the base of the aorta.
   True _____   False _____

The heart is a muscular pump located within the mediastinum and resting upon the diaphragm. It is enclosed by the lungs, thoracic vertebrae, and sternum, and attached at its superior end (the base) are several large blood vessels. Its inferior end extends downward to the left and terminates as a bluntly pointed apex.

The heart and the proximal ends of the attached blood vessels are enclosed by a double-layered pericardium. The innermost layer of this membrane (visceral layer of serous pericardium) consists of a thin covering closely applied to the surface of the heart, whereas the outer layer (parietal layer of serous pericardium with fibrous pericardium) forms a tough, protective sac surrounding the heart. Between the parietal and visceral layers of the pericardium is a space, the pericardial cavity, that contains a small volume of serous (pericardial) fluid.

Functioning as a muscular pump of blood, the heart consists of four chambers: two superior atria and two inferior ventricles. The interatrial septum and the interventricular septum prevent oxygen-rich (oxygenated) blood in the left-side chambers from mixing with oxygen-poor (deoxy-genated) blood in the right chambers. Atrioventricular (AV) valves (mitral and tricuspid) are located between each atrium and ventricle, while semilunar valves (aortic and pulmonary) are at the exits of the ventricles into the two large arteries. The heart valves only allow blood flow in one direction.

Procedure A—The Human Heart

As you study the human heart, use a combination of laboratory manual figures and heart models available in the laboratory to locate structures. This laboratory exercise covers basic information that will be very important as you work on the rest of the cardiovascular system exercises.

1. Study figures 44.1, 44.2, 44.3, 44.4, and 44.5.
2. Examine the human heart model and locate the following features:

   heart
   - base of heart—superior region where blood vessels emerge
   - apex of heart—inferior, rounded end

FIGURE 44.1 Anterior view of the human heart.

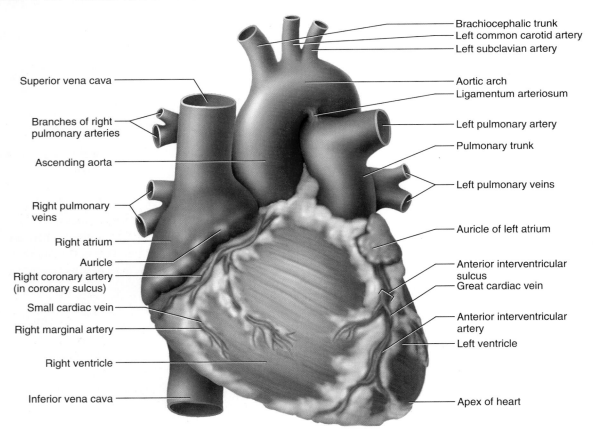

Brachiocephalic trunk
Left common carotid artery
Left subclavian artery

Superior vena cava

Aortic arch
Ligamentum arteriosum

Branches of right pulmonary arteries

Left pulmonary artery

Pulmonary trunk

Ascending aorta

Left pulmonary veins

Right pulmonary veins

Auricle of left atrium

Right atrium

Auricle

Anterior interventricular sulcus

Right coronary artery (in coronary sulcus)

Great cardiac vein

Small cardiac vein

Anterior interventricular artery

Right marginal artery

Left ventricle

Right ventricle

Inferior vena cava

Apex of heart

FIGURE 44.2 Posterior view of the human heart.

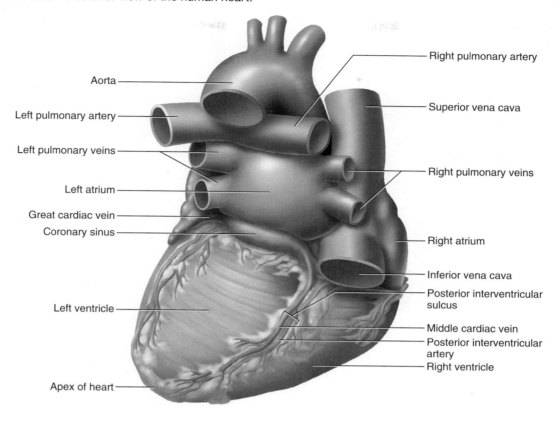

Aorta

Left pulmonary artery

Left pulmonary veins

Left atrium

Great cardiac vein

Coronary sinus

Left ventricle

Apex of heart

Right pulmonary artery

Superior vena cava

Right pulmonary veins

Right atrium

Inferior vena cava

Posterior interventricular sulcus

Middle cardiac vein

Posterior interventricular artery

Right ventricle

FIGURE 44.3 Frontal section of the human heart.

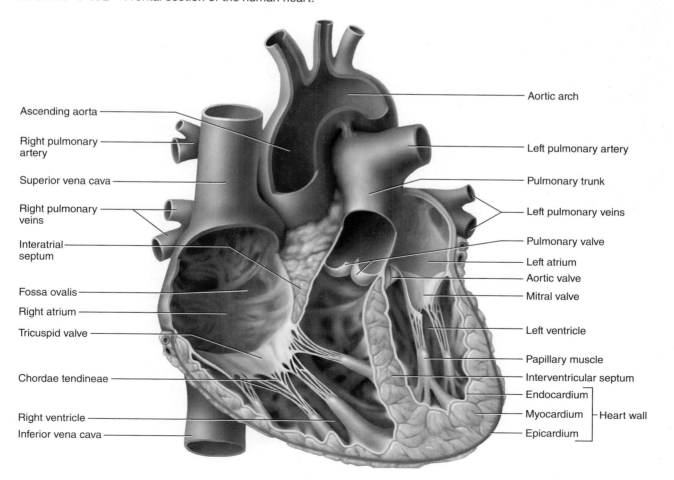

Ascending aorta

Right pulmonary artery

Superior vena cava

Right pulmonary veins

Interatrial septum

Fossa ovalis

Right atrium

Tricuspid valve

Chordae tendineae

Right ventricle

Inferior vena cava

Aortic arch

Left pulmonary artery

Pulmonary trunk

Left pulmonary veins

Pulmonary valve

Left atrium

Aortic valve

Mitral valve

Left ventricle

Papillary muscle

Interventricular septum

Endocardium

Myocardium — Heart wall

Epicardium

FIGURE 44.4 Superior view of the four heart valves (atria removed).

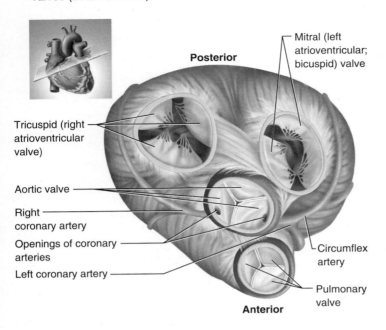

Posterior

Mitral (left atrioventricular; bicuspid) valve

Tricuspid (right atrioventricular valve)

Aortic valve

Right coronary artery

Openings of coronary arteries

Left coronary artery

Circumflex artery

Pulmonary valve

Anterior

FIGURE 44.5 Frontal section of the heart wall and pericardium.

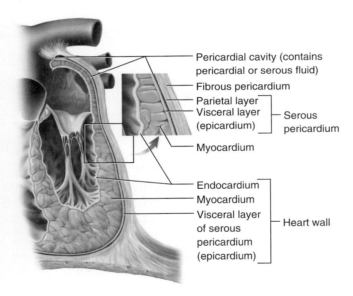

Pericardial cavity (contains pericardial or serous fluid)

Fibrous pericardium

Parietal layer
Visceral layer (epicardium) — Serous pericardium

Myocardium

Endocardium
Myocardium
Visceral layer of serous pericardium (epicardium) — Heart wall

3. Examine figure 44.5 of the pericardium and the heart wall. Observe the arrangement of the structures and note the descriptions of the following features:

pericardium (pericardial sac)
- fibrous pericardium—outer layer
- serous pericardium
  - parietal layer—inner fused lining of fibrous pericardium
  - visceral layer (epicardium)

pericardial cavity—between parietal and visceral layers of serous pericardial membranes

myocardium—composed of cardiac muscle tissue

endocardium—lines heart chambers

4. Use the human heart model and examine the structures, locations, and descriptions of the following:

atria—two superior chambers
- right atrium—receives blood from systemic veins
- left atrium—receives blood from pulmonary veins
- auricles—earlike extensions of atria allowing greater expansions for blood volumes

ventricles—two inferior chambers
- right ventricle—pumps blood into pulmonary circuit
- left ventricle—pumps blood into systemic circuit; has thicker myocardium than right ventricle

atrioventricular orifices—openings between atria and ventricles and guarded by AV valves

atrioventricular valves (AV valves)—located between atria and ventricles
- tricuspid valve (right atrioventricular valve)—has three cusps
- mitral valve (left atrioventricular valve; bicuspid valve)—has two cusps

semilunar valves—regulate flow into two large arteries that exit from ventricles
- pulmonary valve—has three pocketlike cusps at entrance of pulmonary trunk
- aortic valve—has three pocketlike cusps at entrance of aorta

chordae tendineae (tendinous cords)—connect AV valve cusps to papillary muscles

papillary muscles—projections from inferior ventricles; anchor AV valve cusps during ventricular contraction

coronary sulcus (atrioventricular sulcus or groove)—encircles heart between atria and ventricles

interventricular sulci—grooves with blood vessels along septum between ventricles
- anterior sulcus
- posterior sulcus

5. Inspect figure 44.6 as a reference to trace the flow of blood through the heart chambers and the systemic and pulmonary circuits. The arrows within the figure indicate the direction of the blood flow within the heart chambers and the two circuits. Use a dissectible human torso model and the heart model and locate the major blood vessels near the heart. Follow the arrows on figure 44.6 to trace the pathway of blood. Observe the figures and the models to take notice of the use of the red and blue colors that pertain to the relative oxygen levels of the blood within

FIGURE 44.6 The heart pumps blood into two major circuits. The left heart side pumps blood into the systemic circuit, serving all body tissues except the lungs. The right heart side pumps blood into the pulmonary circuit, serving both lungs. Red indicates oxygen-rich blood; blue indicates oxygen-poor blood. The arrows indicate the direction of blood flow.

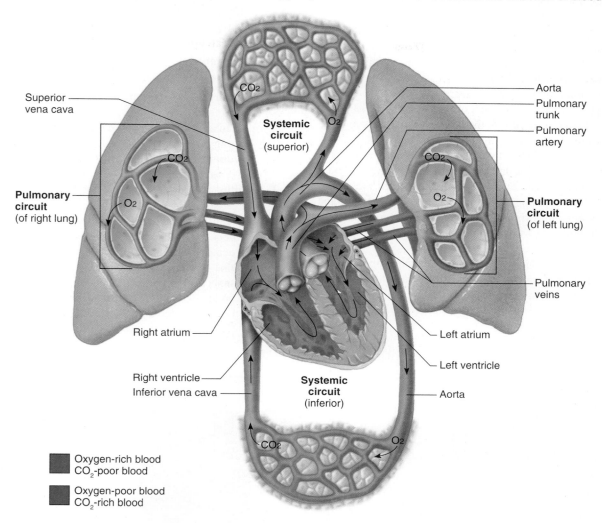

those blood vessels and heart chambers. The red color indicates oxygen-rich (oxygenated) blood; the blue color indicates oxygen-poor (deoxygenated) blood. Compare the two circuits in relation to the oxygen content in the arteries and veins. *Arteries carry blood away from the heart, while veins return blood to the heart. Notice that the arteries of the systemic circuit contain oxygen-rich blood; the arteries of the pulmonary circuit contain oxygen-poor blood.* Locate the following blood vessels associated with the heart and those serving the myocardium:

superior vena cava—returns blood to right atrium from superior part of systemic circuit

inferior vena cava—returns blood to right atrium from inferior part of systemic circuit

pulmonary trunk—main artery of pulmonary circuit

pulmonary arteries—main arteries to lungs

pulmonary veins—return blood from pulmonary circuit into left atrium

aorta—main artery of systemic circuit

left coronary artery—originates from base of aorta; has two main branches (fig. 44.4)

- circumflex artery
- anterior interventricular (descending) artery

right coronary artery—originates from base of aorta; has two main branches (fig. 44.4)

- posterior interventricular artery
- right marginal artery

cardiac veins—drain the myocardium into the coronary sinus; include several main veins

- great cardiac vein
- middle cardiac vein
- small cardiac vein

coronary sinus—for return of blood from cardiac veins into right atrium

6. Complete Parts A and B of Laboratory Assessment 44.

Procedure B—Dissection of a Sheep Heart

The directions and the figures for this procedure are for a sheep heart. The directions and the figures would be comparable if another mammal heart were used for the dissection. Other mammal hearts that might be used include pig, cow, and deer. A cadaver observation would enhance this laboratory experience if it could be arranged.

1. Obtain a preserved sheep heart. Rinse it in water thoroughly to remove as much of the preservative as possible. Also run water into the large blood vessels to force any blood clots out of the heart chambers.

2. Place the heart in a dissecting tray with its ventral surface up (fig. 44.7), and proceed as follows:

   a. Although the relatively thick *pericardial sac* probably is missing, look for traces of this membrane around the origins of the large blood vessels. Review figure 44.5 to help you locate some pericardial sac structures.

   b. Locate the *visceral pericardium,* which appears as a thin, transparent layer on the surface of the heart. Use a scalpel to remove a portion of this layer and expose the *myocardium* beneath. Also note the abundance of fat along the paths of various blood vessels.

This adipose tissue occurs in the loose connective tissue that underlies the visceral pericardium.

   c. Identify the following:

   right atrium

   right ventricle

   left atrium

   left ventricle

   atrioventricular sulcus

   anterior interventricular sulcus

   d. Carefully remove the fat from the anterior interventricular sulcus and expose the blood vessels that pass along this groove. They include a branch of the *left coronary artery* (anterior interventricular artery) and a *cardiac vein.*

3. Examine the dorsal surface of the heart (fig. 44.8), and proceed as follows:

   a. Identify the *atrioventricular sulcus* and the *posterior interventricular sulcus.*

   b. Locate the stumps of two relatively thin-walled veins that enter the right atrium. Demonstrate this connection by passing a slender probe through them. The upper vessel is the *superior vena cava,* and the lower one is the *inferior vena cava.*

FIGURE 44.7 Ventral surface of sheep heart.

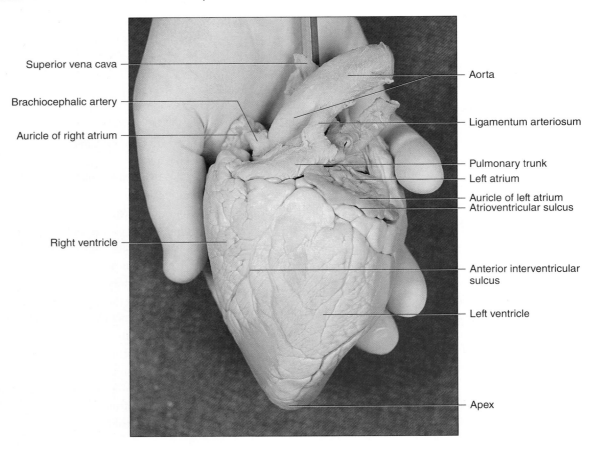

Superior vena cava

Brachiocephalic artery

Auricle of right atrium

Right ventricle

Aorta

Ligamentum arteriosum

Pulmonary trunk

Left atrium

Auricle of left atrium

Atrioventricular sulcus

Anterior interventricular sulcus

Left ventricle

Apex

FIGURE 44.8 Dorsal surface of sheep heart. To open the right atrium, insert a blade of the scissors into the superior (anterior in sheep) vena cava and cut downward.

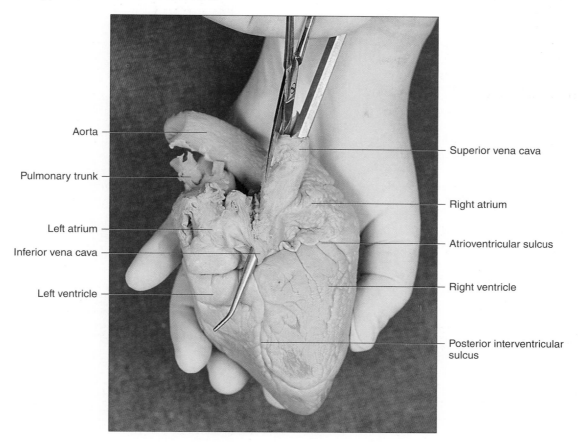

Aorta

Pulmonary trunk

Left atrium

Inferior vena cava

Left ventricle

Superior vena cava

Right atrium

Atrioventricular sulcus

Right ventricle

Posterior interventricular sulcus

4. Open the right atrium. To do this, follow these steps:
   a. Insert a blade of the scissors into the superior vena cava and cut downward through the atrial wall (fig. 44.8).
   b. Open the chamber, locate the *right atrioventricular valve,* and examine its cusps.
   c. Also locate the opening to the *coronary sinus* between the valve and the inferior vena cava.
   d. Run some water through the right atrioventricular valve to fill the chamber of the right ventricle.
   e. Gently squeeze the ventricles, and watch the cusps of the valve as the water moves up against them.
5. Open the right ventricle as follows:
   a. Continue cutting downward through the right atrioventricular valve and the right ventricular wall until you reach the apex of the heart.
   b. Locate the *chordae tendineae* and the *papillary muscles.*
   c. Find the opening to the *pulmonary trunk* and use the scissors to cut upward through the wall of the right ventricle. Follow the pulmonary trunk until you have exposed the *pulmonary valve.*
   d. Examine the valve and its cusps.

6. Open the left side of the heart. To do this, follow these steps:
   a. Insert the blade of the scissors through the wall of the left atrium and cut downward on the ventral portion to the apex of the heart.
   b. Open the left atrium and locate the four openings of the *pulmonary veins.* Pass a slender probe through each opening and locate the stump of its vessel.
   c. Examine the *left atrioventricular valve* and its cusps.
   d. Also examine the left ventricle and compare the thickness of its wall with that of the right ventricle.
7. Locate the aorta, which leads away from the left ventricle, and proceed as follows:
   a. Compare the thickness of the aortic wall with that of a pulmonary artery.
   b. Use scissors to cut along the length of the aorta to expose the *aortic valve* at its base.
   c. Examine the cusps of the valve and locate the openings of the *coronary arteries* just distal to them.
8. As a review, locate and identify the stumps of each of the major blood vessels associated with the heart.
9. Discard or save the specimen as directed by the laboratory instructor.
10. Complete Part C of the laboratory assessment.

Laboratory Assessment

44

Name ______________________

Date ______________________

Section ______________________

The [A] corresponds to the indicated outcome(s) found at the beginning of the laboratory exercise.

Heart Structure

Part A Assessments

Examine the labeled features and identify the numbered features in figures 44.9, 44.10, and 44.11 of the heart.

FIGURE 44.9 Identify the features on this anterior view of the heart region of a cadaver, using the terms provided. [1]

Anatomy & Physiology REVEALED apr.revealed.com

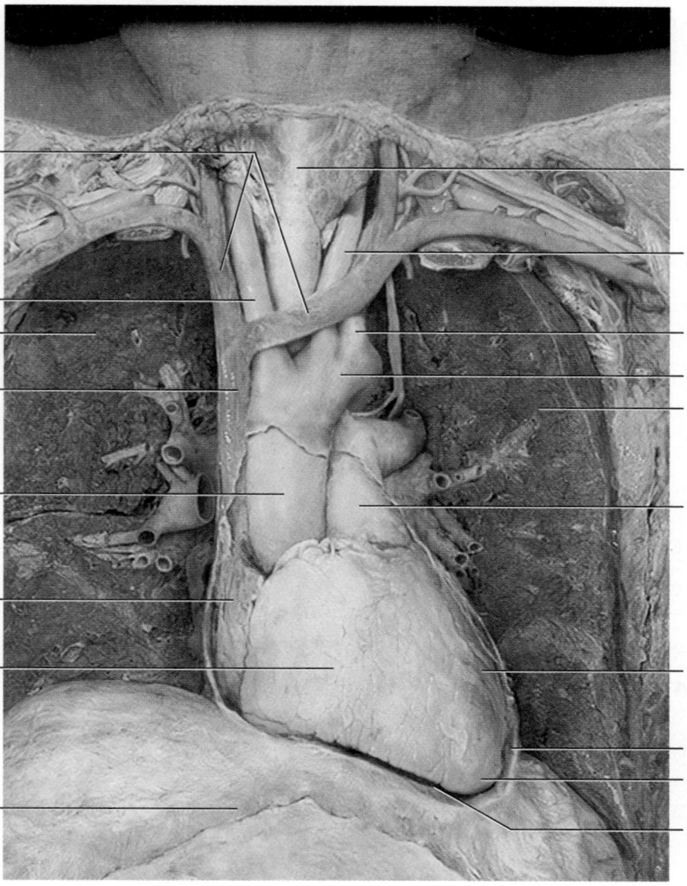

Brachiocephalic veins
brachiocephalic 1
Right lung (cut)
superior vena 2
right atrium 3
4
right vent 5
Diaphragm
Trachea
6 left common carotid
Left subclavian artery
7 pulmonary trunk
Left lung (cut)
8 left atrium
9 left ventricle
Pericardium
10 apex
Pericardial cavity

Terms:

- Aortic arch
- Apex of heart
- Ascending aorta
- Brachiocephalic trunk
- Left common carotid artery
- Left ventricle
- Pulmonary trunk
- Right atrium
- Right ventricle
- Superior vena cava

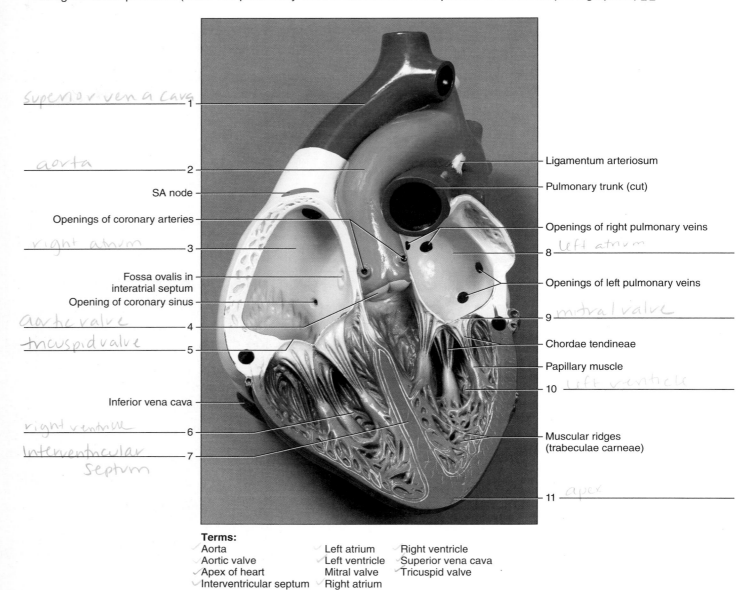

Handwritten labels:
1. superior vena cava
2. aorta
3. right atrium
4. aortic valve
5. tricuspid valve
6. right ventricle
7. Interventricular septum
8. left atrium
9. mitral valve
10. left ventricle
11. apex

Printed labels (left): SA node · Openings of coronary arteries · Fossa ovalis in interatrial septum · Opening of coronary sinus · Inferior vena cava

Printed labels (right): Ligamentum arteriosum · Pulmonary trunk (cut) · Openings of right pulmonary veins · Openings of left pulmonary veins · Chordae tendineae · Papillary muscle · Muscular ridges (trabeculae carneae)

Terms:

✓ Aorta	✓ Left atrium	✓ Right ventricle
✓ Aortic valve	✓ Left ventricle	✓ Superior vena cava
✓ Apex of heart	Mitral valve	✓ Tricuspid valve
✓ Interventricular septum	✓ Right atrium	

Learning Extension Assessment

Use red and blue colored pencils to color the blood vessels in figure 44.11. Use red to illustrate a blood vessel with oxygen-rich blood, and use blue to illustrate a blood vessel with oxygen-poor blood. 2

FIGURE 44.11 Label this frontal section of the human heart. The arrows indicate the direction of blood flow. 🔼

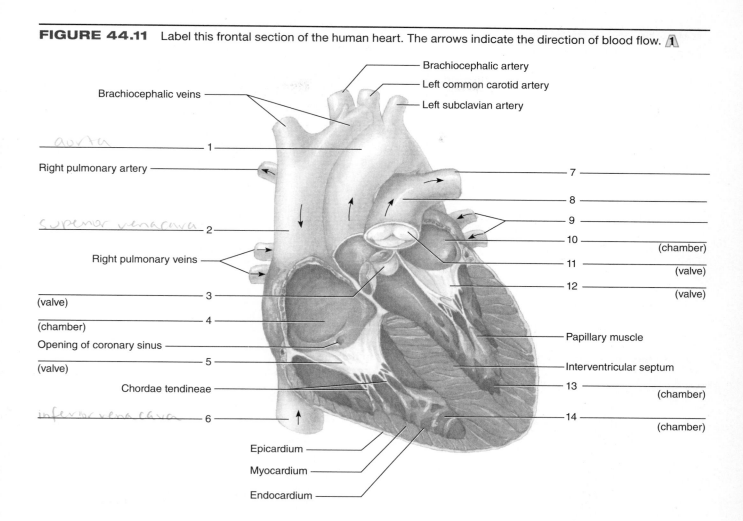

Brachiocephalic veins

Brachiocephalic artery
Left common carotid artery
Left subclavian artery

aorta 1

Right pulmonary artery

7
8
9
10
(chamber)
11
(valve)
12
(valve)

superior venacava 2

Right pulmonary veins

3
(valve)

4
(chamber)

Opening of coronary sinus

5
(valve)

Chordae tendineae

inferior vena cava 6

Papillary muscle

Interventricular septum

13
(chamber)

14
(chamber)

Epicardium
Myocardium
Endocardium

Part B   Assessments

Match the terms in column A with the descriptions in column B. Place the letter of your choice in the space provided. 2️⃣ 3️⃣

Column A
a. Aorta
b. Cardiac vein
c. Coronary artery
d. Coronary sinus
e. Endocardium
f. Mitral valve
g. Myocardium
h. Papillary muscle
i. Pericardial cavity
j. Pericardial sac
k. Pulmonary trunk
l. Tricuspid valve

Column B
_____ **1.** Structure from which chordae tendineae originate

_____ **2.** Prevents blood movement from right ventricle to right atrium

_____ **3.** Membranes around heart

_____ **4.** Prevents blood movement from left ventricle to left atrium

_____ **5.** Gives rise to left and right pulmonary arteries

_____ **6.** Drains blood from myocardium into right atrium

_____ **7.** Inner lining of heart chamber

_____ **8.** Layer largely composed of cardiac muscle tissue

_____ **9.** Space containing serous fluid to reduce friction during heartbeats

_____ **10.** Drains blood from myocardial capillaries

_____ **11.** Supplies blood to heart muscle

_____ **12.** Distributes blood to body organs (systemic circuit) except lungs

Part C Assessments

Complete the following:

1. Compare the structure of the right atrioventricular valve with that of the pulmonary valve. ⚠ _____

   _____

   _____

   _____

   _____

2. Describe the action of the right atrioventricular valve when you squeezed the water-filled right ventricle. ⚠ _____

   _____

   _____

   _____

   _____

3. Describe the function of the chordae tendineae and the papillary muscles. ⚠ _____

   _____

   _____

   _____

   _____

4. What is the significance of the difference in thickness between the wall of the aorta and the wall of the pulmonary trunk? ⚠

   _____

   _____

   _____

   _____

5. List in order the major blood vessels, chambers, and valves through which blood must pass in traveling from a vena cava

   to the aorta. ⚠ ⚠ _____

   _____

   _____

   _____

Critical Thinking Assessment

What is the significance of the difference in thickness of the ventricular walls? ⚠

_____

_____

_____

Laboratory Exercise 45

Cardiac Cycle

Purpose of the Exercise

To review the events of a cardiac cycle, to become acquainted with normal heart sounds, and to record an electrocardiogram.

Materials Needed

Ph.I.L.S. 4.0

For Procedure A—Heart Sounds:
Stethoscope
Alcohol swabs (wipes)

For Procedure B—Electrocardiogram:
Electrocardiograph (or other instrument for recording an ECG)
Cot or table
Alcohol swabs (wipes)
Electrode cream (paste)
Plate electrodes and cables
Self-sticking leads (optional)
Lead selector switch

Learning Outcomes

After completing this exercise, you should be able to

1. Interpret the major events of a cardiac cycle.
2. Associate the sounds produced during a cardiac cycle with the valves closing.
3. Correlate the components of a normal ECG pattern with the phases of a cardiac cycle.
4. Record and interpret an electrocardiogram.
5. Diagram and label the waves of an electrocardiogram (ECG) and correlate with the heart sounds.
6. Integrate the electrical changes and the heart sounds with the cardiac cycle.

Pre-Lab

Carefully read the introductory material and examine the entire lab. Be familiar with the cardiac cycle, heart sounds, cardiac conduction system, and electrocardiograms from lecture or the textbook. Answer the pre-lab questions.

Pre-Lab Questions: Select the correct answer for each of the following questions:

1. The _____ of the conduction system is known as the pacemaker.
   a. SA node   b. AV node
   c. AV bundle   d. left bundle branch
2. The _____ of the conduction system is/are located throughout the ventricular walls.
   a. AV node   b. left bundle branch
   c. right bundle branch d. Purkinje fibers
3. The first of two heart sounds (lubb) occurs when the
   a. AV valves open.
   b. AV valves close.
   c. semilunar valves open.
   d. semilunar valves close.
4. One cardiac cycle would consist of
   a. left chamber contractions followed by right chamber contractions.
   b. right chamber contractions followed by left chamber contractions.
   c. atrial chamber contractions followed by ventricular chamber contractions.
   d. ventricular chamber contractions followed by atrial chamber contractions.
5. The depolarization of ventricular fibers is indicated by the _____ of an ECG.
   a. P wave   b. QRS complex
   c. T wave   d. P-Q interval
6. The dupp sound occurs when the semilunar valves are closing during ventricular diastole.
   True _____ False _____
7. The P wave of an ECG occurs during the repolarization of the atria.
   True _____ False _____

447

A set of atrial contractions while the ventricular walls relax, followed by ventricular contractions while the atrial walls relax, constitutes a *cardiac cycle*. Such a cycle is accompanied by blood pressure changes within the heart chambers, movement of blood in and out of the chambers, and opening and closing of heart valves. These valves closing produce vibrations in the tissues and thus create the sounds associated with the heartbeat. If backflow of blood occurs when a heart valve is closed, it creates a turbulence noise known as a murmur.

The regulation and coordination of the cardiac cycle involves the *cardiac conduction system*. The pathway of electrical signals originates from the *sinoatrial (SA) node* located in the right atrium near the entrance of the superior vena cava. The stimulation of the heartbeat and the heart rate originate from the SA node, so it is often called the *pacemaker*. As the signals pass through the atrial walls toward the *atrioventricular (AV) node*, contractions of the atria followed by relaxations take place. Once the AV node has been signaled, the rapid continuation of the electrical signals occurs through the *AV bundle* and the *right and left bundle branches* within the interventricular septum and terminates via the *Purkinje fibers* throughout the ventricular walls. Once the myocardium of the ventricles has been stimulated, ventricular contractions followed by relaxations occur, completing one cardiac cycle. The sympathetic and para-sympathetic subdivisions of the autonomic nervous system influence the activity of the pacemaker under various conditions. An increased rate results from sympathetic responses; a decreased rate results from parasympathetic responses.

A number of electrical changes also occur in the myocardium as it contracts and relaxes. These changes can be detected by using metal electrodes and an instrument called an *electrocardiograph*. The recording produced by the instrument is an *electrocardiogram*, or *ECG (EKG)*. *Depolarization* and *repolarization* electrical events of the cardiac cycle can be observed and interpreted from the ECG graphic recording. The heart rate can be determined by counting the number of QRS complexes on the ECG during one minute.

Procedure A—Heart Sounds

1. Listen to your heart sounds. To do this, follow these steps:
   a. Obtain a stethoscope and clean its earpieces and the diaphragm by using alcohol swabs.
   b. Fit the earpieces into your ear canals so that the angles are positioned in the forward direction.
   c. Firmly place the diaphragm (or bell) of the stethoscope on the chest over the fifth intercostal space near the apex of the heart (fig. 45.1) and listen to the

FIGURE 45.1 The first sound (lubb) of a cardiac cycle can be heard by placing the diaphragm of a stethoscope over the fifth intercostal space near the apex of the heart. The second sound (dupp) can be heard over the second intercostal space, just left of the sternum. The thoracic regions circled indicate where the sounds of each heart valve are most easily heard.

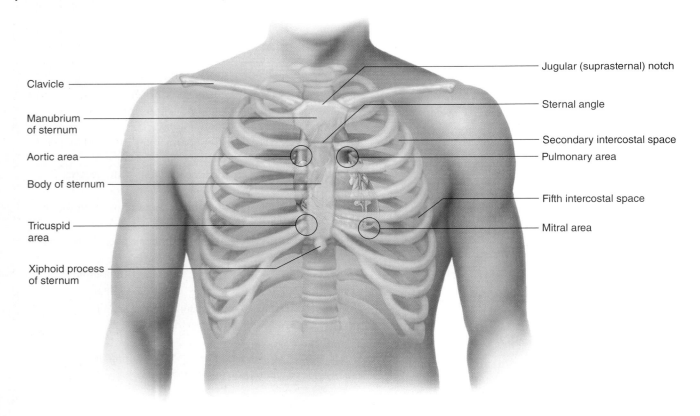

Clavicle

Manubrium of sternum

Aortic area

Body of sternum

Tricuspid area

Xiphoid process of sternum

Jugular (suprasternal) notch

Sternal angle

Secondary intercostal space

Pulmonary area

Fifth intercostal space

Mitral area

sounds. This is a good location to hear the first sound (*lubb*) of a cardiac cycle when the AV valves are closing, which occurs during ventricular *systole* (contraction).

   d. Move the diaphragm to the second intercostal space, just to the left of the sternum and listen to the sounds from this region. You should be able to hear the second sound (*dupp*) of the cardiac cycle clearly when the semilunar valves are closing, which occurs during ventricular *diastole* (relaxation).

   e. It is possible to hear sounds associated with the aortic and pulmonary valves by listening from the second intercostal space on either side of the sternum. The aortic valve sound comes from the right and the pulmonary valve sound from the left. The sound associated with the mitral valve can be heard from the fifth intercostal space at the nipple line on the left. The sound of the tricuspid valve can be heard at the fifth intercostal space just to the right of the sternum (fig. 45.1).

2. Inhale slowly and deeply and exhale slowly while you listen to the heart sounds from each of the locations as before. Note any changes that have occurred in the sounds.

3. Exercise moderately outside the laboratory for a few minutes so that other students listening to heart sounds will not be disturbed. After the exercise period, listen to the heart sounds and note any changes that have occurred in them. *This exercise should be avoided by anyone with health risks.*

4. Complete Parts A and B of Laboratory Assessment 45.

Procedure B—Electrocardiogram

A single heartbeat is a sequence of precisely timed contractions and relaxations of the four heart chambers initiated from the cardiac conduction system (fig. 45.2). The basic rhythm is controlled from the SA node (pacemaker), but it can be modified by the autonomic nervous system. During atrial contraction (systole), the ventricles are in a state of relaxation (diastole) and are filling with blood forced into them from the atria. Atrial contraction is soon followed by ventricular contraction (systole), during which time atrial relaxation (diastole) occurs. As a result of the synchronized ventricular contractions, blood is forced into the two large arteries that carry blood away from the heart. The aorta receives blood from the left ventricle into the systemic circuit; the pulmonary trunk receives blood from the right ventricle into the pulmonary circuit. This entire sequence of events of a single heartbeat is the cardiac cycle.

   Due to the action of the cardiac conduction system, the heart contracts in a coordinated manner. The events of the cardiac conduction system and the cardiac cycle form the basis for the electrocardiogram. The appearance of a normal ECG pattern for one heartbeat is shown in figure 45.3. The ECG components, normal time durations, and corresponding significance of the waves and intervals are

FIGURE 45.2 The cardiac conduction system pathway, indicated by the arrows, from the SA node to the Purkinje fibers (frontal section of heart).

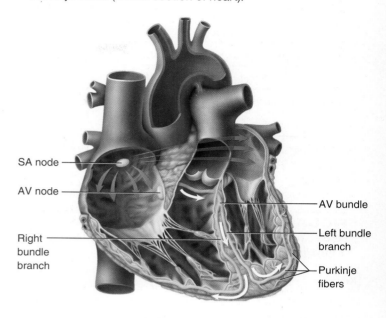

FIGURE 45.3 Components of a normal ECG pattern with a time scale.

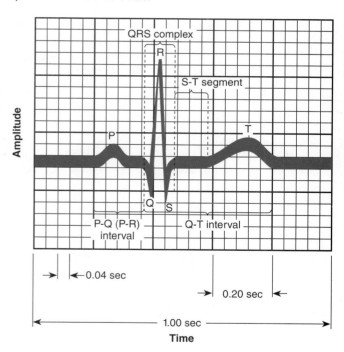

placed in table 45.1. An illustration of the relationship of the ECG components with the depolarization and repolarization of the atria followed by the ventricles is presented in figure 45.4.

1. Study figures 45.2, 45.3, and 45.4, and table 45.1.

2. Complete Part C of the laboratory assessment.

TABLE 45.1 ECG Components, Durations, and Significance

ECG Component	Duration in Seconds	Corresponding Significance
P wave	0.06–0.11	Depolarization of atrial fibers
P-Q (P-R) interval	0.12–0.20	Time for cardiac impulse from SA node through AV node
QRS complex	< 0.12	Depolarization of ventricular fibers
S-T segment	0.12	Time for ventricles to contract
Q-T interval	0.32–0.42	Time from ventricular depolarization to end of ventricular repolarization
T wave	0.16	Repolarization of ventricular fibers (ends pattern)

Note: Atrial repolarization is not observed in the ECG because it occurs at the same time that ventricular fibers depolarize. Therefore it is obscured by the QRS complex.

FIGURE 45.4 Depolarization waves (red) followed by repolarization waves (green) of the atria and ventricles during a cardiac cycle, along with the representative segment of the ECG pattern. The yellow arrows indicate the direction the wave is traveling.

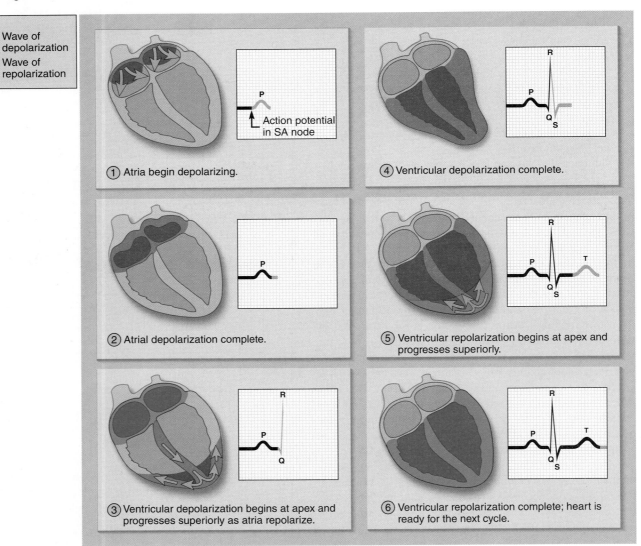

Key
- Wave of depolarization
- Wave of repolarization

1. Atria begin depolarizing.
Action potential in SA node

2. Atrial depolarization complete.

3. Ventricular depolarization begins at apex and progresses superiorly as atria repolarize.

4. Ventricular depolarization complete.

5. Ventricular repolarization begins at apex and progresses superiorly.

6. Ventricular repolarization complete; heart is ready for the next cycle.

3. The laboratory instructor will demonstrate the proper adjustment and use of the instrument available to record an electrocardiogram.

4. Record your laboratory partner's ECG. To do this, follow these steps:

   a. Have your partner lie on a cot or table close to the electrocardiograph, remaining as relaxed and still as possible.

   b. Scrub the electrode placement locations with alcohol swabs (fig. 45.5). Apply a small quantity of electrode cream to the skin on the insides of the wrists and ankles. (Any jewelry on the wrists or ankles should be removed.)

   c. Spread some electrode cream over the inner surfaces of four plate electrodes and attach one to each of the prepared skin areas (fig. 45.5). Make sure there is good contact between the skin and the metal of the electrodes. The electrode plate on the right ankle is the grounding system.

   d. Attach the plate electrodes to the corresponding cables of a lead selector switch. When an ECG recording is made, only two electrodes are used at a time, and the selector switch allows various combinations of electrodes (leads) to be activated. Three standard limb leads placed on the two wrists and the left ankle are used for an ECG. This arrangement has become known as *Einthoven's triangle*,[1] which permits the recording of the potential difference between any two of the electrodes (fig. 45.6).

   The standard leads I, II, and III are called bipolar leads because they are the potential difference between two electrodes (a positive and a negative). Lead I measures the potential difference between the right wrist (negative) and the left wrist (positive). Lead II measures the potential difference between the right wrist and the left ankle, and lead III measures the potential difference between the left wrist and the left ankle. The right ankle is always the ground.

   e. Turn on the recording instrument and adjust it as previously demonstrated by the laboratory instructor. The paper speed should be set at 2.5 cm/second. This is the standard speed for ECG recordings.

   f. Set the lead selector switch to lead I (right wrist, left wrist electrodes), and record the ECG for 1 minute.

   g. Set the lead selector switch to lead II (right wrist, left ankle electrodes), and record the ECG for 1 minute.

   h. Set the lead selector switch to lead III (left wrist, left ankle electrodes), and record the ECG for 1 minute.

---

[1]Willem Einthoven (1860–1927), a Dutch physiologist, received the Nobel prize for physiology or medicine for his work with electrocardiograms.

FIGURE 45.5 To record an ECG, attach electrodes to the wrists and ankles.

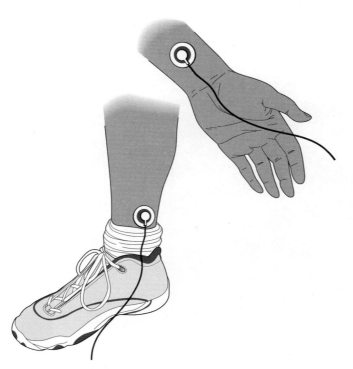

FIGURE 45.6 Standard limb leads for electrocardiograms.

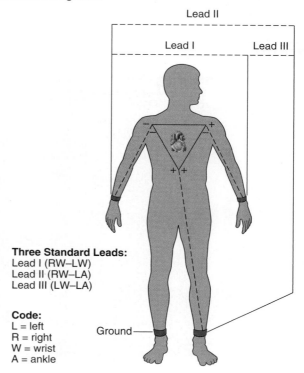

Lead II

Lead I   Lead III

Three Standard Leads:
Lead I (RW–LW)
Lead II (RW–LA)
Lead III (LW–LA)

Code:
L = left
R = right
W = wrist
A = ankle

Ground

i. Remove the electrodes and clean the cream from the metal and skin.

j. Use figure 45.3 to label the ECG components of the results from leads I, II, and III. The P-Q interval is often called the P-R interval because the Q wave is often small or absent. The normal P-Q interval is 0.12–0.20 seconds. The normal QRS complex duration is less than 0.10 seconds.

5. Complete Part D of the laboratory assessment.

Procedure C—Ph.I.L.S. Lesson 26 ECG and Heart Function: The Meaning of Heart Sounds

1. Open Exercise 26, ECG and Heart Function: The Meaning of Heart Sounds.

2. Read the objectives and introduction and take the pre-lab quiz.

3. After completing the pre-lab quiz, read through the wet lab. *Be sure to click open and view the videos that are indicated in red in the wet lab.*

4. The lab exercise will open when you click Continue after completing the wet lab (fig. 45.7).

Setup

5. Check that you can hear heart sounds.

6. View animation on recording when the heart sounds are produced. Notice that the button is pushed down with the first heart sound (lubb) and released with the second heart sound (dupp). Complete the practice test.

7. To start, click the power switch on the virtual computer screen then click the power switch to turn on the Data Acquisition Unit.

8. Electrodes detect electrical activity of the heart that is conducted through the body and transmit **to** the Data Acquisition Unit (input). Connect the electrodes to the Data Acquisition Unit by inserting the blue plug into input 1 of the Data Acquisition Unit and attach electrodes to the volunteer (red electrode to left wrist; black electrode [ground] to right wrist; and green electrode to left ankle).

Producing ECG and Recording Heart Sounds

9. Click Start on the virtual computer screen (blue line is ECG and yellow line is heart sounds).

10. As the line tracing scrolls across the control panel screen, use the mouse to mark the heart sounds. Record four to six cycles.

11. *Print the graph of the ECG and heart sounds by clicking the P in the bottom left of the monitor screen.*

Interpreting Results

12. *With the graph still on the screen,* answer the questions in Part E of the laboratory assessment. If you accidentally closed the graph, click on the Journal panel (red rectangle at bottom right of screen).

13. Complete the Post-Lab Quiz (click open Post-Lab Quiz and Lab Report) by answering the ten questions on the computer screen.

14. Read the conclusion on the computer screen.

15. You may print the Lab Report for ECG and Heart Function: The Meaning of Heart Sounds.

FIGURE 45.7 Opening screen for the laboratory exercise on ECG and Heart Function: The Meaning of Heart Sounds.

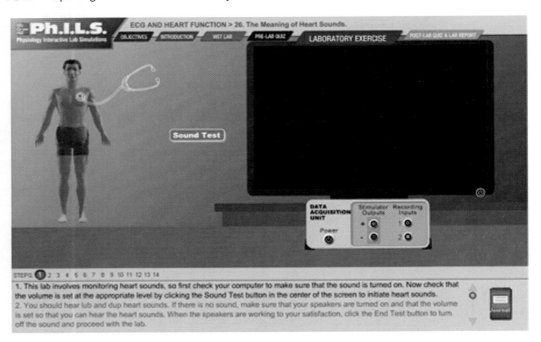

Name _____

Date _____

Section _____

The Ⓐ corresponds to the indicated outcome(s) found at the beginning of the laboratory exercise.

Cardiac Cycle

Part A Assessments

Complete the following statements:

1. The period during which a heart chamber is contracting is called ___systole___. Ⓐ1

2. The period during which a heart chamber is relaxing is called ___diastole___. Ⓐ1

3. During ventricular contraction, the AV valves (tricuspid and mitral valves) are _____. Ⓐ1

4. During ventricular relaxation, the AV valves are _____. Ⓐ1

5. The pulmonary and aortic valves open when the pressure in the _____ exceeds the pressure in the pulmonary trunk and aorta. Ⓐ1

6. The first sound of a cardiac cycle occurs when the _____ are closing. Ⓐ2

7. The second sound of a cardiac cycle occurs when the _____ are closing. Ⓐ2

8. The sound created when blood leaks back through an incompletely closed valve is called a _____. Ⓐ2

Part B Assessments

Complete the following:

1. What changes did you note in the heart sounds when you inhaled deeply? Ⓐ2

_____

_____

_____

_____

2. What changes did you note in the heart sounds following the exercise period? Ⓐ2

_____

_____

_____

_____

Part C Assessments

Complete the following statements:

1. Normally, the _____ node serves as the pacemaker of the heart. **3**

2. The _____ node is located in the inferior portion of the interatrial septum. **3**

3. The fibers that carry cardiac impulses from the interventricular septum into the myocardium are called _____ fibers. **3**

4. A(n) _____ is a recording of electrical changes occurring in the myocardium during a cardiac cycle. **3**

5. The P wave corresponds to depolarization of the muscle fibers of the _____. **3**

6. The QRS complex corresponds to depolarization of the muscle fibers of the _____. **3**

7. The T wave corresponds to repolarization of the muscle fibers of the _____. **3**

8. Why is atrial repolarization not observed in the ECG? **3**

_____

_____

Part D Assessments

1. Attach a short segment of the ECG recording from each of the three leads you used, and label the waves of each. **4**

Lead I

Lead II

Lead III

2. What differences do you find in the ECG patterns of these leads? 🅰 _____

_____

_____

_____

_____

3. How much time passed from the beginning of the P wave to the beginning of the QRS complex (P-Q interval, or P-R interval) in the ECG from lead I? 🅰 _____

4. What is the significance of this P-Q (P-R) interval? 🅰 _____

_____

_____

_____

_____

5. How can you determine the heart rate from an electrocardiogram? 🅰 _____

_____

_____

_____

6. What was your heart rate as determined from the ECG? 🅰 _____

Critical Thinking Assessment

If a person's heart rate is 72 beats per minute, determine the number of QRS complexes that would have appeared on an ECG during the first 30 seconds. 🅰

_____

_____

_____

Part E Ph.I.L.S. Lesson 26, ECG and Heart Function: The Meaning of Heart Sounds Assessments

1. Diagram the graph of the ECG recording from the computer simulation and indicate the heart sounds.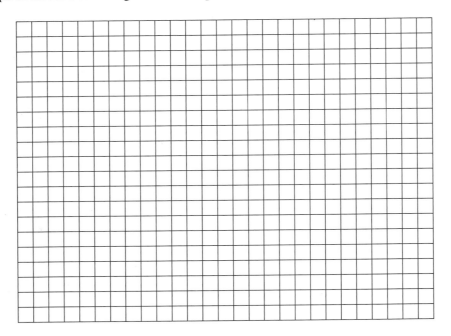

2. Answer the following questions about the graph in question 1. **5**

   a. The QRS wave is correlated with the _____ (first heart sound, lubb, or the second heart sound, dupp)?

   b. The T wave is correlated with the _____ (first heart sound, lubb, or the second heart sound, dupp)?

3. Complete the following: **1 2 4 5 6**

   a. Atrial depolarization, causing the _____ wave of the ECG, has triggered contraction of the atria, and the blood is pumped from the atria into the ventricles. The position of the valves (AV valves open and semilunar valves closed) remains unchanged.

   b. Ventricular depolarization, causing the _____ wave of the ECG, has triggered contraction of the ventricles, and the blood is pumped from the ventricles into the arteries (aorta and pulmonary trunk). The

   _____ valves close, producing the _____ heart sound, and the

   _____ valves remain open.

   c. Ventricular repolarization, causing the _____ wave of the ECG, occurs as the ventricles are relaxing. To prevent the backflow of blood from the arteries into the ventricles, the _____ valves close, producing

   the _____ heart sounds.

   d. At what point do the AV valves reopen? _____

Critical Thinking Assessment

If a person has a heart murmur (caused by the improper opening or closing of a heart valve), the _____
(ECG or the heart sounds) will be unusual as the blood is forced through a more narrow opening (stenotic valve), or the blood will backflow through a valve that does not close properly (incompetent valve). **2**

Electrocardiography: BIOPAC© Exercise

Purpose of the Exercise

To record an ECG and calculate heart rate and rhythm changes that occur with changes in body position and breathing.

Materials Needed

Computer system (Mac OS X 10.4–10.6, or PC running Windows XP, Vista, or 7)
BIOPAC Student Lab software ver. 3.7.7 or above
BIOPAC Acquisition Unit MP36, MP35, MP30 (Windows only) or MP45 (Note: The MP30 is being phased out.) with (AC100A) transformer
BIOPAC serial cable (CBLSERA)
BIOPAC electrode lead set (SS2L)
BIOPAC disposable vinyl electrodes (EL503), 6 electrodes per subject
BIOPAC electrode gel (GEL1) and abrasive pad (ELPAD) or alcohol wipes
Exercise mat, cot, or lab table with pillow
Chair or stool

Safety

▶ The BIOPAC electrode lead set is safe and easy to use and should be used only as described in the procedures section of the laboratory exercise.
▶ The electrode lead clips are color-coded. Make sure they are connected to the properly placed electrode as demonstrated in figure 46.3.
▶ The vinyl electrodes are disposable and are meant to be used only once. Each subject should use a new set.

Learning Outcomes

After completing this exercise, you should be able to

(1) Record the electrical components of an electrocardiogram.

(2) Correlate electrical events as displayed on the ECG with the mechanical events that occur during the cardiac cycle.

(3) Interpret rate and rhythm changes in the ECG associated with changes in body position, activity level, and breathing.

Pre-Lab

Carefully read the introductory material and examine the entire lab content. Be familiar with the cardiac cycle, cardiac conduction system, and electrocardiogram from lecture or the textbook.

The beating of the heart is one of the basic requirements for maintaining life, pumping blood that contains vital oxygen and nutrients to all the active cells of the body. A single heartbeat is a coordinated sequence of events, involving precisely timed contractions and relaxations of the four chambers of the heart. During a heartbeat, the atria, the upper chambers of the heart, contract simultaneously. This action squeezes the blood from the atria into the ventricles, the lower chambers of the heart. During atrial contraction (*systole*), the ventricles are in a state of relaxation (*diastole*) and are filling with blood from the contracting atria. Atrial contraction is soon followed by ventricular contraction, the simultaneous contraction of the two large pumping chambers of the heart. After a short period in which all four heart chambers are relaxed, these events repeat. This entire sequence of events occurs in every heartbeat and is called the *cardiac cycle.*

The heart is a large, cone-shaped pump, consisting mainly of cardiac muscle, one of the three types of muscle tissue in the body. Cardiac muscle differs from skeletal muscle in that it is controlled by specialized cardiac muscle cells right within the heart, rather than by motor neurons. The action of cardiac muscle can be modified by autonomic nerve input, but its basic rhythm is set by specialized cells of the heart, called *pacemaker cells,* found in the small clumps (nodes) of the cardiac conduction system.

The *cardiac conduction system* is a specialized group of cardiac muscle cells that generates and spreads cardiac impulses throughout the heart. The major components of the cardiac conduction system are:

1. **Sinoatrial (SA) node:** This small clump of specialized cardiac muscle cells in the upper wall of the right atrium is called the *pacemaker* of the heart. The SA node sets the rhythm of the heart, since it has the fastest innate firing rate of any conducting cells of the heart. It spreads its impulses across the atria to initiate atrial contraction and to the next component of the cardiac conduction system, the AV node.

2. **Atrioventricular (AV) node:** This tiny clump of conducting cardiac muscle cells in the lower portion of the right atrium receives impulses from the SA node and transports them to the next structure of the conduction system, the AV bundle. The AV node provides a "bottleneck" in the conduction system, which serves to delay the signal from the SA node for about 0.1 seconds before sending it on to the AV bundle. This gives the atria time to complete their contraction and squeeze any remaining blood into the ventricles before the ventricles begin their contraction.

3. **Atrioventricular bundle (AV bundle; bundle of His):** This group of conducting cells at the top of the interventricular septum transports impulses from the AV node to the next structures of the conduction system, the left and right bundle branches. This distributes impulses toward the left and right ventricles.

4. **Left and right bundle branches:** The bundle branches transport cardiac impulses down the interventricu-

lar septum toward the apex of the heart. From these branches, the Purkinje fibers split off.

5. **Purkinje fibers:** Purkinje fibers are large cardiac muscle fibers that split off from the bundle branches and transport cardiac impulses throughout the entire myocardium of the ventricles. These fibers are considered the final component of the cardiac conduction system. From here, the impulses are spread from one contractile cardiac muscle cell to the next via gap junctions in their cell membranes.

Figure 46.1 shows the arrangement of the cardiac conduction system and the locations of its components.

Due to the action of the cardiac conduction system, the heart contracts in a coordinated manner. The atria contract just before the ventricles, providing the ventricles with time to fill completely, before starting their own contraction. This increases the efficiency of the heart as a pump and ensures that the heart contracts as a functional unit almost every time.

The events of the cardiac conduction system and the cardiac cycle form the basis for this lab exercise. An *electrocardiogram* (ECG or EKG) is a recording of the electrical activity derived from the heart during the cardiac cycle. As in the EMG and EEG that you studied in previous lab exercises, the ECG is derived from the sum of the ionic movements that occur during the action potentials of many cardiac muscle cells at the same time. It is not a recording of action potentials in a single cardiac muscle cell. The movement of ions in the extracellular fluids generates current flow, which can be detected by electrodes placed on the skin. ECGs are used not only to observe normal heart action, but also to help

FIGURE 46.1 The components of the cardiac conduction system.

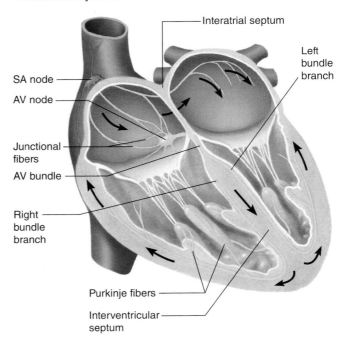

FIGURE 46.2 Components of the ECG.

Source: Courtesy of and © BIOPAC Systems Inc.

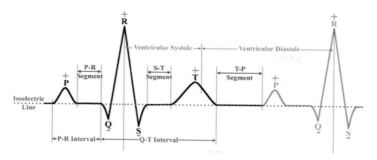

FIGURE 46.3 Electrode attachments for Lead II.

Source: Courtesy of and © BIOPAC Systems Inc.

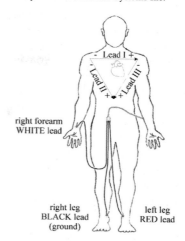

in the diagnosis of many disorders of the cardiac conduction system as well as some types of heart disease.

The ECG is recorded in units of mV (millivolts). The typical appearance of an ECG for a single heartbeat is shown in figure 46.2. It consists of three deflections from the isoelectric line (the baseline, or zero-voltage line, which occurs during the short periods in which no voltage is being recorded):

1. **P wave:** The P wave is a positive (+) deflection from the isoelectric line that results from atrial depolarization. This depolarization is caused by the firing of the SA node, and it leads to the immediate contraction of the atria.
2. **QRS complex:** The QRS complex is a large deflection that results from ventricular depolarization. This depolarization of the ventricles is immediately followed by ventricular contraction (systole). The electrical events of atrial repolarization occur simultaneously with those of ventricular depolarization. Therefore, atrial repolarization is "hidden" inside the QRS complex and cannot be detected separately on an ECG.
3. **T wave:** The T wave is a positive deflection that results from ventricular repolarization. Repolarization of the ventricles is immediately followed by ventricular relaxation (diastole).

In addition to the wave components of the ECG, various segments and intervals can also be measured, all with their own significance in the cardiac cycle. A *segment* is a period of time between the end of a wave and the beginning of the next wave; it lies entirely along the isoelectric line. For example, the P-R segment runs from the end of the P wave to the beginning of the QRS complex; it represents the time required for impulses to be conducted from the AV node to the ventricles. An *interval* is a period of time that contains at least one wave and one segment along the isoelectric line. For example, the P-R interval runs from the beginning of the P wave to the beginning of the QRS complex; it represents the time between the beginning of atrial depolarization and the beginning of ventricular depolarization.

ECGs can be conducted by placing electrodes in a variety of positions on the limbs and chest and measuring the voltage between them. There are several possible combinations of electrodes that can measure the ECG, and the results will vary with electrode placement. The particular arrange-

ment of positive and negative electrodes along with a ground electrode, that is used to conduct the ECG is called a *lead*. In this lab exercise, you are going to use lead II, in which the positive electrode is on the left ankle, the negative electrode on the right wrist, and the ground electrode on the right ankle. Proper electrode placement for lead II is shown in figure 46.3. Typical lead II ECG values are shown in table 46.1.

Procedure A—Setup

1. With your computer turned **ON** and the BIOPAC MP3X or MP45 unit turned **OFF,** plug the electrode lead set (SSL2) into Channel 1. Plug the headphones into the back of the unit if desired to listen to the heartbeat.
2. Turn on the MP3X or MP45 Data Acquisition Unit.
3. Attach the electrodes to the subject as shown in figure 46.3. To help to ensure a good contact, make sure the area to be in contact with the electrodes is clean by wiping with the abrasive pad (ELPAD) or alcohol wipe. A small amount of BIOPAC electrode gel (GEL1) may also be used to make better contact between the sensor in the electrode and the skin. For optimal adhesion, the

TABLE 46.1 Normal Lead II ECG Values

Phase	Duration (seconds)	Amplitude (millivolts)
P wave	0.06–0.11	<0.25
P-R interval	0.12–0.20	
P-R segment	0.08	
QRS complex (R)	<0.12	0.8–1.2
S-T segment	0.12	
Q-T interval	0.36–0.44	
T wave	0.16	<0.5

electrodes should be placed on the skin at least 5 minutes before the calibration procedure.

4. Attach the electrode lead set with the pinch connectors as shown in figure 46.3. Make sure the proper color electrode cable is attached to the appropriate electrode: *white lead* on anterior surface of right forearm at the wrist; *red lead* on medial surface of left leg at the ankle; and *black lead* on medial surface of right leg at the ankle (ground).

5. Have the subject lie down in the supine position on the mat, cot, or table and relax. Make sure to position the electrode cables so that they are not pulling on the electrodes. The cable clips may be attached to the subject's clothing for convenience.

6. Start the BIOPAC student lab program for Electrocardiography I, lesson L05-ECG-1.

7. Designate a unique filename to be used to save the subject's data.

Procedure B—Calibration

1. Make sure the subject is relaxed throughout the calibration procedure. *For all procedures, the subject must be still—no laughing or talking.* Click on **Calibrate.** A dialog box will direct you through the calibration procedure. After reading the box, click **OK.**

2. The calibration will stop automatically (this will take about 8 seconds).

3. The calibration recording should look similar to figure 46.4. If it does not, click **Redo Calibration.** If the baseline is not flat (which is called baseline drift), or there are any large spikes or jitters, click **Redo Calibration.**

Procedure C—Recording

This lab exercise consists of recording the ECG under four different conditions (segments): lying down in a supine position, seated (immediately after sitting up), performing deep breathing, and after exercise. In addition to the subject, you will have to designate a **Director** to coordinate the group members and a **Recorder** to insert event markers and time the exercise.

Segment 1—Supine (Lying Down)

1. With the subject lying down and relaxed, click **Record.** A marker labeled "Supine" will come up automatically. Record for 20 seconds.

FIGURE 46.4 Sample calibration recording.

Source: Courtesy of and © BIOPAC Systems Inc.

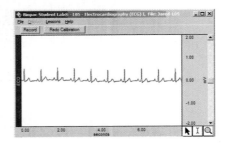

FIGURE 46.5 Segment 1 ECG—supine (lying down).

Source: Courtesy of and © BIOPAC Systems Inc.

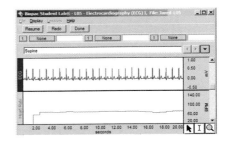

2. Click **Suspend.** The recording should be similar to figure 46.5. If it is not, click **Redo** and repeat step 1.

3. If correct, move on to step 4 (Segment 2).

Segment 2—Seated (After Sitting Up)

4. Have the subject quickly get up and sit in a chair, with forearms resting comfortably on the chair armrests or thighs. Click on **Resume.** A marker labeled "Seated" will come up automatically.

5. Record for 20 seconds (seconds 21–40).

6. Click **Suspend.** The recording should be similar to figure 46.6. If it is not, click **Redo** and repeat steps 4 and 5.

7. If correct, move on to step 8 (Segment 3).

Segment 3—Deep Breathing

8. Have the subject remain seated and prepare to take five deep breaths during the segment 3 recording. Allow the subject to breathe at his/her own resting rate.

9. Click on **Resume.** A marker labeled "Deep Breathing" will come up automatically. Have the subject inhale and exhale deeply for five cycles. At the beginning of each inhalation, the Recorder should press the F4 key. At the beginning of each exhalation, the Recorder should press the F5 key.

10. After five breaths, click **Suspend.** The recording should be similar to figure 46.7. If it is not, click **Redo** and repeat step 9.

11. If correct, move on to step 12 (Segment 4).

FIGURE 46.6 Segment 2 ECG—seated (after sitting up).

Source: Courtesy of and © BIOPAC Systems Inc.

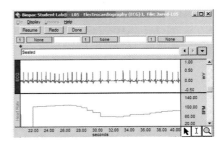

FIGURE 46.7 Segment 3 ECG—deep breathing.

Source: Courtesy of and © BIOPAC Systems Inc.

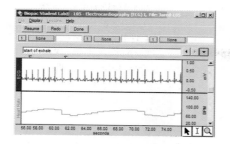

Segment 4—After Exercise

12. Have the subject perform moderate exercise to elevate his/her heart rate, such as jumping jacks or running in place. The electrode cable pinch connectors may be removed during exercising but not the electrodes. Immediately after the subject stops exercising, quickly reattach the pinch connectors; then move on to step 13.
13. Click on **Resume**. A marker labeled "After Exercise" will come up automatically.
14. Record for 60 seconds (seconds 71–130).
15. Click **Suspend**. The recording should be similar to figure 46.8. If it is not, click **Redo** and repeat steps 12–14.
16. If correct, click **Done**.
17. Click **Yes**. Remove the electrodes.

Procedure D—Data Analysis

1. You can either analyze the current file or save it and do the analysis later using the **Review Saved Data** mode and selecting the correct file. Note the channel number designations: **CH 1, ECG (Lead II)**, and **CH 40, Heart Rate.** The data window should look similar to figure 46.9.

Heart Rate

2. Note the settings for the measurement boxes across the top as follows:
   - CH 40—Value
   - CH 1—Delta T (Δ T)
   - CH 1—P-P
   - CH 1—BPM

FIGURE 46.8 Segment 4 ECG—after exercise.

Source: Courtesy of and © BIOPAC Systems Inc.

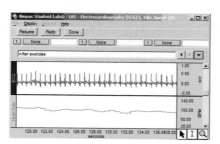

FIGURE 46.9 Data window for the entire ECG exercise.

Source: Courtesy of and © BIOPAC Systems Inc.

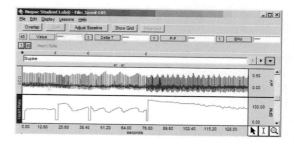

These will provide you with the following measurements:

Value—BPM or beats per minute in the preceding R-R interval

Δ T—the difference in time between the beginning and the end of the selected area

P-P (Peak to Peak)—measures the difference in amplitude between the maximum and minimum values of the selected area

BPM—the beats per minute; time difference between the beginning and the end of the selected area divided into 60 seconds/minute

3. Set up your display window for optimal viewing of three to five successive beats from Segment 1 (0–20 seconds). Do this by clicking anywhere in the horizontal scale (the numbers just above the word "seconds" at the bottom of the data window). A box will pop up in the middle of the screen, called **Horizontal Scale**. Under "Scale Range," change the lower scale to 2 and the upper scale to 6. Then click **OK**. This will spread out the data so you can see ECG recordings of individual heartbeats.
4. To measure heart rate, click on the I-beam and select any point within R-R interval, as shown in figure 46.10. The result will be displayed by CH 40, value. Record the heart rate in table 46.2 of Part A of Laboratory Assessment 46.

FIGURE 46.10 Proper selection of a data point for measuring heart rate.

Source: Courtesy of and © BIOPAC Systems Inc.

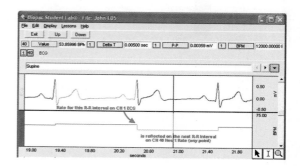

5. Measure heart rate for two more heartbeats within the 0–20 second range. You can scroll through the segment by clicking on the small square under the word "seconds" on the horizontal scale and dragging the square with your cursor. Record these heart rate measurements in table 46.2. Then take the mean of the three measurements.

6. Now scroll to the other three segments of the lab exercise and repeat steps 4 and 5. Record the data in table 46.2 of the lab assessment.

7. Now hide CH 40. Do this by holding down the Ctrl key and clicking on the CH 40 square.

Ventricular Systole and Diastole

8. Zoom in on a single cardiac cycle from Segment 1. Using the I-beam tool, measure the duration of ventricular systole and diastole by selecting the appropriate areas, as shown in figure 46.2. To measure ventricular systole, select the area between the peak of an R wave and the point one-third of the way down the descending part of the T wave. To measure ventricular diastole, select the area between the end point of the systole measurement and the peak of the next R wave. The result will be displayed by CH 1, Delta T. Record your results in table 46.3.

9. Now scroll to Segment 4, "After Exercise," Repeat step 8 for Segment 4, and record your results in table 46.3.

Waves, Intervals, and Segments of the ECG

10. Scroll back to Segment 1 (Supine) and observe the ECG of a single cardiac cycle (heartbeat).

11. Using the I-beam tool, measure the durations of the various waves, intervals, and segments of the ECG, as required to fill out table 46.4. See figure 46.11 for an

FIGURE 46.11 Proper selection of area for measuring the duration of a P wave.

Source: Courtesy of and © BIOPAC Systems Inc.

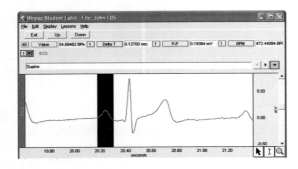

example of how to measure the duration of a P wave. See figure 46.12 for measuring P-R interval. The results for duration measurements will be displayed by CH 1, Delta T. (Refer back to figure 46.2 for proper selection of the areas to measure the intervals and segments.)

12. Measure durations for all waves, intervals, and segments of two more heartbeats in Segment 1 and record the results in table 46.4. Then calculate the means as required in the table.

13. Now scroll to the beginning of Segment 4 (After Exercise).

14. Measure the durations of all waves, intervals, and segments for a single heartbeat close to the beginning of Segment 4. Record the results in table 46.4.

15. Now scroll back to Segment 1 to measure amplitudes for the P wave, QRS complex, and T wave.

16. To measure the amplitude of a P wave, for example, use the I-beam tool to select the area between the beginning of the P wave deflection and the highest point of the P wave. The result will be displayed by CH 1, P-P. (The P-P measurement calculates the difference between the maximum and minimum values in the selected area. Therefore, your selection does not have to be perfect; you can simply include the maximum and minimum points in your selection of the P wave, and the P-P measurement will calculate the amplitude correctly.) Record your results in table 46.4.

17. Measure the wave amplitudes for two other heartbeats within Segment 1. Record the results in table 46.4. Then calculate the means.

18. Exit the program.

19. Complete Part B of the laboratory assessment.

FIGURE 46.12 Proper selection of area for measuring the duration of the P-R interval.

Source: Courtesy of and © BIOPAC Systems Inc.

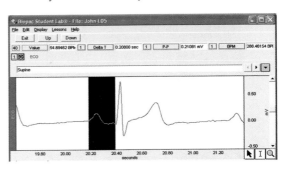

Name _____

Date _____

Section _____

The ⚠ corresponds to the indicated outcome(s) found at the beginning of the laboratory exercise.

Electrocardiography: BIOPAC© Exercise

Part A   Data and Calculations Assessments

Subject Profile

Name _____   Height _____

Age_____   Weight _____

Gender: Male / Female

Complete the following tables with the lesson data indicated and calculate the mean as appropriate. ⚠

1. Heart Rate Measurement

TABLE 46.2

| Segment Condition | Cardiac Cycle 40 Value | | | Mean (calculate) |
	1	2	3	
1: Supine				
2: Seated				
3: Start of inhale				
3: Start of exhale				
4: After exercise				

* If CH 40 was not recorded, use [1] [BPM].

2. Ventricular Systole and Diastole Measurements

TABLE 46.3

| Segment Condition | Duration (ms) 1 Delta T | |
	Ventricular Systole	Ventricular Diastole
1: Supine		
4: After exercise		

3. Waves, Intervals, and Segments of the ECG ⟨2⟩ ⟨3⟩

TABLE 46.4

Segment Condition	Normative Values Based on resting heart rate 75 BPM		Duration (ms) 1 Delta T					Amplitude (mV) 1 P-P			
			Segment 1 Cycle			Seg 1 Mean (calc.)	Seg 4 One cycle	Segment 1 Cycle			Seg 1 Mean (calc.)
			1	2	3			Cycle 1	Cycle 2	Cycle 3	
Waves	Dur. (sec)	Amp (mV)									
P	.07–.18	< .20									
QRS complex	.06–.12	.10–1.5									
T	.10–.25	< .5									
Intervals	Duration (seconds)										
P-R	.12–.20										
Q-T	.32–.36										
R-R	.80										
Segments	Duration (seconds)										
P-R	.02–.10										
S-T	< .20										
T-P	0–.40										

Note: Interpreting ECGs is a skill that requires practice to distinguish between normal variation and those arising from medical conditions. Do not be alarmed if your ECG does not match the "normal values" and references above and in the introduction.

Part B Assessments

Complete the following:

1. Was there a change in heart rate when the subject went from the supine position (Segment 1) to sitting (Segment 2)?

_____

Describe the physiological mechanisms causing this change. ⟨3⟩

_____

_____

_____

_____

2. Was there a difference between the heart rates when the subject was sitting at rest (Segment 2) and just after exercise (beginning of Segment 4)? _____ Explain any differences. ⟨3⟩

_____

_____

_____

3. What changes occurred in the duration of ventricular systole between the supine position (Segment 1) and post-exercise (beginning of Segment 4)? Explain any differences. **3**

_____

_____

_____

_____

_____

4. What changes occurred in the duration of ventricular diastole between the supine position (Segment 1) and post-exercise (beginning of Segment 4)? Explain any differences. **3**

_____

_____

_____

_____

_____

5. What changes occurred in the cardiac cycle during the respiratory cycle? In other words, did you observe a difference in heart rate between inhaling and exhaling (Segment 3)? Explain.

_____

_____

_____

_____

_____

Blood Vessel Structure, Arteries, and Veins

Purpose of the Exercise

To review the structure and functions of blood vessels and the circulatory pathways, and to locate major arteries and veins.

Materials Needed

Compound light
  microscope
Prepared microscope slides:
  Artery cross section
  Vein cross section
Live frog or goldfish
Frog Ringer's solution
Frog board or heavy
  cardboard (with a 1-inch
  hole cut in one corner)

Paper towel
Rubber bands
Masking tape
Dissecting pins
Thread
Dissectible human heart
  model
Human torso model
Anatomical charts of the
  cardiovascular system

For Learning Extension Activity:
Ice   Thermometer
Hot plate

Safety

▸ Wear disposable gloves when handling the live frogs.
▸ Return the frogs to the location indicated after the experiment.
▸ Wash your hands before leaving the laboratory.

Learning Outcomes

After completing this exercise, you should be able to

① Distinguish and sketch the three major layers in the wall of an artery and a vein.

② Interpret the types of blood vessels in the web of a frog's foot.

③ Distinguish and trace the pulmonary and systemic circuits.

④ Locate the major arteries in these circuits on a diagram, chart, or model.

⑤ Locate the major veins in these circuits on a diagram, chart, or model.

Pre-Lab

Carefully read the introductory material and examine the entire lab. Be familiar with the structure of blood vessels and the major arteries and veins of the systemic and pulmonary circuits from lecture or the textbook. Visit www.mhhe.com/martinseries2 for LabCam videos. Answer the pre-lab questions.

Pre-Lab Questions: Select the correct answer for each of the following questions:

1. The middle layer of an artery and vein contains mostly
   a. connective tissue.
   b. simple squamous epithelium.
   c. smooth muscle.
   d. nervous tissue.

2. Arteries carry blood
   a. away from the heart.
   b. toward the heart.
   c. both away from and toward the heart.
   d. only in the systemic circuit.

3. The _____ has the thickest wall.
   a. right atrium   **b.** left atrium
   c. right ventricle   **d.** left ventricle

4. Which of the following arteries is *not* a direct branch of the aorta?
   a. coronary artery   **b.** brachiocephalic trunk
   c. left subclavian   **d.** brachial artery
        artery

5. Which of the following veins is *not* located in the neck?
   a. vertebral vein
   b. brachiocephalic vein
   c. external jugular vein
   d. internal jugular vein

The blood vessels form a closed system of tubes that carry blood to and from the heart, lungs, and body cells. These tubes include *arteries* and *arterioles* that conduct blood away from the heart; *capillaries* in which exchanges of substances occur between the blood and surrounding tissues; and *venules* and *veins* that return blood to the heart.

Arteries and veins are composed of three layers or tunics. The inner layer, *tunica interna*, is composed of an endothelium of simple sqamous epithelial cells and the basement membrane. This establishes a smooth lining next to the lumen of the blood vessel. The middle layer, *tunica media*, is composed of smooth muscle. From various contractions and relaxations of the smooth muscle, the lumen size is modified, which

influences the amount of blood flow through a particular area of the body. The outer layer, *tunica externa*, is composed of connective tissues rich in collagen, providing some support and protection. In general, the arteries have thicker middle and outer layers and elastic laminae, which provide additional support and elasticity to withstand the higher blood pressures of arteries. In contrast, veins have thinner middle and outer layers and also have occasional valves to prevent backflow within these vessels that experience much lower pressures. Relatively small arteries are called arterioles; small veins are termed venules. Capillaries connecting arterioles to venules retain only the endothelium and basement membrane.

The blood vessels of the cardiovascular system can be divided into two major pathways—the *pulmonary circuit* and the *systemic circuit*. Within each circuit, arteries transport blood away from the heart. After exchanges of gases, nutrients, and wastes have occurred between the blood and the surrounding tissues, veins return the blood to the heart.

Frequent variations exist in anatomical structures among humans, especially in arteries and veins. The illustrations in this laboratory manual represent normal (normal means the most common variation) anatomy.

Procedure A—Blood Vessel Structure

1. Reexamine the introductory material. Compare the structure of an artery and a vein, illustrated in figure 47.1. Note that arteries and veins possess the same three

---

FIGURE 47.1   Structure of the wall of an artery and a vein.

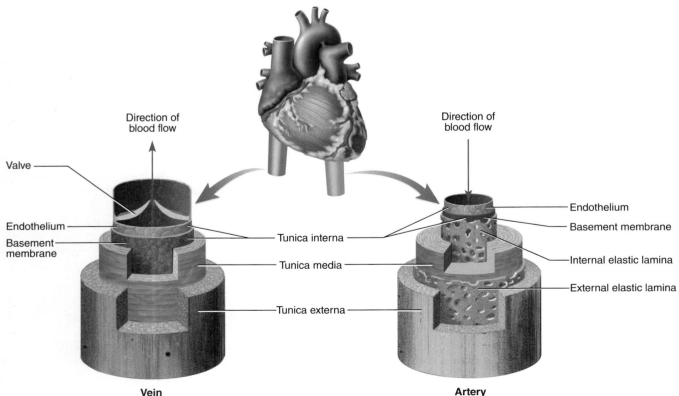

Valve

Endothelium

Basement membrane

Direction of blood flow

Tunica interna

Tunica media

Tunica externa

Vein

Direction of blood flow

Endothelium

Basement membrane

Internal elastic lamina

External elastic lamina

Artery

FIGURE 47.2 Cross section of a small artery (arteriole) and a small vein (venule) (200×).

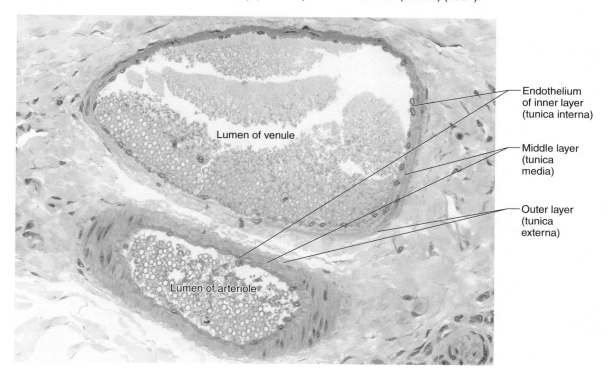

Endothelium
of inner layer
(tunica interna)

Middle layer
(tunica
media)

Outer layer
(tunica
externa)

Lumen of venule

Lumen of arteriole

layers or tunics, but an artery has thicker middle and outer tunics and also contains two elastic laminae. The vein has thinner middle and outer tunics, lacks the two elastic laminae, and has periodic valves.

2. Obtain a microscope slide of an artery cross section, and examine it using low-power and high-power magnification (fig. 47.2). Identify the three distinct layers (tunics) of the arterial wall. The inner layer (tunica interna), is composed of an endothelium (simple squamous epithelium) and appears as a wavy line due to an abundance of elastic fibers that have recoiled just beneath it. The middle layer (tunica media) consists of numerous concentrically arranged smooth muscle cells with elastic fibers scattered among them. The outer layer (tunica externa) contains connective tissue rich in collagenous fibers.

3. Prepare a labeled sketch of the arterial wall in Part A of Laboratory Assessment 47.

4. Obtain a slide of a vein cross section and examine it as you did the artery cross section (fig. 47.2). Note the thinner wall and larger lumen relative to an artery of comparable locations. Identify the three layers of the wall and prepare a labeled sketch in Part A of the laboratory assessment.

5. Complete Part A of the laboratory assessment.

6. Observe the blood vessels in the webbing of a frog's foot. (As a substitute for a frog, a live goldfish could be used to observe circulation in the tail.) To do this, follow these steps:

a. Obtain a live frog. Wrap its body in a moist paper towel, leaving one foot extending outward. Secure the towel with rubber bands, but be careful not to wrap the animal so tightly that it could be injured. Try to keep the nostrils exposed.

b. Place the frog on a frog board or on a piece of heavy cardboard with the foot near the hole in one corner.

c. Fasten the wrapped body to the board with masking tape.

d. Carefully spread the web of the foot over the hole and secure it to the board with dissecting pins and thread (fig. 47.3). Keep the web moist with frog Ringer's solution.

e. Secure the board on the stage of a microscope with heavy rubber bands and position it so that the web is beneath the objective lens.

f. Focus on the web, using low-power magnification, and locate some blood vessels. Note the movement of the blood cells and the direction of the blood flow. You might notice that red blood cells of frogs are nucleated. Identify an arteriole, a capillary, and a venule.

g. Examine each of these vessels with high-power magnification.

h. When finished, return the frog to the location indicated by your instructor. The microscope lenses and stage will likely need cleaning after the experiment.

7. Complete Part B of the laboratory assessment.

FIGURE 47.3 Spread the web of the foot over the hole and secure it to the board with pins and thread.

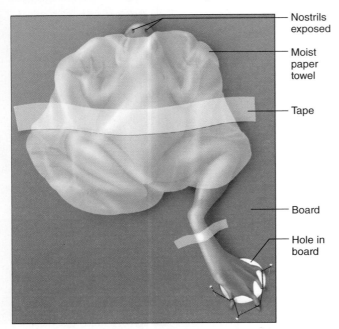

- Nostrils exposed
- Moist paper towel
- Tape
- Board
- Hole in board

Learning Extension Activity

Investigate the effect of temperature change on the blood vessels of the frog's foot by flooding the web with a small quantity of ice water. Observe the blood vessels with low-power magnification and note any changes in their diameters or the rate of blood flow. Remove the ice water and replace it with water heated to about 35°C (95°F). Repeat your observations. What do you conclude from this experiment?

_____

_____

_____

Procedure B—Paths of Circulation

1. Study figure 47.4, which illustrates the pathway of circulation in the pulmonary and systemic circuits. Examine the major blood vessels of the two circuits that are associated near and connected to the heart chambers.

2. Locate the following blood vessels on the available anatomical charts, dissectible human heart model, and the human torso model:

   pulmonary circuit
   - pulmonary trunk   1
   - pulmonary arteries   2
   - pulmonary veins   4

   systemic circuit
   - aorta   1
   - superior vena cava   1
   - inferior vena cava   1

Procedure C—The Arterial System

The pulmonary trunk, connected to the right ventricle, is the main artery of the pulmonary circuit. The aorta, connected to the left ventricle, is the main artery of the systemic circuit. The aorta has a thicker wall than the pulmonary trunk. This allows the aorta to withstand the greater pressure created from the contraction of the very thick myocardium of the left ventricle compared to the right ventricle. In this procedure, trace the pathway of blood vessels as outlined in the various sections. The first branches of the aorta near the aortic valve are the two coronary arteries, which deliver oxygen-rich blood to the myocardium. The brachiocephalic trunk, the left common carotid artery, and the left subclavian artery branch from the aortic arch. One pathway of particular importance to note is the _cerebral arterial circle (circle of Willis)_ of the arteries at the base of the brain. This _arterial anastomosis_ of blood vessels encircles the pituitary gland and the optic chiasma, which allows for alternative routes of blood supply to various parts of the brain.

1. Use figures 47.5, 47.6, 47.7, 47.8, and 47.9 as references to identify the major arteries of the systemic circuit.

Critical Thinking Activity

Why is the left ventricle wall thicker than the right ventricle wall?

_____

_____

_____

_____

_____

_____

_____

FIGURE 47.4 The major blood vessels associated with the pulmonary and systemic circuits. The red colors indicate locations of oxygen-rich (oxygenated) and CO_2-poor blood. The blue colors indicate locations of oxygen-poor (deoxygenated) and CO_2-rich blood.

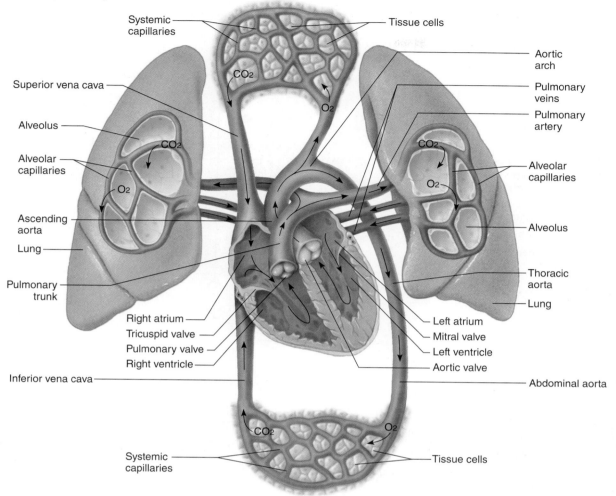

FIGURE 47.5 Arteries supplying the right side of the neck and head. (The clavicle has been removed.)

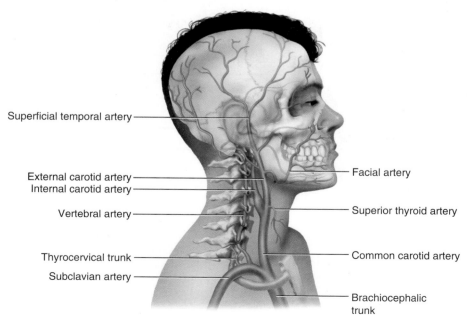

FIGURE 47.6 Inferior view of the brain showing the major arteries of the base of the brain, including the cerebral arterial circle.

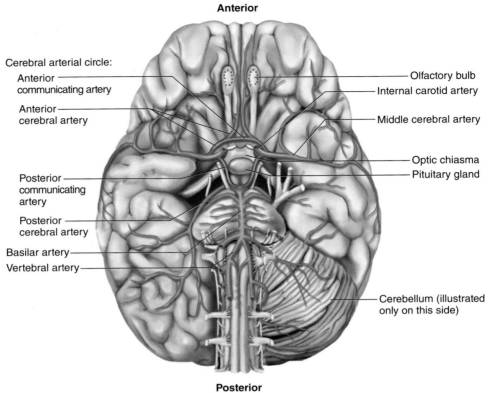

Anterior

Cerebral arterial circle:
Anterior communicating artery
Anterior cerebral artery
Posterior communicating artery
Posterior cerebral artery
Basilar artery
Vertebral artery

Olfactory bulb
Internal carotid artery
Middle cerebral artery
Optic chiasma
Pituitary gland
Cerebellum (illustrated only on this side)

Posterior

FIGURE 47.7 Major arteries of the right shoulder and upper limb.

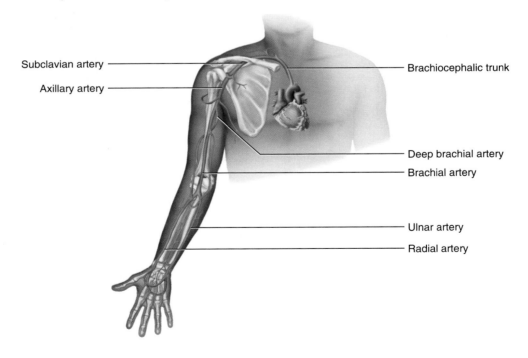

Subclavian artery
Axillary artery

Brachiocephalic trunk

Deep brachial artery
Brachial artery

Ulnar artery
Radial artery

FIGURE 47.8   Major abdominal, mesenteric, and pelvic arteries (anterior view).

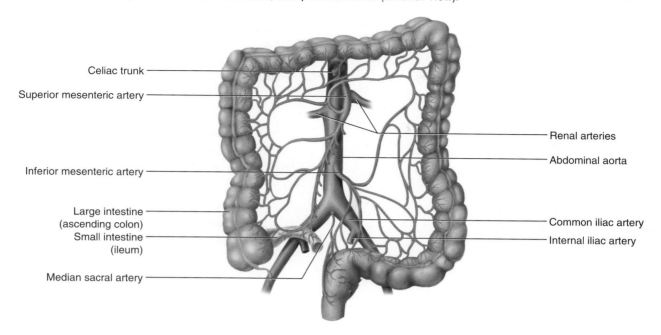

Celiac trunk

Superior mesenteric artery

Inferior mesenteric artery

Large intestine
(ascending colon)
Small intestine
(ileum)

Median sacral artery

Renal arteries

Abdominal aorta

Common iliac artery

Internal iliac artery

2. Locate the following arteries on the charts and human torso model:

aorta
- ascending aorta
- aortic arch (arch of aorta)
- thoracic aorta
- abdominal aorta

branches of the aorta
- coronary artery
- brachiocephalic trunk (artery)
- left common carotid artery
- left subclavian artery
- celiac trunk (artery)
- superior mesenteric artery
- renal artery
- inferior mesenteric artery
- median sacral artery
- common iliac artery

arteries to neck, head, and brain
- vertebral artery
- thyrocervical trunk
- common carotid artery
- external carotid artery
  - superior thyroid artery
  - superficial temporal artery
  - facial artery
- internal carotid artery

arteries of base of brain
- vertebral artery
- basilar artery
- internal carotid artery
- cerebral arterial circle (circle of Willis)
  - anterior cerebral artery
  - anterior communicating artery
  - posterior communicating artery
  - posterior cerebral artery
- middle cerebral artery

arteries to shoulder and upper limb
- subclavian artery
- axillary artery
- brachial artery
- deep brachial artery
- ulnar artery
- radial artery

arteries to pelvis and lower limb
- common iliac artery
- internal iliac artery
- external iliac artery
- femoral artery
- deep artery of thigh (deep femoral artery)
- popliteal artery
- anterior tibial artery
- dorsalis pedis artery (dorsal artery of foot)
- posterior tibial artery

3. Complete Part C of the laboratory assessment.

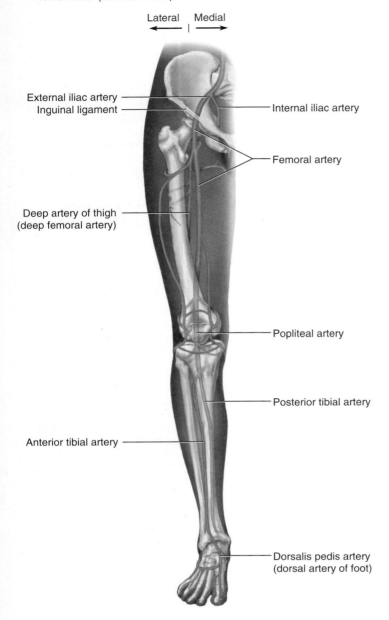

Lateral | Medial

External iliac artery
Inguinal ligament
Internal iliac artery
Femoral artery
Deep artery of thigh (deep femoral artery)
Popliteal artery
Posterior tibial artery
Anterior tibial artery
Dorsalis pedis artery (dorsal artery of foot)

Procedure D—The Venous System

The two vena cavae represent the very large veins returning blood from the superior and inferior portions of the systemic circuit into the right atrium. In this procedure trace the veins as outlined in the various sections, keeping in mind that the flow of blood from smaller veins converges into larger veins.

It should be noted that many of the veins have names corresponding to the arteries that they parallel.

1. Use figures 47.10, 47.11, 47.12, and 47.13 as references to identify the major veins of the systemic circuit.
2. Locate the following veins on the charts and the human torso model:

veins from brain, head, and neck
- dural venous sinus
- external jugular vein
- internal jugular vein
- vertebral vein
- subclavian vein
- brachiocephalic vein
- superior vena cava

veins from upper limb and shoulder
- radial vein
- ulnar vein
- brachial vein
- basilic vein
- cephalic vein
- median cubital vein (antecubital vein)
- axillary vein
- subclavian vein

veins of the abdominal viscera
- hepatic portal vein
- gastric vein
- superior mesenteric vein
- splenic vein
- inferior mesenteric vein
- hepatic vein
- renal vein

veins from lower limb and pelvis
- anterior tibial vein
- posterior tibial vein
- popliteal vein
- femoral vein
- great (long) saphenous vein
- small (short) saphenous vein
- external iliac vein
- internal iliac vein
- common iliac vein
- inferior vena cava

3. Complete Parts D and E of the laboratory report.

FIGURE 47.10 Major veins associated with the right side of the head and neck. (The clavicle has been removed.)

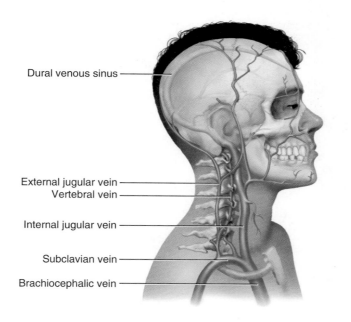

FIGURE 47.11 Anterior view of the veins of the right upper limb. The superficial veins are dark blue and the deep veins are light blue.

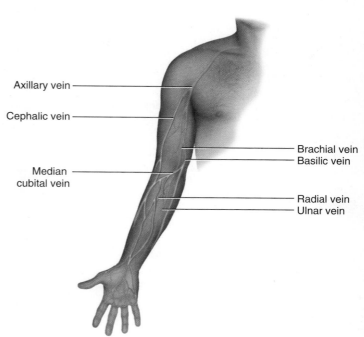

FIGURE 47.12 Major abdominal, hepatic portal, and pelvic veins (anterior view).

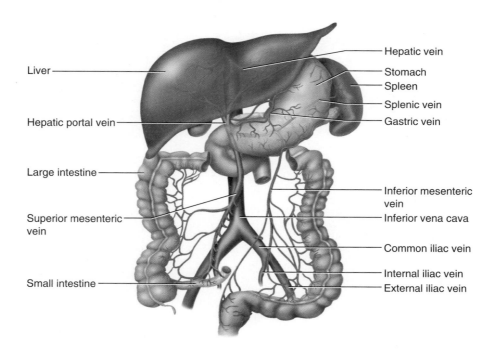

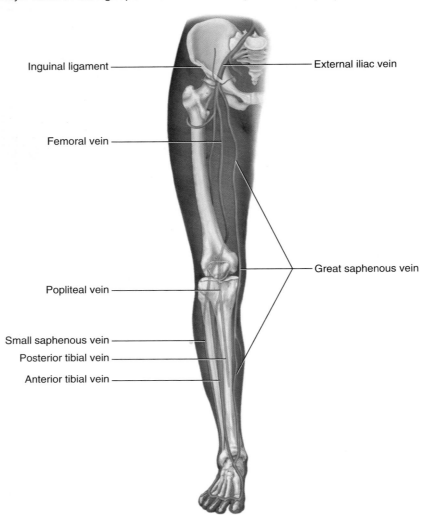

Inguinal ligament

External iliac vein

Femoral vein

Great saphenous vein

Popliteal vein

Small saphenous vein

Posterior tibial vein

Anterior tibial vein

Name _____

Date _____

Section _____

The ⚠ corresponds to the indicated outcome(s) found at the beginning of the laboratory exercise.

Blood Vessel Structure, Arteries, and Veins

Part A   Assessments

1. Sketch and label a section of an arterial wall next to a section of a venous wall. ⚠

Artery wall (_____×)	Vein wall (_____×)

2. Describe the differences you noted in the structures of the arterial and venous walls. Mention each of the three layers of the wall. ⚠ _____

_____

_____

Critical Thinking Assessment

Explain the functional significance of the differences you noted in the structures of the arterial and venous walls. ⚠

_____

_____

Part B Assessments

Complete the following:

1. How did you distinguish between arterioles and venules when you observed the vessels in the web of the frog's foot? 2

_____

_____

2. How did you recognize capillaries in the web? 2 _____

_____

3. What differences did you note in the rate of blood flow through the arterioles, capillaries, and venules? 2 _____

_____

_____

Part C Assessments

Provide the name of the missing artery in each of the following sequences: 3

1. Brachiocephalic trunk, _____, right axillary artery

2. Ascending aorta, _____, descending thoracic aorta

3. Abdominal aorta, _____, ascending colon (right side of large intestine)

4. Brachiocephalic trunk, _____, right external carotid artery

5. Axillary artery, _____, radial artery

6. Common iliac artery, _____, femoral artery

7. Pulmonary trunk, _____, lungs

Part D Assessments

Provide the name of the missing vein or veins in each of the following sequences: 3

1. Right subclavian vein, _____, superior vena cava

2. Anterior tibial vein, _____, femoral vein

3. Internal iliac vein, _____, inferior vena cava

4. Vertebral vein, _____, brachiocephalic vein

5. Popliteal vein, _____, external iliac vein

6. Lungs, _____, left atrium

7. Kidney, _____, inferior vena cava

Part E Assessments

Label the major arteries and veins indicated in figures 47.14 and 47.15.

FIGURE 47.14 Label the major systemic arteries. Ⓐ

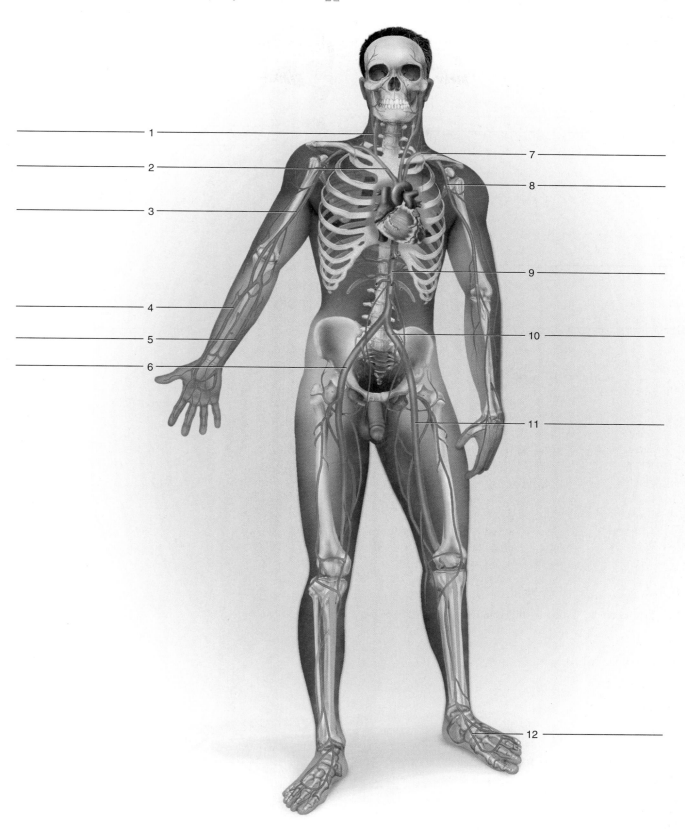

1

2

3

4

5

6

7

8

9

10

11

12

FIGURE 47.15 Label the major systemic veins. 5

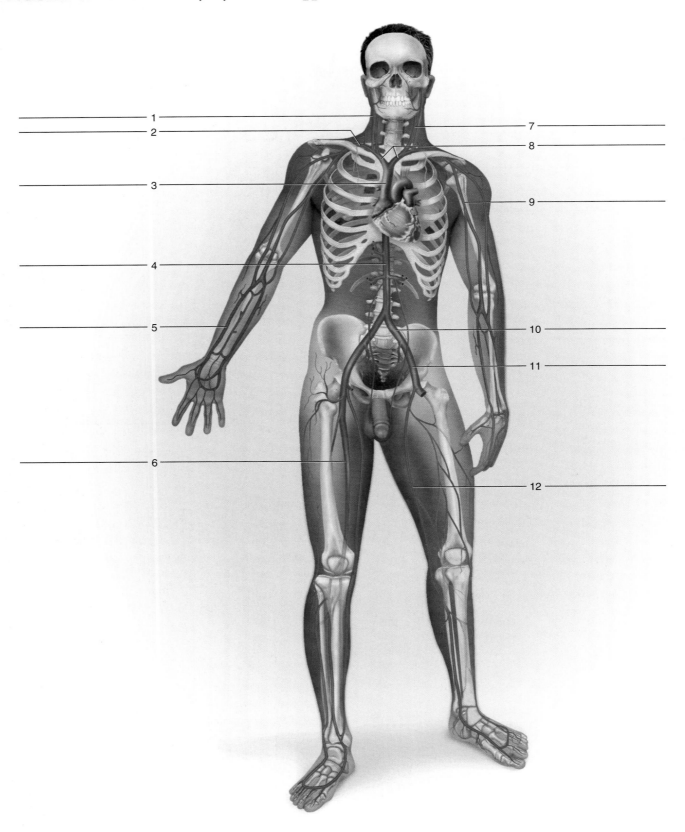

Pulse Rate and Blood Pressure

Purpose of the Exercise

To examine the pulse, determine the pulse rate, measure blood pressure, and investigate the effects of body position and exercise on pulse rate and blood pressure. To investigate the effects of breathing on heart rate and stroke volume.

Materials Needed

Clock with second hand
Sphygmomanometer
Stethoscope
Alcohol swabs (wipes)
Ph.I.L.S. 4.0

For Demonstration Activity:
Pulse pickup transducer or plethysmograph
Physiological recording apparatus

Learning Outcomes

After completing this exercise, you should be able to

① Determine pulse rate and pulse characteristics.

② Test the effects of various factors on pulse rate.

③ Measure and record blood pressure using a sphygmomanometer.

④ Test the effects of various factors on blood pressure.

⑤ Distinguish components of pulse and blood pressure measurements.

⑥ Explain the relationship of the heart rate (and stroke volume) to the breathing cycle.

Pre-Lab

Carefully read the introductory material and examine the entire lab. Be familiar with characteristics of pulse and blood pressure from lecture or the textbook. Answer the pre-lab questions.

Pre-Lab Questions: Select the correct answer for each of the following questions:

1. The _____ is responsible for the pulse wave in the systemic circuit.
    a. right atrium   **b.** right ventricle
    c. left atrium   **d.** left ventricle

2. Which of the following arteries would *not* be palpated on the surface of the body?
    a. facial artery
    b. dorsalis pedis artery
    c. common iliac artery
    d. common carotid artery

3. The _____ is the most common artery palpated to determine heart rate.
    a. temporal artery   **b.** radial artery
    c. popliteal artery   **d.** posterior tibial artery

4. The _____ is the standard artery used to determine blood pressure.
    a. brachial artery   **b.** common carotid artery
    c. femoral artery   **d.** posterior tibial artery

5. Which of the following resting blood pressures would be considered hypertension?
    a. 100/60 mm Hg   **b.** 110/75 mm Hg
    c. 120/80 mm Hg   **d.** 140/90 mm Hg

6. Pulse pressure is the difference between the systolic and diastolic pressures.
    True _____   False _____

7. The blood pressure in a capillary is higher than in a small artery.
    True _____   False _____

8. The stroke volume is the amount of blood pumped by a ventricle in one minute.
    True _____   False _____

The sudden surge of blood that enters the arteries each time the ventricles of the heart contract (systole) causes the elastic walls of these vessels to expand. Then, as the ventricles relax (diastole), the arterial walls recoil. This alternate expanding and recoiling of an arterial wall can be palpated (felt) as a *pulse* in any vessel that is near the surface of the body. Because most of the major arteries are deep in the body, there are a limited number of possible locations to palpate the pulse. If any artery is superficial and firm tissue is just beneath the artery, a pulse can be palpated at that location. The number of pulse expansions per minute correlates with the heart rate or cardiac cycles. The left ventricle is the systemic pumping chamber, and it is therefore responsible for the pulse wave and for the blood pressure in the arteries selected for these assessments.

The force exerted by the blood pressing against the inner walls of arteries creates *blood pressure*. This systemic arterial pressure reaches its maximum, called the *systolic pressure,* during contraction of the left ventricle. Because the left ventricle relaxes when it fills with blood again, the blood pressure then drops to its lowest level, called the *diastolic pressure*. Blood pressure is measured in millimeters of mercury (mm Hg) and is expressed as a systolic pressure over diastolic pressure. A normal resting pressure is considered to be 120/80 mm Hg or slightly lower. If pressures during resting conditions are too high, the condition is called hypertension; if pressures are too low, the condition is called hypotension.

The pulse and the systolic and diastolic blood pressures have some close relationships. The difference between the systolic pressure and the diastolic pressure is called the *pulse pressure*. Obtain the pulse pressure by subtracting the diastolic pressure from the systolic pressure. For example, a person with a blood pressure of 120/80 mm Hg would have a normal pulse pressure of 40 mm Hg. The pulse pressure indicates the force exerted upon the arteries from ventricular contraction and indicates general condition of the cardiovascular system. An expanded pulse pressure could be an indication of atherosclerosis and hypertension. There are many factors that influence blood pressure, so it is important to realize that a single measurement does not provide enough information to make conclusions about health.

After blood has passed through the capillaries, the blood pressure established by the heart is no longer sufficient to return blood to the heart via the veins. The skeletal muscle pump of the limbs and the respiratory pump of the torso assist in venous return. When the skeletal muscles of the limbs contract, they squeeze in on the blood in the veins of the limbs, forcing the blood into the torso (valves in the veins prevent the backflow of blood). In the torso there are two cavities, the thoracic cavity and abdominopelvic cavity, divided by the diaphragm. When the diaphragm contracts (along with the external intercostal muscles), the thoracic cavity volume increases and the pressure decreases. Simultaneously, the volume of the abdominopelvic cavity decreases and the pressure increases. This difference in pressure between the abdominopelvic cavity and the thoracic cavity assists the movement of blood into the veins of the thoracic

cavity and back into the heart. As a result, more blood enters the heart, and the heart rate increases to pump the additional blood. Less blood enters the heart during expiration, since the thoracic cavity volume decreases and its pressure increases.

Procedure A—Pulse Rate

1. Identify some of the pulse locations indicated in figure 48.1.
2. Examine your laboratory partner's radial pulse. To do this, follow these steps:
   a. Have your partner sit quietly, remaining as relaxed as possible.
   b. Locate the pulse by placing your index and middle fingers over the radial artery on the anterior surface near the lateral side of the wrist (fig. 48.2). Do not use your thumb for sensing the pulse because you may feel a pulse coming from an artery in the thumb. The radial artery is most often used because it is easy and convenient to locate.
   c. Note the characteristics of the pulse. That is, could it be described as regular or irregular, strong or weak, hard or soft? The pulse should be regular and the amplitude (magnitude) will decrease as the distance from the left ventricle increases. The strength of the pulse gives some indications of blood pressure. Under high blood pressure, the pulse feels very hard and strong; under low blood pressure, the pulse feels weak and can be easily compressed.

FIGURE 48.1 Locations where an arterial pulse can be palpated.

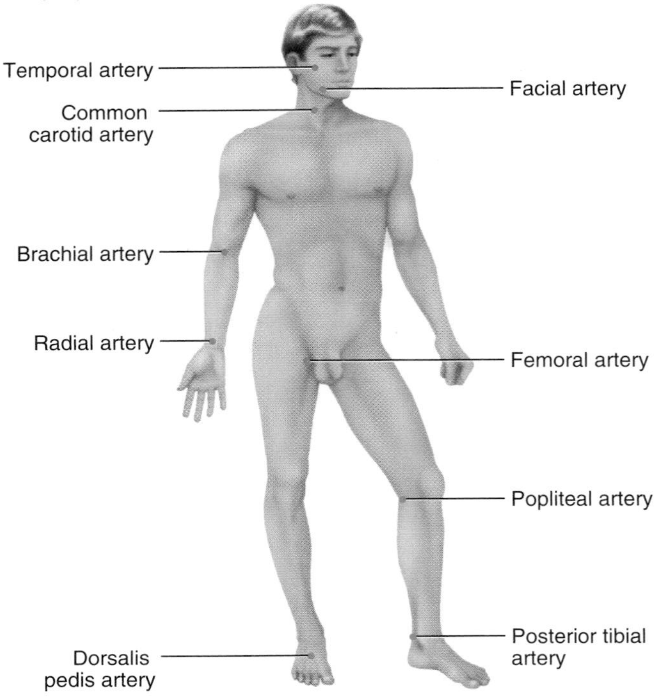

Temporal artery
Facial artery
Common carotid artery
Brachial artery
Radial artery
Femoral artery
Popliteal artery
Posterior tibial artery
Dorsalis pedis artery

FIGURE 48.2 Taking a pulse rate from the radial artery.

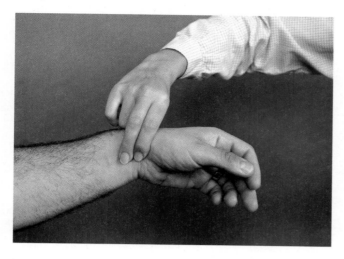

d. To determine the pulse rate, count the number of pulses that occur in 1 minute. This can be accomplished by counting pulses in 30 seconds and multiplying that number by 2. (A pulse count for a full minute, although taking a little longer, would give a better opportunity to detect irregularities.)

e. Record the pulse rate in Part A of Laboratory Assessment 48.

3. Repeat the procedure and determine the pulse rate in each of the following conditions:
   a. immediately after lying down;
   b. 3–5 minutes after lying down;
   c. immediately after standing;
   d. 3–5 minutes after standing quietly;
   e. immediately after 3 minutes of moderate exercise (omit if the person has health problems);
   f. 3–5 minutes after exercise has ended.
   g. Record the pulse rate in Part A of the laboratory assessment.

4. Complete Part A of the laboratory assessment.

Learning Extension Activity

Determine the pulse rate and pulse characteristics in two additional locations on your laboratory partner. Locate and record the pulse rate from the common carotid artery and the dorsalis pedis artery (fig. 48.1).

Common carotid artery pulse rate _____

Dorsalis pedis artery pulse rate _____

Compare the amplitude of the pulse characteristic between these two pulse locations, and interpret any of the variations noted. _____

_____

Demonstration Activity

If the equipment is available, the laboratory instructor will demonstrate how a photoelectric pulse pickup transducer or plethysmograph can be used together with a physiological recording apparatus to record the pulse. Such a recording allows an investigator to analyze certain characteristics of the pulse more precisely than is possible using a finger to examine the pulse. For example, the pulse rate can be determined very accurately from a recording, and the heights of the pulse waves provide information about the blood pressure.

Procedure B—Blood Pressure

Reexamine the introductory material related to blood pressure. Several basic structural and functional factors influence arterial pressure, including *cardiac output, blood volume,* and *peripheral resistance.* Cardiac output is the volume of blood discharged from a ventricle in a minute. The blood ejected from a single ventricle contraction is about 70 mL and is called the *stroke volume.* A 75 beats/minute resting heart rate and a stroke volume of 70 mL would calculate as a cardiac output of 5,250 mL/minute, which is about the average adult blood volume. If either heart rate or stroke volume increases, cardiac output would increase as would blood pressure. Blood volume influences blood pressure. Because blood pressure is directly proportional to blood volume, any volume change can alter blood pressure. Peripheral resistance refers to the amount of friction between the blood and the walls of the blood vessels. Although blood viscosity and blood vessel length influence peripheral resistance, the blood vessel diameter changes are the most significant from vasoconstriction or vasodilation as a result of contraction or relaxation of smooth muscle of the middle tunic. Certain conditions, such as obesity, excess salt intake, inactive lifestyle, stress, smoking, medications, and others can aggravate blood pressures. Be aware of these factors and conditions influencing blood pressure as you take a blood pressure measurement using a sphygmomanometer and a stethoscope.

1. Although blood pressure is present in all blood vessels, the standard location to record blood pressure is the brachial artery. Examine figure 48.3, which indicates various pressure relationships throughout the systemic circuit.

2. Measure your laboratory partner's arterial blood pressure. To do this, follow these steps:
   a. Obtain a sphygmomanometer and a stethoscope.
   b. Clean the earpieces and the diaphragm of the stethoscope with alcohol swabs.
   c. Have your partner sit quietly with upper limb resting on a table at heart level. Have the person remain as relaxed as possible.
   d. Locate the brachial artery at the antecubital space. Wrap the cuff of the sphygmomanometer around the

FIGURE 48.3 Pressure changes in blood vessels of the systemic circuit.

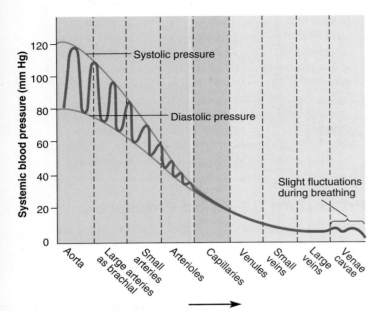

Increasing distance from left ventricle to right atrium

FIGURE 48.4 Blood pressure is commonly measured by using a sphygmomanometer (blood pressure cuff). The use of a column of mercury is the most accurate measurement, but due to environmental concerns, it has been replaced by alternative gauges and digital readouts.

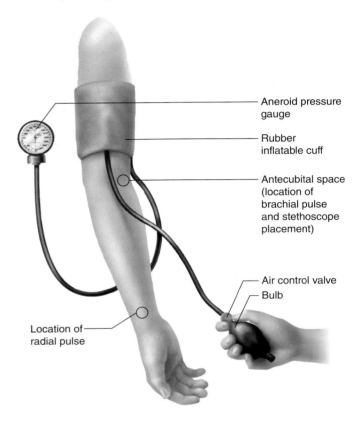

arm so that its lower border is about 2.5 cm above the bend of the elbow. Center the bladder of the cuff in line with the *brachial pulse* (fig. 48.4).

e. Palpate the *radial pulse.* Close the valve on the neck of the rubber bulb connected to the cuff and pump air from the bulb into the cuff. Inflate the cuff while watching the sphygmomanometer and note the pressure when the pulse disappears. (This is a rough estimate of the systolic pressure.) Immediately deflate the cuff. *Do not leave the cuff inflated for more than 1 minute.*

f. Position the stethoscope over the brachial artery. Reinflate the cuff to a level 30 mm Hg higher than the point where the pulse disappeared during palpation (fig. 48.5).

g. Slowly open the valve of the bulb until the pressure in the cuff drops at a rate of about 2 or 3 mm Hg per second.

h. Listen for sounds (*Korotkoff sounds*) from the brachial artery. When the first loud tapping sound is heard, record the reading as the systolic pressure. This indicates the pressure exerted against the arterial wall during systole.

i. Continue to listen to the sounds as the pressure drops and note the level when the last sound is heard. Record this reading as the diastolic pressure, which measures the constant arterial resistance.

j. Release all of the pressure from the cuff.

k. Repeat the procedure until you have two blood pressure measurements from each arm, allowing 2–3 minutes of rest between readings.

l. Average your readings and enter them in the table in Part B of the laboratory assessment.

FIGURE 48.5 Sphygmomanometer and stethoscope placement on arm while taking blood pressure.

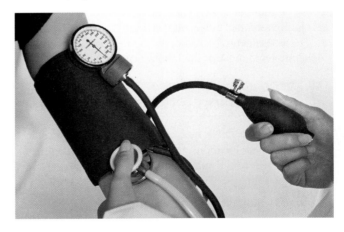

3. Measure your partner's blood pressure in each of the following conditions:
   a. 3–5 minutes after lying down;
   b. 3–5 minutes after standing quietly;
   c. immediately after 3 minutes of moderate exercise (*omit if the person has health problems*);

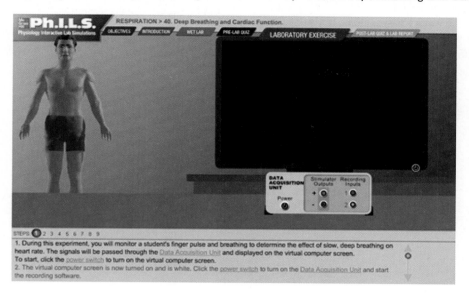

d. 3–5 minutes after exercise has ended.

e. Record the blood pressures in Part B of the laboratory assessment.

4. Complete Parts B and C of the laboratory assessment.

Procedure C—Ph.I.L.S. Lesson 40 Respiration: Deep Breathing and Cardiac Function

1. Open Exercise 40; Respiration: Deep Breathing and Cardiac Function.

2. Read the objectives and introduction and take the pre-lab quiz.

3. After completing the pre-lab quiz, read through the wet lab. *Be sure to click open and view the videos that are indicated in red in the wet lab.*

4. The lab exercise will open when you click Continue after completing the wet lab (fig. 48.6).

Setup

5. To start, click the power switch on the virtual computer screen then click the power switch to turn on the Data Acquisition Unit.

6. The *finger pulse unit* indirectly measures heart rate by detecting pulses of blood that pass through the arteries of the finger, and it transmits *to* the data acquisition unit (input). Connect to the data acquisition unit by inserting the orange plug into input 2 of the data acquisition unit and attach the finger pulse unit to the left hand of the volunteer.

7. The *breathing apparatus* detects airflow into and out of the mouth; it transmits *to* the data acquisition unit (input). Connect to the data acquisition unit by inserting

the blue plug into input 1 of the data acquisition unit and attach the breathing apparatus to the volunteer's mouth.

Recording Pulse and Breathing Pattern

8. Click Start in the upper right of the virtual computer screen. The volume of the air moving in and out of the volunteer's lung is displayed on the upper blue line tracing, the finger pulse recording is on the orange trace, and the heart rate is shown on the lower gray trace.

9. After two complete breathing cycles, click Stop.

10. *Print the graph of the breathing pattern and pulse by clicking the P in the top left of the virtual computer screen.* See figure 48.7 as an example of a final graph.

FIGURE 48.7 Example: A recording to show the relationship between slow breathing (upper trace) and heart function; finger pulse is on the middle tracing, and heart rate is on the lower.

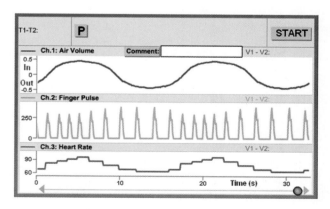

Interpreting Results

11. *With the graph still on the screen,* answer the questions in Part D of the laboratory assessment. If you accidentally closed the graph, click on the Journal panel (red rectangle at bottom right of screen).

12. Complete the Post-Lab Quiz (click open Post Lab Quiz and Lab Report) by answering the ten questions on the computer screen.

13. Read the conclusion on the computer screen.

14. You may print the Lab Report for Respiration: Deep Breathing and Cardiac Function.

Name _____

Date _____

Section _____

The Ⓐ corresponds to the indicated outcome(s) found at the beginning of the
laboratory exercise.

Pulse Rate and Blood Pressure

Part A Assessments

1. Enter your observations of pulse characteristics and pulse rates in the table. Ⓐ1 Ⓐ2

Test Subject	Pulse Characteristics	Pulse Rate (beats/min)
Sitting		74 Bpm
Lying down		67 Bpm
3–5 minutes later		66 Bpm
Standing		69 Bpm
3–5 minutes later		69 Bpm
After exercise		86 Bpm
3–5 minutes later		75 Bpm

2. Summarize the effects of body position and exercise on the characteristics and rates of the pulse. Ⓐ2 _____

Exercise causes HR to increase. When I stood up
my HR decreased and when I lay down it
also decreased.

Part B Assessments

1. Enter the initial measurements of blood pressure in the table. **3**

Reading	Blood Pressure in Right Arm	Blood Pressure in Left Arm
First	117/75	117/76
Second	119/76	118/76
Average	118/75.5	117.5/76

2. Enter your test results in the table. **4**

Test Subject	Blood Pressure
3–5 minutes after lying down	115/72
3–5 minutes after standing	117/74
After 3 minutes of moderate exercise	120/79
3–5 minutes later	118/76

3. Summarize the effects of body position and exercise on blood pressure. **4** _Exercise causes_

HR to increase

4. Summarize any correlations between pulse rate and blood pressure from any of the experimental conditions. **5**

Pulse rate and blood pressure are directly related

Critical Thinking Assessment

When a pulse is palpated and counted, which pressure (systolic or diastolic) would be characteristic at that moment? Explain your answer. **1**

Part C Assessments

Complete the following statements:

1. The maximum pressure achieved during ventricular contraction is called _____ pressure. **5**

2. The lowest pressure that remains in the arterial system during ventricular relaxation is called _____ pressure. **5**

3. The pulse rate is equal to the _____ rate. **5**

4. A pulse that feels full and is not easily compressed is produced by an elevated _____. **5**

5. The instrument commonly used to measure systemic arterial blood pressure is called a _____. **5**

6. Blood pressure is expressed in units of _____. **5**

7. The upper number of the fraction used to record blood pressure indicates the _____ pressure. **5**

8. The _____ artery in the arm is the standard systemic artery in which blood pressure is measured. **5**

9. A person with a blood pressure of 120/80 mm Hg would have a pulse pressure of _____ mm Hg. **5**

Part D Ph.I.L.S. Lesson 40, Respiration: Deep Breathing and Cardiac Function Assessments

Use the printed graph to help you answer the following: **6**

1. During inspiration, the breathing curve is _____ (an upward, a downward) deflection, and during expiration the breathing curve is _____ (an upward, a downward) deflection.

2. The heart rate is determined by the _____ (frequency or amplitude) of the pulses.

3. The stroke volume is determined by the _____ (frequency or amplitude) of the pulses.

4. During inspiration:

    a. The heart rate _____ (increases, decreases, or stays the same).

    b. The stroke volume _____ (increases, decreases, or stays the same).

5. During expiration:

    a. The heart rate _____ (increases, decreases, or stays the same).

    b. The stroke volume _____ (increases, decreases, or stays the same).

6. Cardiac output (CO) is determined by heart rate (HR) and stroke volume (SV): HR x SV = CO. To maintain a consistent cardiac output at rest, as heart rate increases, stroke volume _____, and as heart rate decreases, stroke volume _____. Did you observe this relationship on the graph? _____ Is there a direct or inverse relationship between heart rate and stroke volume when maintaining cardiac output at a set value? _____

Critical Thinking Assessment

Explain the relationship between inspiration, pressure changes in the thoracic cavity, venous return (movement of blood in veins), heart rate, stroke volume, and cardiac output at rest. **2** **6**

_____

_____

_____

_____

_____

Lymphatic System

Purpose of the Exercise

To review the structure of the lymphatic system and to observe the microscopic structure of a lymph node, thymus, spleen, and tonsil.

Materials Needed

Human torso model
Anatomical chart of the lymphatic system
Compound light microscope
Prepared microscope slides:
    Lymph node section
    Human thymus section
    Human spleen section
    Human tonsil section

Learning Outcomes

After completing this exercise, you should be able to

1. Identify the major lymphatic pathways and components of the lymphatic system.

2. Locate and sketch the major microscopic structures of a lymph node, the thymus, the spleen, and a tonsil.

3. Describe the structure and function of a lymph node, the thymus, the spleen, and a tonsil.

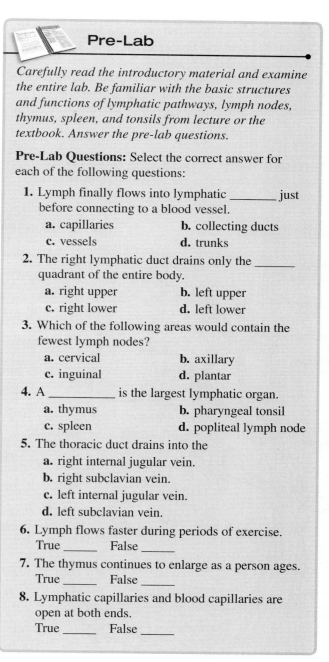

Pre-Lab

Carefully read the introductory material and examine the entire lab. Be familiar with the basic structures and functions of lymphatic pathways, lymph nodes, thymus, spleen, and tonsils from lecture or the textbook. Answer the pre-lab questions.

Pre-Lab Questions: Select the correct answer for each of the following questions:

1. Lymph finally flows into lymphatic _____ just before connecting to a blood vessel.
   - **a.** capillaries
   - **b.** collecting ducts
   - **c.** vessels
   - **d.** trunks

2. The right lymphatic duct drains only the _____ quadrant of the entire body.
   - **a.** right upper
   - **b.** left upper
   - **c.** right lower
   - **d.** left lower

3. Which of the following areas would contain the fewest lymph nodes?
   - **a.** cervical
   - **b.** axillary
   - **c.** inguinal
   - **d.** plantar

4. A _____ is the largest lymphatic organ.
   - **a.** thymus
   - **b.** pharyngeal tonsil
   - **c.** spleen
   - **d.** popliteal lymph node

5. The thoracic duct drains into the
   - **a.** right internal jugular vein.
   - **b.** right subclavian vein.
   - **c.** left internal jugular vein.
   - **d.** left subclavian vein.

6. Lymph flows faster during periods of exercise.
   True _____ False _____

7. The thymus continues to enlarge as a person ages.
   True _____ False _____

8. Lymphatic capillaries and blood capillaries are open at both ends.
   True _____ False _____

The lymphatic system is closely associated with the cardiovascular system and includes a network of vessels that assist in the circulation of body fluids. Fluid in the cardiovascular vessels is blood; fluid in lymphatic vessels is *lymph.* As blood flows through capillary beds, fluids continually filter from the plasma of the blood along with some dissolved nutrients, but leaving most of the plasma proteins behind in the blood. Most of the fluids are reabsorbed back into the blood at the venous ends of the capillary beds; however the amount not reabsorbed becomes part of the interstitial fluids among the tissue cells. Lymphatic vessels provide a necessary pathway for the accumulating tissue fluids to enter lymphatic vessels as lymph and return to the blood of the cardiovascular system (fig. 49.1). Without this system, this fluid would accumulate in tissue spaces, producing *edema.*

The *lymphatic capillaries* are microscopic, blind-ended vessels, in which excess interstitial fluid first enters the lymphatic system. These capillaries drain into *lymphatic collect-*

ing vessels, then into larger *lymphatic trunks,* and finally into one of two *collecting ducts* just before the connections with the subclavian veins (fig. 49.1). As a result of the orientation of valves, only one-way flow of lymph occurs. This happens during skeletal muscle contraction and breathing, causing compression upon the vessels that forces fluid through the lymphatic vessels. The larger *thoracic duct* provides drainage for the entire body except for the right upper quadrant. The thoracic duct connects to the left subclavian vein, while a *right lymphatic duct* drains only the right upper quadrant into the right subclavian vein (fig. 49.2).

Along the pathway of the lymphatic vessels, *lymph nodes* are positioned; they are especially concentrated in areas of major joints like axillary and inguinal regions. The lymph nodes contain concentrations of lymphocytes and macrophages that cleanse the lymph and activate an immune response. Swollen and sore lymph nodes represent areas of infection that occur as our bodies fight off the microorganisms.

Other lymphatic organs are part of the lymphatic system. The red bone marrow is a major production site of white blood cells. The *thymus,* located within the mediastinum, is a location of T lymphocyte maturation. The thymus is proportionately very large in a fetus and in a child, but it begins to atrophy around puberty, and this continues as we age. The *spleen,* the largest lymphatic organ, is located in the left upper quadrant of the abdominopelvic cavity. The spleen is highly vascular and contains large numbers of red blood cells, lymphocytes, and macrophages. Palatine, pharyngeal, and lingual *tonsils* are located in regions of the entrance into the pharynx (throat). Numerous nodules within the tonsils contain concentrations of lymphocytes. The organs of the lymphatic system help to defend the tissues against infections by filtering particles from the lymph and by supporting

FIGURE 49.1 Schematic relationship of fluid movements between the cardiovascular and lymphatic systems.

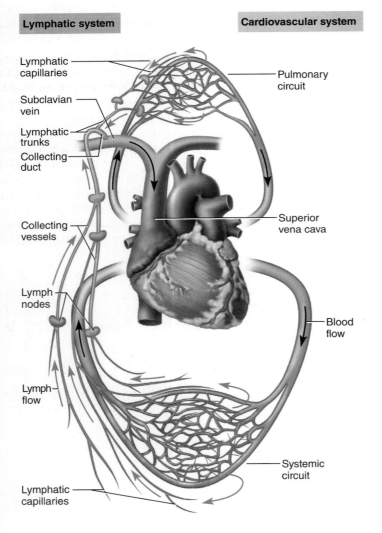

Lymphatic system

Cardiovascular system

- Lymphatic capillaries
- Subclavian vein
- Lymphatic trunks
- Collecting duct
- Collecting vessels
- Lymph nodes
- Lymph flow
- Lymphatic capillaries

- Pulmonary circuit
- Superior vena cava
- Blood flow
- Systemic circuit

FIGURE 49.2 Regions of the body drained by the right lymphatic duct are shaded purple. The unshaded regions of the body are drained by the larger thoracic duct into the left subclavian vein.

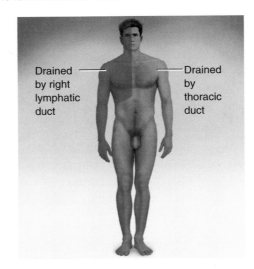

Drained by right lymphatic duct

Drained by thoracic duct

the activities of lymphocytes that furnish immunity against specific disease-causing agents or pathogens.

Procedure A—Lymphatic Pathways

1. Study figures 49.3 and 49.4.
2. Observe the human torso model and the anatomical chart of the lymphatic system, and locate the following features:

   lymphatic collecting vessels—collect lymph from microscopic lymphatic capillaries

   lymph nodes—bean-shaped organs that cleanse lymph and activate immune response

   lymphatic trunks—collect lymph from lymphatic collecting vessels
   - lumbar trunk
   - intestinal trunk
   - bronchomediastinal trunk
   - subclavian trunk
   - jugular trunk

   cisterna chyli—inferior expanded sac of thoracic duct where fatty intestinal lymph collects

   collecting ducts—largest lymphatic vessels; collect lymph from lymphatic trunks
   - thoracic duct—connects to left subclavian vein near left internal jugular vein
   - right lymphatic duct—connects to right subclavian vein near right internal jugular vein

   internal jugular veins

   subclavian veins—receive lymph from collecting ducts

3. Complete Part A of Laboratory Assessment 49.

FIGURE 49.3 Lymphatic vessels, lymph nodes, and lymphatic organs.

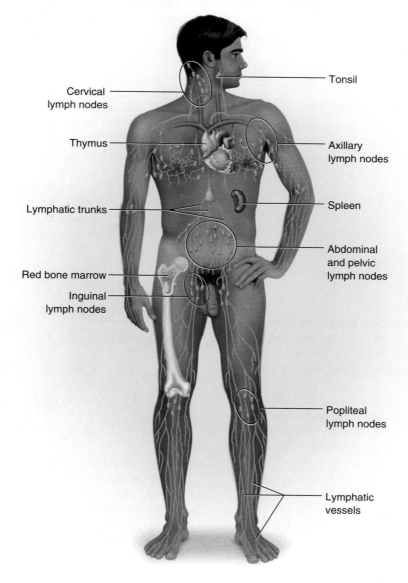

Cervical lymph nodes

Thymus

Lymphatic trunks

Red bone marrow

Inguinal lymph nodes

Tonsil

Axillary lymph nodes

Spleen

Abdominal and pelvic lymph nodes

Popliteal lymph nodes

Lymphatic vessels

FIGURE 49.4 Lymphatic pathways into the subclavian veins. The arrows indicate the direction of lymph drainage.

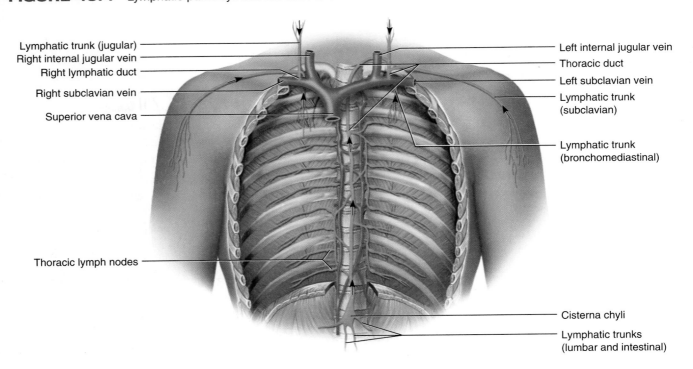

Lymphatic trunk (jugular)
Right internal jugular vein
Right lymphatic duct
Right subclavian vein
Superior vena cava

Left internal jugular vein
Thoracic duct
Left subclavian vein
Lymphatic trunk (subclavian)
Lymphatic trunk (bronchomediastinal)

Thoracic lymph nodes

Cisterna chyli
Lymphatic trunks (lumbar and intestinal)

Procedure B—Lymph Nodes

1. Study the lymph node locations in figures 49.3 and 49.4.
2. Observe the anatomical chart of the lymphatic system and the human torso model, and locate the clusters of lymph nodes in the following regions:

   cervical region
   axillary region
   popliteal region
   inguinal region
   pelvic cavity
   abdominal cavity
   thoracic cavity

3. Palpate the lymph nodes in your cervical region. They are located along the lower border of the mandible and between the ramus of the mandible and the sterno-cleidomastoid muscle. They feel like small, firm lumps.
4. Study figures 49.5 and 49.6.
5. Obtain a prepared microscope slide of a lymph node and observe it, using low-power magnification (fig. 49.6c). Identify the *capsule* that surrounds the node and is mainly composed of collagenous fibers, the *lymph nodules* that appear as dense masses near the surface of the node, and the *lymph sinuses* that appear as narrow spaces where lymph circulates between the nodules and the capsule.

FIGURE 49.5 Diagram of a lymph node showing internal structures, attached lymphatic vessels, and arrows indicating lymph flow.

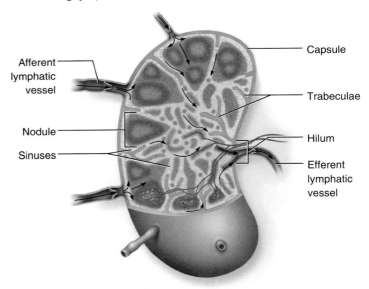

Afferent lymphatic vessel

Nodule
Sinuses

Capsule

Trabeculae

Hilum

Efferent lymphatic vessel

6. Using high-power magnification, examine a nodule within the lymph node. The nodule contains densely packed *lymphocytes*.
7. Prepare a labeled sketch of a representative section of a lymph node in Part B of the laboratory assessment.
8. Complete Part C of the laboratory assessment.

FIGURE 49.6 Lymph nodes: (a) inguinal lymph nodes in a cadaver; (b) photograph of a human lymph node on tip of finger; (c) micrograph of a lymph node (20×).

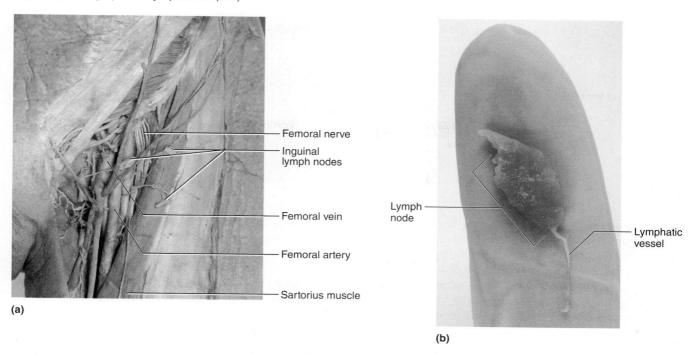

(a)

Femoral nerve
Inguinal lymph nodes
Femoral vein
Femoral artery
Sartorius muscle

Lymph node
Lymphatic vessel

(b)

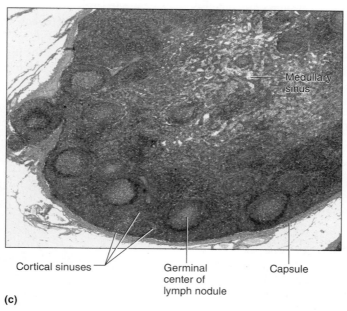

Medullary sinus

Cortical sinuses
Germinal center of lymph nodule
Capsule

(c)

Procedure C—Thymus, Spleen, and Tonsil

1. Locate the thymus, spleen, and tonsils in the anatomical chart of the lymphatic system and on the human torso model.
2. Obtain a prepared microscope slide of human thymus and observe it, using low-power magnification (fig. 49.7). Note how the thymus is subdivided into *lobules* by *septa*

of connective tissue that contain blood vessels. Identify the capsule of loose connective tissue that surrounds the thymus, the outer *cortex* of a lobule composed of densely packed cells and deeply stained, and the inner *medulla* of a lobule composed of loosely packed lymphocytes and epithelial cells and lightly stained.

3. Examine the cortex tissue of a lobule using high-power magnification. The cells of the cortex are composed of densely packed *lymphocytes* among some epithelial cells

FIGURE 49.7 Micrograph of a section of the thymus (15×).

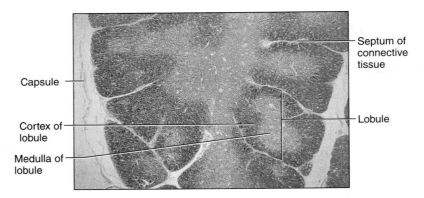

Capsule

Cortex of lobule

Medulla of lobule

Septum of connective tissue

Lobule

and *macrophages.* Some of these cortical cells may be undergoing mitosis, so their chromosomes may be visible.

4. Prepare a labeled sketch of a representative section of the thymus in Part B of the laboratory assessment.

5. Obtain a prepared slide of the human spleen and observe it, using low-power magnification (fig. 49.8). Identify the *capsule* of dense connective tissue that surrounds the spleen. The tissues of the spleen include circular *nodules* of *white pulp* enclosed in a matrix of *red pulp.*

6. Using high-power magnification, examine a nodule of white pulp and red pulp. The cells of the white pulp are mainly lymphocytes. Also, there may be an arteriole centrally located in the nodule. The cells of the red pulp are mostly red blood cells with many lymphocytes and

macrophages. The macrophages engulf and destroy cellular debris and bacteria.

7. Prepare a labeled sketch of a representative section of the spleen in Part B of the laboratory assessment.

8. Obtain a prepared microscope slide of a tonsil and observe it, using low-power magnification (fig. 49.9). Identify the surface epithelium, deep invaginated pits called *tonsillar crypts,* and numerous lymph nodules. Although the tonsillar crypts often harbor bacteria and food debris, infections are usually prevented within these lymphoid organs.

9. Prepare a labeled sketch of a representative section of a tonsil in Part B of the laboratory assessment.

10. Complete Part D of the laboratory assessment.

FIGURE 49.8 Micrograph of a section of the spleen (40×).

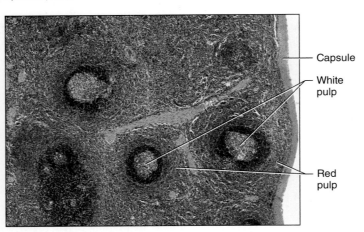

Capsule

White pulp

Red pulp

FIGURE 49.9 Micrograph of a section of the pharyngeal tonsil (5×).

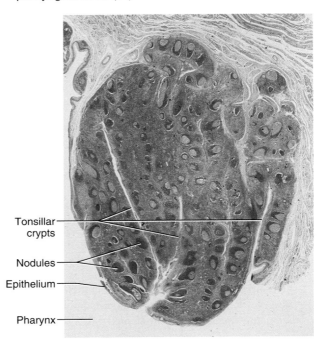

Tonsillar crypts

Nodules

Epithelium

Pharynx

Name _____

Date _____

Section _____

The Ⓐ corresponds to the indicated outcome(s) found at the beginning of the laboratory exercise.

Lymphatic System

Part A Assessments

Complete the following statements:

1. Lymphatic pathways begin as lymphatic _capillaries_ that merge to form lymphatic collecting vessels. **1**

2. Lymph drainage from collecting ducts enters the _Subclavian_ veins. **1**

3. Once tissue (interstitial) fluid is inside a lymph capillary, the fluid is called _lymph_. **1**

4. Lymphatic vessels contain _valves_ that help prevent the backflow of lymph. **1**

5. Lymphatic vessels usually lead to lymph _nodes_ that filter the fluid being transported. **1**

6. The _Thoracic duct_ is the larger and longer of the two lymphatic collecting ducts. **1**

Part B Assessments

Sketch and label the microscopic structures of the following organs: **2**

Lymph node (_____×)	Thymus (_____×)
Spleen (_____×)	Tonsil (_____×)

Part C Assessments

Complete the following statements:

1. Lymph nodes contain large numbers of white blood cells called _lymphocytes_ and macrophages that fight invading microorganisms. **3**

2. The indented region where blood vessels and nerves join a lymph node is called the _hilum_. **3**

3. Lymph _nodules_ that contain germinal centers are the structural units of a lymph node. **3**

4. The spaces within a lymph node are called lymph _sinuses_ through which lymph circulates. **3**

5. Lymph enters a node through a(an) _afferent_ lymphatic vessel. **3**

6. The lymph nodes associated with the lymphatic vessels that drain the lower limbs are located in the _inguinal_ region. **1**

Part D Assessments

Complete the following statements:

1. The thymus is located in the _mediastinum_, anterior to the aortic arch. **1**

2. The _thymus_ is very large in a child and will atrophy in advanced age. **3**

3. The _spleen_ is the largest of the lymphatic organs. **1**

4. A _capsule_ of connective tissue surrounds the spleen. **3**

5. The tiny islands (nodules) of tissue within the spleen that contain many lymphocytes comprise the _white_ pulp. **3**

6. The _red_ pulp of the spleen contains large numbers of red blood cells, lymphocytes, and macrophages. **3**

7. _Venous sinuses_ within the spleen engulf and destroy foreign particles and cellular debris. **3**

8. The lymphoid organs in the pharynx (throat) that are possible infection sites are collectively called _tonsils_. **3**

Respiratory Organs

Purpose of the Exercise

To review the structure and function of the respiratory organs and to examine the tissues of some of these organs.

Materials Needed

Human skull (sagittal section)
Human torso model
Larynx model
Thoracic organs model
Compound light microscope
Prepared microscope slides of the following:
    Trachea (cross section)
    Lung, human (normal)

For Demonstration Activities:

Animal lung with trachea (fresh or preserved)
Bicycle pump
Prepared microscope slides of the following:
    Lung tissue (smoker)
    Lung tissue (emphysema)

⚠ Safety

▶ Wear disposable gloves when working on the fresh or preserved animal lung demonstration.
▶ Wash your hands before leaving the laboratory.

Learning Outcomes

After completing this exercise, you should be able to

① Locate the major organs and structural features of the respiratory system.

② Describe the functions of these organs.

③ Sketch and label features of tissue sections of the trachea and lung.

 Pre-Lab

Carefully read the introductory material and examine the entire lab. Be familiar with the basic structures and functions of the respiratory organs from lecture or the textbook. Visit www.mhhe.com/martinseries2 for LabCam videos. Answer the pre-lab questions.

Pre-Lab Questions: Select the correct answer for each of the following questions:

1. The right lung has _____ lobes; the left lung has _____ lobes.
    a. 3; 3   **b.** 2; 2
    c. 3; 2   **d.** 2; 3

2. Paranasal sinuses are within the following bones *except* the
    a. maxillary bones.   **b.** mandible.
    c. frontal bone.   **d.** sphenoid bone.

3. The alveoli are composed of
    a. simple squamous epithelium.
    b. hyaline cartilage.
    c. smooth muscle.
    d. epithelium with cilia.

4. The _____ adheres to the surface of the lung.
    a. pericardium   **b.** pleural cavity
    c. parietal pleura   **d.** visceral pleura

5. The _____ is the most inferior cartilage of the larynx.
    a. epiglottis   **b.** cricoid cartilage
    c. thyroid cartilage   **d.** corniculate cartilage

6. Which of the following structures increases the surface area and air turbulence the most during breathing?
    a. nasal meatuses   **b.** nasal septum
    c. nasal conchae   **d.** nares (nostrils)

7. Which of the following airway tubes would have the smallest lumens?
    a. alveolar ducts   **b.** segmental bronchi
    c. lobar bronchi   **d.** main bronchi

The respiratory system involves movements of oxygen from our external environment eventually to our cells, and the movements of carbon dioxide produced by our cells until it exits our body. Breathing (pulmonary ventilation) involves the nose, nasal cavity, paranasal sinuses, pharynx, larynx, trachea, and bronchial tree, all serving as passageways for gases into and out of the lungs. The exchange of oxygen and carbon dioxide in the lungs is called *external respiration;* the exchange of oxygen and carbon dioxide in the tissues is called *internal respiration.* The blood transports gases to and from the alveolar sacs and the body cells.

Larger bronchial tubes possess supportive cartilaginous rings and plates so that they do not collapse during breathing. Smooth muscle is part of the walls of these bronchial tubes. If the smooth muscle of these tubes relaxes, the air passages dilate, which allows a greater volume of air movement. The epithelial lining of the respiratory tubes includes occasional goblet cells, which secrete protective mucus. Numerous epithelial cells possess cilia that extend into the mucus. The action of the cilia creates a current of mucus that moves any entrapped debris toward the pharynx. The mucus with any entrapped particles is normally swallowed.

The right lung has three lobes; the left lung has two lobes. A right main bronchus branches into three lobar bronchi, each extending into one of the three right lobes. A left main bronchus branches into two lobar bronchi, each extending into a left lobe. The hilum of the lung represents an area where a bronchus and major blood vessels are located. A *visceral pleura* adheres to the surface of each lung, including the fissures between the lobes. A *parietal pleura* covers the internal surface of the thoracic wall and the superior surface of the diaphragm. A narrow potential space, called the *pleural cavity,* is located between the two pleurae and contains pleural fluid secreted by these membranes. This pleural fluid serves as lubrication during breathing movements, and due to the surface tension, it resists separations of the pleurae.

Procedure A—Respiratory Organs

1. Study figures 50.1 and 50.2.
2. Examine the sagittal section of the human skull and the human torso model. Locate the following features:

   nasal cavity—hollow space behind nose
   - naris (nostril)—external opening into nasal cavity
   - nasal septum—divides nasal cavity into right and left portions
   - nasal conchae—increase surface area and air turbulence during breathing
   - nasal meatuses—air passageways along nasal conchae

   paranasal sinuses—air-filled spaces that open into nasal cavity; lined with mucous membrane
   - maxillary sinus
   - frontal sinus

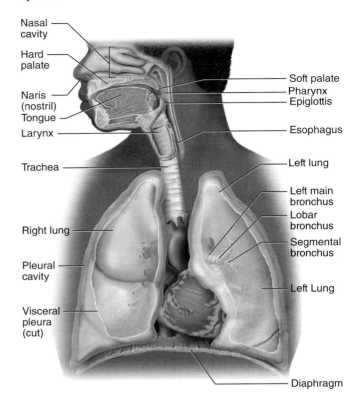

FIGURE 50.1 The major features of the respiratory system.

   - ethmoidal sinus
   - sphenoidal sinus
3. Examine figures 50.1, 50.2, 50.3, 50.4, and 50.5.
4. Observe the larynx model, the thoracic organs model, and the human torso model. Palpate the larynx as you swallow. Locate the following features:

   pharynx—often called throat
   - nasopharynx—air passageway only
   - oropharynx—air and food passageway
   - laryngopharynx—air and food passageway

   larynx—airway enlargement superior to trachea
   - vocal folds—consist of two sets of folds
     - vestibular folds (false vocal cords)—upper folds but do not produce sounds
     - vocal folds (true vocal cords)—vibrate when exhaling to produce sounds
   - thyroid cartilage ("Adam's apple")—largest hyaline cartilage of larynx; covered by part of thyroid gland
   - cricoid cartilage—complete ring of hyaline cartilage of inferior larynx

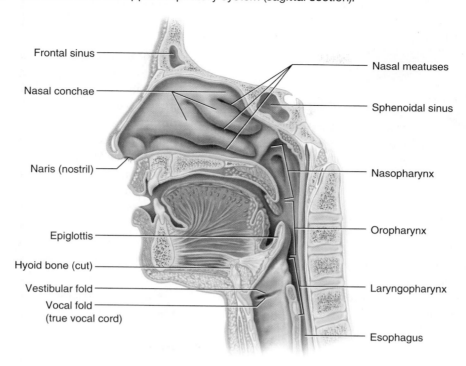

Frontal sinus

Nasal conchae

Naris (nostril)

Epiglottis

Hyoid bone (cut)

Vestibular fold

Vocal fold
(true vocal cord)

Nasal meatuses

Sphenoidal sinus

Nasopharynx

Oropharynx

Laryngopharynx

Esophagus

FIGURE 50.3   Major features of the larynx: (a) anterior view; (b) posterior view.

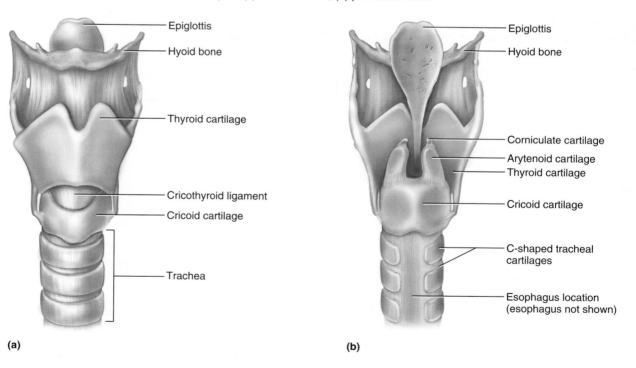

Epiglottis

Hyoid bone

Thyroid cartilage

Cricothyroid ligament

Cricoid cartilage

Trachea

(a)

Epiglottis

Hyoid bone

Corniculate cartilage

Arytenoid cartilage

Thyroid cartilage

Cricoid cartilage

C-shaped tracheal
cartilages

Esophagus location
(esophagus not shown)

(b)

FIGURE 50.4 Superior view of larynx with the glottis open as seen using a laryngoscope.

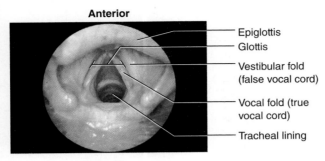

Anterior

- Epiglottis
- Glottis
- Vestibular fold (false vocal cord)
- Vocal fold (true vocal cord)
- Tracheal lining

Posterior

- cricothyroid ligament—connects thyroid and cricoid cartilages
- epiglottis—covers part of larynx opening during swallowing
- epiglottic cartilage—elastic cartilage supporting epiglottis
- arytenoid cartilages—vocal cords attached to these small cartilages
- corniculate cartilages—small cartilages superior to arytenoid cartilages
- glottis—vocal folds and opening between them

trachea—windpipe anterior to esophagus

bronchial tree—branched airways extending throughout lungs
- right and left main (primary) bronchi—arise from trachea
- lobar (secondary) bronchi—branches of main bronchi into lobes of lungs
- segmental (tertiary) bronchi—branches of lobar bronchi into segments of lungs

bronchioles—smaller branches of airways

lung—location of external respiration between alveoli and blood capillaries
- hilum of lung—medial indentation where main bronchus, blood vessels, and nerves enter lung
- lobes—large regions of lungs
  - superior (upper) lobe—located in both lungs
  - middle lobe—only in right lung
  - inferior (lower) lobe—located in both lungs
- lobules—smallest visible subdivisions of a lung

visceral pleura—serous membrane on surface of lung

parietal pleura—serous membrane on inside of thoracic wall and superior surface of diaphragm

pleural cavity—small space between pleurae containing lubricating serous fluid.

5. Complete Part A of Laboratory Assessment 50.

FIGURE 50.5 Features of the lower respiratory system (anterior view).

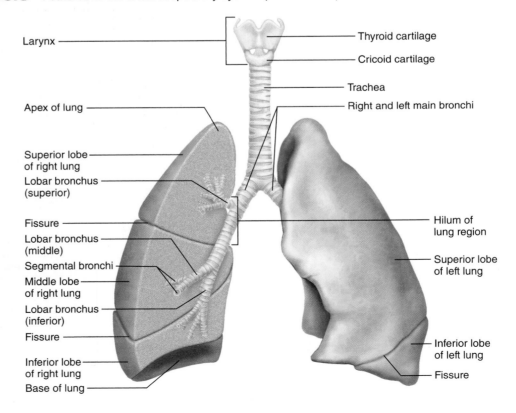

- Larynx
- Thyroid cartilage
- Cricoid cartilage
- Trachea
- Right and left main bronchi
- Apex of lung
- Superior lobe of right lung
- Lobar bronchus (superior)
- Fissure
- Lobar bronchus (middle)
- Segmental bronchi
- Middle lobe of right lung
- Lobar bronchus (inferior)
- Fissure
- Inferior lobe of right lung
- Base of lung
- Hilum of lung region
- Superior lobe of left lung
- Inferior lobe of left lung
- Fissure

Demonstration Activity

Observe the animal lung and the attached trachea. Identify the larynx, major laryngeal cartilages, trachea, and the incomplete cartilaginous rings of the trachea. Open the larynx and locate the vocal folds. Examine the visceral pleura on the surface of a lung. A bicycle pump could be used to demonstrate lung inflation. Section the lung and locate some bronchioles and alveoli. Squeeze a portion of a lung between your fingers. How do you describe the texture of the lung?

_____

_____

_____

Procedure B—Respiratory Tissues

1. Obtain a prepared microscope slide of a trachea, and use low-power magnification to examine it. Notice the inner lining of ciliated pseudostratified columnar epithelium and the deep layer of hyaline cartilage, which represents a portion of an incomplete (C-shaped) tracheal ring (fig. 50.6).
2. Use high-power magnification to observe the cilia on the free surface of the epithelial lining. Locate the wine-glass-shaped goblet cells, which secrete a protective mucus, in the epithelium (fig. 50.7).

FIGURE 50.6 Micrograph of a section of the tracheal wall (60×).

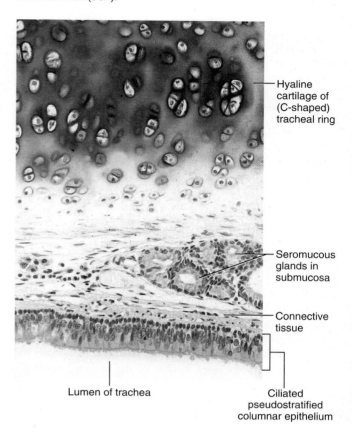

- Hyaline cartilage of (C-shaped) tracheal ring
- Seromucous glands in submucosa
- Connective tissue
- Lumen of trachea
- Ciliated pseudostratified columnar epithelium

FIGURE 50.7 Micrograph of ciliated epithelium in the respiratory tract (275×).

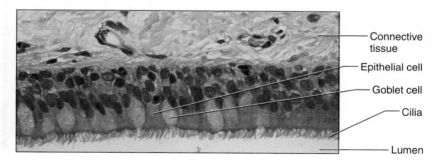

- Connective tissue
- Epithelial cell
- Goblet cell
- Cilia
- Lumen

FIGURE 50.8 Bronchioles, alveoli, and blood vessel network in a lung.

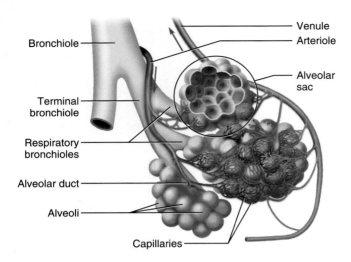

- Bronchiole
- Terminal bronchiole
- Respiratory bronchioles
- Alveolar duct
- Alveoli
- Capillaries
- Venule
- Arteriole
- Alveolar sac

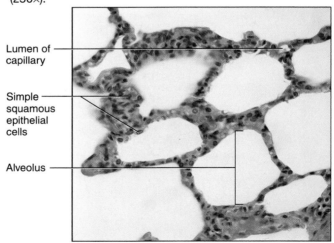

Demonstration Activity

Examine the prepared microscope slides of the lung tissue of a smoker and a person with emphysema, using low-power magnification. How does the smoker's lung tissue compare with that of the normal lung tissue that you examined previously?

_____

_____

How does the emphysema patient's lung tissue compare with the normal lung tissue?

_____

_____

3. Prepare a labeled sketch of a representative portion of the tracheal wall in Part B of the laboratory assessment.
4. Study the internal structures of a lung (fig. 50.8).
5. Obtain a prepared microscope slide of human lung tissue. Examine it, using low-power magnification, and note the many open spaces of the air sacs (alveoli). Look for a bronchiole—a tube with a relatively thick wall and a wavy inner lining. Locate the smooth muscle tissue in the wall of this tube (fig. 50.9). You also may see a section of cartilage as part of the bronchiole wall.
6. Use high-power magnification to examine the alveoli (fig. 50.10). Their walls are composed of simple squamous epithelium. You also may see sections of blood vessels containing blood cells.
7. Prepare a labeled sketch of a representative portion of the lung in Part B of the laboratory assessment.
8. Complete Part C of the laboratory assessment.

FIGURE 50.9 Micrograph of human lung tissue (35×).

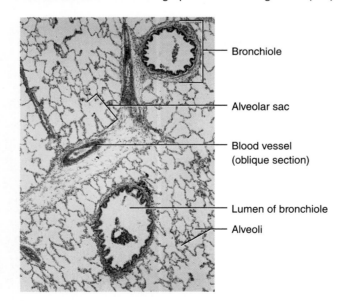

- Bronchiole
- Alveolar sac
- Blood vessel (oblique section)
- Lumen of bronchiole
- Alveoli

FIGURE 50.10 Micrograph of human lung tissue (250×).

- Lumen of capillary
- Simple squamous epithelial cells
- Alveolus

Laboratory Assessment

50

Name ______________________

Date ______________________

Section ______________________

The [A] corresponds to the indicated outcome(s) found at the beginning of the laboratory exercise.

Respiratory Organs

Part A Assessments

1. Match the terms in column A with the descriptions in column B. Place the letter of your choice in the space provided. [1] [2]

Column A	Column B
a. Alveolus	______ 1. Potential space between visceral and parietal pleurae
b. Cricoid cartilage	______ 2. Most inferior portion of larynx
c. Epiglottis	______ 3. Air-filled spaces in skull bones that open into nasal cavity
d. Glottis	______ 4. Microscopic air sac for gas exchange
e. Lung	______ 5. Consists of large lobes
f. Nasal concha	______ 6. Vocal folds, including the opening between them
g. Pharynx	______ 7. Fold of mucous membrane containing elastic fibers responsible for sounds
h. Pleural cavity	______ 8. Increases surface area of nasal mucous membrane
i. Sinus (paranasal sinus)	______ 9. Passageway for air and food
j. Vocal fold (true vocal cord)	______ 10. Partially covers opening of larynx during swallowing

2. Label the structures indicated in figures 50.11, 50.12, and 50.13.

FIGURE 50.11 Label the features of the respiratory system. [1]

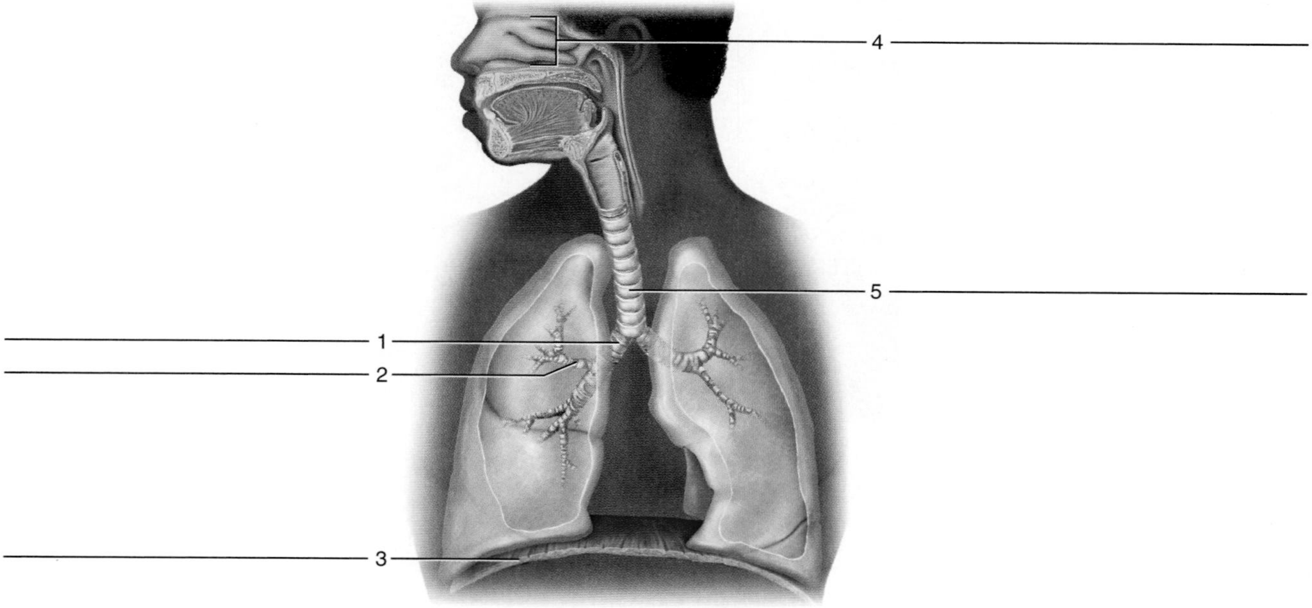

FIGURE 50.12 Label the features of the upper respiratory system (sagittal section). ⚠️

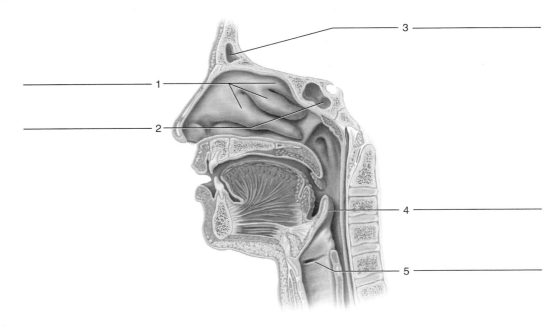

FIGURE 50.13 Label the features of the larynx region of a cadaver (lateral view). ⚠️

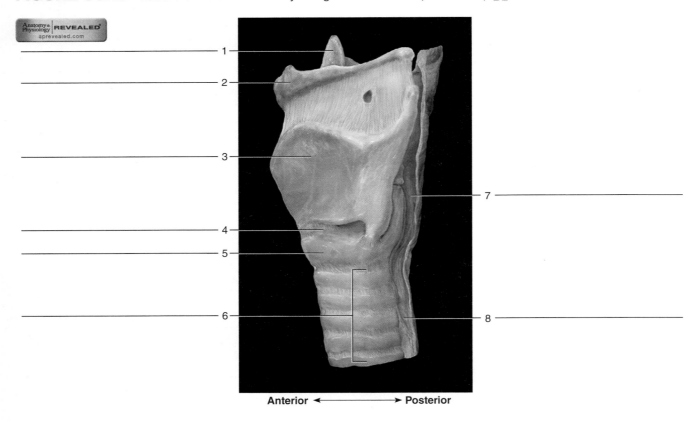

Anterior ← → Posterior

Part B Assessments

Sketch and label a portion of the tracheal wall and a portion of lung tissue. 3

Tracheal wall (_____×)	Lung tissue (_____×)

Part C Assessments

Complete the following:

1. What is the function of the mucus secreted by the goblet cells? 2 _____

   _____

   _____

2. Describe the function of the cilia in the respiratory tubes. 2 _____

   _____

   _____

3. How is breathing affected if the smooth muscle of the bronchial tree relaxes? 2 _____

   _____

   _____

Critical Thinking Assessment

Why are the alveolar walls so thin? 2

_____

_____

_____

Breathing and Respiratory Volumes

Purpose of the Exercise

To review the mechanisms of breathing and to measure or calculate certain respiratory air volumes and respiratory capacities.

Materials Needed

Lung function model
Spirometer, handheld (dry portable)
Alcohol swabs (wipes)
Disposable mouthpieces
Nose clips
Meterstick
Ph.I.L.S. 4.0

For Learning Extension Activity:
Clock or watch with seconds timer

Safety

▶ Clean the spirometer with an alcohol swab (wipe) before each use.
▶ Place a new disposable mouthpiece on the stem of the spirometer before each use.
▶ Dispose of the alcohol swabs and mouthpieces according to directions from your laboratory instructor.

Learning Outcomes

After completing this exercise, you should be able to

① Differentiate between the mechanisms responsible for inspiration and expiration.

② Measure respiratory volumes using a spirometer.

③ Calculate respiratory capacities using the data obtained from respiratory volumes.

④ Match the respiratory volumes and respiratory capacities with their definitions.

⑤ Explain the changes in tidal volume in response to changes in volume of the anatomic dead space.

⑥ Apply concepts covered in this laboratory exercise to the changes that occur when the bronchioles are dilated or constricted by the autonomic nervous system.

Pre-Lab

Carefully read the introductory material and examine the entire lab. Be familiar with inspiration, expiration, and respiratory volumes from lecture or the textbook. Visit www.mhhe.com/martinseries2 for LabCam videos. Answer the pre-lab questions.

Pre-Lab Questions: Select the correct answer for each of the following questions:

1. The size of the thoracic cavity is increased by contractions of all of the following muscles *except* the
   a. diaphragm. b. external intercostals.
   c. pectoralis minor. d. external oblique.

2. A _____ is an instrument to measure air volumes during breathing.
   a. flow meter b. spirometer
   c. lung function model d. capacity meter

3. The _____ is the maximum volume of air that can be exhaled after taking the deepest breath possible.
   a. tidal volume b. expiratory reserve volume
   c. vital capacity d. total lung capacity

4. Tidal volume is estimated to be about
   a. 500 mL. b. 1,200 mL.
   c. 3,600 mL. d. 4,800 mL.

5. A normal resting breathing rate is about _____ breaths per minute.
   a. 5–10 b. 12–15
   c. 16–20 d. 21–30

6. The contraction of the diaphragm increases the size of the thoracic cavity.
   True _____ False _____

7. Vital capacity is the total of tidal volume, expiratory reserve volume, and residual volume.
   True _____ False _____

8. Vital capacities gradually decrease as a person continues to age.
   True _____ False _____

Breathing, or pulmonary ventilation, involves the movement of air from outside the body through the bronchial tree and into the alveoli and the reversal of this air movement to allow gas (oxygen and carbon dioxide) exchange between air and blood. These movements are caused by changes in the size of the thoracic cavity that result from skeletal muscle contractions. The size of the thoracic cavity during *inspiration* is increased by contractions of the diaphragm, external intercostals, internal intercostals (intercartilaginous part), pectoralis minor, sternocleidomastoid, and scalenes. *Expiration* is aided by contractions of the internal intercostals (interosseous part), rectus abdominis, and external oblique, and from the elastic recoil of stretched tissues (fig. 51.1).

The volumes of air that move in and out of the lungs during various phases of breathing are called *respiratory (pulmonary) volumes* and *capacities*. These volumes can be measured by using an instrument called a *spirometer.* Respiratory capacities can be determined by using various combinations of respiratory volumes. However, the values obtained vary with a person's age, sex, height, weight, stress, and physical fitness.

With each breath you take, what volume of air reaches the alveoli of your lungs? You take a normal breath of air through your nose or mouth. The volume of air you breathe in (or out) is called the *tidal volume.* The average tidal volume is 500 mL (imagine the equivalent of a 1/2 liter bottle). This volume of air is moved in through your nose and mouth and through your respiratory passageway, including the nasal cavity, pharynx, larynx, trachea, and bronchial tree, to the alveoli. A certain amount of that air never reaches your alveoli—it is within the respiratory passageway. This air is said to be in the *anatomic dead space.* The average volume of anatomic dead space is 150 mL. If the tidal volume is 500 mL and the volume of the anatomic dead space is 150 mL, what volume of air reaches your alveoli? Understanding the difference between tidal volume and the volume of air that reaches the lungs is important because *only* the air that reaches the alveoli provides oxygen for gas exchange with the blood.

FIGURE 51.1 Respiratory muscles involved in inspiration and forced expiration. Boldface indicates the primary muscles increasing or decreasing the size of the thoracic cavity. Blue arrows indicate the direction of contraction during inspiration that increases the capacity of the thoracic cavity; the green arrows indicate the direction of contraction during forced expiration that decreases the capacity of the thoracic cavity. (The green arrows associated with the diaphragm indicate its ascending direction primarily from the elastic recoil of the stretched tissues during quiet breathing.)

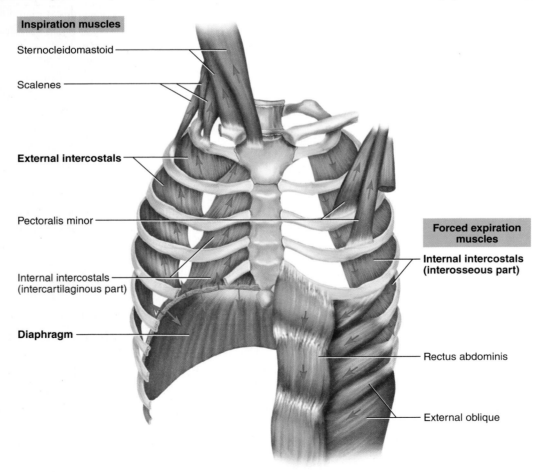

Inspiration muscles
- Sternocleidomastoid
- Scalenes
- **External intercostals**
- Pectoralis minor
- Internal intercostals (intercartilaginous part)
- **Diaphragm**

Forced expiration muscles
- **Internal intercostals (interosseous part)**
- Rectus abdominis
- External oblique

Procedure A—Breathing Mechanisms

The size of the thoracic cavity increases during inspiration by contractions of inspiration muscles, primarily the diaphragm, resulting in increased depth of the thoracic cavity, and external intercostals, widening the thoracic cavity. A combination of synergistic contractions of the sternocleidomastoid, scalenes, pectoralis minor, and the intercartilaginous part of the internal intercostals elevate the sternum and rib cage. During forced expiration, the interosseous part of the internal intercostals depresses the rib cage, narrowing the thoracic cavity. The contractions of the rectus abdominis and the external oblique compress the abdominal viscera, forcing the diaphragm to ascend and thus reducing the size of the thoracic cavity (fig. 51.1).

Air enters the lungs during inspiration and exits the lungs during expiration based upon characteristics of *Boyle's law*. According to Boyle's law, the pressure of a gas is inversely proportional to its volume, assuming a constant temperature. In other words, when the volume increases in a container such as the thoracic cavity during inspiration, pressure decreases and air flows into the lungs. In contrast, when volume decreases in a container such as the thoracic cavity during expiration, pressure increases and air flows out of the lungs. Air moves into and out of the lungs in much the same way as can be demonstrated using a syringe (fig. 51.2). When you pull back on the plunger of the syringe, the volume increases in the barrel of the syringe, which decreases the pressure, resulting in air rushing into the barrel of the syringe. In contrast, pushing the plunger into the barrel of the syringe causes the pressure to increase in the barrel of the syringe, and air flows out of the container.

1. Observe the mechanical lung function model (fig. 51.3). It consists of a heavy plastic bell jar with a rubber sheeting clamped over its wide open end. Its narrow upper opening is plugged with a rubber stopper through which a Y tube is passed. Small rubber balloons are fastened to the arms of the Y. What happens to the balloons when the rubber sheeting is pulled downward?

_____

_____

What happens when the sheeting is pushed upward?

_____

_____

2. Compare the lung function model with figures 51.1 and 51.2. Note the muscles of inspiration that increase the size of the thoracic cavity, and the muscles of expiration that decrease the size of the thoracic cavity. Relate these volume changes to pressure changes and airflow into and out of the lungs.

3. Complete Part A of Laboratory Assessment 51.

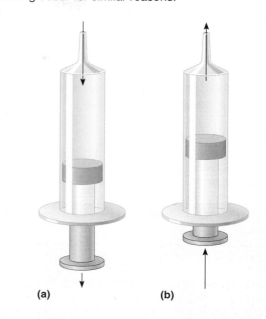

FIGURE 51.2 Moving the plunger of a syringe as in (a) results in air movements into the barrel of the syringe; moving the plunger of the syringe as in (b) results in air movements out of the barrel of the syringe. Air movements in and out of the lung function model and the lungs during breathing occur for similar reasons.

(a)   (b)

FIGURE 51.3 A lung function model.

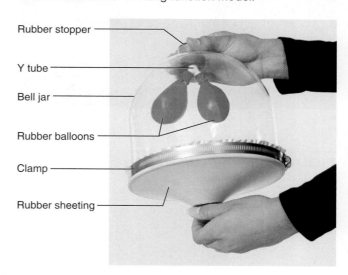

Rubber stopper

Y tube

Bell jar

Rubber balloons

Clamp

Rubber sheeting

Procedure B—Respiratory Volumes and Capacities (This procedure can be in conjunction with Laboratory Exercise 52.)

 Warning

If the subject begins to feel dizzy or light-headed while performing Procedure B, stop the exercise and breathe normally.

1. Get a handheld spirometer (fig. 51.4). Point the needle to zero by rotating the adjustable dial. Before using the instrument, clean it with an alcohol swab and place a new disposable mouthpiece over its stem. The instrument should be held with the dial upward, and *air should be blown only into the disposable mouthpiece* (fig. 51.5). If air tends to exit the nostrils during exhalation, use a nose clip or pinch your nose as a prevention when exhaling into the spirometer. Movement of the needle indicates the air volume that leaves the lungs. The exhalation should be slowly and forcefully performed. Too rapid an exhalation could result in erroneous data or damage to the spirometer.

2. *Tidal volume (TV)* (about 500 mL) is the volume of air that enters (or leaves) the lungs during a *respiratory cycle* (one inspiration plus the following expiration). *Resting tidal volume* is the volume of air that enters (or leaves) the lungs during normal, quiet breathing (fig. 51.6). To measure this volume, follow these steps:
   a. Sit quietly for a few moments.
   b. Position the spirometer dial so that the needle points to zero.

FIGURE 51.5 Demonstration of use of a handheld spirometer. Air should only be blown slowly and forcefully into a disposable mouthpiece. *Use a nose clip or pinch your nose when measuring expiratory reserve volume and vital capacity volume if air exits the nostrils.*

FIGURE 51.4 A handheld spirometer can be used to measure respiratory volumes.

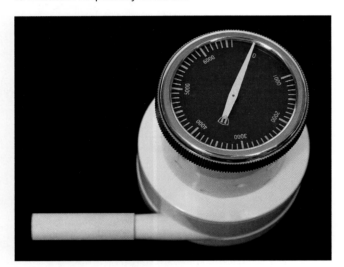

FIGURE 51.6 Graphic representation of respiratory volumes and capacities.

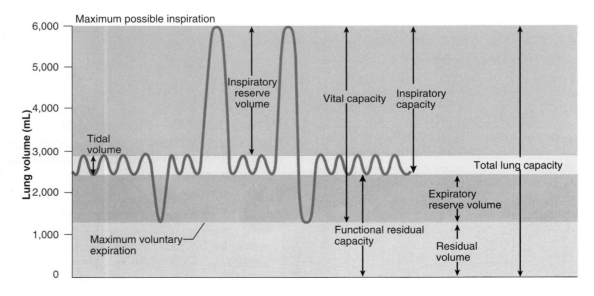

c. Place the mouthpiece between your lips and exhale three ordinary expirations into it after inhaling through the nose each time. *Do not force air out of your lungs; exhale normally.*

d. Divide the total value indicated by the needle by 3 and record this amount as your resting tidal volume on the table in Part B of the laboratory assessment.

3. *Expiratory reserve volume (ERV)* (about 1,200 mL) is the volume of air in addition to the tidal volume that leaves the lungs during forced expiration (fig. 51.6). Chronic obstructive pulmonary disease (COPD) reduces pulmonary ventilation; COPD occurs in emphysema and chronic bronchitis, which are often caused by smoking and other inhaled pollutants. With COPD, the ERV is greatly reduced during physical exertion, and the person becomes easily exhausted from expending more energy just to breathe. To measure this volume, follow these steps:

a. Breathe normally for a few moments. Set the needle to zero.

b. At the end of an ordinary expiration, place the mouthpiece between your lips and exhale all of the air you can force from your lungs through the spirometer. Use a nose clip or pinch your nose to prevent any air from exiting your nostrils.

c. Record the results as your expiratory reserve volume in Part B of the laboratory assessment.

4. *Vital capacity (VC)* (about 4,800 mL) is the maximum volume of air that can be exhaled after taking the deepest breath possible (fig. 51.6). To measure this volume, follow these steps:

a. Breathe normally for a few moments. Set the needle at zero.

b. Breathe in and out deeply a couple of times, then take the deepest breath possible.

c. Place the mouthpiece between your lips and exhale all the air out of your lungs, slowly and forcefully. Use a nose clip or pinch your nose to prevent any air from exiting your nostrils.

d. Record the value as your vital capacity in Part B of the laboratory assessment. Compare your result with that expected for a person of your sex, age, and height listed in tables 51.1 and 51.2. Use the meter-

stick to determine your height in centimeters if necessary or multiply your height in inches times 2.54 to calculate your height in centimeters. Considerable individual variations from the expected will be noted due to parameters other than sex, age, and height, which could include physical shape, health, medications, and others.

5. *Inspiratory reserve volume (IRV)* (about 3,100 mL) is the volume of air in addition to the tidal volume that enters the lungs during forced inspiration (fig. 51.6). Calculate your inspiratory reserve volume by subtracting your tidal volume (TV) and your expiratory reserve volume (ERV) from your vital capacity (VC):

$$IRV = VC - (TV + ERV)$$

6. *Inspiratory capacity (IC)* (about 3,600 mL) is the maximum volume of air a person can inhale following exhalation of the tidal volume (fig. 51.6). Calculate your inspiratory capacity by adding your tidal volume (TV) and your inspiratory reserve volume (IRV):

$$IC = TV + IRV$$

7. *Functional residual capacity (FRC)* (about 2,400 mL) is the volume of air that remains in the lungs following exhalation of the tidal volume (fig. 51.6). Calculate your functional residual capacity (FRC) by adding your expiratory reserve volume (ERV) and your residual volume (RV), which you can assume is 1,200 mL:

$$FRC = ERV + 1,200$$

8. *Residual volume (RV)* (about 1,200 mL) is the volume of air that always remains in the lungs after the most forceful expiration (fig. 51.6). Although it is part of *total lung capacity* (about 6,000 mL), it cannot be measured

Learning Extension Activity

Determine your *minute respiratory volume*. To do this, follow these steps:

1. Sit quietly for a while, and then to establish your breathing rate, count the number of times you breathe in 1 minute. This might be inaccurate because conscious awareness of breathing rate can alter the results. You might ask a laboratory partner to record your breathing rate at some time when you are not expecting it to be recorded. A normal resting breathing rate is about 12–15 breaths per minute.

2. Calculate your minute respiratory volume by multiplying your breathing rate by your tidal volume:

_____ × _____ = _____
(breathing   (tidal   (minute respiratory
rate)   volume)   volume)

3. This value indicates the total volume of air that moves into your respiratory passages during each minute of ordinary breathing.

Critical Thinking Activity

It can be noted from the data in tables 51.1 and 51.2 that vital capacities gradually decrease with age. Propose an explanation for this normal correlation.

_____

_____

_____

_____

TABLE 51.1 Predicted Vital Capacities (in Milliliters) for Females

Age	Height in Centimeters																								
	146	148	150	152	154	156	158	160	162	164	166	168	170	172	174	176	178	180	182	184	186	188	190	192	194
16	2950	2990	3030	3070	3110	3150	3190	3230	3270	3310	3350	3390	3430	3470	3510	3550	3590	3630	3670	3715	3755	3800	3840	3880	3920
18	2920	2960	3000	3040	3080	3120	3160	3200	3240	3280	3320	3360	3400	3440	3480	3520	3560	3600	3640	3680	3720	3760	3800	3840	3880
20	2890	2930	2970	3010	3050	3090	3130	3170	3210	3250	3290	3330	3370	3410	3450	3490	3525	3565	3605	3645	3695	3720	3760	3800	3840
22	2860	2900	2940	2980	3020	3060	3095	3135	3175	3215	3255	3290	3330	3370	3410	3450	3490	3530	3570	3610	3650	3685	3725	3765	3800
24	2830	2870	2910	2950	2985	3025	3065	3100	3140	3180	3220	3260	3300	3335	3375	3415	3455	3490	3530	3570	3610	3650	3685	3725	3765
26	2800	2840	2880	2920	2960	3000	3035	3070	3110	3150	3190	3230	3265	3300	3340	3380	3420	3455	3495	3530	3570	3610	3650	3685	3725
28	2775	2810	2850	2890	2930	2965	3000	3040	3070	3115	3155	3190	3230	3270	3305	3345	3380	3420	3460	3495	3535	3570	3610	3650	3685
30	2745	2780	2820	2860	2895	2935	2970	3010	3045	3085	3120	3160	3195	3235	3270	3310	3345	3385	3420	3460	3495	3535	3570	3610	3645
32	2715	2750	2790	2825	2865	2900	2940	2975	3015	3050	3090	3125	3160	3200	3235	3275	3310	3350	3385	3425	3460	3495	3535	3570	3610
34	2685	2725	2760	2795	2835	2870	2910	2945	2980	3020	3055	3090	3130	3165	3200	3240	3275	3310	3350	3385	3425	3460	3495	3535	3570
36	2655	2695	2730	2765	2805	2840	2875	2910	2950	2985	3020	3060	3095	3130	3165	3205	3240	3275	3310	3350	3385	3420	3460	3495	3530
38	2630	2665	2700	2735	2770	2810	2845	2880	2915	2950	2990	3025	3060	3095	3130	3170	3205	3240	3275	3310	3350	3385	3420	3455	3490
40	2600	2635	2670	2705	2740	2775	2810	2850	2885	2920	2955	2990	3025	3060	3095	3135	3170	3205	3240	3275	3310	3345	3380	3420	3455
42	2570	2605	2640	2675	2710	2745	2780	2815	2850	2885	2920	2955	2990	3025	3060	3100	3135	3170	3205	3240	3275	3310	3345	3380	3415
44	2540	2575	2610	2645	2680	2715	2750	2785	2820	2855	2890	2925	2960	2995	3030	3060	3095	3130	3165	3200	3235	3270	3305	3340	3375
46	2510	2545	2580	2615	2650	2685	2715	2750	2785	2820	2855	2890	2925	2960	2995	3030	3060	3095	3130	3165	3200	3235	3270	3305	3340
48	2480	2515	2550	2585	2620	2650	2685	2715	2750	2785	2820	2855	2890	2925	2960	2995	3030	3060	3095	3130	3160	3195	3230	3265	3300
50	2455	2485	2520	2555	2590	2625	2655	2690	2720	2755	2785	2820	2855	2890	2925	2955	2990	3025	3060	3090	3125	3155	3190	3225	3260
52	2425	2455	2490	2525	2555	2590	2625	2655	2690	2720	2755	2790	2820	2855	2890	2925	2955	2990	3020	3055	3090	3125	3155	3190	3220
54	2395	2425	2460	2495	2530	2560	2590	2625	2655	2690	2720	2755	2790	2820	2855	2885	2920	2950	2985	3020	3050	3085	3115	3150	3180
56	2365	2400	2430	2465	2495	2525	2560	2590	2625	2655	2690	2720	2755	2790	2820	2855	2885	2920	2950	2980	3015	3045	3080	3110	3145
58	2335	2370	2400	2430	2460	2495	2525	2560	2590	2625	2655	2690	2720	2750	2785	2815	2850	2880	2920	2945	2975	3010	3040	3075	3105
60	2305	2340	2370	2400	2430	2460	2495	2525	2560	2590	2625	2655	2685	2720	2750	2780	2810	2845	2875	2915	2940	2970	3000	3035	3065
62	2280	2310	2340	2370	2405	2435	2465	2495	2525	2560	2590	2620	2655	2685	2715	2745	2775	2810	2840	2870	2900	2935	2965	2995	3025
64	2250	2280	2310	2340	2370	2400	2430	2465	2495	2525	2555	2585	2620	2650	2680	2710	2740	2770	2805	2835	2865	2895	2920	2955	2990
66	2220	2250	2280	2310	2340	2370	2400	2430	2460	2495	2525	2555	2585	2615	2645	2675	2705	2735	2765	2800	2825	2860	2890	2920	2950
68	2190	2220	2250	2280	2310	2340	2370	2400	2430	2460	2490	2520	2550	2580	2610	2640	2670	2700	2730	2760	2795	2820	2850	2880	2910
70	2160	2190	2220	2250	2280	2310	2340	2370	2400	2425	2455	2485	2515	2545	2575	2605	2635	2665	2695	2725	2755	2780	2810	2840	2870
72	2130	2160	2190	2220	2250	2280	2310	2335	2365	2395	2425	2455	2480	2510	2540	2570	2600	2630	2660	2685	2715	2745	2775	2805	2830
74	2100	2130	2160	2190	2220	2245	2275	2305	2335	2360	2390	2420	2450	2475	2505	2535	2565	2590	2620	2650	2680	2710	2740	2765	2795

From E. DeF. Baldwin and E. W. Richards, Jr., "Pulmonary Insufficiency I, Physiologic Classification, Clinical Methods of Analysis, Standard Values in Normal Subjects" in *Medicine* 27:243, © by William & Wilkins. Used by permission.

TABLE 52.2 Predicted Vital Capacities (in Milliliters) for Males

Age	Height in Centimeters																								
	146	148	150	152	154	156	158	160	162	164	166	168	170	172	174	176	178	180	182	184	186	188	190	192	194
16	3765	3820	3870	3920	3975	4025	4075	4130	4180	4230	4285	4335	4385	4440	4490	4540	4590	4645	4695	4745	4800	4850	4900	4955	5005
18	3740	3790	3840	3890	3940	3995	4045	4095	4145	4200	4250	4300	4350	4405	4455	4505	4555	4610	4660	4710	4760	4815	4865	4915	4965
20	3710	3760	3810	3860	3910	3960	4015	4065	4115	4165	4215	4265	4320	4370	4420	4470	4520	4570	4625	4675	4725	4775	4825	4875	4930
22	3680	3730	3780	3830	3880	3930	3980	4030	4080	4135	4185	4235	4285	4335	4385	4435	4485	4535	4585	4635	4685	4735	4790	4840	4890
24	3635	3685	3735	3785	3835	3885	3935	3985	4035	4085	4135	4185	4235	4285	4330	4380	4430	4480	4530	4580	4630	4680	4730	4780	4830
26	3605	3655	3705	3755	3805	3855	3905	3955	4000	4050	4100	4150	4200	4250	4300	4350	4395	4445	4495	4545	4595	4645	4695	4740	4790
28	3575	3625	3675	3725	3775	3820	3870	3920	3970	4020	4070	4115	4165	4215	4265	4310	4360	4410	4460	4510	4555	4605	4655	4705	4755
30	3550	3595	3645	3695	3740	3790	3840	3890	3935	3985	4035	4080	4130	4180	4230	4275	4325	4375	4425	4470	4520	4570	4615	4665	4715
32	3520	3565	3615	3665	3710	3760	3810	3855	3905	3950	4000	4050	4095	4145	4195	4240	4290	4340	4385	4435	4485	4530	4580	4625	4675
34	3475	3525	3570	3620	3665	3715	3760	3810	3855	3905	3950	4000	4045	4095	4140	4190	4225	4285	4330	4380	4425	4475	4520	4570	4615
36	3445	3495	3540	3585	3635	3680	3730	3775	3825	3870	3920	3965	4010	4060	4105	4155	4200	4250	4295	4340	4390	4435	4485	4530	4580
38	3415	3465	3510	3555	3605	3650	3695	3745	3790	3840	3885	3930	3980	4025	4070	4120	4165	4210	4260	4305	4350	4400	4445	4495	4540
40	3385	3435	3480	3525	3575	3620	3665	3710	3760	3805	3850	3900	3945	3990	4035	4085	4130	4175	4220	4270	4315	4360	4410	4455	4500
42	3360	3405	3450	3495	3540	3590	3635	3680	3725	3770	3820	3865	3910	3955	4000	4050	4095	4140	4185	4230	4280	4325	4370	4415	4460
44	3315	3360	3405	3450	3495	3540	3585	3630	3675	3725	3770	3815	3860	3905	3950	3995	4040	4085	4130	4175	4220	4270	4315	4360	4405
46	3285	3330	3375	3420	3465	3510	3555	3600	3645	3690	3735	3780	3825	3870	3915	3960	4005	4050	4095	4140	4185	4230	4275	4320	4365
48	3255	3300	3345	3390	3435	3480	3525	3570	3615	3655	3700	3745	3790	3835	3880	3925	3970	4015	4060	4105	4150	4190	4235	4280	4325
50	3210	3255	3300	3345	3390	3430	3475	3520	3565	3610	3650	3695	3740	3785	3830	3870	3915	3960	4005	4050	4090	4135	4180	4225	4270
52	3185	3225	3270	3315	3355	3400	3445	3490	3530	3575	3620	3660	3705	3750	3795	3835	3880	3925	3970	4010	4055	4100	4140	4185	4230
54	3155	3195	3240	3285	3325	3370	3415	3455	3500	3540	3585	3630	3670	3715	3760	3800	3845	3890	3930	3975	4020	4060	4105	4145	4190
56	3125	3165	3210	3255	3295	3340	3380	3425	3465	3510	3550	3595	3640	3680	3725	3765	3810	3850	3895	3940	3980	4025	4065	4110	4150
58	3080	3125	3165	3210	3250	3290	3335	3375	3420	3460	3500	3545	3585	3630	3670	3715	3755	3800	3840	3880	3925	3965	4010	4050	4095
60	3050	3095	3135	3175	3220	3260	3300	3345	3385	3430	3470	3500	3555	3595	3635	3680	3720	3760	3805	3845	3885	3930	3970	4015	4055
62	3020	3060	3110	3150	3190	3230	3270	3310	3350	3390	3440	3480	3520	3560	3600	3640	3680	3730	3770	3810	3850	3890	3930	3970	4020
64	2990	3030	3080	3120	3160	3200	3240	3280	3320	3360	3400	3440	3490	3530	3570	3610	3650	3690	3730	3770	3810	3850	3900	3940	3980
66	2950	2990	3030	3070	3110	3150	3190	3230	3270	3310	3350	3390	3430	3470	3510	3550	3600	3640	3680	3720	3760	3800	3840	3880	3920
68	2920	2960	3000	3040	3080	3120	3160	3200	3240	3280	3320	3360	3400	3440	3480	3520	3560	3600	3640	3680	3720	3760	3800	3840	3880
70	2890	2930	2970	3010	3050	3090	3130	3170	3210	3250	3290	3330	3370	3410	3450	3480	3520	3560	3600	3640	3680	3720	3760	3800	3840
72	2860	2900	2940	2980	3020	3060	3100	3140	3180	3210	3250	3290	3330	3370	3410	3450	3490	3530	3570	3610	3650	3680	3720	3760	3800
74	2820	2860	2900	2930	2970	3010	3050	3090	3130	3170	3200	3240	3280	3320	3360	3400	3440	3470	3510	3550	3590	3630	3670	3710	3740

From E. DeF. Baldwin and E. W. Richards, Jr., "Pulmonary Insufficiency 1, Physiologic Classification, Clinical Methods of Analysis, Standard Values in Normal Subjects" in *Medicine 27:243*, © by William & Wilkins. Used by permission.

with a spirometer. The residual air allows gas exchange and the alveoli to remain open during the respiratory cycle.

9. Complete Parts B and C of the laboratory assessment.

Procedure C—Ph.I.L.S. Lesson 38 Respiration: Altering Airway Volume

1. Open Exercise 38, Respiration: Altering Airway Volume.
2. Read the objectives and introduction and take the pre-lab quiz.
3. After completing the pre-lab quiz, read through the wet lab. *Be sure to click open and view the videos that are indicated in red in the wet lab.*
4. The lab exercise will open when you click Continue after completing the wet lab (fig. 51.7).

Setup

5. To start, click the power switch to turn on the virtual computer screen, then click the power switch to turn on the Data Acquisition Unit.
6. Connect the breathing apparatus by inserting the blue plug into input 1 of the Data Acquisition Unit.
7. Attach the breathing apparatus to the subject's mouth by dragging the other end to the mouth.

Breathing Apparatus Alone

8. Click the Start button. A trace will appear on the control panel. You will see the rhythmic upward deflections of inspiration and the downward deflections of expiration. (This volume of air is tidal volume.) Record three breathing cycles and click Stop.

9. Measure the tidal volume for the first wave on the left. Position the crosshairs (using the mouse) at the top of the wave and click. If not in the correct location, reposition by dragging to a new position. Now position the crosshairs at the bottom of the deflection (after the wave or 0 volts) and click. The result will display underneath the Start button.
10. Click the Journal panel (red rectangle at bottom right of screen) to enter your value into the Journal. A table for tidal volumes (measured without tube and with tube) will appear. There are three trials for each setup.
11. Repeat numbers 9 and 10 (Ph.I.L.S. steps 7 and 8) for the second and third waves.
12. *Print the graph of the tidal volumes by clicking the P in the bottom left of the monitor screen.*
13. Close the Journal window.

Breathing Apparatus with Plastic Tube

14. Insert the plastic tube into the breathing apparatus (insert into end of plastic tube on the right side).
15. Repeat steps 8 through 13.

Interpreting Results

16. *With the Journal still on the screen,* answer the questions in Part D of the laboratory assessment. If you accidentally closed the graph, click on the Journal panel (red rectangle at bottom right of screen).
17. Complete the Post-Lab Quiz (click open Post-Lab Quiz and Lab Report) by answering the ten questions on the computer screen.
18. Read the conclusion on the computer screen.
19. You may print the Lab Report for Respiration: Altering Airway Volume.

FIGURE 51.7 Opening screen for the laboratory exercise on Respiration: Altering Airway Volume.

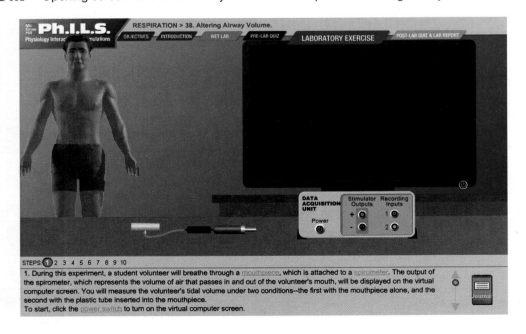

Laboratory Assessment

51

Name _____

Date _____

Section _____

The A corresponds to the indicated outcome(s) found at the beginning of the laboratory exercise.

Breathing and Respiratory Volumes

Part A   Assessments

Complete the following statements:

1. When using the lung function model, what part of the respiratory system is represented by: A

   a. The rubber sheeting? _____

   c. The Y tube? _____

   b. The bell jar? _____

   d. The balloons? _____

2. When the diaphragm contracts, the size of the thoracic cavity _____. A

3. The ribs are raised by contraction of the _____

   muscles, which increases the size of the thoracic cavity. A

4. Muscles that help to force out more than the normal volume of air by pulling the ribs downward and inward include the

   _____. A

5. We inhale when the diaphragm _____. A

Part B   Assessments

1. Test results for respiratory air volumes and capacities: A A

Respiratory Volume or Capacity	Expected Value* (approximate)	Test Result	Percent of Expected Value (test result/expected value × 100)
Tidal volume (resting) (TV)	500 mL		
Expiratory reserve volume (ERV)	1,200 mL		
Vital capacity (VC)	(enter yours from table 51.1 or 51.2)		
Inspiratory reserve volume (IRV)	3,100 mL		
Inspiratory capacity (IC)	3,600 mL		
Functional residual capacity (FRC)	2,400 mL		

*The values listed are most characteristic for a healthy, tall, young adult male. In general, adult females have smaller bodies and therefore smaller lung volumes and capacities. If your expected value for vital capacity is considerably different than 4,800 mL, your other values will vary accordingly.

2. Complete the following:

    a. How do your test results compare with the expected values? _____

    _____

    _____

    _____

    _____

    b. How does your vital capacity compare with the average value for a person of your sex, age, and height?

    _____

    _____

    c. What measurement in addition to vital capacity is needed before you can calculate your total lung capacity?

    _____

3. If your experimental results are considerably different than the predicted vital capacities, propose reasons for the differences. As you write this paragraph, consider factors such as smoking, physical fitness, respiratory disorders, stress, and medications. (Your instructor might have you make some class correlations from class data.)

_____

_____

_____

_____

Part C   Assessments

Match the air volumes in column A with the definitions in column B. Place the letter of your choice in the space provided. ▲4

Column A	Column B
a. Expiratory reserve volume	_____ **1.** Volume in addition to tidal volume that leaves the lungs during forced expiration
b. Functional residual capacity	
c. Inspiratory capacity	_____ **2.** Vital capacity plus residual volume
d. Inspiratory reserve volume	_____ **3.** Volume that remains in lungs after the most forceful expiration
e. Residual volume	
f. Tidal volume	_____ **4.** Volume that enters or leaves lungs during a respiratory cycle
g. Total lung capacity	_____ **5.** Volume in addition to tidal volume that enters lungs during forced inspiration
h. Vital capacity	
	_____ **6.** Maximum volume a person can exhale after taking the deepest possible breath
	_____ **7.** Maximum volume a person can inhale following exhalation of the tidal volume
	_____ **8.** Volume of air remaining in the lungs following exhalation of the tidal volume

Part D Ph.I.L.S. Lesson 38, Respiration: Altering Airway Volume Assessments

1. From the experimental results, complete the following table for the measured tidal volumes comparing breathing apparatus alone and breathing apparatus with a plastic tube. **2**

Trial	Without Tube	With Tube
Trial # 1		
Trial # 2		
Trial # 3		
Average (mean)		

2. Which setup had the larger average tidal volume? _____ **5**

3. To maintain normal alveolar ventilation, if the anatomic dead space is increased, the tidal volume will _____ (increase, decrease, or remain the same). **5**

4. In the setup with the tube attached to the breathing apparatus, did additional air reach the alveoli? _____ **5**

5. Complete the following table for the effect on each variable if a person breathes with the tube. **5**

Volume of Air	Change (increase, decrease, or remain the same)
Anatomic dead space	
Tidal volume	
Alveolar ventilation	

Critical Thinking Assessment **6**

If the bronchioles are dilated (when the sympathetic nervous system is activated, for example), the anatomic dead space _____ (increases, decreases, or remains the same), and tidal volume will _____ _____ (increase, decrease, or remain the same). In patients with emphysema, the anatomic dead space increases (since there is a loss of elasticity of lung tissue and the thoracic cage can not effectively "be pulled back in"). Predict the changes in breathing of a patient with emphysema as he/she tries to compensate for the increase in anatomic dead space. _____

_____

_____

_____

_____

NOTES

Spirometry: BIOPAC© Exercise

Purpose of the Exercise

To measure and calculate pulmonary volumes and capacities using the BIOPAC airflow transducer system.

Materials Needed

Computer system (Mac OS X 10.4–10.6, or PC running Windows XP, Vista, or 7)

BIOPAC Student Lab software ver. 3.7.7 or above

BIOPAC Data Acquisition Unit (MP36, MP35, MP30, or MP45 (Note: The MP30 is being phased out) with (AC100A) transformer

BIOPAC serial cable (CBLSERA)

BIOPAC airflow transducer: either SS11L or SS11LA with removable/autoclavable head

BIOPAC disposable mouthpiece (AFT2) and nose clip (AFT3), one for each subject

BIOPAC bacteriological filter (AFT1) (need at minimum one for calibration; if using SS11L, you will need one filter for each subject)

BIOPAC calibration syringe 0.6 liter (AFT6 or AFT6A + AFT11A) or 2 liter (AFT26)

⚠ Safety

▶ If using nonremovable airflow transducer SS11L, each subject should use a clean filter.

▶ If using removable head airflow transducer SS11LA, make sure each subject has a clean, sterilized head.

▶ Each subject will need a clean, disposable mouthpiece and nose clip.

Learning Outcomes

After completing this exercise, you should be able to

1. Examine experimentally and record and/or calculate selected respiratory volumes and capacities.

2. Compare the measured volumes and capacities with average values.

3. Compare the volumes and capacities of subjects differing in sex, age, height, and weight.

Pre-Lab

Carefully read the introductory material and examine the entire lab content. Be familiar with inspiration, expiration, and respiratory volumes from lecture or the textbook.

Breathing (pulmonary ventilation), one of the most basic processes of life, consists of two phases: inspiration and expiration. *Inspiration* consists of the inhalation of air into the respiratory passages and lungs, which provides vital oxygen (O_2) to the cells of the body. Oxygen is required for the aerobic cellular respiration of food, to break down nutrients into their building blocks, and to generate ATP for energy. Expiration involves the exhalation of air from the lungs into the atmosphere, which rids the body of carbon dioxide (CO_2). Carbon dioxide is produced as a waste product of cellular respiration.

One way that the breathing process can be studied is through the use of a spirometer. There are various types of spirometers, from small handheld dry spirometers, to wet bell varieties, to computer-linked spirometers. But all spirometers work on the same principle—they all measure certain respiratory volumes, based on different amounts of effort in the breathing process. The respiratory volumes used in spirometry are as follows:

1. *Tidal volume (TV):* The volume of air inspired or expired in one normal, quiet breath (about 500 mL).
2. *Inspiratory reserve volume (IRV):* The volume of air that can be inspired forcefully above a tidal inspiration (about 3,100 mL).
3. *Expiratory reserve volume (ERV):* The volume of air that can be expired forcefully beyond a tidal expiration (about 1,200 mL).
4. *Residual volume (RV):* The volume of air remaining in the lungs after forceful expiration (about 1,200 mL).

Tidal respiration does not require any extra effort, but volumes such as IRV and ERV are defined by the extra effort used to measure them. For example, if you are preparing to blow up a balloon, and you inhale the maximum possible amount of air, the amount you inhale above your tidal inspiration will be your IRV. When you actually blow up the balloon, and you exhale the maximum possible amount of air, the amount beyond your tidal expiration will be your ERV. The residual volume cannot be measured with a spirometer, since it always remains in the lungs; for the purpose of this lab exercise, the default setting for RV by the BIOPAC software is 1.0 L.

In addition to respiratory volumes, several respiratory capacities can also be defined, which can be measured in the lab and/or calculated. Respiratory capacities can be calculated by adding two or more volumes together, as follows:

1. *Inspiratory capacity (IC):* The maximal volume of air that can be inspired after a tidal expiration, IC = TV + IRV (about 3,600 mL).
2. *Expiratory capacity (EC):* The maximal volume of air that can be expired after a tidal inspiration, EC = TV + ERV (about 1,700 mL).
3. *Functional residual capacity (FRC):* The volume of air remaining in the lungs after a tidal expiration, FRC = ERV + RV (about 2,400 mL).

4. *Vital capacity (VC):* The maximal volume of air that can be forcefully expired after a forceful inspiration, VC = TV + IRV + ERV (about 4,800 mL).
5. *Total lung capacity (TLC):* The total volume of air that the lungs and respiratory passages can hold, TLC = TV + IRV + ERV + RV or TLC = VC + RV (about 6,000 mL).

It is important to realize that there is quite a bit of individual variation in respiratory volumes and capacities. This variation is due to differences in sex, height, weight, age, and exercise history. Normal averages are given above, but any reading within 20% of the average is still considered to be in the normal range. Using vital capacity as an example, even though there are several factors determining the VC of an individual, VC can be estimated fairly accurately based on sex, height (H) in cm, and age (A) in years, with the following equations:

Male: VC = 0.052 H − 0.022 A − 3.60

Female: VC = 0.041 H − 0.018 A − 2.69

This BIOPAC exercise uses a computer-linked spirometry system to measure respiratory volumes. Each subject will breathe into an airflow transducer; the BIOPAC software will then convert airflow into volume. Several volumes and capacities will be measured and/or calculated in this lab exercise. VC will be both measured and estimated from one of the equations given above; the two values will then be compared. Again, remember that RV cannot be measured through spirometry. Figure 52.1 shows a record of volume changes over time for various respiratory volumes and capacities; this is called a *spirogram.*

Spirometry is used in medicine to help in the diagnosis of specific respiratory disorders, determine the extent of a respiratory illness or disease, and assess the respiratory status of a person recovering from an illness, such as pneumonia or an asthma attack. Certain illnesses are associated with characteristic changes in specific respiratory volumes and/or capacities. For example, in people with obstructive lung disease, such as chronic bronchitis, the FRC and RV may be elevated due to an inability to expel the normal amount of air during expiration and excess inflation of the lungs. In cases of restrictive lung disease, such as pulmonary fibrosis, decreased lung compliance makes it difficult to expand the lungs, which results in a decrease in IC and VC.

Procedure A—Setup

1. With your computer **ON** and the BIOPAC MP3X or MP45 unit turned **OFF,** plug the airflow transducer (SS11L or SSL11LA) into Channel 1.
2. Turn on the MP3X or MP45 Data Acquisition Unit.
3. Connect a bacteriological filter to a calibration syringe, and then connect the calibration syringe/filter assembly to the airflow transducer, as shown in figure 52.2. (Be sure the assembly is inserted on the side labeled "Inlet.")

Source: Courtesy of and © BIOPAC Systems Inc.

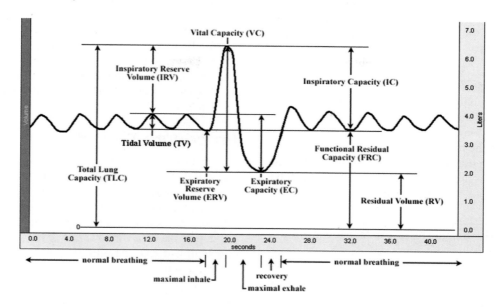

FIGURE 52.2 Calibration setup.

Source: Courtesy of and © BIOPAC Systems Inc.

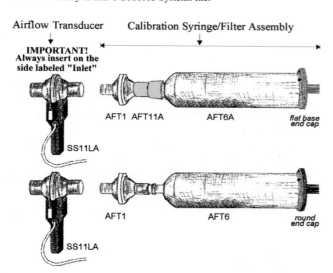

Once you have put the calibration assembly together, always support the calibration syringe by holding it up with your nondominant hand. Do not hold the assembly by the handle of the airflow transducer; this could damage the calibration syringe.

4. Start the BIOPAC student lab program for lesson L12—Pulmonary Function 1.
5. Designate a filename to be used to save the subject's data.

FIGURE 52.3 Sample recording for first calibration stage.

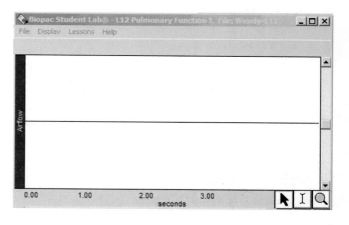

Procedure B—Calibration

1. Pull the calibration syringe plunger all the way out and hold the calibration syringe/filter assembly still and parallel to the floor.
2. The calibration occurs in two stages. During the first stage, simply hold the assembly still and click on the **Calibrate** button. This calibration stage will run for 4 seconds and will shut off automatically. The recording should look similar to Figure 52.3.
3. For the second calibration stage, click **Yes** after reading the alert box. Then push the syringe plunger in and out completely 5 times (10 strokes). This is to mimic restful breathing and should be done at this rate: push in for

1 second, wait for 2 seconds, pull out for 1 second, wait for 2 seconds, and repeat.

4. Click on **End Calibration.** If your calibration data resembles figure 52.4, remove the calibration syringe, and proceed to the **Recording** section. If it does not, **Redo** calibration.

Procedure C—Recording

1. Insert a clean mouthpiece (and filter, if applicable) into the airflow transducer. With nose clip on, have the subject begin breathing into the transducer assembly. See figure 52.5 for the recording setup.

2. Have the subject breathe normally for about 20 seconds, making sure to close his/her lips completely around the mouthpiece. Then click on **Record** and have the subject:
   a. Breathe normally for 5 breaths.
   b. Inhale as deeply as possible once.
   c. Exhale as deeply as possible once.
   d. Breathe normally for 5 breaths.

3. Click on **Stop.**

4. The recording should look similar to figure 52.6. If it does not, click on **Redo.** If similar, click **Done.**

5. Click **Yes.**

Procedure D—Data Analysis

1. You can either analyze the current data file, or save it and do the analysis later, using the **Review Saved Data** mode. Note the channel number designations: CH 1: Airflow (hidden) and CH 2: Volume.

2. Note the settings for the measurement boxes, across the top of the data window: CH 2: P–P, CH 2: Max, CH 2: Min, and CH 2: Delta. Following are explanations of these settings:

P–P:	The difference between the maximum and minimum points in the selected area
Max:	Shows the maximum point within the selected area
Min:	Shows the minimum point within the selected area
Delta:	Calculates the amplitude difference between the last and first points in the selected area

3. Measure the subject's *vital capacity* (VC). Click on the I-beam tool. Using the I-beam cursor, select the area from the highest point of the deep inspiration to the lowest point of the deep expiration. Selection of the P–P area does not have to be perfect; it just has to include the highest and lowest points. See figure 52.7 for an example of proper selection of the area for VC measurement. The result will be displayed by CH 2: P–P. Record your result in Part A of Laboratory Assessment 52.

FIGURE 52.4 Sample recording for second calibration stage.

Source: Courtesy of and © BIOPAC Systems Inc.

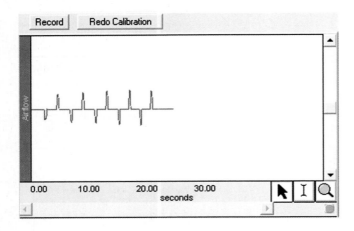

FIGURE 52.5 Recording setup.

Source: Courtesy of and © BIOPAC Systems Inc.

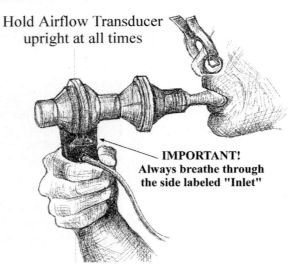

Hold Airflow Transducer upright at all times

IMPORTANT! Always breathe through the side labeled "Inlet"

FIGURE 52.6 Sample subject recording.

Source: Courtesy of and © BIOPAC Systems Inc.

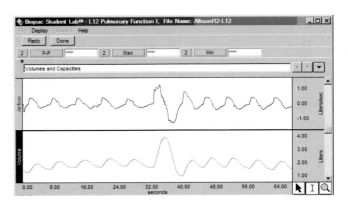

FIGURE 52.7 Proper selection of area for measurement of vital capacity (VC).

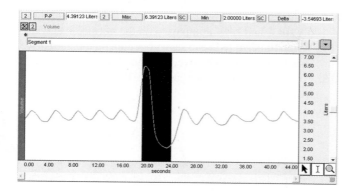

FIGURE 52.9 Proper selection of area for measuring TV during expiration.

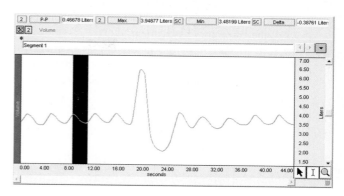

FIGURE 52.8 Proper selection of area for measuring TV during inspiration.

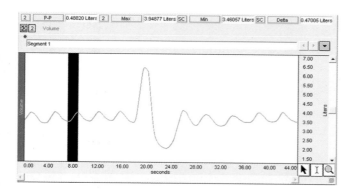

FIGURE 52.10 Proper selection of area for measuring IRV.

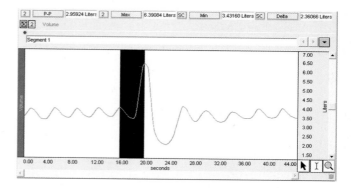

4. Measure the *tidal volume* (TV), during both normal inspiration and expiration. Use the I-beam tool to make these measurements. To measure the TV during inspiration, select the area between the lowest point before the third normal breath and the highest point of the third breath (valley to peak). For TV during expiration, select the area from the highest point of the third breath to the lowest point at the end of the third breath (peak to valley). See figure 52.8 for an example of proper selection of the area for TV measurement during normal inspiration. Figure 52.9 shows proper selection of the area for TV during expiration. The results will be displayed by CH 2: P–P. Record your results in table 52.1 of the laboratory assessment.

5. Measure the *inspiratory reserve volume* (IRV). Using the I-beam cursor, select the area from the peak of the fifth normal breath to the peak of the deep inspiration. See figure 52.10 for an example of proper selection of the area for IRV measurement. The result will be displayed by CH 2: Delta. Record your result in table 52.1.

6. Measure the *expiratory reserve volume* (ERV). Using the I-beam cursor, select the area from the lowest point of the deep expiration to the lowest point after the nor-

mal breath following it (valley to valley). The result will be displayed by CH 2: Delta. Record your result in table 52.1 of the laboratory assessment.

7. Measure the *residual volume* (RV). Using the I-beam cursor, select a small area that includes the lowest point of the deep expiration. The result will be displayed by CH 2: Min. Remember that spirometers cannot actually measure RV. Since the BIOPAC software uses a default setting of 1.0 L for RV, your RV result may be pre-set for 1.0 L. If this is the case, just record 1.0 L as your measurement result for RV in table 52.1 of the laboratory assessment.

8. Now measure the *inspiratory capacity* (IC). Using the I-beam cursor, select the area from the lowest point at the end of the fifth normal breath to the highest point of the deep inspiration (valley to peak). The result will be displayed by CH 2: Delta. Record your result in table 52.1 of the laboratory assessment.

9. Now measure the *expiratory capacity* (EC). Using the I-beam cursor, select the area from the lowest point of the deep expiration to the highest point of the normal breath following it (valley to peak). The result will be displayed by CH 2: Delta. Record your result in table 52.1 of the laboratory assessment.

10. Finally, measure the *total lung capacity* (TLC). Using the I-beam cursor, select a small area that includes the highest point of the deep inspiration. The result will be displayed by CH 2: Max. Record your result in table 52.1 of the laboratory assessment.

11. Now turn to Part A of the laboratory assessment for Laboratory Exercise 52. Complete the calculations for estimated vital capacity and follow the directions to compare measured VC with estimated VC.

12. Now complete the calculations for table 52.1 using the measured results you just obtained. It will be interesting to see how close the measured and calculated results are.

13. Complete table 52.2, to see how the subject's measured TV, IRV, and ERV compare with the normal averages.

14. When finished, you can save your data on the hard drive, a disk, or a network drive, and then **Exit** the system.

15. Answer the questions in Part B of the laboratory assessment.

Name _____

Date _____

Section _____

The A corresponds to the indicated outcome(s) found at the beginning of the laboratory exercise.

Spirometry: BIOPAC© Exercise

Part A   Assessments

1. Vital Capacity

   a. Record the subject's measured vital capacity (VC) as instructed in step 3 of the Data Analysis:

      Subject's measured VC = _____ L.

   b. Estimate the subject's vital capacity (VC) by using the appropriate equation below:

      Male: VC = 0.052 H − 0.022A − 3.60

      Female: VC = 0.041 H − 0.018 A − 2.69

               (H = height in cm, and A = age in years)

      Subject's estimated VC = _____ L.

   c. Perform the following calculation to determine how close the subject's measured VC came to the estimated VC:

      Measured VC / Estimated VC = _____ x 100 = _____ %

2. Measurements and calculations of respiratory volumes and capacities:

TABLE 52.1   Measurements and calculations of subject's respiratory volumes and capacities

Title		Measurement Result			Calculation
Tidal Volume	TV	a = 2	P-P	Cycle 3 inhale:	(a + b) / 2 =
		b = 2	P-P	Cycle 3 exhale:	
Inspiratory Reserve Volume	IRV	2	Delta		
Expiratory Reserve Volume	ERV	2	Delta		
Residual Volume	RV	2	Min		Default = 1 (*Preference setting*)
Inspiratory Capacity	IC	2	Delta		TV + IRV =
Expiratory Capacity	EC	2	Delta		TV + ERV =
Functional Residual Capacity	FRC				ERV + RV =
Total Lung Capacity	TLC	2	Max		IRV + TV + ERV + RV =

3. Comparison of subject's measured volumes with average volumes:

TABLE 52.2 Comparison of subject's measured respiratory volumes with average volumes

Volume Title		Average Volume	Measured Volume

Tidal Volume	TV	Resting subject, normal breathing:
		  TV is approximately 500 mL
		During exercise:
		  TV can be more than 3 liters
			greater than
			equal to
			less than

Inspiratory Reserve Volume	IRV	Resting IRV for young adults is
		  males = approximately 3,300 mL
		  females = approximately 1,900 mL
			greater than
			equal to
			less than

Expiratory Reserve Volume	ERV	Resting ERV for young adults is
		  males = approximately 1,000 mL
		  females = approximately 700 mL
			greater than
			equal to
			less than

Part B Assessments

Complete the following:

1. How do the experimental results compare to the expected or normal values presented in the introduction? **2**

_____

_____

_____

2. Why is the predicted vital capacity dependent upon the height of the subject? **3**

_____

_____

_____

3. Why might vital capacity and expiratory reserve decrease with age? **3**

_____

_____

_____

Control of Breathing

Purpose of the Exercise

To review the muscles and the mechanisms that control breathing and to investigate some of the factors that affect the rate and depth of breathing.

Materials Needed

Clock or watch with seconds timer
Paper bags, small

For Demonstration Activity:
Flasks
Glass tubing
Rubber stoppers, two-hole
Calcium hydroxide solution (limewater)

For Learning Extension Activity:
Pneumograph
Physiological recording apparatus

Learning Outcomes

After completing this exercise, you should be able to

① Locate the respiratory areas in the brainstem.

② Describe the mechanisms that control and influence breathing.

③ Select the respiratory muscles involved in inspiration and forced expiration.

④ Test and record the effect of various factors on the rate and depth of breathing.

Pre-Lab

Carefully read the introductory material and examine the entire lab. Be familiar with control of breathing from lecture or the textbook. Answer the pre-lab questions.

Pre-Lab Questions: Select the correct answer for each of the following questions:

1. Respiratory areas of the brain include all of the following *except* the
   a. brainstem.   b. pons.
   c. medulla oblongata.   d. pineal gland.

2. Breathing rate increases as blood concentrations of
   a. carbon dioxide increase.
   b. carbon dioxide decrease.
   c. hydrogen ions decrease.
   d. oxygen increase.

3. Forced expiration muscles do *not* include the
   a. internal intercostal muscles.
   b. rectus abdominis muscles.
   c. sternocleidomastoid muscles.
   d. external oblique muscles.

4. Peripheral chemoreceptors sensitive to low blood oxygen levels are located in the
   a. heart.
   b. aortic arch and carotid arteries.
   c. aorta only.
   d. carotid arteries only.

5. Normal blood pH is
   a. 6.8.   b. 8.0.
   c. 6.8–8.0.   d. 7.35–7.45.

6. An increase in the duration of inspirations is the normal response from the inflation reflex.
   True _____   False _____

7. The dorsal respiratory group of the medullary respiratory center is involved with stimulation of the diaphragm contractions.
   True _____   False _____

Breathing is controlled from regions of the brainstem called the *respiratory areas,* which control both inspiration and expiration. These areas initiate and regulate nerve impulses that travel to various breathing muscles, causing rhythmic breathing movements and adjustments to the rate and depth of breathing to meet various cellular needs (fig. 53.1). The *medullary respiratory center* is composed of two bilateral groups of neurons in the medulla oblongata. They are called the *dorsal respiratory group* and the *ventral respiratory group.* The dorsal respiratory group primarily stimulates the diaphragm to contract, resulting in inspiration, and

helps process respiratory sensory information involving the cardiovascular system. The ventral respiratory group involves regulation of the basic rhythm of breathing. Neurons in another part of the brainstem, the pons, compose the *pontine respiratory group* (formerly the pneumotaxic center). These neurons may contribute to the rhythm of breathing by modifying the respirations during situations such as exercise and sleep (fig. 53.2).

Various factors can influence the respiratory areas and thus affect the rate and depth of breathing. These factors include stretch of the lung tissues, emotional state, and the presence in the blood of certain chemicals, such as carbon dioxide, hydrogen ions, and oxygen. The breathing rate increases as the blood concentration of carbon dioxide or hydrogen ions increases or as the concentration of oxygen decreases.

The medulla oblongata possesses *central chemoreceptors* that are sensitive to changes in the levels of carbon dioxide and hydrogen ions. If the levels of carbon dioxide or hydrogen ions increase, the chemoreceptors relay this information to the respiratory areas. As a result, the rate and depth of breathing increase and more carbon dioxide is exhaled. When the levels decline to a more normal range, breathing rate decreases to normal. Exercise, breath holding, and hyperventilation are examples that considerably alter the blood levels of carbon dioxide and hydrogen ions.

Low blood oxygen levels are detected by *peripheral chemoreceptors* located in *carotid bodies* in the carotid arteries and *aortic bodies* in the aortic arch (fig. 53.3). This information is relayed to the respiratory areas in the brainstem, and the breathing rate increases. In order to trigger this response, the blood oxygen levels must be very low. Therefore, the blood levels of carbon dioxide and hydrogen ions have a much greater influence upon the rate and depth of breathing.

The depth of breathing is also influenced by the *inflation reflex.* If the lungs are greatly expanded during forceful

FIGURE 53.1 The respiratory areas are located in the pons and the medulla oblongata.

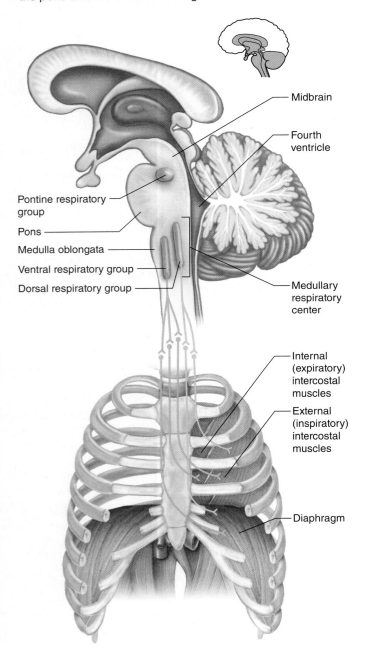

- Midbrain
- Fourth ventricle
- Pontine respiratory group
- Pons
- Medulla oblongata
- Ventral respiratory group
- Dorsal respiratory group
- Medullary respiratory center
- Internal (expiratory) intercostal muscles
- External (inspiratory) intercostal muscles
- Diaphragm

FIGURE 53.2 The medullary respiratory center and the pontine respiratory group control breathing.

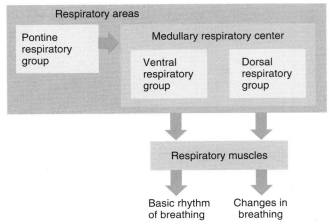

530

FIGURE 53.3 The peripheral chemoreceptors involved in control of breathing. Aortic bodies in the aortic arch and carotid bodies in the carotid arteries are sensitive to decreased blood oxygen levels.

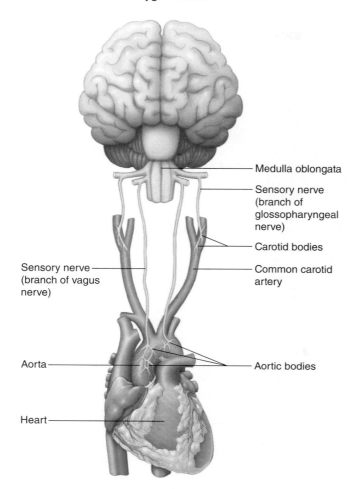

Medulla oblongata

Sensory nerve (branch of glossopharyngeal nerve)

Carotid bodies

Common carotid artery

Sensory nerve (branch of vagus nerve)

Aorta

Aortic bodies

Heart

breathing, stretch receptors within the bronchial tree and the visceral pleura are stimulated. Sensory impulses are conducted to the pontine respiratory group and, as a result, a decrease occurs in the duration of inspirations.

The muscles involved with inspiration and expiration are all voluntary skeletal muscles. Therefore a person has some cerebral conscious influence over the rate and depth of breathing as in talking, singing, and breath holding. However, respiratory areas of the brainstem, chemoreceptors, and certain reflexes represent the primary mechanism of subconscious, automatic control of breathing. The complex coordination of breathing involves considerable integration between respiratory areas of the brain and receptors; however some details of the control of breathing are still obscure.

Procedure A—Control of Breathing

1. Several skeletal muscles contract during breathing. The principal muscles involved during inspiration are the diaphragm and external intercostals. The sternocleido-mastoid, scalenes, and pectoralis minor are synergistic during more forceful inhalation, resulting in greater volumes of air inhaled. During quiet respiration, there is minimal involvement of the expiratory muscles. During forced expiration, the principal muscles are the internal intercostals, but the rectus abdominis and the external oblique muscles can provide extra force (fig. 53.4).

2. Respiratory areas in the brainstem control the cycle of breathing. The dorsal respiratory group of neurons is located in the medulla oblongata. Impulses from the dorsal respiratory group stimulate the muscles of inspiration, especially the diaphragm (fig. 53.1). This results in a normal respiratory rate of 12 to 15 breaths per minute. The ventral respiratory group of neurons will fire during inspiration and expiration to control the appropriate muscles of inspiration and expiration and the basic rhythm of breathing (fig. 53.2).

3. Complete Part A of Laboratory Assessment 53.

Procedure B—Factors Affecting Breathing

Perform each of the following tests, using your laboratory partner as a test subject.

1. *Normal breathing.* To determine the subject's normal breathing rate and depth, follow these steps:
   a. Have the subject sit quietly for a few minutes.
   b. After the rest period, ask the subject to count backwards mentally, beginning with five hundred.
   c. While the subject is distracted by counting, watch the subject's chest movements, and count the breaths taken in a minute. Use this value as the normal breathing rate (breaths per minute).
   d. Note the relative depth of the breathing movements.
   e. Record your observations in the table in Part B of the laboratory assessment.

2. *Effect of hyperventilation.* To test the effect of hyperventilation on breathing, follow these steps:
   a. Seat the subject and *guard to prevent the possibility of the subject falling over.*
   b. Have the subject breathe rapidly and deeply for a maximum of 20 seconds. *If the subject begins to feel dizzy, the hyperventilation should be halted immediately to prevent the subject from fainting from complications of alkalosis. The increased blood pH causes vasoconstriction of cerebral arterioles, which decreases circulation and oxygen to the brain.*
   c. After the period of hyperventilation, determine the subject's breathing rate and judge the breathing depth as before.
   d. Record the results in Part B of the laboratory assessment.

3. *Effect of rebreathing air.* To test the effect of rebreathing air on breathing, follow these steps:
   a. Have the subject sit quietly (approximately 5 minutes) until the breathing rate returns to normal.

FIGURE 53.4 Respiratory muscles involved in inspiration and forced expiration. The blue arrows indicate the direction of muscle contraction during inspiration; the green arrows indicate the direction of muscle contraction during forced expiration. Boldface indicates the primary muscles involved in breathing.

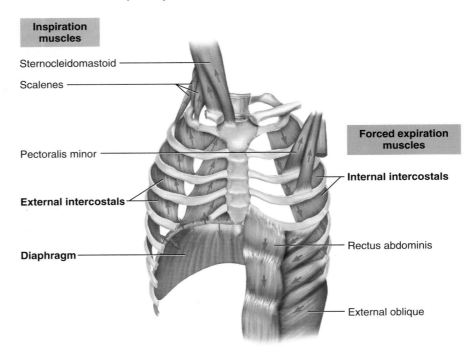

Inspiration muscles

Sternocleidomastoid

Scalenes

Pectoralis minor

External intercostals

Diaphragm

Forced expiration muscles

Internal intercostals

Rectus abdominis

External oblique

Demonstration Activity

When a solution of calcium hydroxide is exposed to carbon dioxide, a chemical reaction occurs, and a white precipitate of calcium carbonate is formed as indicated by the following reaction:

$$Ca(OH)_2 + CO_2 \longrightarrow CaCO_3 + H_2O$$

Thus, a clear water solution of calcium hydroxide (limewater) can be used to detect the presence of carbon dioxide because the solution becomes cloudy if this gas is bubbled through it.

The laboratory instructor will demonstrate this test for carbon dioxide by drawing some air through limewater in an apparatus such as that shown in figure 53.5. Then the instructor will blow an equal volume of expired air through a similar apparatus. (*Note:* A new sterile mouthpiece should be used each time the apparatus is demonstrated.) Watch for the appearance of a precipitate that causes the limewater to become cloudy. Was there any carbon dioxide in the atmospheric air drawn through the limewater?

_____

If so, how did the amount of carbon dioxide in the atmospheric air compare with the amount in the expired air?

_____

_____

FIGURE 53.5 Apparatus used to demonstrate the presence of carbon dioxide in air: (a) atmospheric air is drawn through limewater; (b) expired air is blown through limewater.

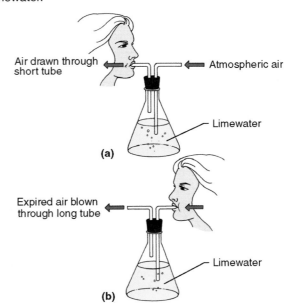

Air drawn through short tube

Atmospheric air

Limewater

(a)

Expired air blown through long tube

Limewater

(b)

b. Have the subject breathe deeply into a small paper bag that is held tightly over the nose and mouth. *If the subject begins to feel light-headed or like fainting, the rebreathing air should be halted immediately to prevent further acidosis and fainting.*

c. After 2 minutes of rebreathing air, determine the subject's breathing rate and judge the depth of breathing.

d. Record the results in Part B of the laboratory assessment.

4. *Effect of breath holding.* To test the effect of breath holding on breathing, follow these steps:

a. Have the subject sit quietly (approximately 5 minutes) until the breathing rate returns to normal.

b. Have the subject hold his or her breath as long as possible. *If the subject begins to feel light-headed or like fainting, breath holding should be halted immediately to prevent further acidosis and fainting.*

c. As the subject begins to breathe again, determine the rate of breathing and judge the depth of breathing.

d. Record the results in Part B of the laboratory assessment.

5. *Effect of exercise.* To test the effect of exercise on breathing, follow these steps:

a. Have the subject sit quietly (approximately 5 minutes) until breathing rate returns to normal.

b. Have the subject exercise by moderately running in place for 3-5 minutes. *This exercise should be avoided by anyone with health risks.*

c. After the exercise, determine the breathing rate and judge the depth of breathing.

d. Record the results in Part B of the laboratory assessment.

6. The importance of the respiratory rate and blood pH relationship is illustrated in figure 53.6. A condition of respiratory acidosis or alkalosis, leading to possible death, can occur from changes in the depth or rate of breathing.

7. Complete Part B of the laboratory assessment.

FIGURE 53.6 The relationship of respiratory rate and blood pH. The normal breathing rate of 12–15 breaths per minute relates to a blood pH within the normal range.

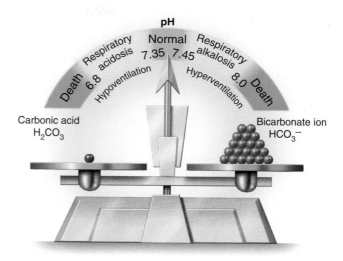

Learning Extension Activity

A *pneumograph* is a device that can be used together with some type of recording apparatus to record breathing movements. The laboratory instructor will demonstrate the use of this equipment to record various movements, such as those that accompany coughing, laughing, yawning, and speaking.

Devise an experiment to test the effect of some factor, such as hyperventilation, rebreathing air, or exercise, on the length of time a person can hold the breath. *After the laboratory instructor has approved your plan,* carry out the experiment, using the pneumograph and recording equipment. What conclusion can you draw from the results of your experiment?

_____

_____

_____

_____

NOTES

Laboratory Assessment

53

Name _____

Date _____

Section _____

The ⚠ corresponds to the indicated outcome(s) found at the beginning of the laboratory exercise.

Control of Breathing

Part A Assessments

Complete the following statements:

1. The respiratory areas are widely scattered throughout the _____ and medulla oblongata of the brainstem. ⚠

2. The _____ respiratory group within the medulla oblongata primarily stimulates the diaphragm. ⚠

3. The _____ respiratory group within the medulla oblongata regulates the basic rhythm of breathing. ⚠

4. Central chemoreceptors are sensitive to changes in the blood concentrations of hydrogen ions and _____. ⚠

5. As the blood concentration of carbon dioxide increases, the breathing rate _____. ⚠

6. As a result of increased breathing, the blood concentration of carbon dioxide is _____. ⚠

7. Peripheral chemoreceptors include aortic bodies and the _____. ⚠

8. The principal muscles of forced expiration are the _____. ⚠

9. The principal muscles of inspiration are the _____ and the external intercostal muscles. ⚠

Part B Assessments

1. Record the results of your breathing tests in the table. ⚠

Factor Tested	Breathing Rate (breaths/minute)	Breathing Depth (+, + +, + + +)
Normal		
Hyperventilation (shortly after)		
Rebreathing air (shortly after)		
Breath holding (shortly after)		
Exercise (shortly after)		

2. Briefly explain the reason for the changes in breathing that occurred in each of the following cases: /2\

    a. Hyperventilation

    _____

    _____

    _____

    _____

    b. Rebreathing air

    _____

    _____

    _____

    _____

    c. Breath holding

    _____

    _____

    _____

    _____

    d. Exercise

    _____

    _____

    _____

    _____

3. Complete the following:

    a. Why is it important to distract a person when you are determining the normal rate of breathing?

    _____

    _____

    b. How can the depth of breathing be measured accurately?

    _____

    _____

Critical Thinking Assessment

Why is it dangerous for a swimmer to hyperventilate in order to hold the breath for a longer period of time? /2\

_____

_____

_____

_____

Digestive Organs

Purpose of the Exercise

To review the structure and function of the digestive organs and to examine the tissues of these organs.

Materials Needed

Human torso model
Head model, sagittal section
Skull with teeth
Teeth, sectioned
Tooth model, sectioned
Paper cup
Compound light microscope
Prepared microscope slides of the following:
    Sublingual gland (salivary gland)
    Esophagus
    Stomach (fundus)
    Pancreas (exocrine portion)
    Small intestine (jejunum)
    Large intestine

Learning Outcomes

After completing this exercise, you should be able to

1 Sketch and label the structures of tissue sections of digestive organs.

2 Locate the major organs and structural features of the digestive system.

3 Match digestive organs with their descriptions.

4 Describe the functions of these organs.

Pre-Lab

Carefully read the introductory material and examine the entire lab. Be familiar with the basic structures and functions of the digestive organs from lecture or the textbook. Answer the pre-lab questions.

Pre-Lab Questions: Select the correct answer for each of the following questions:

1. Which of the following digestive organs possesses villi?
   - **a.** esophagus
   - **b.** stomach
   - **c.** small intestine
   - **d.** large intestine

2. Which is a major function of the oral cavity?
   - **a.** absorb water
   - **b.** mechanical digestion
   - **c.** secrete hydrochloric acid
   - **d.** emulsify fat

3. Bile is produced in the
   - **a.** small intestine.
   - **b.** pancreas.
   - **c.** gallbladder.
   - **d.** liver.

4. The _____ is the exposed part of a tooth.
   - **a.** crown
   - **b.** neck
   - **c.** root
   - **d.** periodontal ligament

5. The _____ sphincter is the muscular valve at the exit of the stomach.
   - **a.** lower esophageal
   - **b.** pyloric
   - **c.** hepatopancreatic
   - **d.** ileocecal

6. The stomach can stretch enough to hold the contents of a large meal because it has
   - **a.** the pyloric antrum.
   - **b.** a fundus region.
   - **c.** gastric folds.
   - **d.** a greater curvature.

7. The _____ lobe is the largest lobe of the liver.
   - **a.** left
   - **b.** quadrate
   - **c.** caudate
   - **d.** right

8. The appendix, cecum, and ascending colon are on the left side of the body.
   True _____   False _____

9. The anal sphincter muscles include an external voluntary sphincter and an internal involuntary sphincter.
   True _____   False _____

The digestive system includes the organs associated with the *alimentary canal* and several accessory structures. The alimentary canal, a muscular tube, passes through the body from the opening of the mouth to the anus. It includes the mouth, pharynx, esophagus, stomach, small intestine, and large intestine. The canal is adapted to move substances throughout its length. It is specialized in various regions to store, digest, and absorb food materials and to eliminate the residues. The accessory organs, which include the salivary glands, liver, gallbladder, and pancreas, secrete products into the alimentary canal that aid digestive functions.

The *oral cavity* is a primary area for mechanical digestion where we masticate the food using the teeth and jaw muscles. Three pairs of salivary glands secrete mucus and salivary amylase into the oral cavity, where chemical digestion begins. The *pharynx* and *esophagus* secrete mucus and serve as passageways for the food and liquids to reach the stomach. The contents are forced along by peristaltic waves.

Within the *stomach,* swallowed contents mix with gastric juice, which contains hydrochloric acid and the enzyme pepsin. The hydrochloric acid creates a pH near 2, serving to destroy most ingested bacteria, and it activates pepsin to begin the chemical digestion of proteins. The gastric folds (rugae) allow the stomach to hold large quantities of food. Only minimal absorption occurs in the stomach. The partially digested contents, called chyme, enter the small intestine periodically through the muscular pyloric sphincter.

Within the *small intestine,* chyme is mixed with bile and pancreatic juice. Bile is produced in the liver and then stored in the gallbladder. Bile will emulsify the fat. Pancreatic juice contains bicarbonate ions necessary to neutralize acidic chyme, and multiple enzymes for the chemical digestion of carbohydrates, fats, and proteins. The small intestine is nearly 6 meters (21 feet) long and possesses villi that increase its surface area, so chemical digestion is completed and most nutrient absorption occurs before the contents reach the large intestine.

Contents not absorbed enter the *large intestine* through the ileocecal valve. Within the large intestine, mucus is secreted and water and electrolytes are absorbed. Bacteria that inhabit the large intestine can break down some remaining residues, producing some vitamins that are absorbed. Feces composed of water, undigested substances, mucus, and bacteria are formed and stored until elimination.

Procedure A—Oral Cavity and Salivary Glands

1. Study figure 54.1 and the list of structures and descriptions. Examine the head model (sagittal section) and a skull. Locate the following structures:

   oral cavity (mouth)—location of mechanical digestion, some chemical digestion

   vestibule—narrow space between teeth and lips and cheeks

   tongue—muscular organ for food manipulation
   - lingual frenulum (frenulum of tongue)— membranous fold connecting tongue and floor of oral cavity
   - papillae—contain taste buds

   palate—roof of oral cavity
   - hard palate—formed by portions of maxillary and palatine bones
   - soft palate—muscular arch without bone
   - uvula—cone-shaped projection

FIGURE 54.1 The features of the oral cavity.

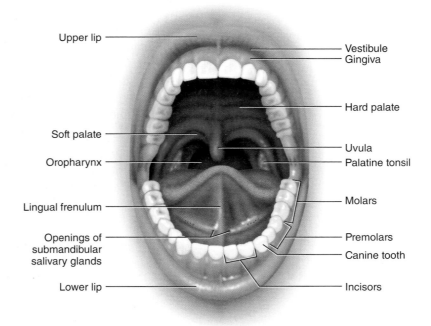

Upper lip
Vestibule
Gingiva
Hard palate
Soft palate
Oropharynx
Uvula
Palatine tonsil
Lingual frenulum
Molars
Openings of submandibular salivary glands
Premolars
Canine tooth
Lower lip
Incisors

palatine tonsils—lymphatic tissues on lateral walls of pharynx

gingivae (gums)—soft tissue that surrounds neck of tooth and alveolar processes

teeth

- incisors
- canines (cuspids)

FIGURE 54.2 Longitudinal section of a molar.

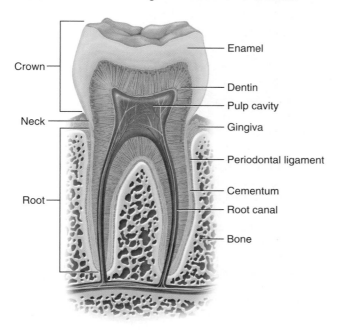

- premolars (bicuspids)
- molars

2. Study figure 54.2 and the list of structures and descriptions. Examine a sectioned tooth and a tooth model. Locate the following features:

crown—tooth portion projection beyond gingivae

- enamel—hardest substance of body on surface of crown
- dentin—living connective tissue forming most of tooth

neck—junction between crown and root

root—deep portion of tooth beneath surface

- pulp cavity—central portion of tooth
- cementum (cement)—thin surface area of bonelike material
- root canal—contains blood vessels and nerves

3. Study figures 54.1 and 54.3. Observe the head of the human torso model and locate the following:

parotid salivary glands—largest glands anterior to ear

- parotid duct (Stensen's duct)

submandibular salivary glands—along mandible

- submandibular duct (Wharton's duct)

sublingual salivary glands—smallest glands inferior to tongue

- sublingual ducts (10–12 ducts)

4. Examine a microscopic section of a sublingual gland using low- and high-power magnification. Note the mucous cells that produce mucus and serous cells that produce enzymes. Serous cells sometimes form caps

---

FIGURE 54.3 The features associated with the salivary glands.

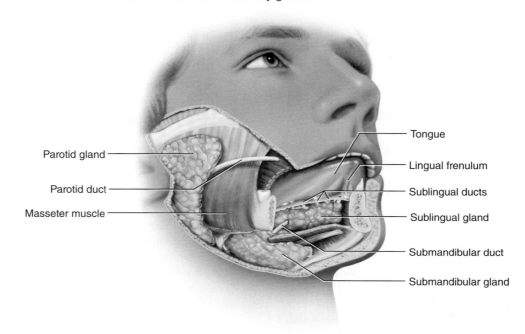

FIGURE 54.4 Micrograph of the sublingual salivary gland (300×).

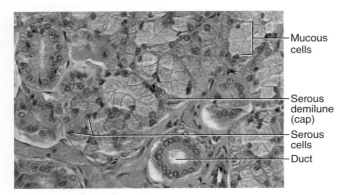

- Mucous cells
- Serous demilune (cap)
- Serous cells
- Duct

FIGURE 54.5 Micrograph of a cross section of the esophagus (10×).

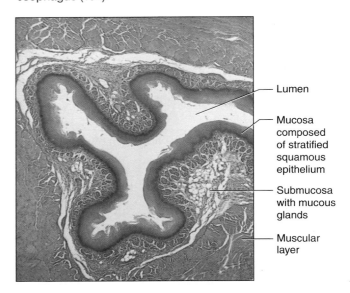

- Lumen
- Mucosa composed of stratified squamous epithelium
- Submucosa with mucous glands
- Muscular layer

(demilunes) around mucous cells. Also note any larger secretory duct surrounded by cuboidal epithelial cells (fig. 54.4).

5. Prepare a labeled sketch of a representative section of a salivary gland in Part A of Laboratory Assessment 54.

Procedure B—Pharynx and Esophagus

1. Reexamine figure 50.2 in Laboratory Exercise 50.
2. Observe the human torso model and locate the following features:

    pharynx—connects nasal and oral cavities with larynx

    - nasopharynx—superior to soft palate; without digestive functions
    - oropharynx—posterior to oral cavity
    - laryngopharynx—extends from inferior oropharynx to esophagus

    epiglottis—deflects food and fluids into esophagus when swallowing

    esophagus—straight collapsible tube from pharynx to stomach

    lower esophageal sphincter (cardiac sphincter; gastroesophageal sphincter)—thickened smooth muscle at junction with stomach

3. Have your partner take a swallow from a cup of water. Carefully watch the movements in the anterior region of the neck. What steps in the swallowing process did you observe?

_____

_____

_____

_____

4. Examine a microscopic section of esophagus wall using low-power magnification (fig. 54.5). The inner lining is composed of stratified squamous epithelium, and there are layers of muscle tissue in the wall. Peristaltic waves occur from contractions of the muscular layer that move contents through the alimentary canal. Locate some mucous glands in the submucosa. They appear as clusters of lightly stained cells.

5. Prepare a labeled sketch of the esophagus wall in Part A of the laboratory assessment.

Procedure C—Stomach

1. Study figure 54.6 and the list of structures and descriptions. Observe the human torso model and locate the following features of the stomach:

    gastric folds (rugae)—allow stomach to expand to hold food and liquid contents

    cardia (cardiac region)—region near lower esophageal sphincter

    fundus of stomach (fundic region)—dome-shaped region superior to opening from esophagus

    body of stomach (body region)—large main region

    pyloric part (pyloric region)—region near passage into small intestine

    - pyloric antrum—funnel-like portion
    - pyloric canal—narrow region
    - pylorus—terminal passage into duodenum

    pyloric sphincter—thick, smooth muscle valve for chyme passage into duodenum

    lesser curvature—on superior surface

    greater curvature—on lateral and inferior surface

2. Examine a microscopic section of stomach wall using low-power magnification. Note how the inner lining of simple columnar epithelium dips inward to form gastric pits. The gastric glands are tubular structures that open into the gastric pits. Near the deep ends of these glands, you should be able to locate some intensely stained

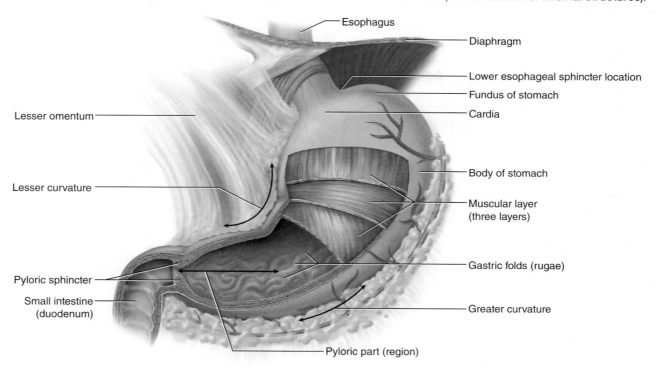

Esophagus
Diaphragm
Lower esophageal sphincter location
Fundus of stomach
Cardia
Lesser omentum
Body of stomach
Lesser curvature
Muscular layer (three layers)
Gastric folds (rugae)
Pyloric sphincter
Small intestine (duodenum)
Greater curvature
Pyloric part (region)

(bluish) chief cells and some lightly stained (pinkish) parietal cells (fig. 54.7). The parietal cells secrete hydrochloric acid; the chief cells produce pepsinogen, the inactive form of pepsin.

3. Prepare a labeled sketch of a representative section of the stomach wall in Part A of the laboratory assessment.

Procedure D—Pancreas and Liver

1. Study figure 54.8 and the list of structures and descriptions. Observe the human torso model and locate the following structures:

pancreas—a retroperitoneal organ; secretes an alkaline mixture of digestive enzymes and bicarbonate ions
- tail of pancreas
- head of pancreas
- pancreatic duct
- accessory pancreatic duct

liver—has four lobes; produces bile
- right lobe—largest lobe
- left lobe—smaller than right lobe
- quadrate lobe—minor lobe near gallbladder
- caudate lobe—minor lobe near vena cava

gallbladder—stores bile

hepatic ducts—bile ducts in liver

common hepatic duct—bile duct from liver to cystic duct

FIGURE 54.7 Micrograph of the mucosa of the stomach wall (60×).

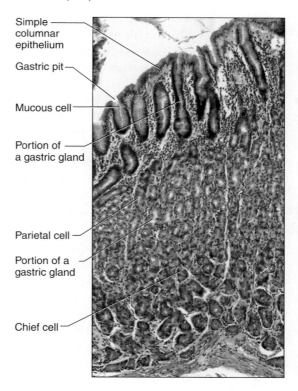

Simple columnar epithelium
Gastric pit
Mucous cell
Portion of a gastric gland
Parietal cell
Portion of a gastric gland
Chief cell

FIGURE 54.8 The features associated with the liver and pancreas.

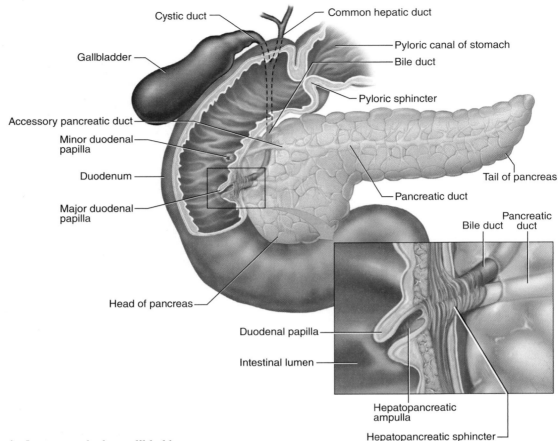

cystic duct—attached to gallbladder

bile duct—duct from cystic duct to duodenum

hepatopancreatic sphincter (sphincter of Oddi)—regulates passage of bile and pancreatic juice

2. Examine the pancreas slide using low-power magnification. Observe the exocrine (acinar) cells that secrete pancreatic juice. See figure 39.12 in Laboratory Exercise 39 for a micrograph of the pancreas.

3. Prepare a labeled sketch of a representative section of the pancreas in Part A of the laboratory assessment.

Procedure E—Small and Large Intestines

1. Study figures 54.9, 54.10, and 54.11 and the list of structures and descriptions. Observe the human torso model and locate each of the following features:

small intestine—tube with small diameter and many loops and coils where chemical digestion and absorption of nutrients occur

- duodenum—first 25 cm (10 inches); shortest portion; located retroperitoneal
- jejunum—next nearly 2.5 m (8 feet)
- ileum—last 3.5 m (12 feet); joins with large intestine

mesentery—fold of peritoneal membrane that suspends and supports abdominal viscera

ileocecal valve (sphincter)—regulates contents entering large intestine

large intestine—tube with large diameter nearly 1.5 m (5 feet) long

- large intestinal wall—has 3 unique features
  - teniae coli—3 longitudinal bands of smooth muscle
  - haustra—pouches created by muscle tone of teniae coli
  - epiploic appendages—fatty pouches on outer surfaces
- cecum—blind pouch inferior to ileocecal valve
- appendix (vermiform appendix)—2–7 cm wormlike blind pouch attached to cecum
- ascending colon—course on right side from cecum to transverse colon
- right colic (hepatic) flexure—right-angle turn
- transverse colon—course from ascending to descending colon
- left colic (splenic) flexure—right-angle turn

FIGURE 54.9 The features of the small intestine and associated structures (anterior view). (*Note:* The small intestine is pulled aside to expose the ileocecal junction.)

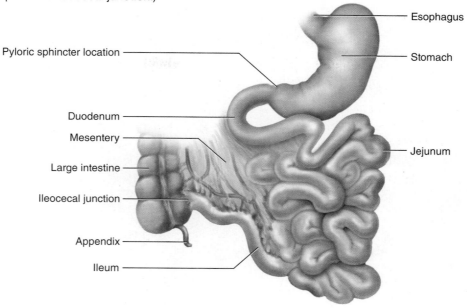

Esophagus

Pyloric sphincter location

Stomach

Duodenum

Mesentery

Jejunum

Large intestine

Ileocecal junction

Appendix

Ileum

FIGURE 54.10 The features of the large intestine (anterior view).

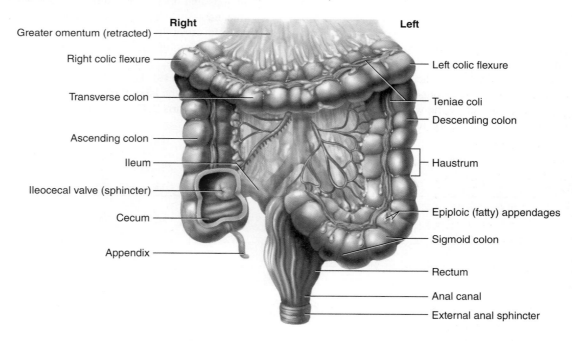

Right   **Left**

Greater omentum (retracted)

Right colic flexure

Left colic flexure

Transverse colon

Teniae coli

Descending colon

Ascending colon

Ileum

Haustrum

Ileocecal valve (sphincter)

Cecum

Epiploic (fatty) appendages

Appendix

Sigmoid colon

Rectum

Anal canal

External anal sphincter

- descending colon—course from transverse to sigmoid colon
- sigmoid colon—S-shaped course to rectum
- rectum—course anterior to sacrum; retains feces until defecation
- anal canal—last few centimeters of large intestine

anal sphincter muscles—encircle anus
  - external anal sphincter—the voluntary sphincter of skeletal muscle
  - internal anal sphincter—the involuntary sphincter of smooth muscle
anus—external opening

FIGURE 54.11 Normal appendix.

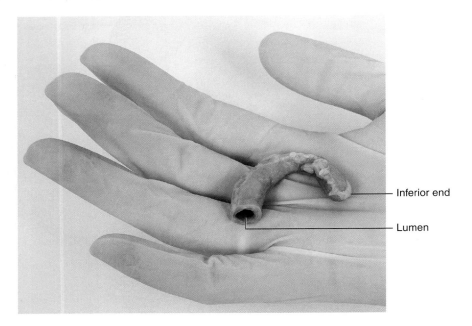

— Inferior end

— Lumen

2. Using low-power magnification, examine a microscopic section of small intestine wall. Identify the mucosa, submucosa, muscular layer, and serosa. Note the villi that extend into the lumen of the tube. Study a single villus using high-power magnification. Note the core of connective tissue and the covering of simple columnar epithelium that contains some lightly stained goblet cells (fig. 54.12). The villi greatly increase the surface area for absorption of digestive products.

3. Prepare a labeled sketch of the wall of the small intestine in Part A of the laboratory assessment.

4. Examine a microscopic section of large intestine wall. Note the lack of villi. Also note the tubular mucous glands that open on the surface of the inner lining and the numerous lightly stained goblet cells. Locate the four layers of the wall (fig. 54.13). The mucus functions as a lubrication and holds the particles of fecal matter together.

5. Prepare a labeled sketch of the wall of the large intestine in Part A of the laboratory assessment.

6. Complete Parts B, C, and D of the laboratory assessment.

 Critical Thinking Activity

How is the structure of the small intestine better adapted for absorption than that of the large intestine?

_____

_____

FIGURE 54.12 Micrograph of the small intestine wall (40×).

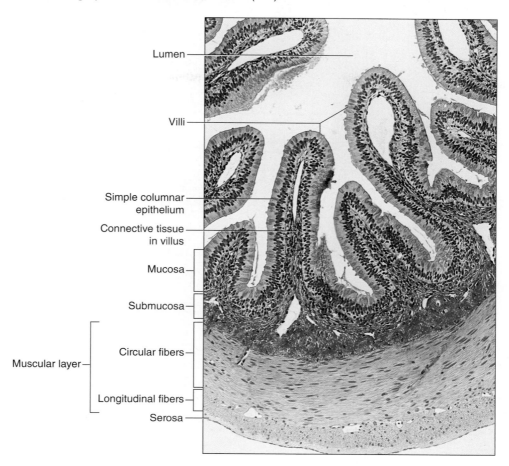

Lumen

Villi

Simple columnar epithelium

Connective tissue in villus

Mucosa

Submucosa

Muscular layer

Circular fibers

Longitudinal fibers

Serosa

FIGURE 54.13 Micrograph of the large intestine wall (64×).

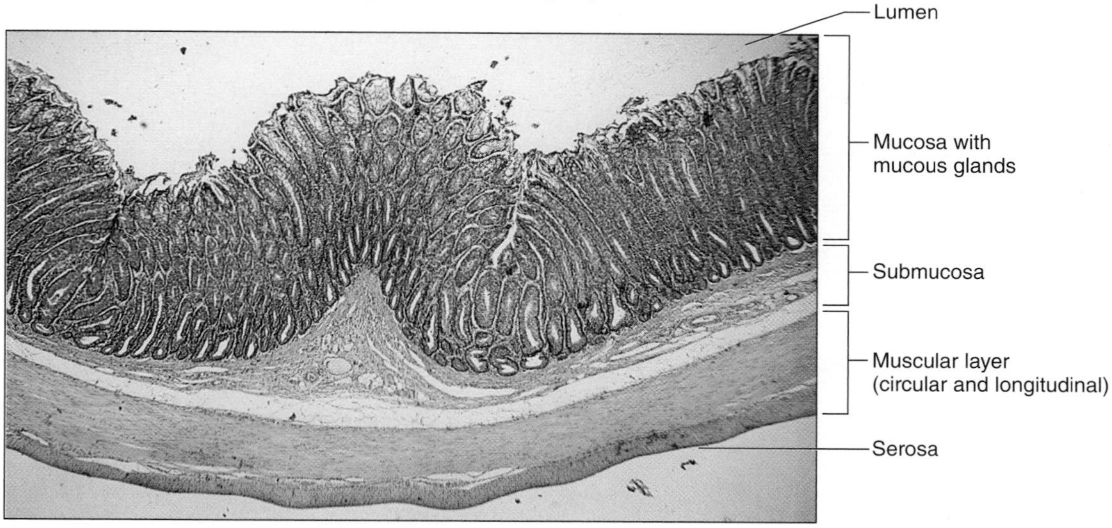

Lumen

Mucosa with mucous glands

Submucosa

Muscular layer (circular and longitudinal)

Serosa

Name _____

Date _____

Section _____

The ⚠ corresponds to the indicated outcome(s) found at the beginning of the laboratory exercise.

Digestive Organs

Part A Assessments

In the space that follows, sketch a representative area of the organ indicated. Label any of the structures observed and indicate the magnification used for each sketch. ⚠

Salivary gland (_____×)	Esophagus (_____×)
Stomach wall (_____×)	Pancreas (_____×)
Small intestine wall (_____×)	Large intestine wall (_____×)

Part B   Assessments

Identify the numbered features in figures 54.14, 54.15, 54.16, and 54.17.

FIGURE 54.14   Label the features of the stomach and nearby regions in this frontal section of a cadaver (anterior view). 🄰

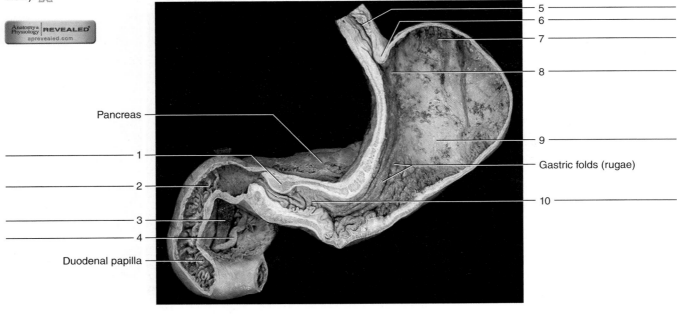

FIGURE 54.15   Label the features associated with the liver and pancreas. 🄰

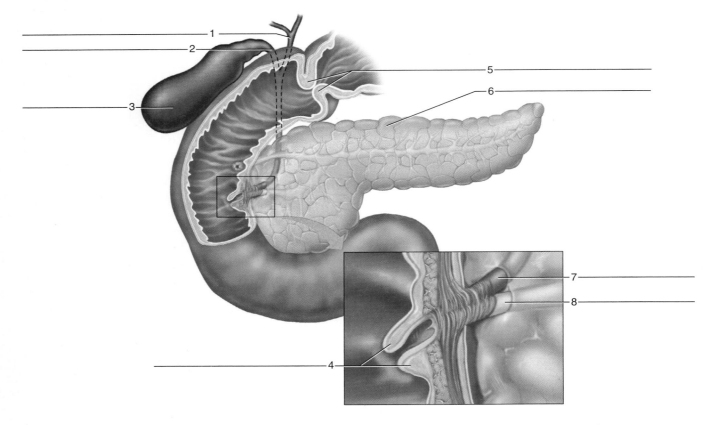

FIGURE 54.16 Label the digestive structures of this abdominopelvic cavity of a cadaver (anterior view). 2

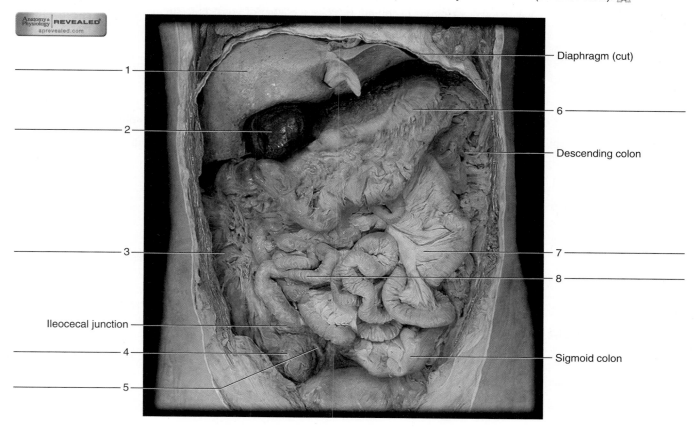

1 _____

2 _____

3 _____

Ileocecal junction _____

4 _____

5 _____

Diaphragm (cut)

6 _____

Descending colon

7 _____

8 _____

Sigmoid colon

FIGURE 54.17 Label the features of the large intestine. 2

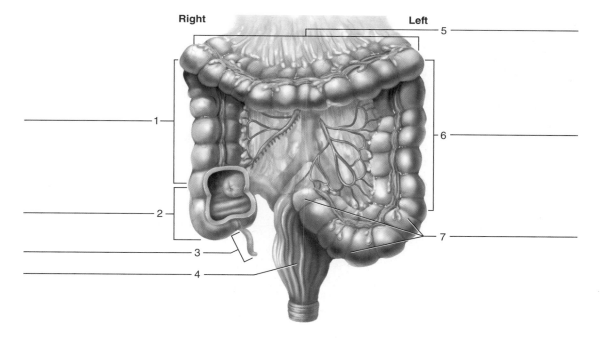

Right   Left

1 _____

2 _____

3 _____

4 _____

5 _____

6 _____

7 _____

Part C Assessments

Match the terms in column A with the descriptions in column B. Place the letter of your choice in the space provided. ⬧3

Column A

a. Cardia
b. Crown
c. Cystic duct
d. Gastric folds
e. Ileum
f. Major duodenal papilla
g. Mucosa
h. Muscular layer
i. Parietal cells
j. Root canal
k. Sublingual gland
l. Vestibule

Column B

_____ 1. Smallest of major salivary glands

_____ 2. Secrete hydrochloric acid into stomach

_____ 3. Last section of small intestine

_____ 4. Region of stomach near lower esophageal sphincter

_____ 5. Contains blood vessels and nerves in a tooth

_____ 6. Responsible for peristaltic waves

_____ 7. Allow stomach to expand

_____ 8. Space between teeth, cheeks, and lips

_____ 9. Attached to gallbladder

_____ 10. Portion of tooth projecting beyond gingivae

_____ 11. Layer nearest lumen of alimentary canal

_____ 12. Common opening region for bile and pancreatic secretions

Part D Assessments

Complete the following:

1. Summarize the functions of the oral cavity. ⬧A _____

_____

_____

2. Summarize the functions of the esophagus. ⬧A _____

_____

3. Name the valve that prevents regurgitation of food from the duodenum back into the stomach. ⬧2 _____

4. Summarize the functions of the stomach. ⬧A _____

_____

_____

5. Name the valve that controls the movement of material between the small and large intestines. ⬧2 _____

6. Summarize the functions of the small intestine. ⬧A _____

_____

_____

_____

7. Summarize the functions of the large intestine. ⬧A _____

_____

_____

Laboratory Exercise 55

Action of a Digestive Enzyme

Purpose of the Exercise

To investigate the action of amylase and the effect of heat on its enzymatic activity.

Materials Needed

0.5% amylase solution*
Beakers (50 and
  500 mL)
Distilled water
Funnel
Pipettes (1 and 10 mL)
Pipette rubber bulbs
0.5% starch solution
Graduated cylinder
  (10 mL)
Test tubes
Test-tube clamps
Wax marker
Iodine-potassium-iodide
  solution

Medicine dropper
Ice
Water bath, 37°C
  (98.6°F)
Porcelain test plate
Benedict's solution
Hot plates
Test-tube rack
Thermometer

***For Alternative
  Procedure Activity:***
Small disposable cups
Distilled water

*The amylase should be free of sugar for best results; a low-maltose solution of amylase yields good results. See Appendix 1.

⚠ Safety

- Use only a mechanical pipetting device (never your mouth). Use pipettes with rubber bulbs or dropping pipettes.
- Wear safety glasses when working with acids and when heating test tubes.
- Use test-tube clamps when handling hot test tubes.
- If an open flame is used for heating the test solutions, keep clothes and hair away from the flame.
- Review all the safety guidelines inside the front cover.
- If student saliva is used as a source of amylase, it is important that students wear disposable gloves and only handle their own materials.
- Use an appropriate disinfectant to wash the laboratory tables before and after the procedures.
- Dispose of chemicals according to appropriate directions.
- Wash your hands before leaving the laboratory.

Learning Outcomes

After completing this exercise, you should be able to

1. Test a solution for the presence of starch or the presence of sugar.
2. Explain the action of amylase.
3. Test the effects of varying temperatures on the activity of amylase.

Pre-Lab

Carefully read the introductory material and examine the entire lab. Be familiar with enzyme structure and activity from lecture or the textbook. Answer the pre-lab questions.

Pre-Lab Questions: Select the correct answer for each of the following questions:

1. The enzyme amylase accelerates the reaction of changing
   a. polysaccharides into monosaccharides.
   b. disaccharides into monosaccharides.
   c. starch into disaccharides.
   d. disaccharides into glucose.

2. Amylase originates from
   a. salivary glands and stomach.
   b. salivary glands and pancreas.
   c. esophagus and stomach.
   d. pancreas and intestinal mucosa.

3. The optimum pH for amylase activity is
   a. 0–14.   b. 2.
   c. 6.8–7.0.   d. 7–8.

4. After an enzyme reaction is completed, the enzyme
   a. is still available.
   b. was structurally changed.
   c. was denatured.
   d. is part of the enzyme-substrate complex.

5. As temperatures decrease, enzyme activity
   a. increases.   b. remains unchanged.
   c. ceases.   d. decreases.

The laboratory exercise is designed to examine the relationship between enzymes and chemical digestion. *Enzymes* are proteins that serve as *biological catalysts*. Acting as a catalyst, an enzyme will modify and rapidly increase the rate of a particular chemical reaction without being consumed in the reaction, which enables the enzyme to function repeatedly. A particular enzyme will catalyze a specific reaction. Various metabolic pathways and chemical digestive processes are accelerated by numerous enzymatic reactions. The **lock-and-key model** of enzyme activity involves an *enzyme* and *substrate* (substance acted upon by the enzyme) temporarily combining as an *enzyme-substrate complex* and ending with a reaction *product* and the original *enzyme* (fig. 55.1).

Enzymes are globular proteins with a three-dimensional structure determined by various chemical bonds including the hydrogen bond. For the enzyme to function as the catalyst, the shape is critical at the active site of the enzyme for the enzyme-substrate complex to take place properly. If the environmental temperature, including the body temperature, rises to a certain level, the three-dimensional shape of the enzyme is modified as hydrogen bonds start breaking and the enzyme becomes *denatured*. Although the primary structure of the amino acid sequence remains unaltered, the secondary and tertiary structures have been altered. A denatured protein could regain its shape unless the three-dimensional shape is destroyed when conditions become too extreme, and the enzyme becomes irreversibly denatured.

The digestive enzyme in salivary secretions is called *salivary amylase*. Pancreatic amylase is secreted among several other pancreatic enzymes. A bacterial extraction of amylase is available for laboratory experiments. This enzyme catalyzes the reaction of changing starch molecules into sugar (disaccharide) molecules, which is the first step in the digestion of complex carbohydrates.

As in the case of other enzymes, amylase is a protein catalyst. Its activity is affected by exposure to certain environmental factors, including various temperatures, pH, radiation, and electricity. As temperatures increase, faster chemical reactions occur as the collisions of molecules happen at a greater frequency. Eventually, temperatures increase to a point that the enzyme is denatured, and the rate of the enzyme activity rapidly declines. As temperatures decrease, enzyme activity also decreases due to fewer collisions of the molecules; however, the colder temperatures do not denature the enzyme. Normal body temperature provides an environment for enzyme activity near the optimum for enzymatic reactions.

The pH of the environment where enzymes are secreted also has a major influence on the enzyme reactions. The optimum pH for amylase activity is between 6.8 and 7.0, typical of salivary secretions. When salivary amylase arrives in the stomach, hydrochloric acid deactivates the enzyme, diminishing any further chemical digestion of remaining starch in the stomach. However, pancreatic amylase is secreted into the small intestine, where an optimum pH for amylase is once more provided. Other enzymes in the digestive system have different optimum pH ranges of activity compared to amylase. For example, pepsin from stomach secretions has an optimum activity around pH 2, while trypsin from pancreatic secretions operates best around pH 7–8.

Procedure A—Amylase Activity

1. Study the lock-and-key model example specific for amylase action on starch digestion (fig. 55.2).
2. Examine the locations where starch is digested into glucose (fig. 55.3).
3. Mark three clean test tubes as *tubes 1, 2,* and *3,* and prepare the tubes as follows (fig. 55.4):

   Tube 1: **Add 6 mL of amylase solution.**

   Tube 2: **Add 6 mL of starch solution.**

   Tube 3: **Add 5 mL of starch solution and 1 mL of amylase solution.**

FIGURE 55.1 Lock-and-key model of an enzyme-catalyzed reaction (**E** + **S** → **E–S** → **P** + **E**). (Many enzyme-catalyzed reactions, as depicted here, are reversible.) In the forward reaction (dark-shaded arrows), (a) the shapes of the substrate molecules fit the shape of the enzyme's active site. (b) When the substrate molecules temporarily combine with the enzyme, a chemical reaction occurs. (c) The result is a product molecule and an unaltered enzyme. The active site changes shape somewhat as the substrate binds, such that formation of the enzyme-substrate complex is more like a hand fitting into a glove, which has some flexibility, than a key fitting into a lock.

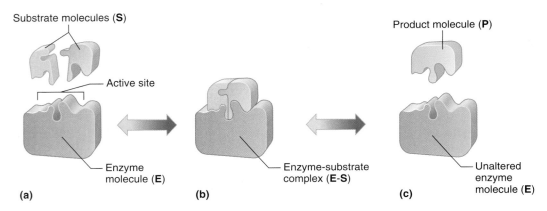

Substrate molecules (**S**)

Active site

Enzyme molecule (**E**)

(a)

Enzyme-substrate complex (**E-S**)

(b)

Product molecule (**P**)

Unaltered enzyme molecule (**E**)

(c)

FIGURE 55.2 Lock-and-key model of enzyme action of amylase on starch digestion.

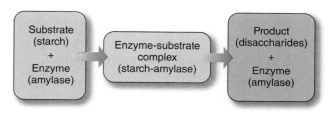

FIGURE 55.3 Flowchart of starch digestion.

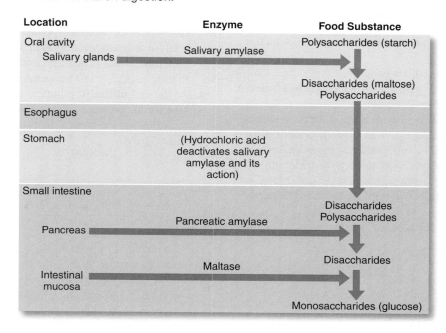

FIGURE 55.4 Test tubes prepared for testing amylase activity.

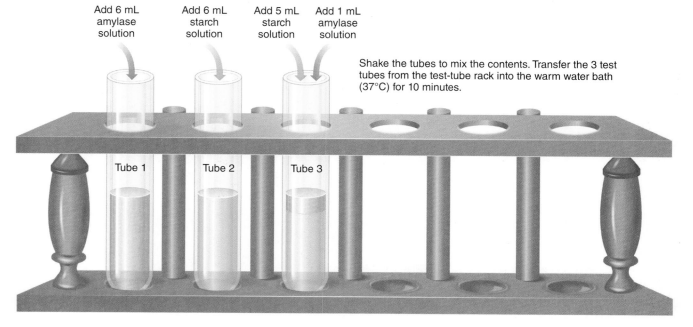

4. Shake the tubes well to mix the contents and place them in a warm water bath, 37°C (98.6°F), for 10 minutes.

5. At the end of the 10 minutes, test the contents of each tube for the presence of starch. To do this, follow these steps:

   a. Place 1 mL of the solution to be tested in a depression of a porcelain test plate.

   b. Next add one drop of iodine-potassium-iodide solution and note the color of the mixture. If the solution becomes blue-black, starch is present.

   c. Record the results in Part A of Laboratory Assessment 55.

6. Test the contents of each tube for the presence of sugar (disaccharides in this instance). To do this, follow these steps:

   a. Place 1 mL of the solution to be tested in a clean test tube.

   b. Add 1 mL of Benedict's solution.

   c. Place the test tube with a test-tube clamp in a beaker of boiling water for 2 minutes.

   d. Note the color of the liquid. If the solution becomes green, yellow, orange, or red, sugar is present. Blue indicates a negative test, whereas green indicates a positive test with the least amount of sugar, and red indicates the greatest amount of sugar present.

   e. Record the results in Part A of the laboratory assessment.

7. Complete Part A of the laboratory assessment.

Procedure B—Effect of Heat

1. Mark three clean test tubes as *tubes 4, 5,* and *6.*

2. Add 1 mL of amylase solution to each of the tubes and expose each solution to a different test temperature for 3 minutes as follows:

   Tube 4: **Place in beaker of ice water (about 0°C [32°F]).**

   Tube 5: **Place in warm water bath (about 37°C [98.6°F]).**

   Tube 6: **Place in beaker of boiling water (about 100°C [212°F]). Use a test-tube clamp.**

3. It is important that the 5 mL of starch solution added to tube 4 be at ice-water temperature before it is added to the 1 mL of amylase solution. Add 5 mL of starch solution to each tube, shake to mix the contents, and return the tubes to their respective test temperatures for 10 minutes.

4. At the end of the 10 minutes, test the contents of each tube for the presence of starch and the presence of sugar by following the same directions as in steps 5 and 6 in Procedure A.

5. Complete Part B of the laboratory assessment.

 Learning Extension Activity

Devise an experiment to test the effect of some other environmental factor on amylase activity. For example, you might test the effect of a strong acid by adding a few drops of concentrated hydrochloric acid to a mixture of starch and amylase solutions. Be sure to include a control in your experimental plan. That is, include a tube containing everything except the factor you are testing. Then you will have something with which to compare your results. *Carry out your experiment only if it has been approved by the laboratory instructor.*

Name _____

Date _____

Section _____

The ⚠ corresponds to the indicated outcome(s) found at the beginning of the laboratory exercise.

Action of a Digestive Enzyme

Part A   Amylase Activity Assessments

1. Test results: ⚠1

Tube	Starch	Sugar
1 Amylase solution		
2 Starch solution		
3 Starch-amylase solution		

2. Complete the following:

   a. Explain the reason for including tube 1 in this experiment. ⚠2 _____

   _____

   _____

   _____

   b. What is the importance of tube 2? ⚠2 _____

   _____

   _____

   _____

   c. What do you conclude from the results of this experiment? ⚠2 _____

   _____

   _____

   _____

Part B Effect of Heat Assessments

1. Test results: **3**

Tube	Starch	Sugar
4 0°C (32°F)		
5 37°C (98.6°F)		
6 100°C (212°F)		

2. Complete the following:

   a. What do you conclude from the results of this experiment? **3** _____

   _____

   _____

   _____

   _____

   _____

   b. If digestion failed to occur in one of the tubes in this experiment, how can you tell if the amylase was destroyed by the factor being tested or if the amylase activity was simply inhibited by the test treatment? **3**

   _____

   _____

   _____

   _____

   _____

Critical Thinking Assessment

What test result would occur if the amylase you used contained sugar? _____ Would your results be as valid? Explain your answer. **2**

_____

_____

_____

_____

Urinary Organs

Purpose of the Exercise

To review the structure of urinary organs, to dissect a kidney, and to observe the major structures of a nephron.

Materials Needed

Human torso model
Kidney model
Preserved pig (or sheep) kidney
Dissecting tray
Dissecting instruments
Long knife
Compound light microscope
Prepared microscope slide of the following:
    Kidney section
    Ureter, cross section
    Urinary bladder
    Urethra, cross section

Safety

▶ Wear disposable gloves when working on the kidney dissection.
▶ Dispose of the kidney and gloves as directed by your laboratory instructor.
▶ Wash the dissecting tray and instruments as instructed.
▶ Wash your laboratory table.
▶ Wash your hands before leaving the laboratory.

Learning Outcomes

After completing this exercise, you should be able to

1. Locate and identify the major structures of a kidney.
2. Identify and sketch the major structures of a nephron.
3. Trace the path of filtrate through a renal nephron.
4. Trace the path of blood through the renal blood vessels.
5. Identify and sketch the structures of a ureter and a urinary bladder wall.
6. Trace the path of urine flow through the urinary system.

Pre-Lab

Carefully read the introductory material and examine the entire lab. Be familiar with the structures and functions of the urinary organs from lecture or the textbook. Answer the pre-lab questions.

Pre-Lab Questions: Select the correct answer for each of the following questions:

1. When comparing the position of the two kidneys,
   a. they are at the same level.
   b. the right kidney is slightly superior to the left kidney.
   c. the right kidney is slightly inferior to the left kidney.
   d. the right kidney is anterior to the left kidney.

2. Cortical nephrons represent about _____ of the nephrons.
   a. 0 %
   b. 100 %
   c. 20 %
   d. 80 %

3. Which of the following does *not* represent one of the processes in urine formation?
   a. secretion of renin
   b. glomerular filtration
   c. tubular reabsorption
   d. tubular secretion

4. The _____ arteries and veins are located in the corticomedullary junction.
   a. arcuate
   b. renal
   c. cortical radiate
   d. peritubular

5. The _____ is the tube from the kidney to the urinary bladder.
   a. urethra
   b. ureter
   c. renal pelvis
   d. renal column

6. The trigone is a triangular, funnel-like region of the
   a. renal cortex.
   b. renal medulla.
   c. urethra.
   d. urinary bladder.

7. The external urethral sphincter is composed of involuntary smooth muscle.
   True _____ False _____

8. Contractions of the detrusor muscle provide the force during micturition (urination).
   True _____ False _____

The two kidneys are the primary organs of the urinary system. They are located in the abdominal cavity, against the posterior wall and behind the parietal peritoneum (retroperitoneal). Masses of adipose tissue associated with the kidneys hold them in place at a vertebral level between T12 and L3. The right kidney is slightly inferior due to the large mass of the liver near its superior border. Ureters force urine by means of peristaltic waves into the urinary bladder, which temporarily stores urine. The urethra conveys urine to the outside of the body.

Each kidney contains over 1 million nephrons, which serve as the basic structural and functional units of the kidney. A glomerular capsule, proximal convoluted tubule, nephron loop, and distal convoluted tubule compose the microscopic, multicellular structure of a relatively long nephron tubule, which drains into a collecting duct. Approximately 80% of the nephrons are cortical nephrons with short nephron loops, while the remaining represent juxtamedullary nephrons, with long nephron loops extending deeper into the renal medulla. An elaborate network of blood vessels surrounds the entire nephron. Glomerular filtration, tubular reabsorption, and tubular secretion represent three processes resulting in urine as the final product.

A variety of functions occur in the kidneys. They remove metabolic wastes from the blood; help regulate blood volume, blood pressure, and pH of blood; control water and electrolyte concentrations; and secrete renin and erythropoietin. Renin functions in regulation of blood pressure and erythropoietin helps regulate red blood cell production.

Procedure A—Kidney Structure

1. Study figures 56.1 and 56.2 and the list of structures and descriptions.
2. Observe the human torso model and the kidney model. Locate the following:

    kidneys—paired retroperitoneal organs

    ureters—paired tubular organs about 25 cm long that transport urine from kidney to urinary bladder

    urinary bladder—single muscular storage organ for urine

    urethra—conveys urine from urinary bladder to external urethral orifice

    renal sinus—hollow chamber of kidney

    renal pelvis—funnel-shaped sac at superior end of ureter; receives urine from major calyces

    • major calyces—2–3 major converging branches into renal pelvis; receive urine from minor calyces

    • minor calyces—small subdivisions of major calyces; receive urine from papillary ducts within renal papillae

    renal medulla—deep region of kidney

    • renal pyramids—6–10 conical regions composing most of renal medulla

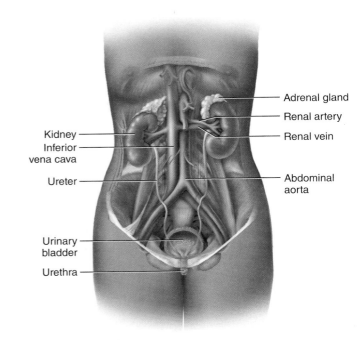

FIGURE 56.1 The urinary system and associated structures.

    • renal papillae—projections at ends of renal pyramids with openings into minor calyces

renal cortex—superficial region of kidney

renal columns—extensions of renal cortical tissue between renal pyramids

nephrons—functional units of kidneys

    • cortical nephrons—80% of nephrons close to surface

    • juxtamedullary nephrons—20% of nephrons close to medulla

3. To observe the structure of a kidney, follow these steps:

    a. Obtain a preserved pig or sheep kidney and rinse it with water to remove as much of the preserving fluid as possible.

    b. Carefully remove any adipose tissue from the surface of the specimen.

    c. Locate the following features:

        fibrous (renal) capsule—protective membrane that encloses kidney

        hilum of kidney—indented region containing renal artery and vein and ureter

        renal artery—large artery arising from abdominal aorta

        renal vein—large vein that drains into inferior vena cava

        ureter—transports urine from renal pelvis to urinary bladder

    d. Use a long knife to cut the kidney in half longitudinally along the frontal plane, beginning on the convex border.

FIGURE 56.2 Frontal section of a pig kidney that has a triple injection of latex (*red* in the renal artery, *blue* in the renal vein, and *yellow* in the ureter and renal pelvis).

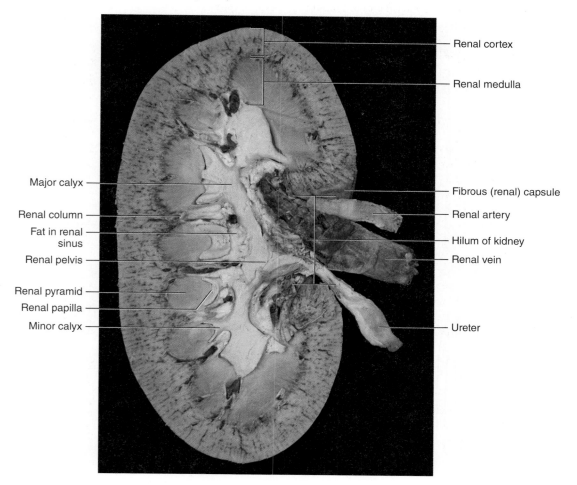

Labels on the figure:
- Renal cortex
- Renal medulla
- Major calyx
- Renal column
- Fat in renal sinus
- Renal pelvis
- Renal pyramid
- Renal papilla
- Minor calyx
- Fibrous (renal) capsule
- Renal artery
- Hilum of kidney
- Renal vein
- Ureter

e. Rinse the interior of the kidney with water, and using figure 56.2 as a reference, locate the following:

renal pelvis
- major calyces
- minor calyces

renal cortex

renal columns

renal medulla
- renal pyramids
- renal papillae

4. Complete Part A of Laboratory Assessment 56.

Procedure B—Renal Blood Vessels and Nephrons

1. Study figures 56.3 and 56.4 illustrating renal blood vessels and structures of a nephron. The arrows within the figures indicate blood flow and tubular fluid flow.
2. Obtain a microscope slide of a kidney section and examine it using low-power magnification. Locate the *renal capsule,* the *renal cortex* (which appears some-what granular and may be more darkly stained than the other renal tissues), and the *renal medulla* (fig. 56.5).
3. Examine the renal cortex using high-power magnification. Locate a *renal corpuscle.* These structures appear as isolated circular areas. Identify the *glomerulus,* the capillary cluster inside the corpuscle, and the *glomerular (Bowman's) capsule,* which appears as a clear area surrounding the glomerulus. A glomerulus and a glomerular capsule compose a *renal corpuscle.* Also note the numerous sections of renal tubules that occupy the spaces between renal corpuscles (fig. 56.5a).
4. Prepare a labeled sketch of a representative section of the renal cortex in Part B of the laboratory assessment.
5. Examine the renal medulla using high-power magnification. Identify longitudinal and cross-sectional views of various collecting ducts. These ducts are lined with simple epithelial cells, which vary in shape from squamous to cuboidal (fig. 56.5b).
6. Prepare a labeled sketch of a representative section of the renal medulla in Part B of the laboratory assessment.
7. Complete Part C of the laboratory assessment.

FIGURE 56.3 Renal blood vessels associated with cortical and juxtamedullary nephrons. The arrows indicate the flow of blood. Blood from the renal artery flows through a segmental artery and an interlobar artery to the arcuate artery. Blood from an arcuate vein flows through an interlobar vein to the renal vein.

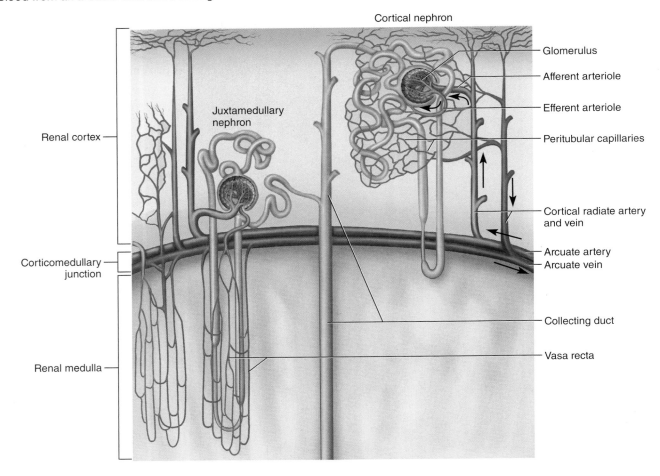

FIGURE 56.4 Structure of a nephron with arrows indicating the flow of tubular fluid. The corticomedullary junction is typical for a cortical nephron.

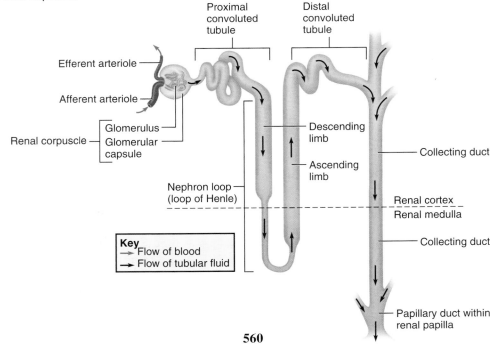

FIGURE 56.5 (a) Micrograph of a section of the renal cortex (220×). (b) Micrograph of a section of the renal medulla (80×).

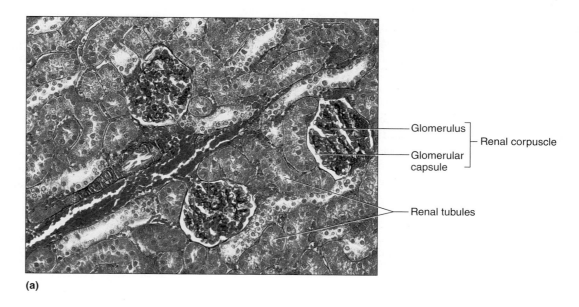

Glomerulus ⎤
⎥ Renal corpuscle
Glomerular capsule ⎦

Renal tubules

(a)

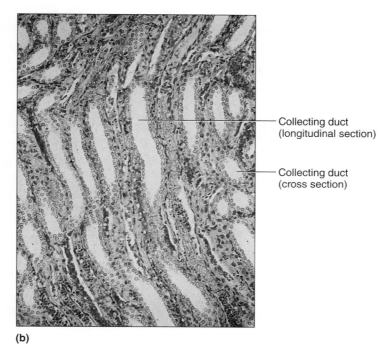

Collecting duct (longitudinal section)

Collecting duct (cross section)

(b)

Procedure C—Ureter, Urinary Bladder, and Urethra

The *ureters* are paired tubes about 25 cm (10 inches) long extending from the renal pelvis to the *ureteral openings (ureteric orifices)* of the *urinary bladder.* Peristaltic waves, created from rhythmic contractions of smooth muscle, occur about every 30 seconds, transporting the urine toward the urinary bladder. The bladder possesses mucosal folds called *rugae* that enable it to distend for temporary urine storage. Coarse bundle layers of smooth muscle compose the somewhat broad *detrusor muscle.* The triangular floor of the bladder, the *trigone,* is bordered by the two ureteral openings and the opening into the urethra, which is surrounded by the *internal urethral sphincter.* The internal urethral sphincter is formed by a thickened region of the detrusor muscle (fig. 56.6).

Although the bladder can hold about 600 mL of urine, the desire to urinate from the pressure of a stretched bladder occurs when the bladder has about 150–200 mL of urine. During *micturition (urination)* the powerful detrusor muscle contracts, and the internal urethral sphincter is forced open. The *external urethral sphincter,* at the level of the urogenital diaphragm, is under somatic control. Because the external urethral sphincter is skeletal muscle, there is considerable voluntary control over the voiding of urine.

The *urethra* extends from the exit of the urinary bladder to the *external urethral orifice.* In the female the urethra is about 4 cm (1.5 inches) long. In the male, the urethra includes passageways for both urine and semen and extends the length of the penis. The male urethra can be divided into three sections: prostatic urethra, membranous (intermediate) urethra, and spongy urethra (fig. 56.6*b*). A total length of

FIGURE 56.6 Ureters and frontal sections of urinary bladder and urethra of (a) female and (b) male.

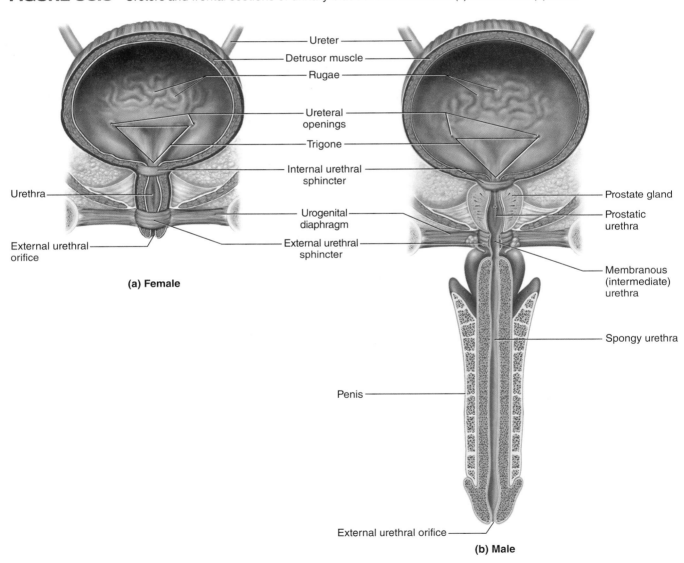

(a) Female

(b) Male

about 20 cm (8 inches) would be characteristic for the urethra of a male.

1. Obtain a microscope slide of a cross section of a ureter and examine it using low-power magnification. Locate the *mucous coat* layer next to the lumen. Examine the middle *muscular coat* composed of longitudinal and circular smooth muscle cells responsible for the peristaltic waves that propel urine from the kidneys to the urinary bladder. The outer *fibrous coat,* composed of connective tissue, secures the ureter in the retroperitoneal position (fig. 56.7).

2. Examine the mucous coat using high-power magnification. The specialized tissue is transitional epithelium, which allows changes in its thickness when unstretched and stretched.

3. Prepare a labeled sketch of a ureter in Part D of the laboratory assessment.

4. Obtain a microscope slide of a segment of the wall of a urinary bladder and examine it using low-power magnification. Examine the *mucous coat* next to the lumen and the *submucous coat* composed of connective tissue just beneath the mucous coat. Examine the *muscular coat* composed of bundles of smooth muscle fibers interlaced in many directions. This thick muscular layer is called the *detrusor muscle* and functions in the elimination of urine. Also note the outer *serous coat* of connective tissue (fig. 56.8).

5. Examine the mucous coat using high-power magnification. The tissue is transitional epithelium, which allows changes in its thickness from unstretched when the bladder is empty to stretched when the bladder distends with urine.

FIGURE 56.7 Micrograph of a cross section of a ureter (75×).

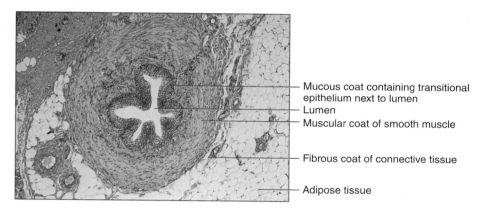

— Mucous coat containing transitional epithelium next to lumen
— Lumen
— Muscular coat of smooth muscle

— Fibrous coat of connective tissue

— Adipose tissue

FIGURE 56.8 Micrograph of a segment of the human urinary bladder wall (6×).

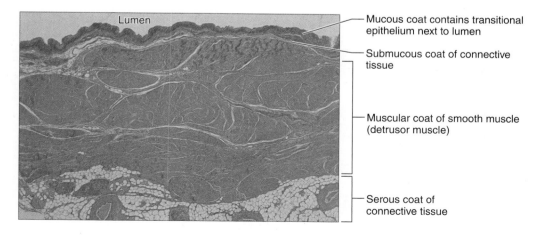

Lumen

— Mucous coat contains transitional epithelium next to lumen

— Submucous coat of connective tissue

— Muscular coat of smooth muscle (detrusor muscle)

— Serous coat of connective tissue

6. Prepare a labeled sketch of a segment of the urinary bladder wall in Part D of the laboratory assessment.

7. Obtain a microscope slide of a cross section of a urethra and examine it using low-power magnification. Locate the muscular coat composed of smooth muscle fibers. Locate groups of mucous glands, called *urethral glands,* in the urethral wall that secrete protective mucus into the lumen of the urethra. Examine the mucous coat using high-power magnification. The mucous membrane is composed of a type of stratified epithelium. The specific type of stratified epithelium varies from transitional, to stratified columnar, to stratified squamous epithelium between the urinary bladder and the external urethral orifice. Depending upon where the section of the urethra was taken for the preparation of the microscope slide, the type of stratified epithelial tissue represented could vary (fig. 56.9).

8. Complete Part E of the laboratory assessment.

FIGURE 56.9 Cross section through the urethra (10×).

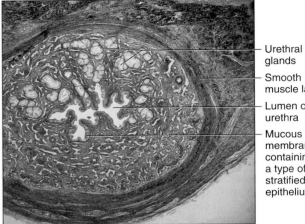

- Urethral glands
- Smooth muscle layer
- Lumen of urethra
- Mucous membrane containing a type of stratified epithelium

Laboratory Assessment

56

Name _____

Date _____

Section _____

The ⚠ corresponds to the indicated outcome(s) found at the beginning of the laboratory exercise.

Urinary Organs

Part A Assessments

1. Label the features indicated in figure 56.10 of a kidney (frontal section).

FIGURE 56.10   Label the structures in the frontal section of a kidney. ⚠

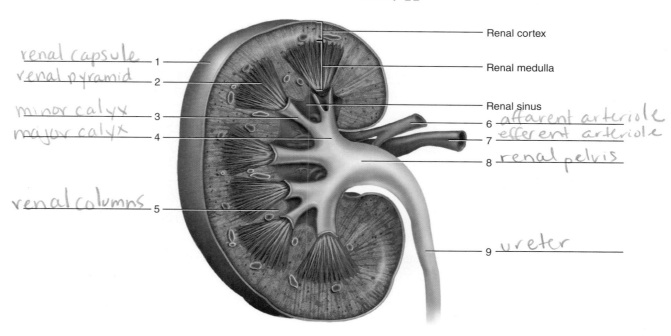

renal capsule 1
renal pyramid 2
minor calyx 3
major calyx 4
renal columns 5

Renal cortex
Renal medulla
Renal sinus
6 afferent arteriole
7 efferent arteriole
8 renal pelvis
9 ureter

2. Match the terms in column A with the descriptions in column B. Place the letter of your choice in the space provided. ⚠

Column A		Column B
a. Calyces	E	**1.** Superficial region around the renal medulla
b. Hilum of kidney	A	**2.** Branches of renal pelvis to renal papillae
c. Nephron	H	**3.** Conical mass of tissue within renal medulla
d. Renal column	F	**4.** Projection with tiny openings into a minor calyx
e. Renal cortex	i	**5.** Hollow chamber within kidney
f. Renal papilla	C	**6.** Microscopic functional unit of kidney
g. Renal pelvis	d	**7.** Located between renal pyramids
h. Renal pyramid	g	**8.** Superior funnel-shaped end of ureter inside the renal sinus
i. Renal sinus	b	**9.** Medial depression for blood vessels and ureter to enter kidney chamber

Part B Assessments

Sketch a representative section of the renal cortex and the renal medulla. Label the glomerulus, glomerular capsule, and sections of renal tubules in the renal cortex. Label a longitudinal section and cross section of a collecting duct in the renal medulla. **2**

Renal cortex (_____×)	Renal medulla (_____×)

Part C Assessments

Complete the following:

1. Distinguish between a renal corpuscle and a renal tubule. **2** _____

_____

_____

2. Number the following structures to indicate their respective positions in relation to the nephron. Assign the number 1 to the structure nearest the glomerulus. **3**

   _5_   Ascending limb of nephron loop

   _7_   Collecting duct

   _4_   Descending limb of nephron loop

   _6_   Distal convoluted tubule

   _2_   Glomerular capsule

   _3_   Proximal convoluted tubule

   _1_   Papillary duct in renal papilla

3. Number the following structures (the list does not include all possible blood vessels) to indicate their respective positions in the blood pathway within the kidney. Assign the number 1 to the vessel nearest the abdominal aorta. ◢4

_____ Afferent arteriole

_____ Arcuate artery

_____ Arcuate vein

_____ Cortical radiate artery

_____ Cortical radiate vein

_____ Efferent arteriole

_____ Glomerulus

_____ Peritubular capillary (or vasa recta)

_____ Renal artery

_____ Renal vein

Part D   Assessments

Sketch a cross section of a ureter and label the three layers and the lumen. Sketch a segment of a urinary bladder and label the four layers and the lumen. ◢5

Ureter (_____ ×)	Urinary bladder (_____ ×)

Part E Assessments

Number the following structures to indicate respective positions in the pathway of urine flow in a male. Assign the number 1 to the structure nearest the papillary duct in a renal papilla. 6

10	External urethral orifice
2	Major calyx
8	Membranous urethra
1	Minor calyx
7	Prostatic urethra
3	Renal pelvis
9	Spongy urethra
4	Ureter
5	Ureteral opening
6	Urinary bladder

Urinalysis

Purpose of the Exercise

To perform the observations and tests commonly used to analyze the characteristics and composition of urine.

Materials Needed

Normal and abnormal simulated urine specimens are suggested as a substitute for collected urine.

Disposable urine-collecting container
Paper towel
Urinometer cylinder
Urinometer hydrometer
Laboratory thermometer
pH test paper

Reagent strips (individual or combination strips such as Chemstrip or Multistix) to test for the presence of the following:

Glucose
Protein
Ketones

Bilirubin
Hemoglobin/occult blood

Compound light microscope
Microscope slide
Coverslip
Centrifuge

Centrifuge tube
Graduated cylinder, 10 mL
Medicine dropper
Sedi-stain

Safety

► Consider using normal and abnormal simulated urine samples available from various laboratory supply houses.
► Wear disposable gloves when working with body fluids.
► Work only with your own urine sample.
► Use an appropriate disinfectant to wash the laboratory table before and after the procedures.
► Place glassware in a disinfectant when finished.
► Dispose of contaminated items as directed by your laboratory instructor.
► Wash your hands before leaving the laboratory.

Learning Outcomes

After completing this exercise, you should be able to

1. Evaluate the color, transparency, and specific gravity of a urine sample.
2. Measure the pH of a urine sample.
3. Test a urine sample for the presence of glucose, protein, ketones, bilirubin, and hemoglobin.
4. Perform a microscopic study of urine sediment.
5. Summarize the results of these observations and tests.

Pre-Lab

Carefully read the introductory material and examine the entire lab. Be familiar with normal urine components from lecture or the textbook. Answer the pre-lab questions.

Pre-Lab Questions: Select the correct answer for each of the following questions:

1. Urine consists of about _____ percent water and _____ percent solutes.
   a. 100; 0   **b.** 95; 5
   c. 90; 10   **d.** 50; 50
2. A urinalysis consists of all of the following *except*
   a. physical characteristics.
   b. chemical analysis.
   c. microscopic examination.
   d. body temperature analysis.
3. Microscopic solids are stained with _____ in order to make identifications possible.
   a. reagents   **b.** iodine
   c. methylene blue   **d.** Sedi-stain
4. Which of the following would be the most typical urinary output in a day?
   a. 0.5 liters   **b.** 0.8 liters
   c. 1.2 liters   **d.** 2.2 liters
5. Which of the following would represent a normal pH range of urine?
   a. 4.6–8.0   **b.** 0–14
   c. 7.35–7.45   **d.** 5–10

Urine is the product of three processes of the nephrons within the kidneys: glomerular filtration, tubular reabsorption, and tubular secretion. As a result of these processes, various waste substances are removed from the blood, and body fluid and electrolyte balance are maintained. Consequently, the composition of urine varies considerably because of differences in dietary intake and physical activity from day to day and person to person. The normal urinary output ranges from 1.0–1.8 liters per day. Typically, urine consists of 95% water and 5% solutes. The volume of urine produced by the kidneys varies with such factors as fluid intake, environmental temperature, relative humidity, respiratory rate, and body temperature.

An analysis of urine composition and volume often is used to evaluate the functions of the kidneys and other organs. This procedure, called *urinalysis,* is a clinical assessment and a diagnostic tool for certain pathological conditions and general overall health. A urinalysis and a complete blood analysis complement each other for an evaluation of certain diseases and general health.

A urinalysis involves three aspects: physical characteristics, chemical analysis, and a microscopic examination. Physical characteristics of urine that are noted include volume, color, transparency, and odor. The chemical analysis of solutes in urine addresses urea and other nitrogenous wastes; electrolytes; pigments; and possible glucose, protein, ketones, bilirubin, and hemoglobin. The specific gravity and pH of urine are greatly influenced by the components and amounts of solutes. An examination of microscopic solids, including cells, casts, and crystals, assists in the diagnosis of injury, various diseases, and urinary infections.

⚠ Warning

While performing the following tests, you should wear disposable latex gloves so that skin contact with urine is avoided. Observe all safety procedures listed for this lab. (Normal and abnormal simulated urine specimens could be used instead of real urine for this lab.)

Procedure A—Chemical Analysis

1. Proceed to the restroom with a clean, disposable container. The first small volume of urine should not be collected because it contains abnormally high levels of microorganisms from the urethra. Collect a midstream sample of about 50 mL of urine. The best collections are the first specimen in the morning or one taken 3 hours after a meal. Refrigerate samples if they are not used immediately.

2. Place a sample of urine in a clean, transparent container. Describe the *color* of the urine. Normal urine varies from light yellow to amber, depending on the presence of urochromes, end-product pigments produced during the decomposition of hemoglobin. Dark urine indicates a high concentration of pigments.

Abnormal urine colors include yellow-brown or green, due to elevated concentrations of bile pigments, and red to dark brown, due to the presence of blood. Certain foods, such as beets or carrots, and various drug substances also may cause color changes in urine, but in such cases the colors have no clinical significance. Enter the results of this and the following tests in Part A of Laboratory Assessment 57.

3. Evaluate the *transparency* of the urine sample (judge whether the urine is clear, slightly cloudy, or very cloudy). Normal urine is clear enough to see through. You can read newsprint through slightly cloudy urine; you can no longer read newsprint through cloudy urine. Cloudy urine indicates the presence of various substances that may include mucus, bacteria, epithelial cells, fat droplets, or inorganic salts.

4. Determine the *specific gravity* of the urine sample, which indicates the solute concentration. Specific gravity is the ratio of the weight of something to the weight of an equal volume of pure water. For example, mercury (at 15°C) weighs 13.6 times as much as an equal volume of water; thus, it has a specific gravity of 13.6. Although urine is mostly water, it has substances dissolved in it and is slightly heavier than an equal volume of water. Thus, urine has a specific gravity of more than 1.000. Actually, the specific gravity of normal urine varies from 1.003 to 1.035. If the specific gravity is too low, the urine contains few solutes and represents dilute urine, a likely result of excessive fluid intake or the use of diuretics. A specific gravity above the normal range represents a higher concentration of solutes, likely from a limited fluid intake. Concentrated urine over an extended time increases the risk of the formation of kidney stones.

   To determine the specific gravity of a urine sample, follow these steps:
   a. Pour enough urine into a clean urinometer cylinder to fill it about three-fourths full. Any foam that appears should be removed with a paper towel.
   b. Use a laboratory thermometer to measure the temperature of the urine.
   c. Gently place the urinometer hydrometer into the urine, *and make sure that the float is not touching the sides or the bottom of the cylinder* (fig. 57.1).
   d. Position your eye at the level of the urine surface. Determine which line on the stem of the hydrometer intersects the lowest level of the concave surface (meniscus) of the urine.
   e. Liquids tend to contract and become denser as they are cooled, or to expand and become less dense as they are heated, so it may be necessary to make a temperature correction to obtain an accurate specific gravity measurement. To do this, add 0.001 to the hydrometer reading for each 3 degrees of urine temperature above 25°C or subtract 0.001 for each 3 degrees below 25°C. Enter this calculated value in the table of the laboratory assessment as the test result.

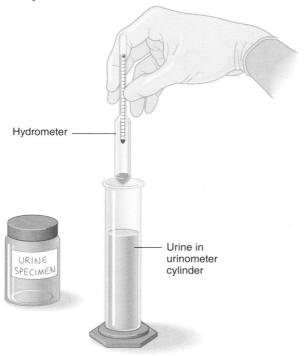

Hydrometer

URINE SPECIMEN

Urine in urinometer cylinder

5. Reagent strips can be used to perform a variety of urine tests. In each case, directions for using the strips are found on the strip container. *Be sure to read them.*

To perform each test, follow these steps:

a. Obtain a urine sample and the proper reagent strip.

b. Read the directions on the strip container.

c. Dip the strip in the urine sample.

d. Remove the strip at an angle and let it touch the inside rim of the urine container to remove any excess liquid.

e. Wait for the length of time indicated by the directions on the container before you compare the color of the test strip with the standard color scale on the side of the container. The value or amount represented by the matching color should be used as the test result and recorded in Part A of the laboratory assessment.

Alternative Procedure

If combination reagent strips (Chemstrip or Multistix) are being used, locate the appropriate color chart for each test being evaluated. Wait the designated time for each color reaction before the comparison is made to the standard color scale.

---

6. Perform the *pH test*. The pH of normal urine varies from 4.6 to 8.0, but most commonly, it is near 6.0 (slightly acidic). The pH of urine may decrease as a result of a diet high in protein, or it may increase with a vegetarian diet. Significant daily variations within the broad normal range are results of concentrations of excesses from variable diets.

7. Perform the *glucose test*. Normally, there is no glucose in urine. However, glucose may appear in the urine temporarily following a meal high in carbohydrates. Glucose also may appear in the urine as a result of uncontrolled diabetes mellitus.

8. Perform the *protein test*. Normally, proteins of large molecular size are not present in urine. However, those of small molecular sizes, such as albumins, may appear in trace amounts, particularly following strenuous exercise. Increased amounts of proteins also may appear as a result of kidney diseases in which the glomeruli are damaged or as a result of high blood pressure.

9. Perform the *ketone test*. Ketones are products of fat metabolism. Usually they are not present in urine. However, they may appear in the urine if the diet fails to provide adequate carbohydrate, as in the case of prolonged fasting or starvation or as a result of insulin deficiency (diabetes mellitus).

10. Perform the *bilirubin test*. Bilirubin, which results from hemoglobin decomposition in the liver, normally is absent in urine. It may appear, however, as a result of liver disorders that cause obstructions of the biliary tract. Urochrome, a normal yellow component of urine, is a result of additional breakdown of bilirubin.

11. Perform the *hemoglobin/occult blood test*. Hemoglobin occurs in the red blood cells, and because such cells normally do not pass into the renal tubules, hemoglobin is not found in normal urine. Its presence in urine usually indicates a disease process, a transfusion reaction, an injury to the urinary organs, or menstrual blood.

12. Complete Part A of the laboratory assessment.

Procedure B—Microscopic Sediment Analysis

1. A urinalysis usually includes a study of urine sediment—the microscopic solids present in a urine sample. This sediment normally includes mucus, certain crystals, and a variety of cells, such as the epithelial cells that line the urinary tubes and an occasional white blood cell. Other types of solids, such as casts or red blood cells, may indicate a disease or injury if they are present in excess. (Casts are cylindrical masses of cells or other substances that form in the renal tubules and are flushed out by the flow of urine.)

To observe urine sediment, follow these steps:

a. Thoroughly stir or shake a urine sample to suspend the sediment, which tends to settle to the bottom of the container.

b. Pour 10 mL of urine into a clean centrifuge tube and centrifuge it for 5 minutes at slow speed (1,500 rpm). Be sure to balance the centrifuge with an even number of tubes filled to the same levels.

FIGURE 57.2 Types of urine sediment. Healthy individuals lack many of these sediments and possess only occasional to trace amounts of others. (*Note:* Shades of white to purple sediments are most characteristic when using Sedi-stain.)

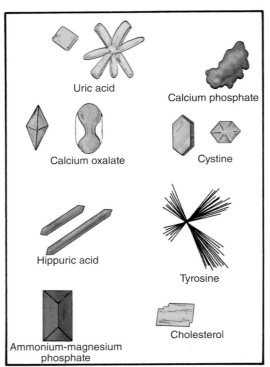

Crystals

Uric acid

Calcium phosphate

Calcium oxalate

Cystine

Hippuric acid

Tyrosine

Ammonium-magnesium phosphate

Cholesterol

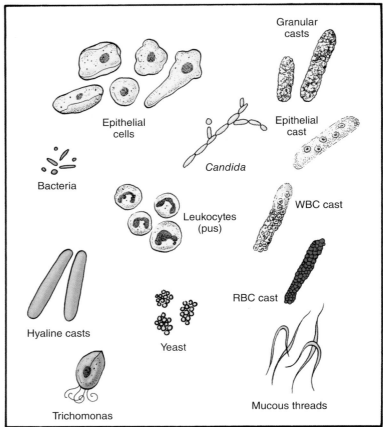

Cells and casts

Epithelial cells

Granular casts

Candida

Epithelial cast

Bacteria

Leukocytes (pus)

WBC cast

Hyaline casts

Yeast

RBC cast

Trichomonas

Mucous threads

c. Carefully decant 9 mL (leave 1 mL) of the liquid from the sediment in the bottom of the centrifuge tube, as directed by your laboratory instructor. Resuspend the 1 mL of sediment.

d. Use a medicine dropper to remove some of the sediment and place it on a clean microscope slide.

e. Add a drop of Sedi-stain to the sample, and add a coverslip.

f. Examine the sediment with low-power (reduce the light when using low power) and high-power magnifications.

g. With the aid of figure 57.2, identify the types of solids present.

2. In Part B of the laboratory assessment, make a sketch of each type of sediment that you observed.

3. Complete Part B of the laboratory assessment.

Name _____

Date _____

Section _____

The A corresponds to the indicated outcome(s) found at the beginning of the laboratory exercise.

Urinalysis

Part A Assessments

1. Enter your observations, test results, and evaluations in the following table: A 1 A 2 A 3

Urine Characteristics	Observations and Test Results	Normal Values	Evaluations
Color	Amber dark yellow	Light yellow to amber	concentrated
Transparency	Cloudy not transparent	Clear	excessive cellular material or protein
Specific gravity (corrected for temperature)		1.003–1.035	
pH	9	4.6–8.0	slightly higher then normal
Glucose	0 - Neg.	0 (negative)	normal
Protein	3.0	0 to trace	abnormal, cause of cloudiness
Ketones	0 - Neg	0	normal
Bilirubin		0	
Hemoglobin/occult blood		0	
(Other)			
(Other)			

2. Summarize the results of the chemical analysis of urine. 5 ______________________________

____

____

____

____

____

____

Critical Thinking Assessment

Why do you think it is important to refrigerate a urine sample if an analysis cannot be performed immediately after collecting it?

____

____

____

____

Part B Assessments

1. Make a sketch for each type of sediment you observed. Label any from those shown in figure 57.2. 4

=RBC
=Yeast
=Renal Cells

2. Summarize the results of the urine sediment analysis. 5 ______________________________

____

____

____

____

Male Reproductive System

Purpose of the Exercise

To review the structure and functions of the male reproductive organs and to examine some of these organs.

Materials Needed

Human torso model
Model of the male reproductive system
Anatomical chart of the male reproductive system
Compound light microscope
Prepared microscope slides of the following:
  Testis section
  Epididymis, cross section
  Ductus deferens, cross section
  Penis, cross section

Learning Outcomes

After completing this exercise, you should be able to

1 Locate and identify the organs of the male reproductive system.

2 Match the structures and functions of these organs.

3 Sketch and label the major features of microscopic sections of the testis, epididymis, ductus deferens, and penis.

Pre-Lab

Carefully read the introductory material and examine the entire lab. Be familiar with the structures and functions of the male reproductive organs from lecture or the textbook. Answer the pre-lab questions.

Pre-Lab Questions: Select the correct answer for each of the following questions:

1. Sperm are produced in the
   a. seminiferous tubules. **b.** interstitial cells.
   c. epididymis. **d.** ductus deferens.
2. The ductus deferens is located within the
   a. testis. **b.** epididymis.
   c. spermatic cord. **d.** penis.
3. The _____ is the common tube for the passage of urine and semen.
   a. ductus deferens **b.** urethra
   c. ureter **d.** ejaculatory duct
4. Which of the following would *not* be visible in a cross section of a penis?
   a. urethra **b.** corpus spongiosum
   c. corpora cavernosa **d.** epididymis
5. The interstitial cells produce
   a. sperm. **b.** testosterone.
   c. nutrients. **d.** semen.
6. The following glands are paired *except* the
   a. prostate. **b.** testis.
   c. seminal vesicle. **d.** bulbourethral.
7. The smooth muscle represents the thickest layer of the wall of the ductus deferens.
   True _____ False _____
8. The rete testis connects directly to the tail of the epididymis.
   True _____ False _____

The organs of the male reproductive system are specialized to produce and maintain the male sex cells, to transport these cells together with supporting fluids to the female reproductive tract, and to produce and secrete male sex hormones. These organs include the *testes* and sets of internal and external genitalia.

The testes originate internally but descend through an inguinal canal prior to birth to a position in the *scrotum,* which provides a lower temperature necessary for sperm production and storage. Within the testes, numerous *seminiferous tubules* produce sperm (spermatozoa) by spermatogenesis, and *interstitial cells* between seminiferous tubules produce testosterone. The interstitial cells then release the testosterone for transport in the bloodstream. Testosterone influences the sex drive and development of the secondary sex characteristics.

The sperm produced in the seminiferous tubules enter a tubular network, the *rete testis,* which joins the tightly coiled *epididymis* through *efferent ductules.* Within head, body, and tail of the epididymis, immature sperm are stored and nourished during their maturation. A muscular *ductus deferens (vas deferens),* part of the *spermatic cord,* passes through the inguinal canal into the pelvic cavity and posterior to the urinary bladder, where it unites with ducts from *seminal vesicles* to form the ejaculatory ducts. An alkaline fluid containing nutrients and prostaglandins is secreted by the seminal vesicles. The ejaculatory ducts unite with the urethra within the *prostate gland.* Additional secretions are added from the prostate gland and two *bulbourethral glands.* The combination of sperm and the secretions from the accessory glands is semen.

The urethra, a common tube to convey semen and urine, passes through the *penis* within the *corpus spongiosum* to the external urethral orifice. The corpus spongiosum plus two *corpora cavernosa* provide vascular spaces that become engorged with blood during an erection. A loose fold of skin, forming a cuff over the glans penis, is called the *prepuce,* and is frequently removed in a procedure called a circumcision shortly after birth.

Procedure A—Male Reproductive Organs

1. Study figures 58.1 and 58.2 and the list of structures and descriptions.
2. Observe the human torso model, the model of the male reproductive system, and the anatomical chart

---

FIGURE 58.1 The major structures of the male reproductive system, sagittal view.

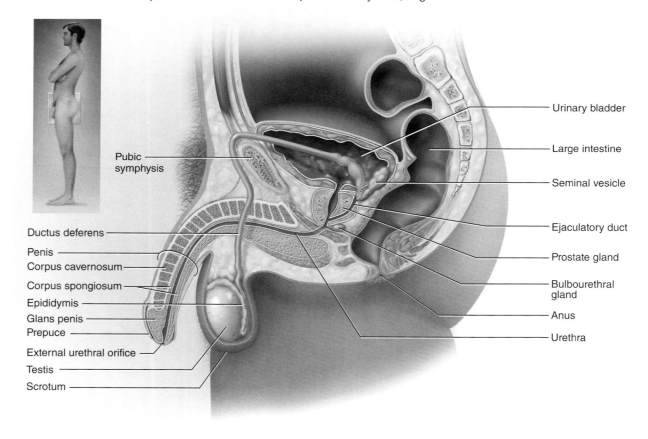

- Pubic symphysis
- Ductus deferens
- Penis
- Corpus cavernosum
- Corpus spongiosum
- Epididymis
- Glans penis
- Prepuce
- External urethral orifice
- Testis
- Scrotum
- Urinary bladder
- Large intestine
- Seminal vesicle
- Ejaculatory duct
- Prostate gland
- Bulbourethral gland
- Anus
- Urethra

FIGURE 58.2 The testis and associated structures (a) from a cadaver and (b) in an illustration showing some internal structures.

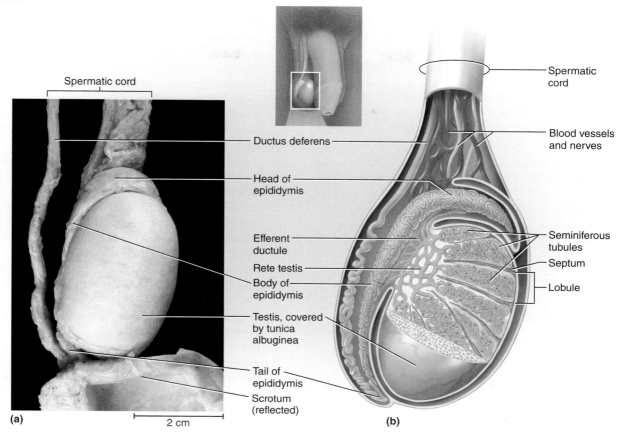

Spermatic cord

Ductus deferens

Head of epididymis

Efferent ductule

Rete testis

Body of epididymis

Testis, covered by tunica albuginea

Tail of epididymis

Scrotum (reflected)

(a)

2 cm

Spermatic cord

Blood vessels and nerves

Seminiferous tubules

Septum

Lobule

(b)

of the male reproductive system. Locate the following features:

testes—paired organs suspended by spermatic cord within scrotum
- lobules—subdivisions of testes
- seminiferous tubules—long, tiny coiled tubes; site of spermatogenesis
- interstitial cells (Leydig cells)—endocrine cells between seminiferous tubules; produce and secrete testosterone
- rete testis—network of channels from seminiferous tubules to efferent ductules
- efferent ductules—several ducts from rete testis to epididymis

inguinal canal—contains the spermatic cord

spermatic cord—contains ductus deferens, blood vessels, nerves, and lymphatic vessels

epididymis—site of sperm maturation and storage; tube about 6 meters (18 feet) if uncoiled

ductus deferens (vas deferens)—muscular tube that transports and stores sperm

ejaculatory duct—tube within prostate gland from ductus deferens to urethra

seminal vesicles (seminal glands)—paired glands; produce most of seminal fluid

prostate gland—produces seminal fluid that helps to activate sperm; single gland

bulbourethral glands—tiny paired glands; secretions neutralize any residual acidic urine

scrotum—protective covering of testes and epididymis

penis—male copulatory organ
- corpora cavernosa—paired erectile tissue cylinders; engorge with blood during an erection
- corpus spongiosum—single erectile tissue cylinder; engorges with blood during an erection
- tunica albuginea—fibrous capsule around testes and corpora erectile tissues
- glans penis—expanded distal end of penis
- external urethral orifice (external urethral meatus)—external opening for urine and semen
- prepuce (foreskin)—fold of skin around glans penis; removed if a surgical circumcision is performed

3. Complete Part A of Laboratory Assessment 58.

Procedure B—Microscopic Anatomy

1. Obtain a microscope slide of a human testis section and examine it using low-power magnification (fig. 58.3). Locate the thick *tunica albuginea (fibrous capsule)* on the surface and the numerous sections of *seminiferous tubules* inside.

2. Focus on some of the seminiferous tubules using high-power magnification (fig. 58.4). Locate the *basement membrane* and the layer of *spermatogonia* just beneath the basement membrane. Identify some *sustentacular cells* (supporting cells; Sertoli cells), which have pale, oval-shaped nuclei, and some *spermatogenic cells,* which have smaller, round nuclei. Spermatogonia give rise to spermatogenic cells that are in various stages of spermatogenesis as they are forced toward the lumen. Near the lumen of the tube, find some darkly stained, elongated heads of developing sperm (spermatozoa). In the spaces between adjacent seminiferous tubules, locate some isolated *interstitial cells* (Leydig cells) of the endocrine system. Interstitial cells produce the hormone testosterone, transported by the blood.

3. Prepare a labeled sketch of a representative section of the testis in Part B of the laboratory assessment.

4. Obtain a microscope slide of a cross section of *epididymis* (fig. 58.5). Examine its wall using high-power magnification. Note the elongated, *pseudostratified columnar epithelial cells* that compose most of the inner lining. These cells have nonmotile stereocilia (microvilli) on their free surfaces. Also note the thin layer of smooth muscle and connective tissue surrounding the tube.

5. Prepare a labeled sketch of the epididymis wall in Part B of the laboratory assessment.

6. Obtain a microscope slide of a cross section of *ductus deferens,* and examine it with scan and low-power magnifications (fig. 58.6*a*). Note the small lumen lined with an epithelium and the thick *smooth muscle* layer composing most of the structure of the duct. Three layers of smooth muscle fibers are distinctive: a middle circular layer between two layers of longitudinal muscle fibers. The peristaltic contractions of the muscular layer transport sperm along the passageway toward the ejaculatory duct and urethra. Sperm storage occurs within the lumen of the ductus deferens and the epididymis. An outer covering of the ductus deferens is called the *adventitia.* Use high-power magnification and examine the epithelium layer composed of pseudostratified columnar epithelial cells (fig. 58.6*b*).

FIGURE 58.3 Micrograph of a human testis (1.7×).

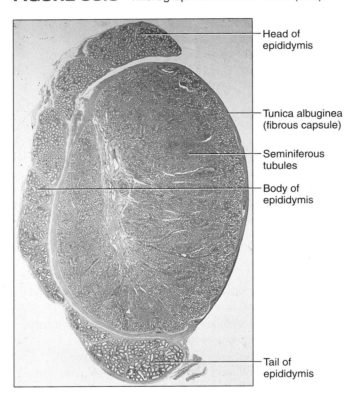

- Head of epididymis
- Tunica albuginea (fibrous capsule)
- Seminiferous tubules
- Body of epididymis
- Tail of epididymis

FIGURE 58.4 Micrograph of a seminiferous tubule (250×).

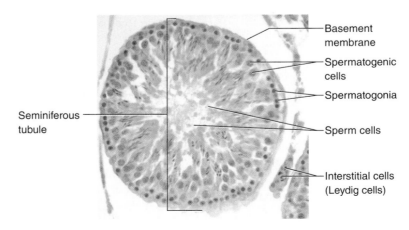

- Seminiferous tubule
- Basement membrane
- Spermatogenic cells
- Spermatogonia
- Sperm cells
- Interstitial cells (Leydig cells)

FIGURE 58.5 Micrograph of a cross section of a human epididymis (200×).

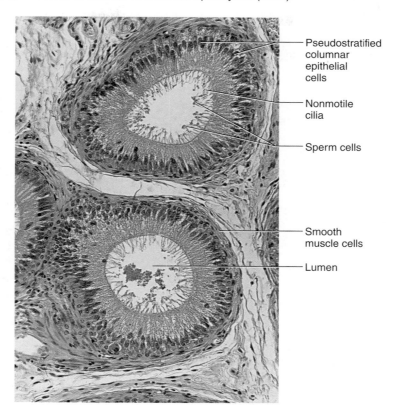

Pseudostratified columnar epithelial cells

Nonmotile cilia

Sperm cells

Smooth muscle cells

Lumen

FIGURE 58.6 Ductus deferens. (a) Micrograph of a cross section of the ductus deferens (40×). (b) Micrograph of the wall of the ductus deferens (400×).

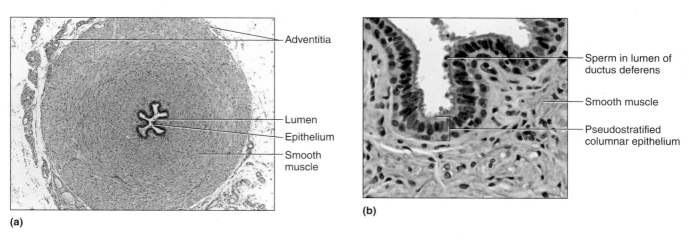

Adventitia

Lumen
Epithelium
Smooth muscle

(a)

Sperm in lumen of ductus deferens

Smooth muscle

Pseudostratified columnar epithelium

(b)

FIGURE 58.7 Micrograph of a cross section of the body of the penis (5×).

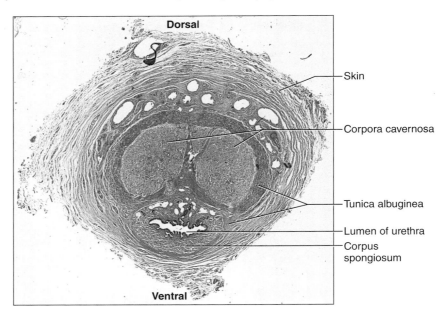

Dorsal

Skin

Corpora cavernosa

Tunica albuginea

Lumen of urethra

Corpus spongiosum

Ventral

7. Prepare a labeled sketch of a representative section of a ductus deferens in Part B of the laboratory assessment.

8. Study figure 58.7 and the list of structures and descriptions. Obtain a microscope slide of a *penis* cross section, and examine it with scan and low-power magnifications. Identify the following features:

  corpora cavernosa—two dorsal erectile tissue cylinders; engorge with blood during an erection

  corpus spongiosum—single erectile tissue cylinder around urethra; engorges with blood during an erection

  tunica albuginea—fibrous capsule around corpora

  urethra—passageway for urine and semen

  skin—surrounds entire penis

9. Prepare a labeled sketch of a penis cross section in Part B of the laboratory assessment.

10. Complete Part B of the laboratory assessment.

Name _____

Date _____

Section _____

The 🄐 corresponds to the indicated outcome(s) found at the beginning of the laboratory exercise.

Male Reproductive System

Part A   Assessments

1. Label the structures in figures 58.8 and 58.9.

FIGURE 58.8   Label the major structures of the male reproductive system in this sagittal view. 🄐

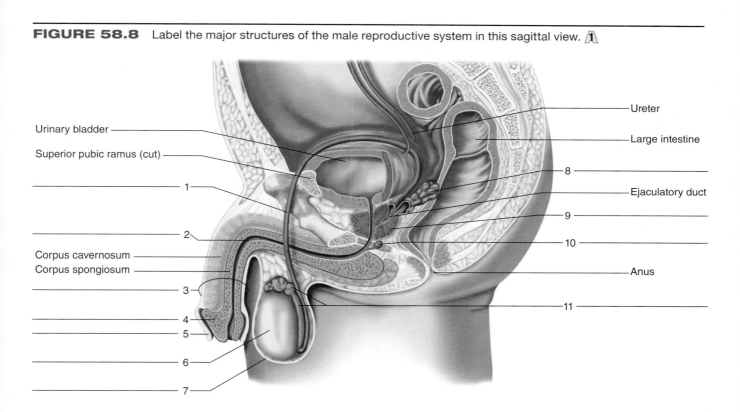

Urinary bladder

Superior pubic ramus (cut)

1

2

Corpus cavernosum

Corpus spongiosum

3

4

5

6

7

Ureter

Large intestine

8

Ejaculatory duct

9

10

Anus

11

FIGURE 58.9 Label the diagram of (a) the sagittal section of a testis and (b) a cross section of a seminiferous tubule by placing the correct numbers in the spaces provided.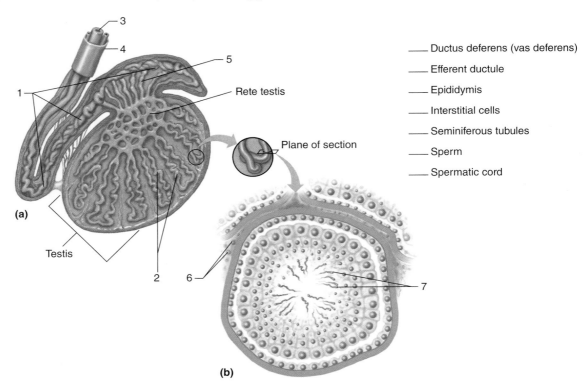

_____ Ductus deferens (vas deferens)

_____ Efferent ductule

_____ Epididymis

_____ Interstitial cells

_____ Seminiferous tubules

_____ Sperm

_____ Spermatic cord

Rete testis

Plane of section

(a)

Testis

(b)

2. Match the terms in column A with the descriptions in column B. Place the letter of your choice in the space provided. /2\

Column A	Column B
a. Bulbourethral glands	_____ **1.** Process by which sperm are formed
b. Ductus deferens	_____ **2.** Common tube for urine and semen
c. Ejaculatory duct	_____ **3.** Sperm duct located within spermatic cord
d. Epididymis	_____ **4.** Produce male hormones
e. Interstitial cells	_____ **5.** Location of spermatogenesis
f. Penis	_____ **6.** Located within prostate gland and can propel semen into urethra
g. Prepuce	_____ **7.** Contains sperm and secretions from accessory glands
h. Prostate	_____ **8.** Tiny paired glands that secrete into urethra
i. Semen	_____ **9.** Maturation and storage site within a coiled tube along testis
j. Seminiferous tubules	_____ **10.** Surrounds the urethra and secretes a seminal fluid
k. Spermatogenesis	_____ **11.** Contains three columns of erectile tissue
l. Urethra	_____ **12.** Removed in a surgical circumcision

Part B Assessments

1. In the space that follows, sketch a representative area of the organ indicated. Label any of the structures that were observed, indicate the magnification used for each sketch. 3

Testis (_____×)	Epididymis (_____×)
Ductus deferens (_____×)	Penis (_____×)

2. Describe the function of each of the following: 2

   a. Scrotum

   _____

   _____

   b. Spermatogenic cell

   _____

   _____

   c. Interstitial cell (Leydig cell)

   _____

   _____

   d. Epididymis

   _____

   _____

   e. Corpora cavernosa and corpus spongiosum

   _____

   _____

NOTES

Laboratory Exercise 59

Female Reproductive System

Purpose of the Exercise

To review the structure and functions of the female reproductive organs and to examine some of their features.

Materials Needed

Human torso model
Model of the female reproductive system
Anatomical chart of the female reproductive system
Compound light microscope
Prepared microscope slides of the following:
  Ovary section with maturing follicles
  Uterine tube, cross section
  Uterine wall section

For Demonstration Activity:
Prepared microscope slides of the following:
  Uterine wall, early proliferative phase
  Uterine wall, secretory phase
  Uterine wall, early menstrual phase

Learning Outcomes

After completing this exercise, you should be able to

1. Locate and identify the organs of the female reproductive system.
2. Match the structures and functions of these organs.
3. Sketch and label the major features of microscopic sections of the ovary, uterine tube, and uterine wall.

Pre-Lab

Carefully read the introductory material and examine the entire lab. Be familiar with the structures and functions of the female reproductive system from lecture or the textbook. Answer the pre-lab questions.

Pre-Lab Questions: Select the correct answer for each of the following questions:

1. Oogenesis originates within the
   a. follicle.   b. uterine tube.
   c. uterus.   d. corpus albicans.

2. Early cleavage division occurs within the
   a. follicle.   b. corpus luteum.
   c. uterine tube.   d. uterus.

3. Which of the following is *not* a structure of the uterus?
   a. cervix   b. fimbriae
   c. endometrium   d. myometrium

4. The _____ are the milk-producing parts of the mammary glands.
   a. areolar glands   b. lactiferous ducts
   c. adipose cells   d. alveolar glands

5. How many openings are located on a human nipple?
   a. 1   b. 2–3
   c. 5–10   d. 15–20

6. Which of the following structures occupies the most anterior position of the external genitalia?
   a. hymen   b. clitoris
   c. external urethral   d. vaginal orifice
      orifice

7. The broad ligament is the largest ligament involved in holding female reproductive organs in position.
   True _____   False _____

8. Primordial follicles are concentrated in the medulla region of an ovary.
   True _____   False _____

The organs of the female reproductive system are specialized to produce and maintain the female sex cells, to transport these cells to the site of fertilization, to provide a favorable environment for a developing offspring, to move the offspring to the outside, and to produce female sex hormones.

These organs include the *ovaries,* which produce the egg cells and female sex hormones, and sets of internal and external genitalia (accessory organs). The internal genitalia include the uterine tubes, uterus, and vagina. The external genitalia include the labia majora, labia minora, clitoris, and vestibular glands.

Within the cortex region of an ovary, numerous follicles are in some stage of development to produce ova (eggs) by the process called oogenesis (meiosis). Numerous *primordial follicles* formed during prenatal development. During the period from puberty to menopause, reproductive cycles occur as individual follicles complete maturation. A *primary follicle* contains a *primary oocyte* that undergoes oogenesis. The primary follicle develops into a *secondary follicle,* and finally a *mature tertiary follicle* shortly before ovulation of a *secondary oocyte.* The ruptured follicle becomes a *corpus luteum,* which secretes estrogens and progesterone, and eventually degenerates to form a *corpus albicans,* composed of connective tissue.

The *uterine tube (oviduct; fallopian tube)* has finger-like fimbriae partially around the ovary. After ovulation, the secondary oocyte is conveyed through the uterine tube by the action of cilia and peristaltic waves. If fertilization occurs, early cleavage divisions take place during the passage to the *uterus* during the next several days. The uterus has three regions: fundus, body, and cervix. There are three layers to its wall: perimetrium, myometrium, and endometrium. Development of the embryo and the fetus take place within the uterus until birth. The *vagina* has a mucosal lining, and near the vaginal orifice a membrane (hymen) partially closes the orifice unless broken, often during the first intercourse.

The external genitalia (vulva) are within a diamond-shaped perineum with boundaries established by the pubis, ischial tuberosities, and coccyx. Larger rounded folds of skin, labia majora, protect the labia minora and other external genitalia. Between the labia minora is a space (vestibule) that encloses the urethral and vaginal openings. A clitoris at the anterior vulva has similar structures to a penis, including a prepuce, glans, and corpora cavernosa.

The *breasts* contain the mammary glands, which are comprised of many lobes, ducts, ligaments, and adipose tissue. At puberty, estrogens stimulate breast development, and at the birth of a baby, various hormones promote alveolar gland development and milk production. Lactiferous ducts terminate in 15–20 openings on the nipple in the center of the pigmented areola.

Procedure A—Female Reproductive Organs

1. Study figures 59.1, 59.2, 59.3, and 59.4 of the internal reproductive organs and the list of structures and descriptions.

2. Observe the human torso model, the model of the female reproductive system, and the anatomical chart of the female reproductive system. Locate the following features:

ovaries—paired organs that produce ova and sex hormones
- medulla—internal region
- cortex—outer region

ligaments—hold reproductive organs in place
- broad ligament—largest ligament attaches to uterine tube, ovary, and uterus
- suspensory ligament of ovary—holds superior end of ovary to pelvic wall
- ovarian ligament—attaches ovary to uterus
- round ligament of uterus—band within broad ligament from lateral uterus to the connective tissue of external genitalia

uterine tubes (oviducts; fallopian tubes)—paired tubes about 10 cm (4 inches) long; site of fertilization and early cleavage development
- infundibulum—expanded open end near ovary
- fimbriae—slender extensions on end of infundibulum

uterus—houses embryonic and fetal development during pregnancy
- fundus of uterus—wide superior curvature
- body of uterus—large middle section
- cervix of uterus—inferior cylindrical end
  - cervical canal—narrow passage within cervix
  - external os—opening into vagina
- uterine wall—thick and composed of three layers
  - endometrium—inner mucosal layer with simple columnar epithelium and numerous tubular glands
  - myometrium—thick middle layer of smooth muscle; contracts during childbirth
  - perimetrium (serosa; serous coat)—outer layer

rectouterine pouch—inferior recess of peritoneal cavity between rectum and uterus

vagina—tube about 10 cm (4 inches) long; receives penis during sexual intercourse; birth canal
- fornices—recesses between vaginal wall and cervix
- vaginal orifice—external opening
- hymen—thin fold of mucosa that partially closes vaginal orifice
- mucosal layer—inner vaginal wall of stratified squamous epithelium

FIGURE 59.1 The structures of the female reproductive system.

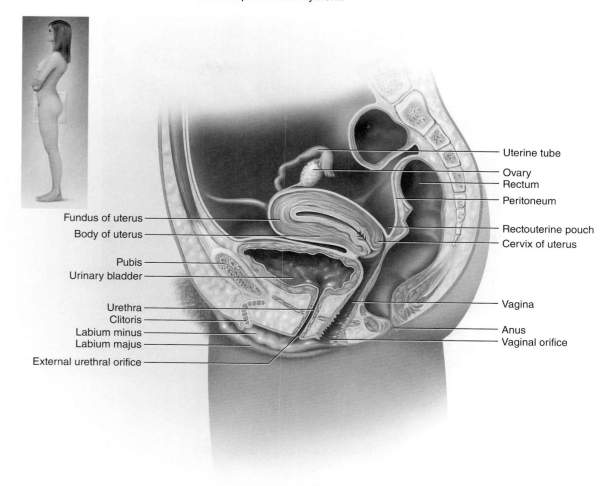

Uterine tube
Ovary
Rectum
Peritoneum

Fundus of uterus
Body of uterus

Rectouterine pouch
Cervix of uterus

Pubis
Urinary bladder

Urethra
Clitoris
Labium minus
Labium majus

Vagina

Anus
Vaginal orifice

External urethral orifice

FIGURE 59.2 The female reproductive organs (posterior view). The left-side organs are shown in section and the ligaments are shown on the right side.

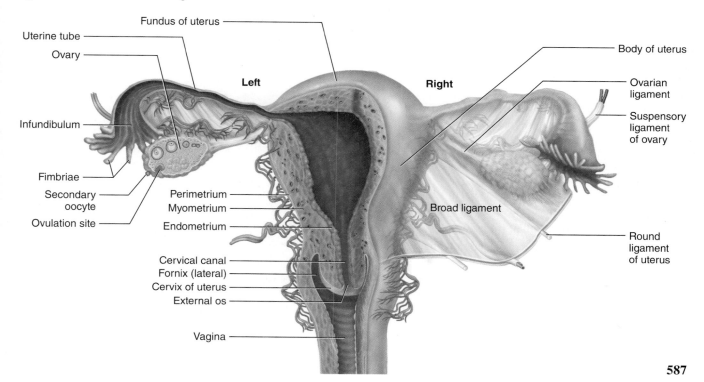

Fundus of uterus

Uterine tube
Ovary

Body of uterus

Left

Right

Ovarian ligament

Suspensory ligament of ovary

Infundibulum

Fimbriae
Secondary oocyte
Ovulation site

Perimetrium
Myometrium
Endometrium

Broad ligament

Round ligament of uterus

Cervical canal
Fornix (lateral)
Cervix of uterus
External os

Vagina

FIGURE 59.3 The female reproductive organs of a cadaver (anterior view).

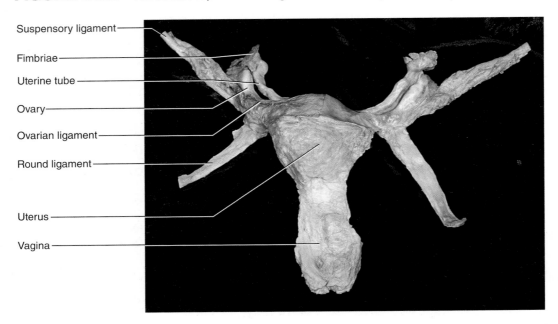

Suspensory ligament
Fimbriae
Uterine tube
Ovary
Ovarian ligament
Round ligament
Uterus
Vagina

FIGURE 59.4 The structures associated with the female perineum. The external genitalia (vulva) are within the perineum.

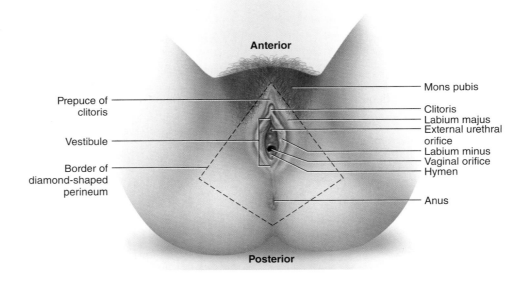

Anterior

Prepuce of clitoris
Vestibule
Border of diamond-shaped perineum

Mons pubis
Clitoris
Labium majus
External urethral orifice
Labium minus
Vaginal orifice
Hymen
Anus

Posterior

3. Study figures 59.4 and 59.5 and the list of structures and descriptions of the external genitalia and the breasts.

external genitalia (vulva; pudendum)

- mons pubis—rounded elevation with adipose tissue anterior to pubic symphysis
- labia majora—folds of skin with hair that enclose and protect structures between them
- labia minora—lesser hair-free folds between labia majora
- vestibule—region between labia minora
- vestibular glands—located on each side of vagina; secrete a lubricating fluid
- clitoris—small projection at anterior end of vestibule; homologous to penis

breasts—paired elevations containing mammary glands

- nipple—protruding region in center of areola containing openings of the lactiferous ducts

588

FIGURE 59.5 The structures of a lactating breast (anterior view).

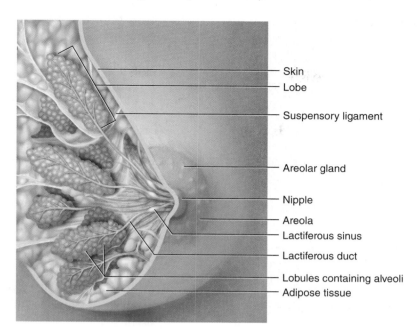

- Skin
- Lobe
- Suspensory ligament
- Areolar gland
- Nipple
- Areola
- Lactiferous sinus
- Lactiferous duct
- Lobules containing alveoli
- Adipose tissue

- areola—circular pigmented skin around nipple
- alveolar glands—compose the milk-producing parts of mammary glands that develop during pregnancy
- lactiferous ducts—drain the 15–20 lobes of the breast
- adipose tissue—compose much of internal structure of the breast

4. Complete Part A of Laboratory Assessment 59.

Procedure B—Microscopic Anatomy

1. Obtain a microscope slide of an ovary section with maturing follicles, and examine it with low-power magnification (fig. 59.6). Locate the outer layer, or *cortex,* composed of densely packed cells with developing follicles, and the inner layer, or *medulla,* which largely consists of loose connective tissue.

FIGURE 59.6 Micrograph of the ovary (80×).

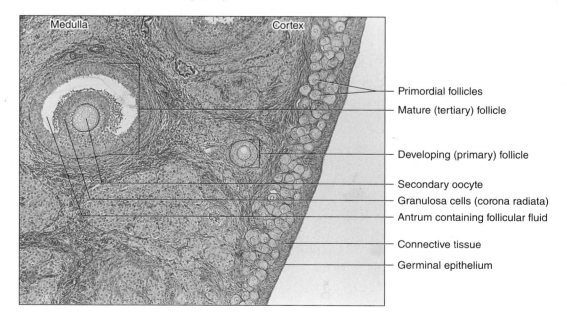

Medulla   Cortex

- Primordial follicles
- Mature (tertiary) follicle
- Developing (primary) follicle
- Secondary oocyte
- Granulosa cells (corona radiata)
- Antrum containing follicular fluid
- Connective tissue
- Germinal epithelium

2. Focus on the cortex of the ovary using high-power magnification (fig. 59.7). Note the thin layer of small cuboidal cells on the free surface. These cells comprise the *germinal epithelium*. Also locate some *primordial follicles* just beneath the germinal epithelium. Each follicle consists of a single, relatively large *primary oocyte* with a prominent nucleus and a layer of simple squamous epithelial cells.

3. Use low-power magnification to search the ovarian cortex for maturing follicles in various stages of development (fig. 59.6). Locate and compare primordial follicles, *primary follicles, secondary follicles,* and *mature (tertiary) follicles.* Developing follicles have a covering from a single layer to a double layer of folliclular cells. The type of epithelial tissue cells depends upon the stage of development. Mature (tertiary) follicles have a covering of several layers of granulosa cells.

4. Prepare labeled sketches in Part B of the laboratory assessment to illustrate the changes that occur in a follicle as it matures.

5. Obtain a microscope slide of a cross section of a uterine tube. Examine it using low-power magnification. The shape of the lumen is very irregular because of folds of the mucosa layer.

6. Focus on the inner lining of the uterine tube using high-power magnification (fig. 59.8). The lining is composed of *simple columnar epithelium,* and some of the epithelial cells are ciliated on their free surfaces.

7. Prepare a labeled sketch of a representative region of the wall of the uterine tube in Part B of the laboratory assessment.

8. Obtain a microscope slide of the uterine wall section (fig. 59.9). Examine it using low-power magnification, and locate the following:

> **endometrium**—inner mucosal layer
>
> **myometrium**—middle, thick muscular layer
>
> **perimetrium**—outer serosal layer

9. Prepare a labeled sketch of a representative section of the uterine wall in Part B of the laboratory assessment.

10. Complete Part C of the laboratory assessment.

FIGURE 59.7 Micrograph of the ovarian cortex (200×).

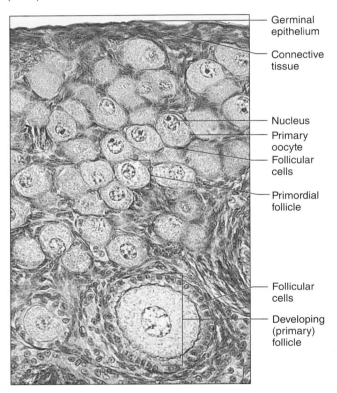

- Germinal epithelium
- Connective tissue
- Nucleus
- Primary oocyte
- Follicular cells
- Primordial follicle
- Follicular cells
- Developing (primary) follicle

FIGURE 59.8 Micrograph of a cross section of the uterine tube (800×).

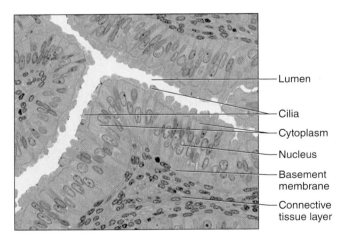

- Lumen
- Cilia
- Cytoplasm
- Nucleus
- Basement membrane
- Connective tissue layer

FIGURE 59.9 Micrograph of the uterine wall (10×).

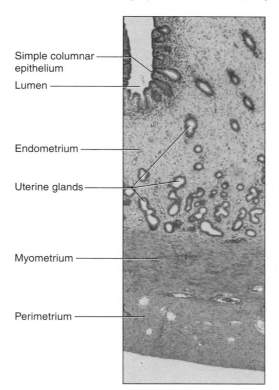

Simple columnar epithelium

Lumen

Endometrium

Uterine glands

Myometrium

Perimetrium

Demonstration Activity

Observe the slides in the demonstration microscopes. Each slide contains a section of uterine mucosa taken during a different phase in the reproductive cycle. In the *early proliferative phase,* note the simple columnar epithelium on the free surface of the mucosa and the many sections of tubular uterine glands in the tissues beneath the epithelium. In the *secretory phase,* the endometrium is thicker, the uterine glands appear more extensive, and they are coiled. In the *early menstrual phase,* the endometrium is thinner because its surface functional layer has been shed, while retaining the basal layer. Also the uterine glands are less apparent, and the spaces between the glands contain many leukocytes. The basal layer of the endometrium renews the functional layer after menstruation ends.

Name _____

Date _____

Section _____

The ⚠ corresponds to the indicated outcome(s) found at the beginning of the laboratory exercise.

Female Reproductive System

Part A   Assessments

1. Identify the numbered structures in figures 59.10 and 59.11.

FIGURE 59.10   Label the major structures of the female reproductive system in this sagittal view. ⚠

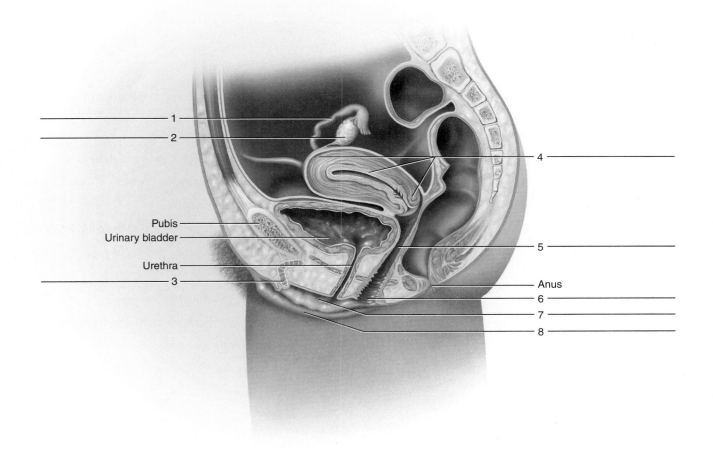

FIGURE 59.11 Label the female reproductive structures in this posterior view.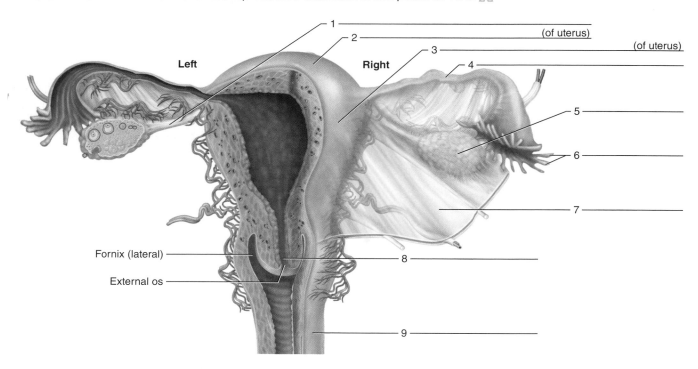

1 ———————————
2 ——————————— (of uterus)
3 ——————————— (of uterus)
Left   Right
4 ———————————
5 ———————————
6 ———————————
7 ———————————
Fornix (lateral) ———————
8 ———————————
External os ———————
9 ———————————

2. Match the terms in column A with the descriptions in column B. Place the letter of your choice in the space provided. 2

Column A

a. Cervix
b. Cilia
c. Endometrium
d. Fimbriae
e. Fundus
f. Hymen
g. Lactiferous duct
h. Myometrium
i. Ovulation
j. Perineum
k. Suspensory ligament
l. Vagina

Column B

_____ 1. Rounded end of uterus near uterine tubes

_____ 2. Smooth muscle that contracts with force during childbirth

_____ 3. Thin membrane that partially closes vaginal orifice

_____ 4. Surface region around external genitalila

_____ 5. Process by which a secondary oocyte is released from the ovary

_____ 6. Portion of uterus extending into superior portion of vagina

_____ 7. Inner mucosal lining of uterus

_____ 8. Transports milk from milk-producing lobes to nipple

_____ 9. Help move secondary oocyte through the uterine tube toward uterus

___D___ 10. Fingerlike projections of uterine tube near ovary

_____ 11. Birth canal during birth process

_____ 12. Holds ovary to pelvic wall

Part B Assessments

In the space that follows, sketch a representative area of the organ indicated. Label any of the structures observed and indicate the magnification used for each sketch. ⒊

Primordial or primary follicle (_____×)	Secondary or mature follicle (_____×)
Uterine tube (_____×)	Wall of uterus (_____×)

Part C Assessments

Identify the numbered features in figure 59.12 of a sectioned ovary that represent structures present at some stage during a reproductive cycle.

FIGURE 59.12 Label this ovary by placing the correct numbers in the spaces provided. The arrows indicate development over time; follicles do not migrate through an ovary.

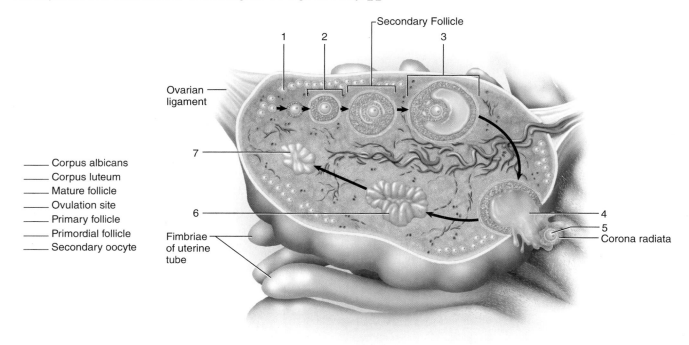

_____ Corpus albicans
_____ Corpus luteum
_____ Mature follicle
_____ Ovulation site
_____ Primary follicle
_____ Primordial follicle
_____ Secondary oocyte

Fertilization and Early Development

Purpose of the Exercise

To review the process of fertilization, to observe sea urchin eggs being fertilized, and to examine embryos in early stages of development.

Materials Needed

Sea urchin egg suspension*
Sea urchin sperm suspension*
Compound light microscope
Depression microscope slide
Coverslip
Medicine droppers
Models of human embryos
Prepared microscope slide of the following:
   Sea urchin embryos (early and late cleavage)

For Learning Extension Activity:
Vaseline
Toothpick

For Demonstration Activity:
Preserved mammalian embryos

*See the Instructor's Manual for a source of materials.

Learning Outcomes

After completing this exercise, you should be able to

1. Sketch the early developmental stages of a sea urchin.
2. Distinguish the major features of human embryos.
3. Describe fertilization and the early developmental stages of a human.

Pre-Lab

Carefully read the introductory material and examine the entire lab. Be familiar with fertilization and early embryonic development from lecture or the textbook. Answer the pre-lab questions.

Pre-Lab Questions: Select the correct answer for each of the following questions:

1. The ovum and sperm unite and form a diploid cell called the
   a. pronucleus.   **b.** blastocyst.
   c. morula.   **d.** zygote.

2. The _____ is the development stage that implants in the endometrium.
   a. morula   **b.** blastocyst
   c. zygote   **d.** two-cell stage

3. Which of the following is *not* an extraembryonic membrane?
   a. umbilical cord   **b.** amnion
   c. chorion   **d.** yolk sac

4. The placenta is comprised of
   a. only the myometrium.
   b. only the decidua basalis.
   c. only the chorionic villi.
   d. combinations of the decidua basalis and chorionic villi.

5. The embryonic stage of human development exists
   a. from fertilization until implantation.
   b. through the first four weeks of development.
   c. through the first eight weeks of development.
   d. from fertilization until birth.

6. The zona pellucida gel layer surrounds the secondary oocyte and early development stages through the morula.
   True _____   False _____

7. A single completed oogenesis sequence produces four functional ova (eggs).
   True _____   False _____

8. A single completed spermatogenesis sequence produces four functional sperm.
   True _____   False _____

Gametes are produced in testes and ovaries by a distinctive process of nuclear division called *meiosis* that reduces a single *diploid (2n)* cell to possibly four *haploid (n)* cells. Meiosis is a double division that includes a first meiotic division followed by a second meiotic division. In males the process is called *spermatogenesis,* and four functional haploid sperm result from each completed sequence. In females the process is called *oogenesis,* and one functional ovum is formed along with *polar bodies.* During puberty some *primary oocytes* continue through the first meiotic division, producing a large *secondary oocyte* and a small first polar body. (The secondary oocyte is often referred to as the ovum, although a second meiotic division must still occur after ovulation of the secondary oocyte.) If fertilization occurs, the secondary oocyte undergoes the second meiotic division, producing a fertilized ovum, the *zygote,* and a second tiny polar body that degenerates. The first small polar body from the first meiotic division, either degenerates or undergoes the second meiotic division, resulting in two degenerating polar bodies.

Ovulation occurs when a secondary oocyte, surrounded by a gel layer called the *zona pellucida,* is expelled from the ovary into the uterine tube. Before fertilization can occur, semen must be deposited in the vagina; some of the sperm must pass through the uterus into the uterine tubes and contact a secondary oocyte. Upon penetration of a sperm, the second meiotic division takes place, forming a second polar body. Fertilization is the process by which the pronuclei of the ovum (egg) and sperm come together and combine their chromosomes, forming a single diploid cell called a zygote.

Shortly after fertilization, the zygote undergoes cell division (mitosis and cytokinesis) to form two cells. These two cells become four, then in turn divide into eight, and so forth. A solid cluster of cells forms, the *morula,* which is still surrounded by the zona pellucida. The cells of the morula continue cell cycles and the hollow *blastocyst* develops. The blastocyst contains an *inner cell mass,* which develops into the embryo, and an outer *trophoblast* layer, which develops into the extraembryonic membranes and a portion of the *placenta.* These early *cleavage* divisions, resulting in an increasing number of cells that are smaller than the zygote, continue for most of the first week of development. The blastocyst implants into the endometrium about 6–7 days after fertilization. The early stages of development in some animals, such as sea urchins, are similar to those in humans.

In human development, the *embryonic stage* lasts through the first eight weeks, followed by the *fetal stage* until birth. The blastocyst develops into a *gastrula* containing three primary germ layers: the *ectoderm, endoderm,* and *mesoderm,* which give rise to particular tissues and body systems. The extraembryonic membranes include the amnion, yolk sac, allantois, and chorion. The *amnion* extends around the embryo and is filled with amniotic fluid, providing a protective cushion and constant environment. The *yolk sac* forms early blood cells and a portion of the embryonic digestive tube; however, the placental region, not the yolk, provides the nutritive functions. The *allantois* gives rise to the umbilical blood vessels and forms part of the urinary bladder. The *chorion* extends around the embryo and the other membranes, and it eventually fuses with the amnion to form the amniochorionic membrane.

The *placenta* is comprised of maternal and fetal components. The maternal component is a region of the *endometrium (decidua basalis);* the fetal portion contains an elaborate series of *chorionic villi* formed from the chorion. An examination of the placenta upon birth (the afterbirth) reveals a rough surface of the maternal portion and a smooth surface with the attached umbilical cord of the fetal part.

Procedure A—Fertilization

1. Study figure 60.1.
2. Although it is difficult to observe fertilization in mammals since the process occurs internally, it is possible to view forms of external fertilization. For example, egg and sperm cells can be collected from sea urchins, and the process of fertilization can be observed microscopically. To make this observation, follow these steps:
   a. Place a drop of sea urchin egg-cell suspension in the chamber of a depression slide and add a coverslip.
   b. Examine the egg cells using low-power magnification.
   c. Focus on a single egg cell with high-power magnification and sketch the cell in Part A of Laboratory Assessment 60.
   d. Remove the coverslip and add a drop of sea urchin sperm-cell suspension to the depression slide. Replace the coverslip and observe the sperm cells with high-power magnification as they cluster around the egg cells. This attraction is stimulated by gamete secretions.
   e. Observe the egg cells with low-power magnification once again and watch for the appearance of *fertilization membranes.* Such a membrane forms as soon as an egg cell is penetrated by a sperm cell; it looks like a clear halo surrounding the egg cell.
   f. Focus on a single fertilized egg cell and sketch it in Part A of the laboratory assessment.

FIGURE 60.1 Fertilization and early embryonic development in the uterine tube and implantation of blastocyst stage in endometrium of uterus.

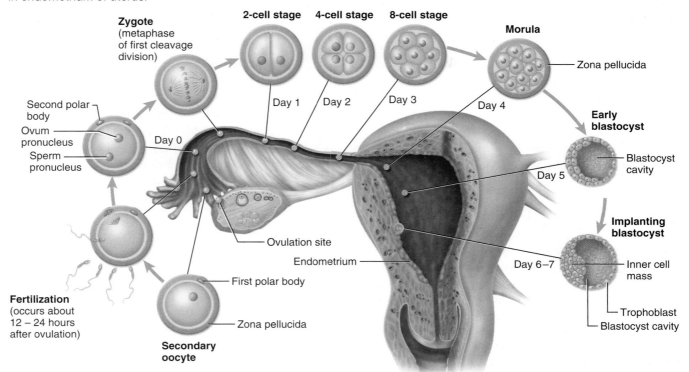

 ## Learning Extension Activity

Use a toothpick to draw a thin line of Vaseline around the chamber of the depression slide containing the fertilized sea urchin egg cells. Place a coverslip over the chamber, and gently press it into the Vaseline to seal the chamber and prevent the liquid inside from evaporating. Keep the slide in a cool place so that the temperature never exceeds 22°C (72°F). Using low-power magnification, examine the slide every 30 minutes, and look for the appearance of two-, four-, and eight-cell stages of developing sea urchin embryos.

Procedure B—Sea Urchin Early Development

1. Obtain a prepared microscope slide of developing sea urchin embryos. This slide contains embryos in various stages of cleavage. Search the slide using low-power magnification, and locate embryos in two-, four-, eight-cell, and morula stages. Observe that cleavage results in an increase of cell numbers; however, the cells get progressively smaller.
2. Prepare a sketch of each stage in Part B of the laboratory assessment.

Procedure C—Human Early Development

1. Study figures 60.2 and 60.3 and the list of structures and descriptions.
2. Observe the models of human embryos and identify the following features:

blastocyst—implantation stage of embryo into the endometrium
- blastocyst cavity (blastocoel)—hollow part
- inner cell mass (embryoblast)—develops into the actual embryo
- trophoblast—develops into the extraembryonic membranes and part of the placenta

primary germ layers—primitive tissues that develop into all organs
- ectoderm—outer layer; develops into epidermal structures and the nervous system
- endoderm—inner layer; gives rise to linings of digestive, respiratory, and urinary organs
- mesoderm—middle layer; develops into bones, muscles, dermis, and connective tissues

chorion—forms fetal portion of placenta

chorionic villi—branched projections of chorion

amnion—membrane that encloses embryo

amniotic fluid—protective fluid between amnion and embryo

yolk sac—produces early blood cells and gonadal tissues of embryo

allantois—forms part of umbilical cord

umbilical cord—suspends embryo in amniotic fluid; attached to the placenta
- umbilical arteries (2)—transport oxygen-poor blood of fetus to maternal portion of placenta
- umbilical vein (1)—transports oxygen-rich blood from maternal portion of placenta to embryo

placenta—site of exchange of nutrients, gases, and wastes between embryonic and maternal blood

3. Complete Parts C and D of the laboratory assessment.

FIGURE 60.2 Structures associated with this 12-week-old fetus: (a) fetus, membranes, and uterus; (b) fetal and maternal vasculature of placenta.

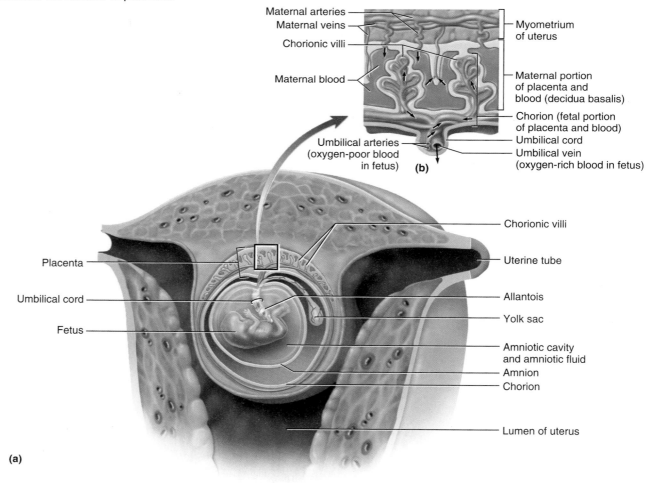

FIGURE 60.3 Human placenta (afterbirth): (a) fetal surface; (b) maternal (uterine) surface. At full term, the placenta is about 18 cm in diameter, 4 cm thickness, and weighs about 450 g (one pound).

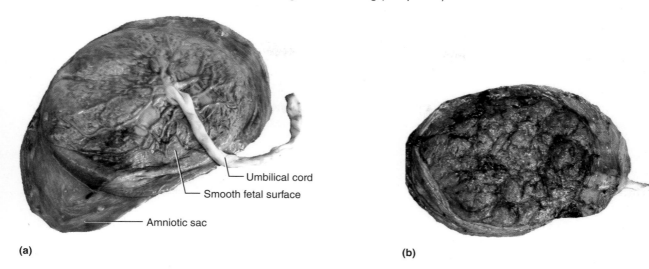

Umbilical cord

Smooth fetal surface

Amniotic sac

(a)

(b)

Demonstration Activity

Observe the preserved mammalian embryos that are on display. In addition to observing the developing external body structures, identify such features as the chorion, chorionic villi, amnion, yolk sac, umbilical cord, and placenta. What special features provide clues to the types of mammals these embryos represent?

_____

_____

_____

NOTES

Name _____

Date _____

Section _____

The ⚠ corresponds to the indicated outcome(s) found at the beginning of the laboratory exercise.

Fertilization and Early Development

Part A Assessments

Prepare sketches of the following: ⚠

Single sea urchin egg (_____×)	Fertilized sea urchin egg (_____×)

Part B Assessments

Prepare sketches of the following: ⚠

Two-cell sea urchin embryo (_____×)	Four-cell sea urchin embryo (_____×)
Eight-cell sea urchin embryo (_____×)	Morula sea urchin embryo (_____×)

Part C Assessments

Match the terms in column A with the descriptions in column B. Place the letter of your choice in the space provided. **2**

Column A	Column B
a. Blastocyst	_____ **1.** Gel layer around oocyte and early cleavage stages
b. Chorionic villi	_____ **2.** Hollow ball of cells
c. Endometrium	_____ **3.** Develops into extraembryonic membranes
d. Gastrula	_____ **4.** Solid ball of about sixteen cells
e. Morula	_____ **5.** Forms fetal portion of placenta
f. Polar bodies	_____ **6.** Forms maternal portion of placenta
g. Trophoblast	_____ **7.** Contains the ectoderm, endoderm, and mesoderm
h. Zona pellucida	_____ **8.** Small cells formed during meiosis that degenerate

Part D Assessments

Complete the following statements:

1. The umbilical cord contains three blood vessels, two of which are _____. **3**

2. _____ fluid protects the embryo from jarred movements and provides a watery environment for development. **3**

3. The cell resulting from fertilization is called a(an) _____. **3**

4. Tiny cells that form during meiotic divisions of oogenesis and then degenerate are called _____. **3**

5. _____ is the phase of development during which cellular divisions result in smaller and smaller cells. **3**

6. Fertilization and early embryonic development occur within the lumen of a _____. **3**

7. A human offspring is called a(an) _____ until the end of the eighth week of development. **3**

8. After the eighth week, a developing human is called a(an) _____ until the time of birth. **3**

9. Summarize your observations of any similarities noted between sea urchin and human early development stages. **3**

_____

_____

_____

_____

Genetics

Purpose of the Exercise

To observe some selected human traits, to use pennies and dice to demonstrate laws of probability, and to solve some genetic problems using a Punnett square.

Materials Needed

Pennies (or other coins)
Dice
PTC paper
Astigmatism chart
Ichikawa's or Ishihara's color plates for color blindness test

Many examples of human genetics in this laboratory exercise are basic, external features to observe. Some traits are based upon simple Mendelian genetics. As our knowledge of genetics continues to develop, traits such as tongue roller and free earlobe may no longer be considered examples of simple Mendelian genetics. We may need to continue to abandon some simple Mendelian models. New evidence may involve polygenic inheritance, effects of other genes, or environmental factors as more appropriate explanations. Therefore, the analysis of family genetics is not always feasible or meaningful.

Learning Outcomes

After completing this exercise, you should be able to

1. Examine and record twelve genotypes and phenotypes of selected human traits.

2. Demonstrate the laws of probability using tossed pennies and dice and interpret the results.

3. Predict genotypes and phenotypes of complete dominance, codominance, and sex-linked problems using Punnett squares.

Pre-Lab

Carefully read the introductory material and examine the entire lab. Be familiar with genetic terminology and genetic problems from lecture or the textbook. Answer the pre-lab questions.

Pre-Lab Questions: Select the correct answer for each of the following questions:

1. The term _____ is used for a person possessing two identical alleles.
   a. homozygous   **b.** heterozygous
   c. genotype   **d.** phenotype

2. The term _____ is used for the appearance of the person as a result of the way the genes are expressed.
   a. homozygous   **b.** heterozygous
   c. genotype   **d.** phenotype

3. The _____ is a genetic tool to simulate all possible genetic combinations of offspring.
   a. genome   **b.** Punnett square
   c. gamete formation   **d.** gene pool

4. The allele is termed _____ if one allele determines the phenotype.
   a. homozygous   **b.** heterozygous
   c. dominant   **d.** recessive

5. Which of the following represents a dominant human trait?
   a. straight hairline   **b.** attached earlobe
   c. dimples   **d.** blood type O

6. An attached earlobe is considered a recessive human trait.
   True _____   False _____

7. A diploid chromosome number of 46 and a haploid number of 23 are normal for humans.
   True _____   False _____

8. The possession of freckles is considered a recessive human trait.
   True _____   False _____

Genetics is the study of the inheritance of characteristics. The *genes* that transmit this information are coded in segments of DNA in chromosomes. Homologous chromosomes possess the same gene at the same *locus*. These genes may exist in variant forms, called *alleles*. If a person possesses two identical alleles, the person is said to be *homozygous* for that particular trait. If a person possesses two different alleles, the person is said to be *heterozygous* for that particular trait. The particular combination of these gene variants (alleles) represents the person's *genotype;* the appearance or physical manifestation of the individual that develops as a result of the way the genes are expressed, represents the person's *phenotype.*

If one allele determines the phenotype by masking the expression of the other allele in a heterozygous individual, the allele is termed *dominant*. The allele whose expression is masked is termed *recessive*. A heterozygous person is known as a *carrier* when a recessive gene for a genetic disease "skips" generations and the condition reappears. If the heterozygous condition determines an intermediate phenotype, the inheritance represents *incomplete dominance.*

If more than two allelic forms of a gene exist in the gene pool, it represents an example of *multiple alleles*. Three alleles exist related to the inheritance of the human ABO blood types. However, different alleles are *codominant* if both are expressed in the heterozygous condition.

Some characteristics inherited on the sex chromosomes result in phenotype frequencies that might be more prevalent in males or females. Such characteristics are called *sex-linked* (X-linked or Y-linked) characteristics.

As a result of meiosis during the formation of ova (eggs) and sperm, a mother and father each transmit an equal number of chromosomes (the haploid number 23) to form the zygote (diploid number 46). An offspring will receive one allele from each parent. These gametes combine randomly in the formation of each offspring. Hence, the *laws of probability* can be used to predict possible genotypes and phenotypes of offspring. A genetic tool called a *Punnett square* simulates all possible combinations (probabilities) that can occur in offspring genotypes and resulting phenotypes.

Procedure A—Human Genotypes and Phenotypes

A complete set of genetic instructions in one human cell constitutes one's *genome*. The human genome contains about 2.9 billion base pairs representing approximately 20,500 protein-encoding genes. These instructions represent our genotypes and are expressed as phenotypes sometimes clearly observable on our bodies. Some of these traits are listed in table 61.1 and are discernible in figure 61.1.

A dominant trait might be homozygous or heterozygous; only one capital letter is used along with a blank for the possible second dominant or recessive allele. For a recessive trait, two lowercase letters represent the homozygous recessive genotype for that characteristic. Dominant does not always correlate with the predominance of the allele in the gene pool; dominant means one allele will determine the appearance of the phenotype.

1. **Tongue roller/nonroller:** The dominant allele (R) determines the person's ability to roll the tongue into a U-shaped trough. The homozygous recessive condition (*rr*) prevents this tongue rolling (fig. 61.1). Record your results in the table in Part A of Laboratory Assessment 61.
2. **Freckles/no freckles:** The dominant allele (F) determines the appearance of freckles. The homozygous recessive condition (*ff*) does not produce freckles (fig. 61.1). Record your results in the table in Part A of the laboratory assessment.

TABLE 61.1   Examples of Some Common Human Phenotypes

Dominant Traits and Genotypes	Recessive Traits and Genotypes
Tongue roller ($R\_\_$)	Nonroller (*rr*)
Freckles ($F\_\_$)	No freckles (*ff*)
Widow's peak ($W\_\_$)	Straight hairline (*ww*)
Dimples ($D\_\_$)	No dimples (*dd*)
Free earlobe ($E\_\_$)	Attached earlobe (*ee*)
Normal skin coloration ($M\_\_$)	Albinism (*mm*)
Astigmatism ($A\_\_$)	Normal vision (*aa*)
Natural curly hair ($C\_\_$)	Natural straight hair (*cc*)
PTC taster ($T\_\_$)	Nontaster (*tt*)
Blood type A ($I^A\_\_$), B ($I^B\_\_$), or AB ($I^A I^B$)	Blood type O (*ii*)
Normal color vision ($X^C X^C$), ($X^C X^c$), or ($X^C Y$)	Red-green color blindness ($X^c X^c$) or ($X^c Y$)

FIGURE 61.1 Representative genetic traits comparing dominant and recessive phenotypes: (a) tongue roller; (b) nonroller; (c) freckles; (d) no freckles; (e) widow's peak; (f) straight hairline; (g) dimples; (h) no dimples; (i) free earlobe; (j) attached earlobe.

Dominant Traits

Recessive Traits

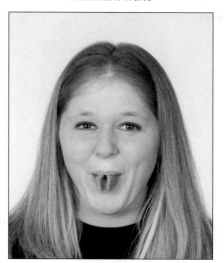

(a) Tongue roller

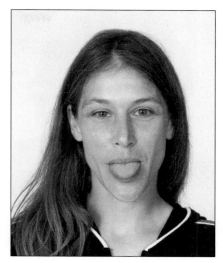

(b) Nonroller

(c) Freckles

(d) No freckles

(e) Widow's peak

(f) Straight hairline

FIGURE 61.1 *Continued.*

Dominant Traits   **Recessive Traits**

(g) Dimples

(h) No dimples

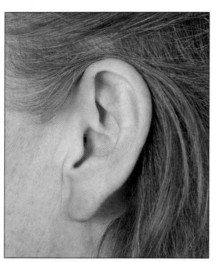

(i) Free earlobe

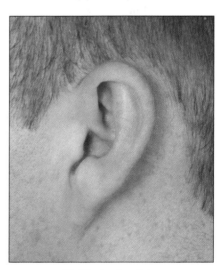

(j) Attached earlobe

3. **Widow's peak/straight hairline:** The dominant allele (*W*) determines the appearance of a hairline above the forehead that has a distinct downward point in the center called a widow's peak. The homozygous recessive condition (*ww*) produces a straight hairline (fig. 61.1). A receding hairline would prevent this phenotype determination. Record your results in the table in Part A of the laboratory assessment.

4. **Dimples/no dimples:** The dominant allele (*D*) determines the appearance of a distinct dimple in one or both cheeks upon smiling. The homozygous recessive condition (*dd*) results in the absence of dimples (fig. 61.1). Record your results in the table in Part A of the laboratory assessment.

5. **Free earlobe/attached earlobe:** The dominant allele (*E*) codes for the appearance of an inferior earlobe that hangs freely below the attachment to the head. The homozygous recessive condition (*ee*) determines the earlobe attaching directly to the head at its inferior border (fig. 61.1). Record your results in the table in Part A of the laboratory assessment.

6. **Normal skin coloration/albinism:** The dominant allele (*M*) determines the production of some melanin, producing normal skin coloration. The homozygous recessive condition (*mm*) determines albinism due to the inability to produce or use the enzyme tyrosinase in pigment cells. An albino does not produce melanin in the skin, hair, or the middle tunic (choroid, ciliary body, and

iris) of the eye. The absence of melanin in the middle tunic allows the pupil to appear slightly red to nearly black. Remember that the pupil is an opening in the iris filled with transparent aqueous humor. An albino human has pale white skin, flax-white hair, and a pale blue iris. Record your results in the table in Part A of the laboratory report.

7. **Astigmatism/normal vision:** The dominant allele (A) results in an abnormal curvature to the cornea or the lens. As a consequence, some portions of the image projected on the retina are sharply focused, and other portions are blurred. The homozygous recessive condition (aa) generates normal cornea and lens shapes and normal vision. Use the astigmatism chart and directions to assess this possible defect described in Laboratory Exercise 36. Other eye defects, such as nearsightedness (myopia) and farsightedness (hyperopia), are different genetic traits due to genes at other locations. Record your results in the table in Part A of the laboratory assessment.

8. **Curly hair/straight hair:** The dominant allele (C) determines the appearance of curly hair. Curly hair is somewhat flattened in cross section, as the hair follicle of a similar shape served as a mold for the root of the hair during its formation. The homozygous recessive condition (cc) produces straight hair. Straight hair is nearly round in cross section from being molded into this shape in the hair follicle. In some populations (Caucasians) the heterozygous condition (Cc) expresses the intermediate wavy hair phenotype (incomplete dominance). This trait determination assumes no permanents or hair straightening procedures have been performed. Such hair alterations do not change the hair follicle shape, and future hair growth results in original genetic hair conditions. Record your unaltered hair appearance in the table in Part A of the laboratory assessment.

9. **PTC taster/nontaster:** The dominant allele (T) determines the ability to experience a bitter sensation when PTC paper is placed on the tongue. About 70% of people possess this dominant gene. The homozygous recessive condition (tt) makes a person unable to notice the substance. Place a piece of PTC (phenylthiocarbamide) paper on the upper tongue surface and chew it slightly to see if you notice a bitter sensation from this harmless chemical. The nontaster of the PTC paper does not detect any taste at all from this substance. Record your results in the table in Part A of the laboratory assessment.

10. **Blood type A, B, or AB/blood type O:** Blood type inheritance is an example of multiple-allele inheritance with codominant alleles. There are three alleles (I^A, I^B, and i) in the human population affecting RBC membrane structure. These alleles are located on a single pair of homologous chromosomes, so a person could possess either two of the three alleles or two of the same allele. All of the possible combinations of these alleles of genotypes and the resulting phenotypes are depicted in table 61.2. The expression of the blood type

TABLE 61.2 Genotypes and Phenotypes (Blood Types)

Genotypes	Phenotypes (Blood Types)
$I^A I^A$ or $I^A i$	A
$I^B I^B$ or $I^B i$	B
$I^A I^B$ (codominant)	AB
ii	O

AB is a result of both codominant alleles located in the same individual. Possibly you have already determined your blood type in Laboratory Exercise 43 or have it recorded on a blood donor card. (If simulated blood-typing kits were used for Laboratory Exercise 43, those results would not be valid for your genetic factors.) Record your results in the table in Part A of the laboratory assessment.

11. **Sex determination:** A person with sex chromosomes XX displays a female phenotype. A person with sex chromosomes XY displays a male phenotype. Record your results in the table in Part A of the laboratory assessment.

12. **Normal color vision/red-green color blindness:** This condition is a sex-linked (X-linked) characteristic. The alleles for color vision are also on the X chromosome, but absent on the Y chromosome. As a result, a female might possess both alleles (C and c), one on each of the X chromosomes. The dominant allele (C) determines normal color vision; the homozygous recessive condition (cc) results in red-green color blindness. However, a male would possess only one of the two alleles for color vision because there is only a single X chromosome in a male. Hence a male with even a single recessive gene for color blindness possesses the defect. Note all the possible genotypes and phenotypes for this condition (table 61.1). Review the color vision test in Laboratory Exercise 36 using the color plates in figure 36.5 and Ichikawa's or Ishihara's book. Record your results in the table in Part A of the laboratory assessment.

13. Complete Part A of the laboratory assessment.

Procedure B—Laws of Probability

The laws of probability provide a mathematical way to determine the likelihood of events occurring by chance. This prediction is often expressed as a ratio of the number of results from experimental events to the number of results considered possible. For example, when tossing a coin there is an equal chance of the results displaying heads or tails. Hence the probability is one-half of obtaining either a heads or a tails (there are two possibilities for each toss). When all of the probabilities of all possible outcomes are considered for the result, they will always add up to a 1. To predict the probability of two or more events occurring in succession, multiply the probabilities of each individual event. For example, the probability of tossing a die and displaying a 4 two times in a row is 1/6 × 1/6 = 1/36 (there are six possibilities for each toss). Each toss in a sequence is an *independent event* (chance has no memory). The same laws apply when parents have multiple children (each fertilization is an independent event). Perform the following experiments to demonstrate the laws of probability:

1. Use a single penny (or other coin) and toss it 20 times. Predict the number of heads and tails that would occur from the 20 tosses. Record your prediction and the actual results observed in Part B of the laboratory assessment.

2. Use a single die (*pl.* dice) and toss it 24 times. Predict the number of times a number below 3 (numbers 1 and 2) would occur from the 24 tosses. Record your prediction and the actual results in Part B of the laboratory assessment.

3. Use two pennies and toss them simultaneously 32 times. Predict the number of times two heads, a heads and a tails, and two tails occur. Record your prediction and the actual results in Part B of the laboratory assessment.

4. Use a pair of dice and toss them simultaneously 32 times. Predict the number of times for both dice coming up with odd numbers, one die an odd and the other an even number, and both dice coming up with even numbers. Record your prediction and the actual results in Part B of the laboratory assessment.

5. Obtain class totals for all of the coins and dice tossed by adding your individual results to a class tally location as on the blackboard.

6. A Punnett square can be used for a visual representation to demonstrate the probable results for two pennies tossed simultaneously. For the purpose of a genetic comparison, an *h* (heads) will represent one "allele" on the coin; a *t* (tails) will represent a different "allele" on the coin.

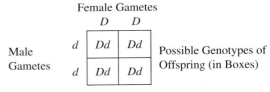

7. Complete Part B of the laboratory assessment.

Procedure C—Genetic Problems

1. A Punnett square can be constructed to demonstrate a visual display of the predicted offspring from parents with known genotypes. Recall that in complete dominance, a dominant allele is expressed in the phenotype, as it can mask the other recessive allele on the homologous chromosome pair. During meiosis, the homologous chromosomes with their alleles separate (Mendel's Law of Segregation) into different gametes. An example of such a cross might be a homozygous dominant mother for dimples (*DD*) has offspring with a father homozygous reces-

sive (*dd*) for the same trait. The results of such a cross, according to the laws of probability, would be represented by the following Punnett square:

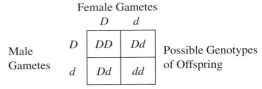

Results: Genotypes: 100% *Dd* (all heterozygous)

Phenotypes: 100% dimples

In another example, assume that both parents are heterozygous (*Dd*) for dimples. The results of such a cross, according to the laws of probability, would be represented by the following Punnett square:

Female Gametes

		D	d
Male Gametes	D	DD	Dd
	d	Dd	dd

Possible Genotypes of Offspring

Results: Genotypes: 25% *DD* (homozygous dominant); 50% *Dd* (heterozygous); 25% *dd* (homozygous recessive) (1:2:1 genotypic ratio)

Phenotypes: 75% dimples; 25% no dimples (3:1 phenotypic ratio)

2. Work the genetic problems 1 and 2 in Part C of the laboratory assessment.

3. The ABO blood type inheritance represents an example of multiple alleles and codominance. Review table 61.2 for the genotypes and phenotypes for the expression of this trait. A similar technique completed previously can be used to predict the offspring of parents of known genotypes. In this example, assume the genotype of the mother is $I^A I^B$, and the father is *ii*. The results of such a cross would be represented by the following Punnett square:

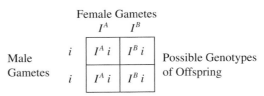

Results: Genotypes: 50% $I^A i$ (heterozygous for A); 50% $I^B i$ (heterozygous for B) (1:1 genotypic ratio)

Phenotypes: 50% blood type A; 50% blood type B (1:1 phenotypic ratio)

Note: In this particular cross, all of the children would have blood types unlike either parent.

4. Work the genetic problems 3 and 4 in Part C of the laboratory assessment.

5. Review the inheritance of red-green color blindness, an X-linked characteristic, in table 61.1. A similar technique used to identify complete dominance can be performed to predict the offspring of parents of known genotypes. In this example, assume the genotype of the mother is heterozygous $X^C X^c$ (normal color vision, but a carrier for the color blindness defect), and the father is $X^C Y$ (normal color vision; no allele on the Y chromosome). The results of such a cross would be represented by the following Punnett square:

		Female Gametes	
		X^C	X^c
Male Gametes	X^C	$X^C X^C$	$X^C X^c$
	Y	$X^C Y$	$X^c Y$

Possible Genotypes of Offspring

Results: Genotypes: 25% $X^C X^C$; 25% $X^C X^c$; 25% $X^C Y$; 25% $X^c Y$ (1:1:1:1 genotypic ratio)

Phenotypes for sex determination: 50% females; 50% males (1:1 phenotypic ratio)

Phenotypes for color vision: Females 100% normal color vision (however, 50% are heterozygous carriers for color blindness). Males 50% normal; 50% with red-green color blindness (1:1 phenotypic ratio)

Phenotypes for sex determination and color vision combined: 50% normal females; 25% normal males; 25% males with color blindness (2:1:1 phenotypic ratio)

Note: In X-linked inheritance, males with color blindness received the recessive gene from their mothers.

6. Complete Part C of the laboratory assessment.

Laboratory Assessment

61

Name _____

Date _____

Section _____

The △ corresponds to the indicated outcome(s) found at the beginning of the laboratory exercise.

Genetics

Part A   Assessments

1. Enter your test results for genotypes and phenotypes in the table. Circle your particular phenotype and genotype for each of the twelve traits. △

Trait	Dominant Phenotype	Genotype	Recessive Phenotype	Genotype
Tongue movement	Roller	$R\_\_$	Nonroller	rr
Freckles	Freckles	$F\_\_$	No freckles	ff
Hairline	Widow's peak	$W\_\_$	Straight	ww
Dimples	Dimples	$D\_\_$	No dimples	dd
Earlobe	Free	$E\_\_$	Attached	ee
Skin coloration	Normal (some melanin)	$M\_\_$	Albinism	mm
Vision	Astigmatism	$A\_\_$	Normal	aa
Hair shape*	Curly	$C\_\_$	Straight	cc
Taste	PTC taster	$T\_\_$	Nontaster for PTC	tt
Blood type	A, B, or AB	$I^A\_\_$; $I^B\_\_$; or $I^A I^B$	O	ii
Sex		XX or XY		
Color vision	Normal	$X^C X\_\_$ or $X^C Y$	Red-green color blindness	$X^c X^c$ or $X^c Y$

*In some populations (Caucasians) the heterozygous condition (Cc) results in the appearance of wavy hair, which actually represents an example of incomplete dominance for this trait.

2. Choose at least three dominant phenotypes that you circled, and analyze the genotypes for those traits. If it is feasible to observe your biological parents and siblings for any of these traits, are you able to determine if any of your dominant genotypes are homozygous dominant or heterozygous? _____ If so, which ones? _____

   Explain the rationale for your response. _____

   _____

   _____

   _____

   _____

   _____

Part B   Assessments

1. Single penny tossed 20 times and counting heads and tails: ⟨2⟩

   Probability (prediction): ___/20 heads ___/20 tails

   (Note: Traditionally, probabilities are converted to the lowest fractional representation.)

   Actual results: _____ heads _____ tails

   Class totals: _____ heads _____ tails

2. Single die tossed 24 times and counting the number of times a number below 3 occurs: ⟨2⟩

   Probability: ____/24 number below 3 (numbers 1 and 2)

   Actual results: ____ number below 3

   Class totals: ____ number below 3 ____ total tosses by class members

3. Two pennies tossed simultaneously 32 times and counting the number of two heads, a heads and a tails, and two tails: ⟨2⟩

   Probability: ____/32 of two heads ____/32 of a heads and a tails ____/32 of two tails

   Actual results: ____two heads ____heads and tails ____two tails

   Class totals: ____two heads ____heads and tails ____two tails

4. Two dice tossed simultaneously 32 times and counting the number of two odd numbers, an odd and an even number, and two even numbers: ⟨2⟩

   Probability: ____/32 of two odd numbers ____/32 of an odd and an even number ____/32 of two even numbers

   Actual results: ____ two odd numbers ____ an odd and an even number ____ two even numbers

   Class totals: ____ two odd numbers ____ an odd and an even number ____ two even numbers

5. Use the example of the two dice tossed 32 times and construct a Punnett square to represent the possible combinations that could be used to determine the probability (prediction) of odd and even numbers for the resulting tosses. Your construction should be similar to the Punnett square for the two coins tossed that is depicted in Procedure B of the laboratory exercise. ⟨2⟩

6. Complete the following:

   a. Are the class totals closer to the predicted probabilities than your results? _____ Explain your response. ⟨2⟩

   _____

   _____

   b. Does the first toss of the penny or the first toss of the die have any influence on the next toss? _____

   Explain your response. ⟨2⟩

   _____

   c. Assume a family has two boys or two girls. They wish to have one more child, but hope for the child to be of the opposite sex from the two they already have. What is the probability that the third child will be of the opposite sex?

   _____ Explain your response. ⟨2⟩

   _____

d. What is the probability (prediction) that a couple without children will eventually have four children, all girls?

_____ Explain your response. **2**

_____

_____

Part C Assessments

For each of the genetic problems, (a) determine the parents' genotypes, (b) determine the possible gametes for each parent, (c) construct a Punnett square, and (d) record the resulting genotypes and phenotypes as ratios from the cross. Problems 1 and 2 involve examples of complete dominance; problems 3 and 4 are examples of codominance; problem 5 is an example of sex-linked (X-linked) inheritance.

   1. Determine the results from a cross of a mother who is heterozygous (*Rr*) for tongue rolling with a father who is homozygous recessive (*rr*). **3**

   2. Determine the results from a cross of a mother and a father who are both heterozygous for freckles. **3**

   3. Determine the results from a mother who is heterozygous for blood type B and a father who is homozygous dominant for blood type A. **3**

4. Determine the results from a mother who is heterozygous for blood type A and a father who is heterozygous for blood type B.

5. Color blindness is an example of X-linked inheritance. Hemophilia is another example of X-linked inheritance, also from a recessive allele (*h*). The dominant allele (*H*) determines whether the person possesses normal blood clotting. A person with hemophilia has a permanent tendency for hemorrhaging due to a deficiency of one of the clotting factors (VIII—antihemophilic factor). Determine the offspring from a cross of a mother who is a carrier (heterozygous) for the disease and a father with normal blood coagulation.

Critical Thinking Assessment

Assume that the genes for hairline and earlobes are on different pairs of homologous chromosomes. Determine the genotypes and phenotypes of the offspring from a cross if both parents are heterozygous for both traits. (1) First determine the genotypes for each parent. (2) Determine the gametes, but remember each gamete has one allele for each trait (gametes are haploid). (3) Construct a Punnett square with 16 boxes that has four different gametes from each parent along the top and the left edges. (This is an application to demonstrate Mendel's Law of Independent Assortment.) (4) List the results of genotypes and phenotypes as ratios.

Preparation of Solutions

Amylase solution, 0.5%

Place 0.5 g of bacterial amylase in a graduated cylinder or volumetric flask. Add distilled water to the 100 mL level. Stir until dissolved. The amylase should be free of sugar for best results; a low-maltose solution of amylase yields good results. (Store amylase powder in a freezer until mixing this solution.)

Benedict's solution

Prepared solution is available from various suppliers.

Caffeine, 0.2%

Place 0.2 g of caffeine in a graduated cylinder or volumetric flask. Add distilled water to the 100 mL level. Stir until dissolved.

Calcium chloride, 2.0%

Place 2.0 g of calcium chloride in a graduated cylinder or volumetric flask. Add distilled water to the 100 mL level. Stir until dissolved.

Calcium hydroxide solution (limewater)

Add an excess of calcium hydroxide to 1 L of distilled water. Stopper the bottle and shake thoroughly. Allow the solution to stand for 24 hours. Pour the supernatant fluid through a filter. Store the clear filtrate in a stoppered container.

Epsom salt solution, 0.1%

Place 0.5 g of Epsom salt in a graduated cylinder or volumetric flask. Add distilled water to the 500 mL level. Stir until dissolved.

Glucose solutions

1. *1.0% solution.* Place 1 g of glucose in a graduated cylinder or volumetric flask. Add distilled water to the 100 mL level. Stir until dissolved.
2. *10% solution.* Place 10 g of glucose in a graduated cylinder or volumetric flask. Add distilled water to the 100 mL level. Stir until dissolved.

Iodine-potassium-iodide (IKI solution)

Add 20 g of potassium iodide to 1 L of distilled water, and stir until dissolved. Then add 4.0 g of iodine, and stir again until dissolved. Solution should be stored in a dark stoppered bottle.

Methylene blue

Dissolve 0.3 g of methylene blue powder in 30 mL of 95% ethyl alcohol. In a separate container, dissolve 0.01 g of potassium hydroxide in 100 mL of distilled water. Mix the two solutions. (Prepared solution is available from various suppliers.)

Monosodium glutamate (MSG) solution, 1%

Place 1 g of monosodium glutamate in a graduated cylinder or volumetric flask. Add distilled water to the 100 mL level. Stir until dissolved.

Physiological saline solution

Place 0.9 g of sodium chloride in a graduated cylinder or volumetric flask. Add distilled water to the 100 mL level. Stir until dissolved.

Potassium chloride, 5%

Place 5.0 g of potassium chloride in a graduated cylinder or volumetric flask. Add distilled water to the 100 mL level. Stir until dissolved.

Quinine sulfate, 0.5%

Place 0.5 g of quinine sulfate in a graduated cylinder or volumetric flask. Add distilled water to the 100 mL level. Stir until dissolved.

Ringer's solution (frog)

Dissolve the following salts in 1 L of distilled water:

6.50 g sodium chloride
0.20 g sodium bicarbonate
0.14 g potassium chloride
0.12 g calcium chloride

Sodium chloride solutions

1. *0.9% solution.* Place 0.9 g of sodium chloride in a graduated cylinder or volumetric flask. Add distilled water to the 100 mL level. Stir until dissolved.
2. *1.0% solution.* Place 1.0 g of sodium chloride in a graduated cylinder or volumetric flask. Add distilled water to the 100 mL level. Stir until dissolved.
3. *3.0% solution.* Place 3.0 g of sodium chloride in a graduated cylinder or volumetric flask. Add distilled water to the 100 mL level. Stir until dissolved.
4. *5.0% solution.* Place 5.0 g of sodium chloride in a graduated cylinder or volumetric flask. Add distilled water to the 100 mL level. Stir until dissolved.

Starch solutions

1. *0.5% solution.* Add 5 g of cornstarch to 1 L of distilled water. Heat until the mixture boils. Cool the liquid, and pour it through a filter. Store the filtrate in a refrigerator.
2. *1.0% solution.* Add 10 g of cornstarch to 1 L of distilled water. Heat until the mixture boils. Cool the liquid, and pour it through a filter. Store the filtrate in a refrigerator.
3. *10% solution.* Add 100 g of cornstarch to 1 L of distilled water. Heat until the mixture boils. Cool the liquid, and pour it through a filter. Store the filtrate in a refrigerator.

Sucrose, 5% solution

Place 5.0 g of sucrose in a graduated cylinder or volumetric flask. Add distilled water to the 100 mL level. Stir until dissolved.

Wright's stain

Prepared solution is available from various suppliers.

Assessment of Laboratory Assessments

Many assessment models can be used for laboratory assessments. A rubric, which can be used for performance assessments, contains a description of the elements (requirements or criteria) of success to various degrees. The term *rubric* originated from *rubrica terra,* which is Latin for the application of red earth to indicate anything of importance. A rubric used for assessment contains elements for judging student performance, with points awarded for varying degrees of success in meeting the learning outcomes. The content and the quality level necessary to attain certain points are indicated in the rubric. It is effective if the assessment tool is shared with the students before the laboratory exercise is performed.

Following are two sample rubrics that could easily be modified to meet the needs of a specific course. Some of the elements for these sample rubrics may not be necessary for every laboratory exercise. The generalized rubric needs to contain the possible assessment points that correspond to learning outcomes for a specific course. The point value for each element may vary. The specific rubric example contains performance levels for laboratory assessments. The elements and the point values could easily be altered to meet the value placed on laboratory assessments for a specific course.

Assessment: Generalized Laboratory Assessment Rubric

Element	Assessment Points Possible	Assessment Points Earned
1. Figures are completely and accurately labeled.		
2. Sketches are accurate, contain proper labels, and are of sufficient detail.		
3. Colored pencils were used extensively to differentiate structures on illustrations.		
4. Matching and fill-in-the-blank answers are completed and accurate.		
5. Short-answer/discussion questions contain complete, thorough, and accurate answers. Some elaboration is evident for some answers.		
6. Data collected are complete, accurately displayed, and contain a valid explanation.		

TOTAL POINTS: POSSIBLE _____ EARNED _____

Assessment: Specific Laboratory Assessment Rubric

Element	Excellent Performance (4 points)	Proficient Performance (3 points)	Marginal Performance (2 points)	Novice Performance (1 point)	Points Earned
Figure labels	Labels completed with ≥ 90% accuracy.	Labels completed with 80%–89% accuracy.	Labels completed with 70%–79% accuracy.	Labels < 70% accurate.	
Sketches	Accurate use of scale, details illustrated, and all structures labeled accurately.	Minor errors in sketches. Missing or inaccurate labels on one or more structures.	Sketch is not realistic. Missing or inaccurate labels on two or more structures.	Several missing or inaccurate labels.	
Matching and fill-in-the-blanks	All completed and accurate.	One to two errors or omissions.	Three to four errors or omissions.	Five or more errors or omissions.	
Short-answer and discussion questions	Answers are complete, valid, and contain some elaboration. No misinterpretations are noted.	Answers are generally complete and valid. Only minor inaccuracies were noted. Minimal elaboration exists.	Marginal answers to the questions and contains inaccurate information.	Many answers are incorrect or fail to address the topic. There may be misinterpretations.	
Data collection and analysis	Data are complete and displayed with a valid interpretation.	Only minor data missing or a slight misinterpretation exists.	Some omissions. Not displayed or interpreted accurately.	Data are incomplete or show serious misinterpretations.	

TOTAL POINTS EARNED _____

Correlation of Laboratory Exercises and Ph.I.L.S. 4.0 Lab Simulations

Laboratory Exercise	Ph.I.L.S. 4.0
Fundamentals of Human Anatomy and Physiology	
1 Scientific Method and Measurements 2 Body Organization, Membranes, and Terminology 3 Chemistry of Life 4 Care and Use of the Microscope	
Cells	

5 Cell Structure and Function
   Ph.I.L.S. 4.0 Lesson 2 Metabolism: Size and Basal Metabolic Rate
6 Movements Through Cell Membranes
   Ph.I.L.S. 4.0 Lesson 1 Osmosis and Diffusion: Varying Extracellular Concentration
7 Cell Cycle
	4 Cyanide and Electron Transport
	2 Size and Basal Metabolic Rate
	1 Varying Extracellular Concentration

Tissues	
8 Epithelial Tissues 9 Connective Tissues 10 Muscle and Nervous Tissues	
Integumentary System	
11 Integumentary System	
Skeletal System	
12 Bone Structure and Classification 13 Organization of the Skeleton 14 Skull 15 Vertebral Column and Thoracic Cage 16 Pectoral Girdle and Upper Limb 17 Pelvic Girdle and Lower Limb 18 Fetal Skeleton 19 Joint Structure and Movements	
Muscular System	
20 Skeletal Muscle Structure and Function Ph.I.L.S. 4.0 Lesson 5 Skeletal Muscle Function: Stimulus-Dependent Force Generation 21 Electromyography: BIOPAC Exercise 22 Muscles of the Head and Neck 23 Muscles of the Chest, Shoulder, and Upper Limb 24 Muscles of the Vertebral Column, Abdominal Wall, and Pelvic Floor 25 Muscles of the Hip and Lower Limb	5 Stimulus-Dependent Force Generation 6 Weight and Contraction 7 The Length-Tension Relationship 8 Principles of Summation and Tetanus 9 EMG and Twitch Amplitude

Continued

Laboratory Exercise	Ph.I.L.S. 4.0
Surface Anatomy	
26 Surface Anatomy	
Nervous System	
27 Nervous Tissue and Nerves	10 Resting Potential and External [K+]
	11 Resting Potential and External [Na+]
	12 The Compound Action Potential
	13 Conduction Velocity and Temperature
	14 Refractory Periods
	15 Measuring Ion Currents
	16 Facilitation and Depression
	17 Temporal Summation of EPSPs
	18 Spatial Summation of EPSPs
28 Spinal Cord, Spinal Nerves, and Meninges	
29 Reflex Arc and Reflexes	
30 Brain and Cranial Nerves	
31 Electroencephalography: BIOPAC Exercise	
32 Dissection of the Sheep Brain	
General and Special Senses	
33 General Senses	
34 Smell and Taste	
35 Eye Structure	
36 Visual Tests and Demonstrations	
37 Ear and Hearing	
38 Ear and Equilibrium	
Endocrine System	
39 Endocrine Structure and Function	19 Thyroid Gland and Metabolic Rate
Ph.I.L.S. 4.0 Lesson 19 Endocrine Function: Thyroid Gland and Metabolic Rate	
40 Diabetic Physiology	20 Insulin and Glucose Tolerance
Cardiovascular System	
41 Blood Cells	
42 Blood Testing	34 pH & Hb-Oxygen Binding
Ph.I.L.S. 4.0 Lesson 34 Blood: pH & Hb-Oxygen Binding	35 DPG and Hb-Oxygen Binding
43 Blood Typing	36 Blood Typing
44 Heart Structure	21 Thermal and Chemical Effects
	23 Starling's Law of the Heart
45 Cardiac Cycle	22 Refractory Period of the Heart
Ph.I.L.S. 4.0 Lesson 26 ECG and Heart Function: The Meaning of Heart Sounds	24 Heart Block
	26 The Meaning of Heart Sounds
46 Electrocardiography: BIOPAC Exercise	25 ECG and Exercise
	27 ECG and Finger Pulse
	28 Electrical Axis of the Heart
	29 ECG and Heart Block
	30 Abnormal ECGs
47 Blood Vessel Structure, Arteries, and Veins	
48 Pulse Rate and Blood Pressure	31 Cooling and Peripheral Blood Flow
Ph.I.L.S. 4.0 Lesson 40 Respiration: Deep Breathing and Cardiac Function	32 Blood Pressure and Gravity
	33 Blood Pressure and Body Position
	40 Deep Breathing and Cardiac Function
Lymphatic System	
49 Lymphatic System	

Continued

Laboratory Exercise	Ph.I.L.S. 4.0
Respiratory System	
50 Respiratory Organs	
51 Breathing and Respiratory Volumes	37 Altering Body Position
Ph.I.L.S. 4.0 Lesson 38 Respiration: Altering Airway Volume	38 Altering Airway Volume
52 Spirometry: BIOPAC Exercise	
53 Control of Breathing	3 Respiratory Quotient
	39 Exercise-Induced Changes
	40 Deep Breathing and Cardiac Function
Digestive System	
54 Digestive Organs	41 Glucose Transport
55 Action of a Digestive Enzyme	
Urinary System	
56 Urinary Organs	42 Anti-Diuretic Hormone
57 Urinalysis	
Reproductive Systems and Development	
58 Male Reproductive System	
59 Female Reproductive System	
60 Fertilization and Early Development	
61 Genetics	
Fetal Pig Dissection Exercises (similar for cat and rat versions)	
62 Fetal Pig Dissection: Musculature	
63 Fetal Pig Dissection: Cardiovascular System	
64 Fetal Pig Dissection: Respiratory System	
65 Fetal Pig Dissection: Digestive System	
66 Fetal Pig Dissection: Urinary System	
67 Fetal Pig Dissection: Reproductive Systems	
Supplemental Laboratory Exercises*	
S-1 Skeletal Muscle Contraction	6 Weight and Contraction
Ph.I.L.S. 4.0 Lesson 8 Skeletal Muscle Function: Principles of Summation and Tetanus	8 Principles of Summation and Tetanus
S-2 Nerve Impulse Stimulation	12 The Compound Action Potential
Ph.I.L.S. 4.0 Lesson 12 Action Potentials: The Compound Action Potential	
S-3 Factors Affecting the Cardiac Cycle	
Ph.I.L.S. 4.0 Lesson 21 Frog Heart Function: Thermal and Chemical Effects	21 Thermal and Chemical Effects

Credits

Photography Ltd.; 38.5: © John D. Cunningham/Visuals Unlimited.

Laboratory Exercise 39

Figure 39.2: © Biophoto Associates/Photo Researchers, Inc.; 39.3: © John D. Cunningham/Visuals Unlimited; 39.4: © Science VU/Visuals Unlimited; 39.6: © Robert Calentine/Visuals Unlimited; 39.8: © Biophoto Associates/Photo Researchers, Inc.; 39.10: © Ed Reschke; 39.12: © Dr. Kent M. Van De Graaff.

Laboratory Exercise 40

Figure 40.1–40.2: © The McGraw-Hill Companies, Inc./Al Telser, photographer; 40.3: © Dr. Frederick Skarva/Visuals Unlimited.

Laboratory Exercise 41

Figure 41.2–41.5: © The McGraw-Hill Companies, Inc./Al Telser, photographer.

Laboratory Exercise 42

Figure 42.2a–d: © The McGraw-Hill Companies, Inc./James Shaffer, photographer; 42.3–42.4: © The McGraw-Hill Companies, Inc./Womack Photography Ltd.

Laboratory Exercise 43

Figure 43.2: © Jean-Claude Revy/ISM/Phototake; 43.4: © Image Source/Getty Images RF.

Laboratory Exercise 44

Figure 44.7–44.8: © The McGraw-Hill Companies, Inc./J. Womack Photography; 44.9: © The McGraw-Hill Companies, Inc./APR; 44.10: © The McGraw-Hill Companies, Inc./J. Womack Photography.

Laboratory Exercise 47

Figure 47.2: © Biophoto Associates/Photo Researchers, Inc.

Laboratory Exercise 48

Figure 48.2: © Image Source/Getty Images RF; 48.5: © Getty Images RF.

Laboratory Exercise 49

Figure 49.6a: © The McGraw-Hill Companies, Inc./Rebecca Gray, photographer/Don Kincaid, dissections; 49.6b: © The McGraw-Hill Companies, Inc./Womack Photography Ltd.; 49.6c: © John Watney/Photo Researchers, Inc.; 49.7: © John D. Cunningham/Visuals Unlimited; 49.8–49.9: © Biophoto Associates/Photo Researchers, Inc.

Laboratory Exercise 50

Figure 50.4: © Collection CNRI/Phototake; 50.6: © Victor P. Eroschenko; 50.7: © Biophoto Associates/Photo Researchers, Inc.; 50.9: © imagingbody.com; 50.10: © Dwight Kuhn; 50.13: © The McGraw–Hill Companies, Inc./APR.

Laboratory Exercise 51

Figure 51.3–51.5: © The McGraw-Hill Companies, Inc./J. Womack Photography.

Laboratory Exercise 54

Figure 54.4: © The McGraw-Hill Companies, Inc./Dennis Strete, photographer; 54.5–54.7: © Ed Reschke; 54.11: © The McGraw-Hill Companies, Inc./J. Womack Photography; 54.12: © Dennis Strete; 54.13: © Ed Reschke/Peter Arnold Images; 54.14–54.16: © The McGraw-Hill Companies, Inc./APR.

Laboratory Exercise 56

Figure 56.2: © The McGraw-Hill Companies, Inc./J. Womack Photography; 56.5a: © Biophoto Associates/Photo Researchers, Inc.; 56.5b: © Manfred Kage/Peter Arnold Images; 56.7: © Per Kjeldsen; 56.8: © John D. Cunningham/Visuals Unlimited; 56.9: © Ed Reschke.

Laboratory Exercise 58

Figure 58.2a: © The McGraw-Hill Companies, Inc./Dennis Strete, photographer; 58.3: © Biophoto Associates/Photo Researchers, Inc.; 58.4–58.5: © Ed Reschke; 58.6a–b: © The McGraw-Hill Companies, Inc./Al Telser, photographer; 58.7: © Michael Peres.

Laboratory Exercise 59

Figure 59.3: © The McGraw-Hill Companies, Inc./Rebecca Gray, photographer/Don Kincaid, dissections; 59.6: © Ed Reschke/Peter Arnold Images; 59.7: © Ed Reschke; 59.8: © 1989 Michael Peres; 59.9: © McGraw-Hill Higher Education, Inc./Carol D. Jacobson, PhD., Dept. of Veterinary Anatomy, Iowa State University.

Laboratory Exercise 60

Figure 60.3a–b: © Dr. Kurt Benirschke.

Laboratory Exercise 61

Figure 61.1a–j: © The McGraw-Hill Companies, Inc./J. Womack Photography.

Index

Page numbers followed by *f*
indicate figures; *t*, tables.

A

AB blood type, 430*t*
Abdominal, 16*f*
Abdominal aorta, 395*f*, 471*f*,
473*f*, 558*f*
Abdominal cavity, 10, 10*f*
Abdominal lymph nodes, 493*f*
Abdominal wall muscles, 243–248
Abdominopelvic cavity, 10, 10*f*, 14
Abdominopelvic quadrants, 12*f*,
14, 15*f*
Abdominopelvic regions, 14, 15*f*
Abducens nerve, 312*f*, 312*t*
sheep, 328*f*
Abduction, 188*f*, 190*t*
Abductor pollicis brevis
muscle, 235*t*
Abductor pollicis longus
muscle, 234*f*
A blood type, 430*t*
Accessory nerve, 312*f*, 312*t*
sheep, 328*f*
Accommodation pupillary
reflex, 366
Accommodation test, 363–364
Acetabulum, 167*f*, 168
Acid, 24, 24*f*
Acidophil cells, 391, 392*f*
Acinar cells, 395*f*
ACL. *See* Anterior cruciate
ligament (ACL)
Acoustic meatus, 133*f*–134*f*, 135,
137*t*, 267*f*, 372
Acromial, 16*f*
Acromioclavicular joint, 268*f*–269*f*
Acromion, 157, 157*f*, 232*f*,
267*f*–269*f*
Actin filament, 198*f*
Action potential, 292, 318
Adam's apple, 267*f*, 392*f*
Adduction, 188*f*, 190*t*
Adductor brevis muscle, 255*f*, 256*t*
Adductor longus muscle,
255*f*, 256*t*
Adductor magnus muscle,
255*f*, 256*t*
Adipose tissue, 88*f*, 90*t*, 104*f*
Adjustment knobs of microscope,
32*f*, 34
Adrenal cortex, 393, 394*f*
Adrenal glands, 391*f*, 393–394,
394*f*, 558*f*
Adrenal medulla, 393, 394*f*
Afferent arteriole, 560*f*
Afferent neurons, 280, 280*t*, 298

Agonist, 199*t*
Agranulocytes, 409*t*
Albinism, 606*t*, 608–609
Alimentary canal, 538
Allantois, 598, 600, 600*f*
Alleles, 606
Alpha cells, 395*f*, 403, 403*f*
Alpha waves, 318, 318*f*, 319*t*
Alveolar arch, 135
Alveolar capillaries, 471*f*
Alveolar duct, 504*f*
Alveolar processes, 132*f*, 135–136
Alveolar sac, 504*f*
Alveolus, 123, 125*f*, 471*f*, 504*f*
Amnion, 600, 600*f*
Amniotic cavity, 600*f*
Amniotic fluid, 600, 600*f*
Amniotic sac, 601*f*
Amoeba, 46
Amphiarthroses, 184, 184*t*
Amplitude, brain wave, 318
Ampullae, 384*f*
Amylase, 552
Anal canal, 543*f*
Anal sphincter, 543*f*
Anal triangle, 244, 247*f*
Analysis of data, 2
Anaphase, 68, 69*f*–70*f*
Anatomical planes, 14, 15*f*
Anatomical position, 10
Anatomic dead space, 510
Anconeus muscle, 234*f*, 234*t*
Ankle-jerk reflex, 299
Antagonist, 199*t*
Antagonistic pairs, 199*f*
Antebrachial, 16*f*
Anterior axillary fold, 268*f*
Anterior cerebral artery, 472*f*
Anterior communicating
artery, 472*f*
Anterior corticospinal tract, 291*f*
Anterior cruciate ligament (ACL),
186, 187*f*
Anterior fontanel, 176*f*
Anterior funiculus, 291*f*
Anterior horn, 291*f*, 293*f*
Anterior inferior iliac spine, 167*f*
Anterior interventricular
artery, 436*f*
Anterior interventricular
sulcus, 436*f*
sheep, 440*f*
Anterior median fissure, 291*f*
Anterior reticulospinal tract, 291*f*
Anterior sacral foramina, 149, 149*f*
Anterior scalene muscle, 223*t*
Anterior spinocerebellar tract, 291*f*
Anterior spinothalamic tract, 291*f*

Anterior superior iliac spine, 167*f*,
254*f*, 268*f*, 271*f*
Anterior tibial artery, 474*f*
Anterior tibial vein, 476*f*
Anterolateral system, 291*f*
Antigens, 430
Anus, 576*f*
Aorta, 395*f*, 436*f*–437*f*, 439, 439*f*,
471*f*, 558*f*
sheep, 440*f*–441*f*
Aortic arch, 436*f*–437*f*, 471*f*
Aortic area, 448*f*
Aortic bodies, 530, 531*f*
Aortic valve, 437*f*–438*f*, 471*f*
Apical cell surface, 44*f*
Apocrine sweat glands, 102, 103*f*
Aponeurosis, 245*f*–246*f*
Appendicular skeleton, 122
Appendix, 543*f*
Aqueous humor, 352
Arachnoid mater, 290, 293,
293*f*–294*f*
Arbor vitae, 309*f*, 311*f*. *See also*
White matter
Arcuate artery, 560*f*
Arcuate vein, 560*f*
Areola, 589, 589*f*
Areolar connective tissue, 88*f*, 90*t*
Areolar glands, 589, 589*f*
Arm, 155–159
Arm of microscope, 32*f*
Arrector pili muscle, 104*f*
Arterial anastomosis, 470
Arteries, 467–477
Arterioles, 468, 504*f*
Articular cartilage, 112, 114*f*, 116,
186, 186*f*
Articular facet, 147*f*
Articulations, bone, 123
Arytenoid cartilage, 501*f*, 502
Ascending aorta, 436*f*–437*f*, 471*f*
Ascending colon, 473*f*, 543*f*
Association areas, 306, 310*f*
Association neurons, 280
Aster, 69*f*–70*f*
Astigmatism, 363, 606*t*, 609
Astigmatism test, 363, 363*f*
Astrocyte, 280, 281*t*, 284*f*–285*f*
Atlas, 146, 146*f*–147*f*
Atrioventricular bundle, 448,
449*f*, 458
Atrioventricular node, 448,
449*f*, 458
Atrioventricular orifices, 438
Atrioventricular sulcus, sheep,
440*f*–441*f*
Atrioventricular valves, 436,
438, 438*f*

Auditory acuity test, 372
Auditory association area, 310*f*
Auditory ossicles, 132, 372, 373*f*
Auditory tube, 372, 373*f*
Auricle, 372, 373*f*, 436*f*
sheep, 440*f*
Axial skeleton, 122
Axilla, 268*f*
Axillary, 16*f*
Axillary artery, 472*f*
Axillary fold, 268*f*–269*f*
Axillary lymph nodes, 493*f*
Axillary vein, 475*f*
Axis, 146, 146*f*–147*f*
Axon, 280, 281*f*–282*f*, 283,
283*f*–286*f*
Axon collateral, 282*f*
Axon hillock, 282*f*–283*f*

B

Babinski reflex, 299
Balance, 372, 381–385
Ball-and-socket joint, 184*t*, 185*f*
Bárány test, 383–385
Basal cell surface, 44*f*
Basal metabolic rate, 46–47
Basal nuclei of cerebellum,
307*f*, 310
Base, 24, 24*f*
Basement membrane, 44*f*,
103*f*–104*f*
Base of microscope, 32*f*
Basilar artery, 472*f*
Basilar membrane, 374*f*
Basilic vein, 475*f*
Basophil cells, 391, 392*f*, 409*t*,
410*f*, 411*t*
B blood type, 430*t*
Benedict's test for sugar, 25–26
Beta cells, 395*f*, 402, 403*f*
Beta waves, 318, 318*f*, 319*t*
Biceps brachii muscle, 199*f*, 230*f*,
234*t*, 255*f*, 268*f*–269*f*
Biceps femoris muscle, 257*t*,
258*f*–259*f*
Biceps reflex, 299, 300*f*
Bile duct, 395*f*, 542, 542*f*
BIOPAC
electrocardiography, 457–462
electroencephalography,
317–321
electromyography, 205–213
spirometry, 521–526
Bipolar neuron, 280*t*, 281*f*
Bitter taste, 340
Biuret test, 25
Bladder, 12*f*, 558, 558*f*, 562,
563*f*, 576*f*